BATTLESHIPS

BATTLESHIPS

Axis and Neutral Battleships in World War II

BY

WILLIAM H. GARZKE, JR.

ROBERT O. DULIN, JR.

LINE DRAWINGS BY

THOMAS G. WEBB

ALAN RAVEN

NAVAL INSTITUTE PRESS

ANNAPOLIS, MARYLAND

Reprinted with corrections and updates 1990

Library of Congress Cataloging-in-Publication Data
Garzke, William H.
Battleships : axis and neutral battleships in World War II.
Bibliography: p.
Includes index.
1. Battleships. 2. Battle cruisers. I. Dulin, Robert O. II. Title.
V815.G373 1985 359.8′3 85-21579
ISBN 0-87021-101-3

Printed in the United States of America on acid-free paper ♾

All photographs not otherwise credited are
official U.S. Navy releases.

10 9 8 7 6 5

Table of Contents

APPENDIXES

Preface

Battleships and battlecruisers are warships of a bygone age, but for the first half of the twentieth century they were regarded as the ultimate weapon of sea power by the major navies of the world. Their large guns, heavy protection, and large displacement accorded them the privilege of being termed "capital ships." Until World War II these capital ships were considered to be the basic measure of relative power, but the airplane and aircraft carrier, by then major components in the naval task force concept, brought about their decline and near disappearance from the naval scene. At this writing (1985), two American battleships, the *Iowa* and *New Jersey*, have been reactivated, with a third ship, the *Missouri*, in the shipyard. Planning is well advanced for the reactivation of the fourth *Iowa*-class battleship, the *Wisconsin*.

During most of the first half of the twentieth century, until air power made them obsolete, the number, characteristics, and availability of capital ships fundamentally influenced foreign policy and naval strategy. From the onset of the dreadnought era, government leaders, naval officers, civilian engineers and technicians, and laymen debated the complex problems of determining the suitable number of capital ships, establishing their characteristics, and finding the best way to deploy and operate battle fleets, the basis of which were the battleships.

Several significant developments during World War I greatly influenced the course of capital-ship design:

- The development of the submarine as an effective combatant type forced the designers of capital ships to provide reasonable protection against torpedo attack.
- Similarly, the development of aircraft prompted interest in antiaircraft gunnery and in heavier deck armor to resist bombs.
- The naval Battle of Jutland was an important factor in the design and operation of capital ships before World War II.

In the years following World War I, the supremacy of the battleship was challenged by the advocates of air power. They insisted, accurately but prematurely, that the primacy of the capital ship was doomed by air power. In the United States the media sensationalized the success of U.S. Army aviators under the leadership of General William "Billy" Mitchell in bombing and sinking a moored destroyer, light cruiser, and three obsolete battleships (most notably the ex-German *Ostfriesland*). This inflamed the controversy. Battleships were not defenseless and certainly not without any protection, but they did eventually succumb to this new and dangerous foe. For the next two decades, periodicals and books that forecast the demise of the battleship under air attack, whether at sea or in port, proliferated. Inevitably, these developments influenced the design of new capital ships.

The Washington Naval Treaty of 1922 and the subsequent London Naval Treaty of 1936 had a profound effect on capital-ship design. These treaties established two basic limitations on capital-ship characteristics—the displacement was limited to a maximum of 35,000 tons "standard," and the maximum bore of the main-battery guns was set at 406 mm. These treaty limitations accelerated technological improvements that promised to save weight in the warships and still retain adequate offensive and defensive qualities. Even so, *all* navies building capital ships were to violate the treaty limitations to some degree. The London Naval Treaty of 1930 played a minor role in the development of the capital ship, although the discussions highlighted the growing closeness of the United States and Great Britain in naval policy, and the accelerating estrangement of Japan. Henceforth, Japan would be perceived as a probable enemy by the United States and Great Britain.

In Europe, the rise of the fascist dictatorship of Benito Mussolini in Italy and the subsequent ascendancy of Adolf Hitler in Germany heralded an increased emphasis on military expansion on the part of both dictatorships. The German decision to build armored ships in 1926, several years prior to the rise of Nazi Germany, had great consequences in capital-ship design. In 1929, prompted by concern about the raider potential of the new "pocket battleships," France initiated the design of a much-larger battlecruiser, designed to overtake and overwhelm the German commerce raiders. Four years later, the Italians authorized the construction of two battleships of 35,000 tons armed with 381-mm guns, prompting the French Navy to begin design studies of a similar battleship type. The following year, the Anglo-German Naval Treaty of 1935 formally consented to let the Germans expand their naval forces up to a total displacement in surface ships equalling 35 percent of that of the Royal Navy, thereby repudiating the inhibitions imposed on Germany by the Versailles Treaty of 1919. By this time, the Royal Navy was also active in new capital-ship design studies.

By 1935 the second battleship construction program of the twentieth century had reached a fever pitch. However, the number of ships would be fewer than in the period preceding World War I, but the displacements involved were much greater. By 1950, when the last battleship (the *Jean Bart*) was finally completed, new capital ships had been built by the United States, Japan, France, the United Kingdom, Germany, and Italy. Several capital ships were actually laid down in the Soviet Union, and capital-ship construction was formally projected in the Netherlands and Spain.

Even now, an air of fascination and mystery surrounds these giant warships, and some have been accorded near-legendary places in naval history. The German battleship *Bismarck* has been written about in many publications, and her exploits are well known, but the details of her design, construction, and eventual destruction have been incomplete. The Japanese *Yamato, Musashi,* and *Shinano* were the largest warships of their type built in their era and have attracted much attention in the past few years. In the past, authors were limited by the absence of official sources and documents; however, patient and diligent research has eliminated errors of the past.

This study provides a definitive technical history of the design, construction, and operation of all battleships and battlecruisers, of vintage subsequent to 1930, including ships completed, laid down, or only projected. This is the final volume of a three-volume set. Volume I, *United States Battleships in World War II*, and Volume II, *Allied Battleships in World War II*, have already been published. This book will deal with the capital ships of Spain and the Axis powers—Germany, Italy, and Japan.

Although there is extensive literature describing the operations of the various battleships and battlecruisers, the available data on their technical characteristics and the development of their designs have been rather limited and contradictory, with virtually no effort having been made to describe them in detail from a technical standpoint.

It is believed that these noteworthy capital ships have been given a definitive treatment here in a study that presents relatively complete plans as well as extensive technical data covering their characteristics and performance. The operational careers of the ships are chronicled, followed by detailed analyses of outstanding incidents as they pertain to design adequacy, particularly from the standpoint of damage resistance. Research over a period of more than two decades has made possible extensive use of authoritative (and frequently official) source material, some of which has never before been published. These sources were also supplemented by an exhaustive correspondence with many of those individuals who were intimately involved in the design and the operations of the ships. In some cases, this correspondence was the sole source of information.

During the research, writing, and drawing of plans for this volume, many individuals and offices in the United States, Europe, and Japan offered valuable assistance.

The chapter concerning the Japanese *Yamato*-class battleships was a direct outgrowth of a prize-winning essay written by then Midshipman Robert Dulin at the U.S. Naval Academy during 1960–1961. As in the earlier paper, by far the most important contributor to the present analysis was former naval constructor Lieutenant Commander Shizuo Fukui, Imperial Japanese Navy. Lieutenant Commander Fukui conducted an extensive correspondence with, and interviewed, many of the principal personnel involved in the design process of these ships, including such men as Vice Admiral Fukuda, Captain Matsumoto, and Captain Makino. All of these gentlemen had key roles in the design development of these ships, including the subsequent conversion of the incomplete *Shinano* to an aircraft carrier. Mr. Fukui also answered many questions on the ships' armament, hull form, arrangement, propulsion, structure, and battle damage. A number of official U.S. Navy Naval Technical Mission to Japan (NAVTECHJAP) reports provided useful starting points for research. U.S. Navy Rear Admirals Sonnesheim and Holtzworth were also interviewed by the authors. Japanese-language action reports were used in the development of the narratives regarding the sinkings of the *Yamato* and *Musashi*. Interviews with Mr. Shigeru Makino were most helpful in providing the authors with a better understanding of the design development of these ships and the important design changes accomplished or contemplated after their delivery to the Japanese Navy. Admiral Suteo Ishida, Japanese Maritime Self-Defense Force, while serving as naval attaché in Washington, was extremely helpful to the authors by translating Japanese works about the ships and helping to arrange contacts with many of the people listed above.

The chapters on the German *Scharnhorst, Bismarck*, "H"-, and "O"-class ships reflect the unparalleled cooperation of surviving members of the construction office of the Oberkommando der Kriegsmarine (OKM)—High Command of the German Navy. By far the most generous German contributors to these four chapters were Captain Albrecht Schnarke (former gunnery officer of the *Tirpitz*) and Dr. Jürgen Rohwer (German military historian). Both gentlemen reviewed the text in detail and personally arranged for the German translations of draft versions. The German text was reviewed by a number of men who had participated in the design of these ships. Messrs. Schnarke and Rohwer collated the comments and, in the case of Mr. Schnarke, revised the text accordingly. The latter also provided data and explanations on German radar and ordnance from his files in addition to a number of photographs from his own collection. Mr. Richard Van Hooff, a practicing naval architect, then translated the German texts and made suggestions as to important revisions. In additon, Mr. Garzke's father-in-law, Professor Andrew Vincze, a retired professor of mechanical engineering at the City University of New York (CUNY) and an expert in marine engineering, did much of the early translating and provided guidance in the design and arrangement of German steam and diesel propulsion plants. These very generous efforts provided final checks on the years of research by the authors.

Dr. Jürgen Rohwer also placed the authors in contact with many individuals who participated in the design and construction of German battleships and battlecruisers of the World War II era, including Professor-Doctor Erwin Strohbusch, Admiral Karl Witzell, Admiral Gerhard Wagner, Admiral Werner Fuchs, and naval constructors Otto Reidel, Wilhelm Hadeler, and Hermann Burckhardt. Operational insights were provided by senior officers Helmuth Giessler (*Scharnhorst*), Baron Burkard Freiherr von Müllenheim-Rechberg (*Bismarck*), Wilhelm Schmidt (*Bismarck*), Gerhard Jünack (*Bismarck*), and Admiral Wolfgang Kähler (*Gneisenau*). The authors supplemented this effort with extensive correspondence with Captain Karl Bidlingmaier (*Prinz Eugen*) who had done extensive research on the *Bismarck* and *Tirpitz*.

A very detailed and exhaustive correspondence was conducted with Mr. Otto Reidel, who had an intimate knowledge of German battleship design and had been interviewed at length after World War II by NAVTECHEUR (Naval Technical Mission to Europe) personnel of the U.S. Navy. He provided the authors with key insights into the background of the major technical and political influences that affected the design decisions regarding these ships. He also reviewed the text with Messrs. Ludwig and Seefisch, both former German naval constructors in the Construction Office and both familiar with the designs and important technical features. To aid in this analysis, declassified U.S. Navy NAVTECHEUR reports were reviewed with Mr. Reidel regarding their accuracy and completeness. Mr. Nathan Okun contributed information on German gunnery and armor and was helpful in putting together technical information regarding the protection schemes of the *Bismarck* and *Scharnhorst* classes. It is believed that, thanks to this unprecedented cooperation, the chapters concerned with German capital-ship design are the most complete that can be written today.

The chapter on the Italian *Vittorio Veneto* class was developed with the full cooperation of the Historical Office of the Italian Navy. This assistance included the provision of damage reports for the various ships, including an analysis of the sinking of the *Roma*. The manuscript was also reviewed by that office.

Full plans of these battleships were made available to the authors, greatly assisting the evaluation and analysis of their design. Plans of earlier studies have contributed to a better understanding of the design evolution of these ships.

Although the Historical Office of the Italian Navy made by far the most important contributions to the development of this chapter, a number of individuals and organizations were most helpful and cooperative, including Messrs. Augusto Nani, Giuseppi Sitzia, Cantieri Ansaldo, and Cantieri Riuniti dell' Adriatico (CRDA). A special note of thanks is due to the office of the Italian Naval Attaché, Washington, whose staff was unfailingly courteous and helpful during almost twenty years of correspondence and research regarding these ships.

The authors first learned of Spanish efforts in capital-ship design and construction from a former Spanish naval officer, Ricardo de Sobrino. Mr. de Sobrino actively cooperated with the authors in translating the limited available published material on the subject, in addition to suggesting further contacts for research.

Other individuals who contributed to the development of this chapter include: Ricardo Gomez Usatore, J. Taibo, and Arturo Genova Sotil.

The illustrations for this book are worthy of special comment. Official plans at times vary in detail from the ships as built, are sometimes poorly preserved, and in some cases have become so illegible as to be considered incomplete. As a result, Mr. Thomas Webb and Mr. Alan Raven had to work in close cooperation with the authors and at times supplement the reference plans with numerous photographs to verify the plans. Particularly difficult was the profile of the *Yamato* of April 1945. Mr. Webb studied all the available photographs of the ship, which are not of high quality, and has drawn what the authors consider the most accurate plan of the ship at the time of her loss. Mr. Webb provided the profiles of the *Yamato, Scharnhorst, Bismarck*, and "H" classes, while Mr. Raven did the profiles for the battlecruiser "O" and the *Vittorio Veneto* classes, as well as some drawings of the *Tirpitz* and the *Gneisenau*. A number of paintings have been specially prepared for this volume by Mr. Richard Allison, a very talented artist with extensive experience with marine and naval subjects.

The authors are particularly grateful for the substantial help provided by official sources. The authors and the illustrators are solely responsible for any errors.

After Dr. Robert Ballard of the Woods Hole Oceanographic Institution discovered the wreck of the *Bismarck*, the authors were retained as technical consultants for a series of clients such as the National Geographic Television, Madison Press (for his book on the discovery of the *Bismarck*), and a number of periodicals such as *National Geographic Magazine*, the German Magazine *GEO*, and the *London Daily Express Magazine* which had articles by Dr. Ballard on his *Bismarck* exploration.

The authors were aided in updating this reprint by one of the *Bismarck*'s survivors, Josef Statz, who was a Seaman, Second Class, in the Damage Control Central. He was able to escape from that space, up the conning tower tube, and through the Forward Conning Tower, which had been destroyed by 356- and 406-mm shells. Mr. Statz observed much damage to a once-proud and beautiful warship and was able to exit the ship on the much damaged port side amidships. Mr. Statz has stated that he came forward so that the true agony of the *Bismarck* could be told for posterity. This in combination with Dr. Ballard's photography has provided the authors with much insight into the damage the *Bismarck* received on the morning of 27 May 1941. Burkard Baron von Müllenheim-Rechberg and a German-American naval architect, Wolfgang Reuter, assisted the authors in compiling this testimony.

Deck Designations

United States Navy	*Italian*	*German*	*French*	*Japanese*	*Dutch*	*Royal Navy*
Tank Top	Stiva	Stauung	Toles de Ballast	Tank Top	Binnenboden	Tank Top
3rd Platform		Unteres Platformdeck		1st Hold Deck		
2nd Platform	Copertino Intermedio	Mittleres Platformdeck		2nd Hold Deck		Lower Platform
1st Platform	Copertino Superiore	Oberes Platformdeck	Platforme	Platform Deck	Vierde Tusschendek	Upper Platform
Fourth Deck		Panzerdeck	Deuxième Faux Pont	Lower Deck	Derde Tusschendek	Lower Deck
Third Deck	II Corridoio	Zwischendeck	Premier Faux Pont	Middle Deck	Tweede Pantserdek	Middle Deck
Second Deck	I Corridoio	Batteriedeck	Pont Principal	Upper Deck	Eerste Tusschendek	Main Deck
Main Deck	Coperta	Oberdeck	Premier Pont	Flying Deck	Bovendek	Upper Deck
01 Level	Castello	Aufbandeck	Pont du Chateau		Slopendek	Shelter Deck
02 Level	Tug A	Unteres Bruckendeck	Platforme 2		Dek H	Boat Deck
03 Level	Tug B	Unteres Mastdeck	Platforme 3		Dek K	No. 2 Platform
04 Level	Tug C	Oberes Mastdeck	Platforme 4		Dek L	Signal Deck
05 Level	Tug D	Admiralbrucke	Platforme 5		Dek M	Lower Bridge
06 Level	Tug E		Platforme 6		Dek N	Upper Bridge

BATTLESHIPS

CHAPTER ONE

Introduction

Heavy gun battleships and battlecruisers, long considered the ultimate embodiment of sea power, have now almost disappeared from the oceans of the world, the victims of technological progress. The final evolution of these ships was highly accelerated during World War II, where in a short time they reached their zenith—as illustrated by the *Hood-Prince of Wales-Bismarck* battle in May 1941. Some had considered battleships of doubtful value after the *Littorio, Caio Duilio,* and *Conte di Cavour* were torpedoed in a daring carrier strike on Taranto in November 1940; however, it was not until December 1941, when the *Prince of Wales* and *Repulse* were sunk by Japanese aircraft in the South China Sea and the American battlefleet was severely damaged at Pearl Harbor, that the importance of the battleship was seriously questioned. Nonetheless, the last battleships and battlecruisers were fast, heavily armed, and well protected.

The modern battleship of the 1930–40 era was an evolutionary warship, benefitting from the battleship and battlecruiser experience in design and operation from the period of 1905 to 1922. The most influential factors in battleship design development were the lessons learned from the naval battles of World War I, technological progress, and the naval limitation treaties drawn up during the interwar years. Generally, the modern battleship combined high speed with increased gun power and protection. The value of this combination of attributes was keenly shown during the Battle of Jutland in 1916, when the German battlecruisers, fast and well protected, but with smaller-caliber main batteries than their British counterparts, were able to escape from under the guns of a superior British fleet, while heavily damaging some of their British opponents. The limitations of the Washington Treaty affected Allied battleships more than those built by Axis powers, but various naval, geographical, and economic limitations ultimately influenced all Axis powers building such capital ships. Despite such artificially imposed limits, modern Axis battleships evolved into massive warships combining high speed with considerable offensive power as well as the ability to sustain heavy punishment. The primary goal was to design a small number of powerful ships that could absorb punishment, yet still be able to deliver a decisive blow to an opponent. This was particularly important for Axis capital ships because they would be fighting numerically superior foes; such requirements forced considerable growth in their physical dimensions and displacements. With the advent of the naval limitation treaties, the limits on displacement were extreme;* this, in turn, restricted the amount

*Washington Conference (1921–22), London Naval Conference (1930), and the Second London Naval Conference (1935–36). The 1927 Geneva Disarmament Conference resulted in no limitations on capital ships, but did result in British efforts to reduce the gun size and battleship displacement from 1927–35. For all practical purposes, all treaty limitations expired on 31 December 1936.

This view of the reconstructed Italian battleship Andrea Doria *vividly shows the topside clutter that resulted from a mixed-caliber secondary battery, to the detriment of the arcs of fire for the antiaircraft battery. (Italian Navy)*

of armor that could be used on a modern battleship. Axis powers recognized the restrictiveness of the treaty limitations in terms of speed, gun power, and protection. They elected to ignore them. Thus, the *Bismarck-, Vittorio Veneto-*, and *Yamato*-class ships were much larger than their Allied contemporaries built to treaty limits.

By the end of World War I, the fast-battleship concept had been accepted by all the major naval powers except the United States. In fact, France, Italy, and the United States were the only major nations not possessing battlecruisers. Japan, Italy, Germany, Russia, and the United Kingdom were all constructing ships that could reasonably be termed fast battleships. Although few of these ships were completed, they established the trend for the next major naval building epoch two decades later, when improved technology facilitated the construction of well-protected fast battleships.

Washington Naval Treaty. At the end of World War I, the U.S. Navy was well started on an all-out expansion program so extensive that it threatened the traditional supremacy of the Royal Navy. As a result, the British initiated new capital-ship design projects during 1920–21. These ships would have prompted even further response from the United States and Japan. The American program also threatened the new position of power that had been established in the Pacific by Japan, which was likewise embarked on a major construction program. As a matter of fact, the new Japanese capital ships would have been individually superior to their American and British contemporaries. Furthermore, the Japanese would have introduced the 457-mm gun in some of these ships. Such ships probably would have brought about a further escalation in gun size. Finally, because of the enormous financial strains imposed by these huge naval expansion programs, the United States proposed a naval disarmament conference. This offer was accepted by the British government and then by Japan, Italy, and France. On 12 November 1921, representatives met in Washington, D.C., for the first session of the Washington Naval Conference, which was to produce a naval agreement that would lastingly affect warship design and construction. The agreement provided a quantitative and qualitative reduction in capital-ship construction and was the most important and successful arms limitation treaty drawn up during the twentieth century.

It is worthwhile to note those provisions of the Washington Naval Treaty, signed on 6 February 1922, that influenced capital ship design:

- British and American fleets were limited to 580,450 and 500,360 tons respectively, with Japan at 60 percent of U.S. strength (301,320 tons), France at 221,170 tons, and Italy at 182,000 tons. (This was the famous "5-5-3 ratio" so bitterly disputed by the Japanese.) All displacements were in tons of 2,240 pounds, 1,016.05 kilograms, and the comparative displacements were calculated on the basis of the "standard" displacements, defined later.*
- Capital ships were limited to 35,000 tons standard displacement, with 406-mm guns the largest permissible.
- No more than 3,000 tons could be added to any modernized battleship or aircraft carrier, and primary batteries could not be increased in size during such modernization.

---

*Metric-English measurements. The comparison of characteristics of ships built by different nations has long been complicated by the use of diverse measurement systems. In this text, measurements for all ships are in the metric system; all tabular data are presented in both the metric and English measurements. This may at times result in too-precise values, where a general value in one system is converted into the other, and the exact conversion is retained in order to best indicate the actual value under discussion.

- Standard displacement was established as the displacement of the vessel complete, fully manned, equipped and ready for sea, including ammunition, provisions, fresh water for her crew, and miscellaneous stores and implements of every description to be carried in war, but *not including* fuel or reserve feed water.
- Treaty lifetime of a battleship was to cover 20 years.

The displacement and armament restrictions made the introduction of efficient weight-saving techniques in ship equipment and armament mandatory. Designers were forced to devise and develop new concepts and procedures, many of which later became fundamental when capital ship construction was resumed before World War II. The Washington Naval Treaty, despite the ten-year naval "holiday" it enforced, altered the course of capital-ship design and exerted the most profound influence on battleship technology since the *Dreadnought.*

The first "Treaty" battleships. In view of the new Japanese and American battleships nearing completion, the Washington Naval Treaty permitted the Royal Navy to build the battleships *Nelson* and *Rodney,* which were laid down on 28 December 1922. In addition, France and Italy were each permitted the construction of new battleships up to 70,000 tons (total displacement), since both countries had been unable to complete new ships during the period 1916–1921. In addition, the British were allowed to complete the battlecruiser *Hood,* but she was armed with 380-mm guns. With the United States and Japan now having new battleships armed with 406-mm guns, the British were allowed to arm their two new battleships, the *Nelson* and *Rodney,* with 406-mm guns.

These ships were novel because they were the first to be built under the restrictions of the Treaty, and as a result many innovations in these ships were incorporated into later battleships. Essentially, the *Nelson* and *Rodney* were modifications of the 1921 battlecruiser design, with the full-load displacement reduced from 53,900 tons to 38,400 tons, waterline length reduced by 160 feet, and shaft horsepower slashed from 160,000 to 45,000. The arrangement of all three turrets forward gave the superstructure the appearance of having been pushed far aft of the conventional position. Armor protection was concentrated in the citadel, with a thick, but still too shallow, internal main side belt inclined from the vertical for increased effective thickness. This concentration left the forward portion of the hull and the sides above the armor deck unprotected. In addition, the British steel industry had developed a new "D" steel with improved yield strength. This made it possible to reduce substantially the structural steel weight in comparison with other British battleships or battlecruisers built or designed several years before. Such efforts to save weight on armor protection and structure would become characteristic of all displacement-limited modern battleship designs.

The *Nelson* and *Rodney,* the first battleships built under the restrictions of the Washington Treaty, were not auspicious heralds of the modern battleship era. Many of the new designs that were being studied in other navies included a *Nelson*-style arrangement and, in the case of the French Navy, resulted in the adoption of the all-main-battery forward arrangement of the *Dunkerque* and *Richelieu* classes. In fact, early conceptual designs of the Japanese *Yamato* class featured such arrangements to save weight.

Germany. One of the most significant developments in warship design occurred in Germany during the interwar years. The very restrictive terms of the Versailles Treaty limited Germany from building warships of greater than 10,000 tons displacement, and the largest ships in the fleet were eight obsolete pre-dreadnought battleships of the

Deutschland and *Braunschweig* classes. Under the treaty terms, these old ships could be replaced 20 years after launch, which meant that Germany could begin replacing these ships in 1925. Skilled technicians and design teams had been dispersed by the terms and sanctions in the treaty. Germany also experienced severe economic hardships, which brought about a further erosion of her shipbuilding industry.

Against this background, design studies for replacements of the pre-dreadnoughts began in the early 1920s; these early designs were attempts to establish what type of warship could be constructed with a 10,000-ton displacement. A 1923 study showed that an armament of four 380-mm guns could be carried, but immense sacrifices in speed, endurance, or protection were necessary. In other words, a good ship was possible, but on a small scale. Design work continued from 1923–1926 and came to several possible solutions; these are summarized in table 1-1.

TABLE 1-1

Design Requirements for Armored Ships

Displacement (standard)	10,000 tons (10,160 mt)
Armament (main battery)	four 15″ (380-mm) twin, or six 12″ (305-mm) twin, or six 11″ (283-mm) twin
Protection (side armor belt)	3.94″ – 9.84″ (100–250 mm)
Speed	18, 21, or 27 knots

From the initial planning stages of the naval construction program, which commenced in 1927 when Admiral Hans Zenker made his final decision on the characteristics of the new "panzerschiffe," there were difficult decisions to be made on the future course of German warship design.* Because of the unwillingness of the Weimar Republic to provide funding and the restrictive nature of the Versailles Treaty, to build a fleet Admiral Zenker had to choose from among the following alternatives:

- A defensive coastal fleet of monitors and minesweepers.
- A limited offensive fleet of small, powerful light and heavy cruisers and several destroyer flotillas.

Admiral Zenker decided on a limited offensive fleet and chose as its capital ship the armored ship, which would be a small but powerful ship with limited protection. The new ship was to feature a 26–27-knot speed (this to elude battleships), 100-mm side armor protection (this barely suitable against 203-mm guns), and six 283-mm guns in two triple turrets (this to outgun heavy cruisers). The light protection and heavy armament represented a compromise solution and best fit the type of fleet Admiral Zenker chose. The armament would overwhelm any heavy cruiser in service or projected, and its concentration in a multibarrel turret reduced ship size and armor protection, but still permitted a maximum speed of 28 knots. These ships, which would ultimately achieve 28.5 knots in the *Admiral Graf Spee*, were faster than all battleships with heavier armament, but were outgunned and slower than the British battlecruisers *Renown, Repulse*, and *Hood.* However, from 1923–1938 the German Naval Command never seriously considered the United Kingdom a likely adversary.

The armored ships *Deutschland* (later renamed *Lützow*), *Admiral Scheer*, and *Admiral Graf Spee* aroused the interest of all naval powers, but most particularly the

*Panzerschiffe means armored ships, literally translated, and will be used throughout this text.

French Navy. When Germany announced the construction of the *Deutschland,* the French Navy began studying a ship that would neutralize her, and this resulted in an intensive design effort on a "croiseur de combat" or "pocket battleship chaser." Ultimately, all this effort culminated in the construction of the battlecruiser *Dunkerque* in 1932. When the French effort became known, studies were begun in Italy, and these ultimately led to the extensive reconstruction of several old battleships. The characteristics of the armored ship *Deutschland* are tabulated in table 1-2.

TABLE 1-2

Characteristics of the Deutschland-Class Armored Ships

Standard displacement*	11,396 tons (11,396 mt) *Deutschland*
	11,785 tons (11,974 mt) *Admiral Graf Spee*
Full load displacement*	15,486 tons (15,735 mt) *Deutschland*
	15,779 tons (16,032 mt) *Admiral Graf Spee*
Waterline length	596.1 feet (181.7 m)
Beam	67.91 feet (20.7 m) *Deutschland*
	71.19 feet (21.7 m) *Admiral Graf Spee*
Draft (Standard displ.)	19.06 feet (5.81 m) *Admiral Graf Spee*
(Full load)	24.08 feet (7.34 m) *Admiral Graf Spee*
Armament	
Main	Six 11.14"/54.5 (283-mm) guns in two twin turrets.
Secondary	Eight 5.9"/55 (150 mm) in single mounts
Antiaircraft	Three 3.46"/45 (88 mm) in single mounts
	(Six in *Admiral Graf Spee*)
Torpedo	Eight 21.0" (533 mm) in quadruple mounts
Speed	28.0 knots (*Deutschland*)
	28.3 knots (*Admiral Scheer*)
	28.5 knots (*Admiral Graf Spee*)
Protection	2.36" (60 mm) side belt—*Admiral Scheer/Deutschland*
	3.15" (80 mm) side belt—*Admiral Graf Spee*
	1.57" (40 mm) deck—*Admiral Scheer/Deutschland*
	1.77" (45 mm) deck—*Admiral Graf Spee*

*German calculations of displacements are based upon the specific gravity of water in the Baltic Sea. Instead of the specific gravity of 64 pounds per cubic foot of salt water used in American and British calculations of displacement, the Germans used a specific gravity of 1.015, which would yield 63.336 pounds per cubic foot. All displacement figures for specified drafts for German ships have been adjusted to be consistent with American and British usage.

The armored ships were an ingenious design brought about by severe displacement limitations.* Although they carried a very heavy armament of 283-mm guns on a cruiser displacement, the ships had very little armor protection. This became a factor in the River Plate battle in December 1939, since the protection of the *Graf Spee* was only marginally effective against the 203-mm guns of the *Exeter,* which had more armor-deck protection than the German armored ship. Aside from the weak protection, the armored ships had a large cruising range and relatively good speed for that period. The long range of their main battery, combined with these attributes, made these ships very suitable for operations against merchant ships. Moreover, their design and construc-

*The armored ships are not considered in this study. Discussions with German naval officers and naval constructors have indicated that these ships were of the heavy cruiser type and do not belong in a study of battleships and battlecruisers.

The Kirishima *was one of four battlecruisers built to a British design before World War I. She was severely damaged by nine shell hits from the USS* Washington *during a savage night battle off Guadalcanal in November 1942. She was scuttled on 15 November 1942. (United States Naval Institute)*

The characteristic pagoda-type forward bridge tower is vividly shown in this striking view of the battleship Ise. *The Japanese adopted this type of superstructure during the modernization of all their capital ships in the 1920s and 1930s. (United States Naval Institute)*

The top view of the battleship Yamashiro *in 1938 and the bottom photo of the* Fuso *vividly show the arrangement complications resulting from the decision to mount six twin turrets with 356-mm guns. (James C. Fahey Collection)*

tion introduced many innovations that would later be used in German battleships and battlecruisers, particularly in the use of high-strength steels, improved armor steels, electric arc welding, and the introduction of a successful diesel propulsion plant.

Italy. In the 1920s the Italian government debated the use of the 70,000 tons allocated to her by the Washington Treaty. During this time, however, Italy also experienced a severe economic crisis, and the government did not have the financial resources to build battleships. Also, with France undecided on whether to proceed with battleship construction, Italy was not going to initiate any action. When France did begin the construction of the *Dunkerque*, however, earlier Italian design projects were revived and new concepts sought. At first, the concept of the "pocket battleship" appeared attractive to the Italian Navy, especially since the United Kingdom was advocating smaller ships and gun sizes. But the cost to build such new ships was prohibitive, and the number of units that could be constructed was limited. After much deliberation, Supermarina (the Italian Naval Command) decided to modernize four of its older battleships, the *Guilio Cesare, Conte Di Cavour, Andrea Doria*, and *Caio Duilio.* This modernization transformed old and obsolete battleships into somewhat more modern warships. Major changes were:

- Extensive rework of the hull with the addition of the Pugliese underwater protection system and 10.8 meters of length (at the bow).
- A complete replacement of the propulsion plant with modern boilers, turbines, and gears. A new propulsion plant of greater power and a hull of longer length meant an increase in speed from 21.5 to 27 knots.
- Removal of center turret of 305-mm guns and changing of all remaining guns to 320 mm. The casemate secondary guns of 120 mm were replaced with 120-mm twin turrets (*Cavour* and *Cesare*) or 135-mm triple turrets (*Doria* and *Duilio*). New antiaircraft weapons were also added.
- Complete new superstructure with stacks and propulsion plant concentrated instead of separated.
- Increase of horizontal armor protection from 97 mm to 135 mm. The side armor remained unchanged.

It would appear that this course of action was due to Italy's limited industrial resources and the navy's unwillingness to commit itself to smaller battleship construction when the major naval powers were not. Political developments in the Mediterranean area brought increasing tensions between France, Great Britain, and Italy. The Italian naval command recognized that the Italian Navy could be confronted with a combined British-French fleet. Therefore, the Italian Navy finally decided to build two 35,000-ton battleships in 1934. These new battleships were much larger than any warship the Italians had built previously, but they were also significant because the Italians would have considerable influence on the design of foreign battleships when Italy later provided assistance to the Russians and the Spanish.

Japan. Japan emerged from World War I as a victorious nation, although she fought no major naval engagements. While the war raged in Europe, Japan began the development of her 8/8 fleet construction program—that is, eight new battleships and eight new battlecruisers. These ships would have completely outclassed any European battleship since they were to be armed with either the 410-mm or the 457-mm gun. The characteristics of these ships were as follows:

Nagato and *Mutsu*	
Displacement	33,800 tons (34,342 mt)*
Armament	eight 410-mm guns in four twin turrets twenty 140-mm guns in casemates
Speed	26.5 knots
Tosa and *Kaga*	
Displacement	39,900 tons (40,541 mt)*
Armament	ten 410-mm guns in five twin turrets twenty 140-mm guns in casemates
Speed	26.5 knots
Amagi, Akagi, Takao, and *Atago*	
Displacement	41,200 tons (41,861 mt)*
Armament	ten 410-mm guns in five twin turrets sixteen 140-mm guns in casemates
Speed	29.75 knots
Owari, Kii, Numbers 11 and 12	
Displacement	42,600 tons (43,284 mt)*
Armament	ten 410-mm guns in five twin turrets sixteen 140-mm guns in casemates
Speed	30.0 knots
Numbers 13-16	
Displacement	47,500 tons (48,263 mt)*
Armament	eight 457-mm guns in four twin turrets sixteen 140-mm guns in casemates
Speed	30.0 knots

The construction of the ships of the 8/8 fleet program was well underway when the Washington Conference was convened in 1921. The completion of this program would have given the Japanese a most powerful battle fleet and would have led to large battleship building programs in the United States and Great Britain. Under the terms of the treaty, the Japanese agreed to complete the *Nagato* and *Mutsu*, while the *Akagi* and *Amagi* were to be completed as carriers. However, the *Amagi* was severely damaged in the 1923 Japanese earthquake and the *Kaga* replaced her. The *Tosa* was used in explosion tests and was finally sunk as a target ship. The construction of the battleships to be armed with 457-mm guns was never begun.

Japan adhered to the 1921 Washington Treaty and signed the 1930 London Naval Treaty. In 1934, however, Japan resigned from the League of Nations over the Manchurian Affair and later renounced her obligations regarding the naval treaties she had signed. This paved the way for the design and construction of the world's largest battleships—the *Yamato* class.

---

*Normal Displacement - In the Imperial Japanese Navy the English ton (2,240 lbs) was used only in the case of standard displacement (Washington Treaty) and in the older normal displacement. The older normal displacement was the English system of measure for a ships condition, that is 25% fuel, 75% ammunition, and 50% fresh water. This was used in the design of ships before 1923. Since that time, the Japanese Navy adopted a new normal displacement that they termed "Trial Condition," and this corresponded to that used in the U.S. Navy. This condition used 67% fuel and other consumables and full load (100%) ammunition. In all conditions after 1924, the units were in metric tons, although standard displacement remained in English units.

The London Naval Conferences. After the unsuccessful 1927 Disarmament Conference in Geneva, the prospects for further limitation on battleship size seemed doubtful. The new British government of Prime Minister Ramsay MacDonald called for a new conference in London to reach an agreement on further naval limitations. When this conference concluded its work, the only additional limitations on capital-ship design were that no carrier decks were permitted on battleships and battlecruisers; however, Great Britain did formally accept the right of the United States to possess a fleet equal to that of the Royal Navy. The "holiday" in capital-ship construction was extended to 31 December 1936, and it was agreed that another naval limitation conference should be convened in 1935.

The London Naval Conference of 1935-36 resulted in the 1936 London Treaty, which was signed by Britain, France, and the United States in March. The treaty reaffirmed the 35,000-ton standard displacement limitation of the earlier treaties, but the main battery was further limited to 356 mm. As the Japanese officially had withdrawn from the conference, the treaty stipulated that the gun caliber limit would revert to 406 mm if the Japanese failed to ratify the treaty by 1 April 1937. If any navy built ships exceeding the treaty limits, signatories to the treaty were given the right of escalation to a new standard displacement limit of 45,000 tons. Although Italy and Japan were not signatories of this agreement, they did send observers to the conference. German observers were also present, but they did not sign the document, since the Anglo-German Naval Treaty of 1935 required Germany to observe all naval treaties that had been or would be concluded.

Post-Treaty Capital-Ship Design. When Germany began the construction of her first armored ship, France authorized the construction of the *Dunkerque* in 1931. Italy, after many design studies, authorized the reconstruction of four old battleships in 1932-1933 and then decided to build two 35,000-ton battleships in 1934. The race was on again! Within a few years, capital-ship construction programs were active in all major navies. The new ships, with the exception of several defined as battlecruisers, all displaced at least 35,000 tons and had powerful armament, good protection, and speeds of at least 27 knots. The heavy gun was still considered the most effective naval weapon in 1935, although aircraft and submarines had advocates who insisted on claiming their supremacy. Naval authorities in Japan, Germany, Italy, Russia, France, the United Kingdom, and the United States showed their confidence in more modern battleships by starting the construction of powerful new ships with the best attainable combination of armor, long-range guns, improved antiaircraft batteries, good speed, and extensive subdivision and compartmentation.

The new ships represented the culmination of a long trend in capital-ship design by merging the speed of the battlecruiser with the gun power and protection of the battleship. The new, fast battleships approached the characteristics of the ideal battlecruiser as first proposed in 1905 by William Hovgaard.* Despite the high speed, technological advances permitted the incorporation of improved protection and more powerful guns in the new capital ships, which had these tactical and strategic characteristics:

- Higher maximum and cruising speeds, which improved mobility and gave added flexibility to the battlefleet.

*Transactions of the Society of Naval Architects and Marine Engineers, 1905 edition, New York, N.Y.

- Larger and more powerful guns, concentrating greater relative power in a single ship.
- Better gun platform stability and considerably improved gunfire control systems. The wartime incorporation of radar systems accentuated the contrast.
- Greater displacement, contributing to the ability to absorb more punishment.
- Great cost and large crews, which combined to make the loss of such ships tantamount to a national disaster.

The preceding discussion covers general trends in capital-ship design. There follows a discussion of the more technical aspects of ship characteristics and design necessary for full appreciation of the various ships described in this study.

Armament. Two basic developments led to the most apparent differences between battleships and battlecruisers of the World War I era and later capital ships: the newer ships had more powerful main battery guns,* with greater maximum ranges (largely the result of increased maximum elevation limits in turrets) capable of penetrating heavy armor at long range; and increased antiaircraft batteries emphasized the growing threat of air power that would eventually make the battleship obsolescent.

Improvements in heavy naval guns were largely evolutionary in nature, but the cumulative effect was of major consequence. Generally, newer guns fired much heavier projectiles than earlier weapons of the same caliber. Muzzle velocities were often less than for similar earlier weapons; this reduced bore erosion noticeably. The reduced velocity, however, was made up by increasing the mass of the projectile. Improved turret designs increased maximum elevation, thereby considerably increasing range, while the heavier projectiles gave better armor penetration.

Turrets. Twin turrets were used in the majority of European World War I capital ships, although several classes carried triple turrets, and the French *Normandie*-class battleships were designed to mount quadruple turrets. For ships designed after the naval holiday, the treaty displacement limitations, coupled with pressures for higher speed and improved horizontal and underwater protection, led to many weight-saving expedients. Among these were major changes in the ammunition-handling arrangements, redesign of turret rotating and elevating machinery to provide more efficient and lighter components, the use of higher-strength steels in the stationary support structure as well as the turrets themselves, and the use of welding. However, more weight savings were possible in the adoption of the triple or quadruple turret because more guns could be carried per total turret weight and the armor citadel could be reduced in length. These considerations led to the selection of quadruple turrets for the British *King George V* class, and for the French battlecruisers of the *Dunkerque* class and battleships of the *Richelieu* class. With the particularly stringent displacement limits on the French battlecruiser design, even the secondary armament of 130-mm guns on the *Dunkerque*-class ships featured quadruple turrets.

German rejection of the triple turret in the *Bismarck* and *Tirpitz*, despite its success in the *Scharnhorst* and *Gneisenau*, emphasized a widely held conviction that twin turrets permitted more effective gunnery with greater dispersion of the main battery and relatively little added displacement. All later German battleships had twin turrets,

---

*See Appendix C for detailed data on all guns.

The Nagato*, shown here, and the* Mutsu *were completed after the signing of the 1922 Washington Naval Treaty. Until the completion of the* Yamato *in 1941, these two ships were the most powerful in the Japanese Navy and mounted eight 410-mm guns in four twin turrets. The unusual "S"-shaped funnel was designed to keep stack gases clear of the forward superstructure. During a subsequent conversion, a more conventional trunked single funnel was provided. The* Nagato *survived the war only to be expended in the Bikini atomic tests in 1946. (United States Naval Institute)*

The Baden *and* Bayern *were the only German battleships of the World War I era to mount 380-mm guns. These ships were inferior to the British* Queen Elizabeth *class. The* Bayern *is shown here. These ships were the basis for the design of the* Bismarck *and* Tirpitz *some 25 years later. (R. O. Dulin Collection)*

The three "pocket battleships" of the Deutschland *class were the first major warships with an all-diesel propulsion. The ships were designed to outrun battleships and overwhelm faster cruisers and smaller combatants. They were desperately vulnerable to the few battlecruisers and fast battleships that had both sufficient speed to overtake them and adequate protection and gunpower to overwhelm them. The ships mounted six 283-mm guns in triple turrets, the first such mountings of large guns in the post–World War I German Navy. At the top is the* Graf Spee. *The* Deutschland, *later renamed* Lützow, *is in the middle, and the* Admiral Scheer *is at the bottom. (Blohm & Voss)*

despite the almost universal adoption of triple or quadruple turrets in other navies. The British battleship *Vanguard* mounted obsolescent 381-mm guns in twin turrets held in reserve for older World War I battleships, but she represented a design and material compromise accepted in an effort to expedite the completion of a battleship before the projected 1944 completion of the first *Lion*-class ship.

The relative power of heavy guns of similar vintage was generally proportional to their bores, with some exceptions. The U.S. *Iowa*-class ships mounted 406-mm guns that were superb weapons for that caliber, while the Japanese *Yamato* and *Musashi* mounted 460-mm guns that were at best of mediocre quality. As a result, for similar battle ranges the penetrative capabilities of the two guns were essentially equal. This is a classic example of the difficulties inherent in comparing the attributes of different classes of capital ships.

Cartridge and bag ammunition. Although powder charges for smaller caliber guns had long been contained in metallic cartridges, the use of such cartridges for heavy naval guns was not widespread. The German and Austro-Hungarian navies had pioneered the use of such heavy cartridges before World War I. This practice was continued in German battleship gun designs of World War II.

In certain respects, the cartridge has a considerable advantage. A magazine loaded with powder charges in cartridges instead of bags in lightweight canisters is much less likely to receive disabling damage from a shell hit. In addition, the cartridge contributes to the sealing of the breech opening, thus making possible the adoption of sliding wedge-type mechanisms, improving the potential for higher rates of fire. When cartridges are used, there is much less likelihood of flareback-type disasters caused by smoldering remnants from a previous powder charge.

On the other hand, cartridges for heavy guns are, of necessity, heavy and awkward to handle, forcing the adoption of complicated handling equipment if the potentialities of higher rates of fire offered by the simplified breech mechanism are to be realized. Bag ammunition is normally segmented, thus permitting the handling equipment to be much less heavy and complicated. Despite having had two years after the end of World War I to examine the *Baden*, one of the most modern German dreadnoughts, the Royal Navy declined the opportunity to follow the German example.

Dual-purpose guns. The United States, the United Kingdom, France, and Japan developed new dual-purpose guns effective against either light surface ships or aircraft. They felt that there was no necessity for a separate antiaircraft battery. A dual-purpose gun was selected with the intention of providing sufficient performance to engage either air or surface targets. Control was to have been arranged to split the battery for antisurface and antiaircraft purposes, with the important advantage that *all* of the secondary battery would be available for either air or destroyer attack.

The Germans and Italians, on the other hand, had not developed such guns and preferred to use a mixed-caliber secondary battery. These navies considered the repulse of the fast-moving destroyer as a vital concern to their capital ships, as potential opponents had superiority in surface ships. Despite expense, extra manpower, weight, erosion, space, and other objections, they advocated the gun with the flattest possible trajectory and the best possible position for resisting destroyer or light cruiser attacks. It was believed that the antiaircraft armament could have a shorter range.

The merits of the dual-purpose battery as compared to the mixed caliber battery can be debated at length. Briefly, the former permitted a more numerous single-caliber battery that would in all likelihood never be needed to engage surface and air targets simultaneously. Dual-purpose mounts were relatively economical in weight and space, while mixed-caliber batteries of necessity required greater space and weight, yet provided fewer barrels to engage either surface or air targets. On the other hand, proponents of the mixed-caliber battery argued that the necessity for a high rate of fire limited dual-purpose batteries to a bore too small to be effective against cruisers and destroyers. Normally, heavier guns in the mixed battery were about 150 mm, with dual-purpose being 127–134 mm. The French Navy, however, was an exception. The *Jean Bart* and *Richelieu* had dual-purpose 152-mm guns, but these had a much slower rate of fire than the British 134-mm and the American 127-mm guns. The point was a valid one, if in fact surface ships of cruiser and destroyer size were encountered. This was a case where there was no "right" answer—the key was the operational employment of the ships under discussion. Although the 150-mm gun was chosen for the *Bismarck* and *Tirpitz*, the ascendancy of aircraft by 1944 finally forced the Germans to develop a flak-type shell for their 150-mm and 380-mm guns for the *Tirpitz*, even though these guns' rate of fire, train, and elevation were not well suited for antiaircraft defense. The Japanese, who had used a dual-purpose 127-mm gun in the *Yamato* and *Musashi*, felt it was necessary to use a special flak shell in their 460-mm guns in 1944! The U.S. Navy, the Italian Navy, and the Royal Navy also experimented with AA shells for their capital-ship main-battery guns.

In any event, the heavy antiaircraft fire of Axis ships generally was sufficient, although more guns would have been desirable. British aircraft were slow, unlike Japanese or American aircraft that were faster and more maneuverable, and this situation masked the weaknesses of German and Italian battleship antiaircraft defenses. Under World War II combat conditions, prewar conceptions of the need for machine gun batteries proved to be hopelessly inadequate.

Antiaircraft machine guns. At the outbreak of World War II a typical capital ship had perhaps two dozen antiaircraft machine guns of two or three calibers. By the end of the war it was common for such ships to carry as many as one hundred machine guns. For example, the Japanese *Yamato* carried 24 25-mm guns at her completion in December 1941, but by 1945 she had been equipped with 150 of these guns. Invariably, many of these guns were mounted in open deck areas and on top of turrets, which made them subject to the blast effects of the larger guns. It was anticipated, however, that the light and heavier guns would not be used simultaneously.

The light machine guns were to provide close-in (less than 5,000 meters) coverage with a large volume of fire. At first, the 37-mm machine gun was considered a good weapon, but the development of the 20-mm gun at the beginning of World War II ultimately led to its use on German battleships in great numbers. The Japanese, however, continued to use their 25-mm gun until the end of the war when the gun became obsolete because its relatively light, low-velocity shells were inadequate against the new, faster, and more heavily armored aircraft. This is why the large batteries of 25-mm guns on the *Musashi* and *Yamato* were relatively ineffective against American carrier planes. The Germans and Italians used the 20-mm gun in their battleships as well as the heavier 37-mm machine gun, which was to cover the medium range. The Germans developed a highly effective 20-mm quadruple mount that gave excellent performance on the *Prinz Eugen*, *Scharnhorst*, and *Gneisenau* during the Channel Dash in February 1942. These guns, however, were hopelessly ineffective in 1944 against air attacks by Lancaster bombers on the *Tirpitz*. By 1944 it was important that an antiaircraft gun have a shell heavy enough and a range sufficient to engage and destroy an aircraft *before* it could release its projectile.

In the placement of the light battery, there were two primary concerns: the problem of providing complete arcs of fire for antiaircraft coverage of the ship, and the problem of how to reduce the blast interference from the main and secondary guns. The latter was virtually impossible to avoid. The Japanese opted for an unusual solution and moved their boat and airplane stowage to the fantail below the main deck, where they were free from blast. The superstructure was now free of such equipment and an ideal location for an antiaircraft battery. Another acute problem was the necessity to stow the ready-service ammunition where it would be accessible to the gun mount. In general, these problems intensified as the war continued when the air threat worsened, and additional guns were needed to counter the threat. These were installed as weight and space permitted.

Gunfire control and radar. Along with the improved ballistic properties of guns, of equal importance was the improvement in the fire control equipment. Even before the advent of radar, such progress had been made that one American naval authority commented:

> By the early 1930s, battleships were demonstrating their ability to fire effectively 356-mm and 406-mm projectiles at unseen, maneuvering targets at ranges greater than 24 kilometers in salvos of from 8 to 12 guns, remotely controlled and fired from a single key . . . .

The Royal Navy pioneered in the service introduction of radar for search and fire control, although Germany, France, and the United States had active programs for its development. Although radar systems were quite primitive at the time of the *Bismarck* chase in May 1941, the British use of radar played an important role in the destruction of the German battleship. Although the first radar sets were of the search type, later developments would see radar teamed with mechanical analog fire-control computers that substantially increased the effective ranges of naval guns against both surface and air targets and permitted for the first time the accurate engagement of targets otherwise obscured by darkness, fog, or smoke.

The proliferation of antiaircraft batteries and the increasing complexity and sophistication of gunfire control systems contributed to the inexorable growth in topside clutter on modern capital ships. In addition to two search radar installations, the German battleship *Bismarck* carried three main-battery fire control directors, two foward and one aft, plus four secondary battery directors, two on either beam. All this equipment, of course, contributed greatly to gunnery accuracy, but it also made the ships more vulnerable to damage, particularly in key gunfire control systems.

Antiaircraft VT fuzes. In addition, the development of the proximity or VT (variable time) fuze for antiaircraft guns in the United States and United Kingdom considerably increased their capital ships' chances of a "kill." A direct hit was no longer necessary to damage an aircraft; the development of the VT fuze represented a vast improvement in antiaircraft defense. The failure of the Axis navies to develop such devices meant that their guns had to rely on mechanical time fuzes for their projectiles. These were set at the gun mount and were based on approximate data.

Torpedoes. The use of torpedo tubes on capital ships resulted from some of the close skirmishes fought during the Battle of Jutland. That is why the *Nelson* and *Rodney* were so equipped. Some early battleship designs also featured such armament, but this was in the days before radar when a potential opponent could be covered by smoke, haze, or weather conditions. Torpedo tubes and torpedo magazines were difficult to protect adequately in weight-sensitive designs. Although most capital ships completed during World War II had no torpedo tubes installed, several German ships (*Scharnhorst,*

Gneisenau, and *Tirpitz*) were fitted after their completion with deck-mounted torpedo tubes to assist in commerce raiding. This resulted from a report filed with the German Naval High Command by Admiral Gunther Lütjens after the Atlantic operations with the *Gneisenau* and *Scharnhorst* in February–March 1941. He recommended their installation as a means of saving shell ammunition when such ships were involved in commerce raiding. Battleships "H" and "O," however, were to be completed with torpedo tubes, since their original designs specified such installations.

Armor Protection. Battleship armor protection, traditionally, had been designed to withstand attack by guns equal in caliber to that of the main battery guns. Battlecruisers were given protection on the basis of widely varying standards, ranging from essentially a battleship scale of protection (*Scharnhorst* and *Strasbourg*) down to the requirement that the citadel be able to withstand shellfire from heavy cruisers (Battleship "O").

Early in the dreadnought era, side-armor thickness was the primary index of adequate protection, because the relatively close ranges anticipated in combat made the likelihood of shells hitting the deck negligible. As ordnance refinements increased effective gun ranges, there was a growing awareness of the need to provide deck armor systems. Eventually the concept of the immunity zone as a criterion for the adequacy of the armor suit of a ship was widely accepted.

The immunity (or protected) zone was a range band for a specified gun in which the armored portions of the target ship were designed to resist penetration by a designated projectile. At or below the lower range limit of the zone, side or vertical armor would have been penetrated, while at or beyond the upper limit, deck or horizontal armor would be penetrated. These range limits were normally determined for a ship floating upright (no pitch, roll, trim, or list) with a specified target angle, usually 90 degrees. Under actual conditions of a seaway, the likelihood of roll, pitch, and yaw complicates the calculation of the real immunity zone. Inevitable variations in armor and projectile quality, as well as ballistic performance, introduce further uncertainties. As a result, the most prudent tactical maneuver was to maintain position near the midpoint of the calculated immunity zone.

Calculations of potential side-armor penetration are critically affected by the target angle. This target angle is the angle formed between the gun and target line as well as the centerline of the target ship, measured in the horizontal plane. For optimum penetration, this angle is 90 degrees.

It is important to remember that a 90-degree target angle does *not* mean that the projectile will strike vertical side armor perpendicular to its surface. The angle of impact of the projectile is dictated by a combination of the target angle, the angle of fall of the shell, and the inclination of the armor plate from the vertical. As optimum armor penetration results from a normal impact angle, warship designers often attempted to enhance the tendency of projectiles to ricochet by inclining the armor from the vertical (top edge farther outboard), thereby increasing the effective thickness of the armor.

The concept of the immunity zone gave the capital-ship designer a useful analytical device for evaluating the most efficient way to improve armor protection. The greater the net width of the immunity zone, the greater the tactical flexibility of the ship. By calculating the added weight of improved armor protection and comparing this with the change in width of the immunity zone for each variation in protection, the designer can select the more efficient way of improving the width of the immunity zone.

For example, for a ship with a theoretical immunity zone of 18,300 to 27,400 meters against the American 356-mm/50-caliber gun, assume it is decided to evaluate the effect of adding 98 tons of armor protection. Over the relatively great area of the deck armor, this is equivalent to a modest increase in thickness, which adds 270 meters to the necessary firing range for penetration. (Remember, long-range plunging fire obtains added deck-armor penetrative capability as range is increased.) Applying the same weight to side armor, a somewhat greater increase in thickness is possible, corresponding to a reduced firing range for side-armor penetration of 460 meters. In this instance, increasing the side-armor protection is clearly the most effective way of improving the width of the immunity zone.

Unfortunately, life is never quite so simple. Such analyses as that presented above ignore the relative probability of the side of the ship being hit compared with the likelihood of the deck being struck. As improved technology made long-range gunnery engagements ever more likely, the relative probability of the deck-armor system being hit increased tremendously, lessening the importance of heavy side-armor systems. This principle was grasped by the Germans and was one of the major reasons why the main side belts of German capital ships were thinner than those of most last-generation dreadnoughts. As in many instances of warship design, the designer is forced to a difficult compromise between contradictory requirements.

Belt armor. Prior to the dreadnought era, belt armor had been fixed vertically on the exterior sides of the hull. A few capital ships at the end of World War I were completed with inclined armor systems in an effort to improve the armor resistance by increasing the tendency of projectiles to ricochet harmlessly from the inclined armor. A projectile striking an armor plate obliquely meets a greater thickness of armor, but the effective thickness of the plate increases well beyond this factor. For example, a plate inclined 15 degrees from the vertical provides resistance equivalent to that of a vertical plate of some 30 percent greater thickness. On the other hand, a projectile approaching an inclined armor plate obliquely meets a smaller target width than if the plate were vertical. The inclined armor system must be wider to present the same effective area to an oncoming projectile. The ballistic data in table 1-3 refers to the 381-mm guns carried by the *Vittorio Veneto*-class ships.

At a firing range of 20,000 meters, an inclined plate (15 degrees from the vertical) provides the resistance of a vertical plate of 30 percent greater thickness, yet must be only 10.6 percent larger than a vertical plate of comparable effective area. For similar resistance, the inclined plate saves about 10 percent of the weight required by a vertical belt system. As the firing range increases, the inclined plate gradually loses its advantage, but it was the closer ranges that were of importance to designers. At longer ranges, the horizontal protection is much more likely to be hit than the side protection.

TABLE 1-3

The Effect of Projectile Angle of Fall on Required Plating Width for Inclined Side Armor

Range (meters)	*Angle of Fall*	*Vertical Plate* Target Width*	*Inclined Plate (15°) Comparable Width*†
20,000	13.4 °	9.73	11.06 (+10.6%)
24,000	18.05°	9.51	11.35 (+13.5%)
28,000	23.4 °	9.18	11.71 (+17.1%)
34,000	34.33°	8.26	12.67 (+26.7%)

*Vertical plate width, 10 meters

†Necessary inclined plate width, to provide equal target to that provided by 10-meter vertical plate.

The thick armor belt was carried only a few meters below the design waterline because shell damage was less probable and such massive weight could not be tolerated in weight-limited designs. Italian and German battleships had relatively shallow underwater belts, while the Japanese were more concerned with underwater hits in the design and arrangement of their side belts. The Japanese were aware of damage to main side belts due to underwater shell hits during tests on the *Tosa* (see Appendix A, Volume 1). It was demonstrated that this type of shell hit resulted in more flooding and local structural damage rather than the deep penetrative type of damage associated with above-water shell hits. Frequently, armor plates were driven back so that openings would occur in the structural backing and hull plating below the bottom edge of the armor. A remedy adopted in almost all modern battleships was to provide a thin armor belt or armored torpedo bulkhead (in those ships with vertical armor on the exterior plating) below the main side belt. Projectiles with underwater trajectories would be sufficiently slowed so that armor of 45-80 mm thickness would provide adequate protection. The structural backing of this bulkhead was important, and in French, American, and German battleships, welded construction was used to increase the resistance to leakage.

Germany and Great Britain were the only countries that equipped their new capital ships of the World War II era with vertical armor protection on the exterior of the ship; the British belt was carried deeper underwater. For similar protection, such vertical armor systems weigh more than the inclined type, but had the advantage of keeping shells from damaging the outer plating, causing flooding of wing compartments and reducing transverse stability. The armor arrangements of the new German capital ships of the *Scharnhorst* and *Bismarck* classes were quite similar to those given the *Ersatz Yorck*-class battlecruisers of World War I, when such practices were deliberately employed, but they also featured new composition armor plate of greater effectiveness. In addition, the Italian and German naval constructors used a lighter citadel belt above the heavier main side belt. This was done because these ships would have to fight an enemy with superior numbers of ships at close range, and protection was necessary to prevent cruisers and destroyers from damaging uptakes and other vital spaces above the main deck. Any splinter-type damage at this level, coupled with a list to the damaged side, could lead to serious flooding above the armor deck, with a consequent loss of stability and large free-surface effects. This additional belt was an important reason why the *Vittorio Veneto*- and *Bismarck*-class ships were heavier than their Allied contemporaries.

Horizontal protection. The usual scheme of horizontal armor protection in pre-1930 battleships consisted of a single deck of sufficient thickness to defeat a projectile at a specified range. In some instances, a splinter deck was added one deck below. Italian and German naval constructors believed that the heavy deck, known as the "protective" or "armor" deck, should be positioned below the weather deck—in Italian battleships two decks and in German three deck heights (one deck height = 2.5 – 2.75 meters). The latter would mean that the lower armor deck would be some 5 to 7 meters below non-ballistic material and at the level of the design waterline. In Japan, an opposite viewpoint was taken. The Japanese *Yamato* was designed to stop the armor penetration at the upper armor-deck level. For the Germans or Italians to use the Japanese philosophy would have meant the addition of more belt armor at the level of the citadel belt, which meant a substantial increase in displacement and a much larger ship. However, the airplane had become an important factor in naval warfare by 1936, particularly in attacks with armor-piercing bombs. This was demonstrated in one action in the Spanish Civil War when one of the armored ships received serious damage

This model is of the German battlecruiser Mackensen. *German battlecruisers were ancestors of the fast battleship, combining battlecruiser speed with the protection and armament of a battleship. The* Mackensen *was launched in 1917. Never completed, the ship would have had a main-battery armament of eight 350-mm guns. (Courtesy of Horst Feistel)*

The Dante Alighieri *was the first dreadnought to be armed with a triple turret in her main battery. Note the interesting arrangement of funnels and the amidships turrets. (United States Naval Institute)*

The Italians totally reconstructed four battleships that had been completed during the World War I era, including a total change in the side protection to include the Pugliese system, removal of the amidships main turret, and the substitution of a powerful new main battery of ten 320-mm guns, with twin turrets superfiring over triples. Included in this conversion was a total revamping of the propulsion plant—new boilers and turbines. The top view shows the Conte di Cavour *in 1940. (Courtesy of A. Fraccaroli) The bottom view is a 1937 photo of the* Conte di Cavour. *(Italian Navy)*

from bomb hits. During the 20-year life of the capital ship built at that time, it was assumed that accurate level-bombing would eventually be possible from heights beyond antiaircraft gun ranges, and the development of accurate bomb sights probably would improve the accuracy of bombing from high altitudes.

Based upon these considerations, the Germans and Italians felt it wise to use a deck of 40-50 mm of special-treatment steel that would be capable of defeating light case demolition bombs. Armor-piercing bombs would penetrate it, but since this deck was at least 5 meters above the armor deck, the armor-piercing bomb in theory should be brought to explosion at the level of the main armor deck. Fuze action would have been initiated as the bomb passed through the upper armor deck. Later during World War II, larger bombs with more explosive filler were developed with even greater fuze delays. The armor protection of a modern battleship was quite effective against conventional bombs of 100–1,000 kilograms. In fact, as the bomb sizes grew in World War II, thicker armor decks were required. A 1944 German battleship design study specified a total thickness of 350 mm of deck armor, as compared to 176 mm of armor on the *Bismarck*, completed only a few years earlier. Such massive protection could only be provided in giant ships, and the full load displacement planned for "H-44" was 141,500 tons as compared to 50,900 tons for the *Bismarck*.

The Germans had concluded that a ship at sea would be very difficult to hit with high-level bombers because of its maneuverability, and dive bombers of the 1930 era could not carry large enough bombs to penetrate the horizontal protection of a *Bismarck*-type ship. In addition, even the probability of hits from long-range shell fire with improved fire control was considered miniscule if the proper tactics were employed. The Germans considered the greatest danger at medium and short range (10,000 to 25,000 meters). Therefore, inclined deck armor was used to supplement the main side belts.

Aircraft bombs evolved during the course of World War II into the most important factor affecting the design of deck armor systems, particularly with the progressively greater lifting capability of modern aircraft that also permitted the introduction of heavier and more powerful bombs. Several factors influenced the penetrative performance of aerial bombs:

- Altitude of release
- Plane speed and angle of dive (if any)
- Weight of bomb
- Ballistic properties of the bomb
- Angle of impact and striking velocity
- Structural adequacy of the bomb

The rapid increase in the required deck armor protection made the problem insoluble and made battleships obsolete. The only ships that could carry such massive armor systems would have to be prohibitively large and expensive. The race was lost, and modern aircraft with heavy armor-piercing bombs could defeat any practical armor system. Additionally, aircraft more frequently employed semi-armor-piercing or high-explosive bombs, in order to take advantage of the substantial possibility of major underwater damage caused by near-miss bombs with heavy explosive charges. This was demonstrated by several incidents of bomb damage to the *Tirpitz* in 1944.

Similarly, although to a somewhat lesser degree, refined fire control systems greatly increased the chance of long-range shell fire hitting the deck armor. This, coupled with the greater striking angles of projectiles, increased the possibility of deck armor penetration as the more nearly normal impact angle of the long-range shell is coupled with increasing striking velocities because of gravity effects. At very long ranges, the striking velocity actually *increases* as range increases.

Two design approaches were adopted by Axis battleship designers in their efforts to counter the effects of modern ordnance. The Japanese accepted the "all-or-nothing" concept of protection in the *Yamato*, with the heaviest possible protection concentrated over the most essential areas and the remainder of the hull and superstructure essentially unprotected by armor. The Germans and Italians used limited armor protection (splinter belts) in the bow area to reduce shell damage to unprotected plate and consequent flooding. The experiences of the *Seydlitz* during the Battle of Jutland showed the wisdom of providing some armor protection outside the main citadel. Whatever the details of the armor arrangement, it was evident that much of the ship outside the armor citadel would remain vulnerable, since massive armor protection could not be afforded in terms of displacement and cost.

Conning towers. Armor protection of conning towers followed two courses in modern battleship design. The main battery director systems of World War I-type battleships and battlecruisers had a high degree of protection, and their operators were located behind heavy armor. A considerable number of optical rangefinders were provided, any one of which could control the main battery. With the advent of heavy antiaircraft batteries, this system was extended to control these guns, whether of the dual-purpose or the single-purpose battery. This led to the need for larger plotting rooms and more computing elements. It was, however, important to provide protection for the additional personnel, and placing them within the armored citadel would provide the advantage of increased protection. It also offered the highly desirable feature of removing the greater part of control personnel from blast interference, which could be so great as to be almost demoralizing when both main and secondary guns were firing. This type of arrangement decreased top weights, including the weight of a heavy conning tower and other splinter protection aloft. This system was adopted in the Royal Navy in its pre–World War II battleships, but not in Axis battleships. All other navies opted for the heavy armored conning tower, but some of the plotting and computation spaces were moved below to the armor box.

In many instances, conning tower armor shielded gunfire control systems, as well as ship control personnel. More desirable gunfire control positions high in the superstructure could only be given very limited armor protection because of stability limitations.

Turrets. Of all armor plate comprising turret armor (not including the barbette), only the roof plate was dependent upon a simple ballistic calculation. The face plate must be increased over its calculated required thickness an indeterminate amount to compensate for the weakening effect of the gun port openings. Side armor plate requirements simply could not be calculated, and the rear plate was generally determined by the requirements of turret balance.

The roof armor thickness given turrets was dependent upon a ballistic calculation. Due to the unfavorable slope of the forward portion over the gun chamber (approximately 5 degrees) and the greater possibility of a shell hit with normal impact, the turret roof plate in this area was frequently 20–50-mm thicker than the flat plate on the after part of the turret. The *Yamato* class had a sloping roof over the entire turret top. The slope of the forward portion was necessary to permit the depression of the main guns, which varied between 3–5 degrees, and so that headroom might be provided in the turret officers' booth and rangefinder. Each turret carried its own rangefinder in case of damage to those on the control towers.

Main-battery turret machinery and ammunition storage and handling facilities were protected by the barbette armor. The barbettes were constructed of heavy face-hardened armor and were dependent upon a ballistic calculation.* The lower portion of the barbette, however, which was below the main armor deck, was not constructed of face-hardened armor and was reduced in thickness. Only splinter protection was necessary in this area.

As far as resistance to penetration is concerned, it is pertinent to note that a single plate of a given thickness provided considerably more resistance than two plates of the same total thickness. Armor consisting of two layers of plating was intermediate in its resistance. Hence, it was highly desirable to concentrate as much protection as possible in a single plate.

In reviewing the armor systems of modern battleships, it appears that all battleships built after the Washington Treaty featured the "all-or-nothing" concept of armor protection, while armor was stripped from nonessential portions of the ship. In German and Italian capital ships, this concept was slightly modified to provide for splinter protection in the bow region. The concept of "all-or-nothing" armor protection was first introduced in the USS *Nevada* and was used by the British in the design of the *Nelson* and *Rodney* to reduce armor weight. With the advent of the 1922 Washington Treaty, the limit on displacement was severe, and this in turn restricted the amount of armor that could be used on a modern battleship design. Axis powers, recognizing the restrictiveness of the treaty in terms of speed, gun power, and protection, elected ultimately to ignore the treaty. Therefore, the *Bismarck-*, *Vittorio Veneto-*, and *Yamato*-class ships were much larger than their Allied contemporaries built to treaty limits.

A tough, yet flexible, structure was vital to the effectiveness of the armor systems. The armor was designed to defeat projectiles by breaking them up, deflecting them, or resisting the effects of their detonation. Although a good understanding of explosive phenomena was important to the design of effective armor systems, it was vital in the design and construction of underwater protection systems.

Underwater protection systems. An imperfect knowledge of underwater explosive phenomena hindered the development of efficient side protection systems for World War II capital ships. Frequently, very crude empirical relationships derived from explosion tests were extrapolated to permit the design of new systems. Although even the most massive capital ship could never approach the ideal of the "unsinkable ship," it was possible to develop systems that would absorb several torpedo hits and enable a

---

* *Face-hardened armor*—Thick armor plates, used in way of armor systems, turrets, and the like. The face-hardening treatment (such as the famed "Krupp Cemented Armor") was designed to make the exterior surface of the armor particularly hard, while retaining the relatively tough, ductile properties of the backing armor steel. The surface hardening was intended to contribute to the breaking up of oncoming armor-piercing projectiles. Earlier armor steels, hardened throughout, proved to be relatively brittle (subject to cracking). The combined structure of the case-hardened armor steel retained the desirable exterior hardness, while adding the necessary ability to yield ductilely to heavy impact loadings without fracture. This type of armor steel is now rarely used in warship construction.

Special-treatment steel—Relatively tough and ductile, suited for structural as well as ballistic protection applications. Special-treatment steel can be shaped relatively readily. Most horizontal armor systems were made of this steel. Maximum thicknesses were about 130 mm. Resistance was provided through the toughness and ductility of the steel, combined with the tendency of projectiles to ricochet. This type of steel remains an important factor in warship construction.

ship to remain in combat. Capital ships generally conformed rather well to this basic requirement, although increased torpedo explosive charges reduced the effectiveness of earlier designs.

Appendix A of Volumes 1 and 2 gave representative examples of the types of experimentation and ordnance trials that profoundly influenced capital-ship design of the World War II era. The specific examples cited were:

- A series of Royal Navy tests on the battleship *Monarch.*
- A series of Royal Navy tests on the German battleship *Baden.*
- The famous tests done on the *Marlborough, Emperor of India,* and the Job 74 mockup of a modern capital ship that affected Royal Navy design philosophy on bomb, shell, and torpedo protection for the *King George V, Lion, Vanguard,* and *Ark Royal* classes.
- Ordnance trials on the USS *Washington* (BB 47).
- Japanese experiments on the incomplete battleship *Tosa.*

The Germans did perform several caisson tests to confirm the principles of their underwater protection schemes for the *Bismarck,* "H," and *Scharnhorst* designs in the early thirties. These tests involved sections of an old predreadnought that was reconstructed to test properties of riveting versus welding, the correct spacing between the side shell and torpedo bulkhead, structural continuity, etc. The Italians used an old tanker, the *Brennero,* to confirm the principles of their new Pugliese underwater protection system in 1930.

Implicit in the design of side protection systems is the acknowledgment that it is impossible to fully protect the ship against the effects of a contact or powerful near-miss detonation. Tests indicated that side protection systems had to be at least 4 meters in breadth to provide adequate protection to the vitals. This is due to the physics of the underwater explosion. On detonation, the solid explosive converts into a powerful gas bubble that expands in the direction of least resistance. If the detonation is sufficiently close to a ship, that direction is through the shell plating into the hull. A well-designed side protective system would permit the absorption of the destructive effects or venting of the gases from the detonation without flooding beyond the holding bulkhead.

Any effective side protection system requires the following basic features:

- The exterior shell plating of the hull must detonate the explosive charge with the minimum possible fragmentation. This suggested the minimum possible hull plating thicknesses be used.
- The protective layer must allow the initial dissipation of energy by permitting the free expansion of gases resulting from the detonation. This led to the provision of venting plates designed to vent these gases into the air or within the ship.
- The protective layer must absorb and contain the remaining explosive energy.
- The inboard boundary (the holding or torpedo protective bulkhead) must remain intact after the explosion of even a large charge and should be located as far away from the explosion as possible.
- Individual bulkheads within the side protection system must have sufficient clearance to permit their maximum plastic and elastic deformation to the point of rupture without contacting the bulkhead inboard.
- The torpedo defense bulkheads should be carried a deck higher than the side protection system in order to lessen stress concentrations as the bulkheads deflect under explosive loadings, thereby lessening the likelihood of flooding. This also contributes to a partial venting of the explosive gas bubble.

The Giulio Cesare *is shown here at Taranto in August 1942. This ship was typical of the four radically reconstructed Italian dreadnoughts that survived World War I. These reconstructed ships were perhaps the most extensively modified capital ships of the between-war era. (Italian Navy)*

The Leonardo da Vinci *was one of six Italian battleships to enter service during the World War I era. The ship was destroyed by a magazine explosion on 2 August 1916. The hulk was later scrapped in 1921. (United States Naval Institute)*

These two photographs vividly show the differences between the Caio Duilio *before and after her reconstruction. The upper photograph (United States Naval Institute), taken during 1916, shows the amidship turret and split stack arrangement. The lower photograph (Italian Navy) shows the ship at Taranto in May 1942 with this turret removed and the stacks brought closer together. The ship had her entire propulsion plant renewed and was given modern boilers and turbines to give her increased power and speed.*

Conventional side protection systems incorporated a series of longitudinal compartments with dewatering and counterflooding capabilities for effective damage control. Compartments were designed to be kept either liquid-loaded or void; specific arrangements varied from design to design. The inboard boundaries of such systems were framed by a protective bulkhead, designed to remain watertight against underwater attack. Although later systems were more sophisticated, the design of World War II-era capital ships did make them quite resistant to underwater attack.

Much of the damage caused by an underwater explosion results from the fragmentation effect from the disintegrated ship structure that is propelled inboard at high velocity by the forces of the blast. It would appear that structure in the vicinity of an underwater protection system should either be very light, thus producing a minimum of dangerous fragments (a principle used by French engineers in the *Richelieu* class) or be sufficiently heavy so that the ship structure does not fragment at all (principle followed with vertical side belts in the *King George V* and *Bismarck* battleships). In most treaty battleship designs, the thickest bulkhead in the underwater protection system was the torpedo (inboard) bulkhead.

When the whole underwater protection system is subjected to a contact or non-contact explosion, the entire system behaves similarly to a beam on two supports with a mid-span load. The system will bend away from the explosive load, so it is important to provide sufficient supporting structure at the top and bottom of the system. Such principles were not rigorously followed in the *King George V, Scharnhorst,* and *Vittorio Veneto* designs. These ships exhibited significant inboard flooding of the torpedo bulkhead, which meant that the supporting structure was not sturdy enough to support the system when it deflected.

Important to the success of any underwater protection system is the structural arrangement of the longitudinal and transverse bulkheads. If the stiffeners for these bulkheads are too rigid and closely spaced, they will in fact break the bulkhead membrane into a series of small panels, each of which has a smaller capability to deflect elastically than would the larger panel. In some more modern systems, the transverse bulkheads in the underwater protection system are deliberately contoured, to enhance deformation. There is some reason to believe that this may have been a contributory cause to the inboard flooding in the *Prince of Wales.*

Early underwater protection systems featured coal as protection, placing it close to the skin of the ship where it would absorb the energy of the blast by being pulverized and acting as a first-rate fragment absorber. The German battleships *Baden* and *Bayern* featured such protection. Coal was replaced by oil in all warship designs after 1920. At first, designers of underwater protection systems believed that oil represented an enormous fire hazard if subjected to a torpedo hit. It was later recognized, however, that no matter how hot an explosion was, oil needed oxygen to support combustion, and whatever air would be in the tanks at the outset of the explosion would not be there afterward as the blast would consume virtually all the oxygen.

In their torpedo tests on the incomplete hull of the battleship *Tosa*, the Japanese discovered that all of the important fragments from a torpedo explosion resulted from disintegrated structure. These fragments moved inboard at an included angle of 65°. The fragments retained their velocity over the breadth of the *Tosa*'s protective system. The Japanese decided that a liquid medium would be one way to slow down the fragments as well as to distribute the force of the blast more evenly over the surface of the tank walls. Other countries soon discovered the same principle in tests that they conducted.

A typical side protection system such as that in the USS *Maryland*, completed in 1921, incorporated five elastic bulkheads, with the inboard one designed to remain intact and watertight after an underwater explosion.* The outboard void compartment permitted unresisted initial expansion of the gas bubble. Inboard were three compartments designed to be kept liquid loaded with fuel oil or water, to permit the inboard-bounding bulkheads to absorb much of the explosive energy as they were deformed and ultimately ruptured. The torpedo bulkhead was located inboard of a void compartment that was carefully kept at such a depth that the next outboard bulkhead would not strike the protective bulkhead before rupturing. The final dissipation of energy was achieved by liquid turbulence in this last compartment. The total breadth as well as the depth of the torpedo protection system were critical factors in its effectiveness, hence the advantage of large beams in protective system design.

Unfortunately, the transverse depth required in an effective torpedo protection system prohibited its installation at the extremities of the ship. Generally the system extended the length of the armored citadel—about 55 percent of the waterline length. This left the bow and stern, and such vital equipment as rudders and propellers, exposed. Sir Eustace Tennyson d'Eyncourt, the designer of the *Nelson* and *Rodney*, found no practical means of protecting the rudders and propellers and called them the "Achilles heel" of the battleship. This weakness was dramatically demonstrated during the *Bismarck*'s last engagement (24–27 May 1941) and the near loss of the *Vittorio Veneto* during the Battle of Cape Matapan.

One of the most difficult problems in designing a torpedo defense system for the modern capital ship, aside from the constantly accelerating size of explosives, was the provision of adequate torpedo protection near the ends of the armor box, particularly with very fine hulls with prismatic coefficients less than 0.61.† Because of the emphasis on speed, large boiler and engine rooms were necessary, although these were less in volume than those of battleships designed from 1915–1925. With magazines forward and aft of these spaces, it was necessary to provide reasonably full sections at the quarter points, particularly at half draft. With somewhat reduced beam and magazines in the vicinity, adequate side protection near the ends of the armor box was of great importance. Furthermore, the floodable length‡ is least in the vicinity of these quarter points, which is yet another important reason for obtaining the best possible torpedo protection in these areas.

In addition to the problem of these "quarter points," there was the passage of shafts through the torpedo bulkheads aft. This was a difficult structural and arrangement problem for either a three-shaft or four-shaft arrangement in the high-speed battleship. The center shaft in the three-shaft and inner shafts in the four-shaft arrangements pass through the hull well aft and clear of the region that requires torpedo protection, but the outboard shafts had to pass through the torpedo bulkheads with a long diagonal intersection. It appears that the three-shaft arrangement created more of a problem with the structural continuity of the longitudinal bulkheads in the torpedo protection system because the outboard shafts had to pass through them near the "quarter point," whereas the outboard shafts in the four-shaft arrangement did not. The integrity of the

*Manning and Schumacher, *Principles of Naval Architecture and Warship Construction*, Plate IV, "Midship section of USS *West Virginia*," Annapolis: U.S. Naval Institute, 1928.
†The ratio of the actual hull displacement volume to the product of the waterline length and the area of the ship's midship section below the waterline. The lower this ratio, the more fine the ship's hull form.
‡Floodable length—The maximum distance between two main transverse watertight bulkheads ensuring no more than the permissible sinking of the ship, that is, to the margin line. This line, determined by the Safety of Life at Sea Convention of 1929, stipulates that it be some 50 mm below the deck to which watertight bulkheads extend.

torpedo bulkhead system at these points was approximately maintained by incorporating the shaft alley bulkhead as a partial torpedo bulkhead. A number of battle damage incidents revealed substantial weaknesses in capital-ship protection in the stern. It appears probable that the torpedo protection of all capital ships was somewhat defective in the stern.

Subdivision. Although highly important in restricting flooding, subdivision is at best an imperfect means of protecting a ship. Normally, fragmentation damage resulting from a bomb, shell, or torpedo hit is sufficient to riddle numerous watertight bulkheads in the vicinity of the hit. As a consequence, considerable flooding often resulted even when the damaged ship had excellent subdivision. Despite this shortcoming, the subdivision of all warships was a most important factor in their resistance to damage.

In high-speed capital ships, the protection of bow and stern areas became very important, as the *Bismarck* and *Musashi* episodes will reveal. As sufficient armor protection was impossible, the most extensive practical subdivision was essential to minimize flooding after damage. The death throes of the Japanese battleship *Musashi* will be mute testimony to this. The *Yamato*-class battleships had very large compartments forward and aft of the armor box, and when they were damaged, large trims. Not only did this reduce speed, but it made the ships very difficult to steer. This vulnerability was met in other navies by keeping the armored citadel large enough to permit the ship to remain upright and stable with both extremities flooded, provided the citadel within the torpedo-protective bulkheads remained intact. In the *Richelieu* class, a large compartment just forward of the armor box was filled with a water exclusion material to prevent bow trims in the damaged condition.

Most capital ships discussed in these three volumes had a variety of multilayer sandwich-type underwater protection systems. Liquid loading of these systems also varied. Some designers were convinced that outboard void compartments were not desirable, because when the ship was damaged, these outboard voids filled with large amounts of water, which caused large lists to develop. Other designers used oil in the outboard compartment to minimize flooding and reduce lists. The smaller the list, the less the need for counterflooding and the greater the amount of reserve buoyancy.

Another important factor in subdivision is damage control. By this is meant the limitation, or reduction to the minimum, of the adverse effect on the ship's fighting efficiency of damage received from any source that causes water to enter the hull. Damage control was exceptionally well practiced by the Germans, as they recognized the importance of keeping their ships on an even keel without excessive trim. The spirit and the attitude of this philosophy can be seen from an excerpt taken from their Damage Control Regulations (Leckregeln) of World War I:

> The control of damage in battle supersedes all other requirements and demands above all the faculty of decision and prompt, reliable action. A damage-control service organized and trained from this point of view must also operate perfectly in cases of accidents occurring in time of peace. The preservation of the buoyancy of the ship is under all circumstances the first and most important requirement, every demand that can be made upon the vessel in time of peace and war being dependent upon this.

Note the emphasis placed upon keeping the vessel afloat. Not only was rigorous training of crews practiced in the German Navy, but diligent and exhaustive preparation for damage control was accomplished by the designers of the ships. That is why German ships were hard to sink.

During the design period of the modern capital ship, subdivision was also considered necessary to control the spread of noxious gases, which proved of little importance in World War II, as such gas attacks never materialized.

Radar. When most capital ships of the World War II era were designed, the requirements for electronics equipment played a very small role. Optical rangefinders were more important factors in gun operation, and their placement aboard ship was carefully thought out. By the end of the war, however, the efficiency of electronic gear, particularly radar, was a major factor in evaluating a ship's performance.

An important feature of battleship design in the pre-radar era was the necessity for gunfire control towers—usually one forward and another aft—where the main and secondary battery optical rangefinders were mounted. It was also important that these towers be high enough to permit uninterrupted lines of sight at least 23–30 meters above the design waterline. Two main battery and four secondary armament directors were deemed essential in modern battleship design. One of the most difficult problems was the placement of the after gunfire control position, due to the interference of stack gases. An ideal solution would have been placement of the aft control tower forward of the stack, but this was difficult to accomplish in a ship with main guns arrayed forward and aft and the propulsion plant amidships. It was not possible to arrange the boilers aft and the engines forward since the boiler heights and the necessary shaft clearances made this difficult if not impracticable.

In many battleship designs, the stack was carried at least 4 meters above the aft main battery director and the aft control tower moved as far aft as possible. Studies of the Battle of Jutland showed that stack-gas interference could be minimized as much as possible with the correct location of rangefinders and control towers. In the British *Lion*-class battleships (1938 design), the aft control tower was mounted on the after turret. These solutions were meant to minimize but could not eliminate the gas interference problem. The decision to keep the after control tower small, which reduced the eddying problem of stack gases, inevitably caused the directors for the secondary armament to be mounted on structures adjacent to the stack or outboard to provide good lines of sight.

Protection of these spaces was limited to splinter armor, usually of 10—20 mm thicknesses. Due to the large weights that would have been involved and the adverse effect on the vertical center of gravity and ultimately the metacentric height (GM), greater protection was not possible.

Electronics. Search-radar equipment developed rapidly to the point where long-range contacts were commonplace. The classic surface engagement of the Battle of Surigao Strait (25 October 1944) resulted in the near annihilation of the Japanese southern force in a night action during which American ships tracked them by radar and used radar fire-control equipment.

Air-search radar proved to be of enormous value in the antiaircraft defense of surface ships. By often detecting enemy aircraft at ranges in excess of fifty miles, it permitted the vectoring of fighters to meet the attack and alerted antiaircraft batteries in ample time to put up effective fire.

As used by the Royal Navy and the U.S. Navy, the combination of radar-directed antiaircraft guns and the variable-time fuze was highly effective in antiaircraft defense, although by no means a complete defense against determined attack. Such a combination was not developed by the Axis navies and put their ships at a severe disadvantage.

The increasing complexity of naval warfare forced a major growth in the communications facilities of all ships, leading to the installation of literally dozens of transmitters and receivers. The net effect was the proliferation of electronic spaces in the superstructures, to the detriment of habitability and battle-worthiness. Ironically, the electronics installations essential to the combat effectiveness of a ship are perhaps

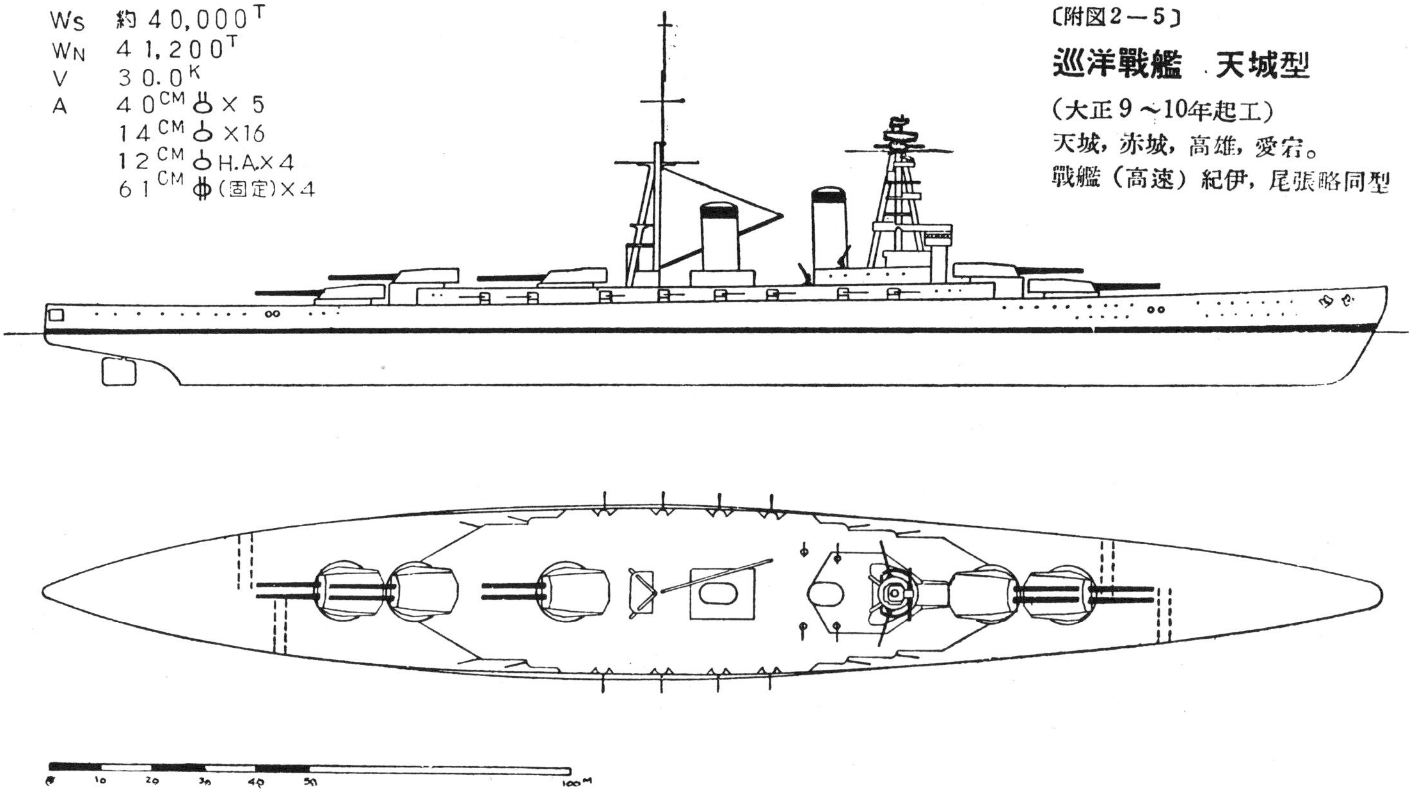

This is a sketch of the Amagi-class battlecruisers, two of which were to be completed as aircraft carriers under the terms of the Washington Naval Treaty of 1922. As battlecruisers these ships would have had a standard displacement of 40,000 tons, a 30-knot speed, and an armament of ten 410-mm guns in five twin turrets, 140-mm guns in casemates, and four 120-mm antiaircraft guns in single mounts. Characteristics of the period in which these ships were to have operated, they were to be armed with four twin 610-mm torpedo tubes. (Courtesy S. Fukui)

These ships were the ancestors of the Yamato. The Number 13–16 battleships that Japan envisioned as part of her 8/8 Fleet Program had the most powerful armament of the period (457-mm) and a speed of 30 knots. They never advanced beyond the design stage; Japan agreed not to construct them under the terms of the Washington Treaty. (Courtesy S. Fukui)

The Derfflinger *was one of three battlecruisers built to this design. With an adequate maximum speed of about 26.5 knots, a very heavy suit of armor for a battlecruiser design, and a powerful main battery of eight 305-mm guns, the ships of this class were very formidable adversaries. The* Derfflinger *has been credited with the destruction of the Royal Navy's battlecruisers* Queen Mary *and* Invincible *during the Battle of Jutland, 31 May 1916. (Foto Drüppel)*

Perhaps the most famous German capital ship of the World War I era in the United States, the Ostfriesland *became famous as the target ship sunk by U.S. Army planes under General William ("Billy") Mitchell in 1921. The four battleships of the* Helgoland *class were the first German battleships to mount the 305-mm gun. The wing turrets reflected a design philosophy of the time to have guns in reserve on the disengaged side—obtained at the price of inordinate complications in machinery arrangement and enhanced vulnerability to battle damage. (Foto Drüppel)*

the most susceptible to damage of all vital systems on board. Consequently, a ship with electronics gear by necessity located in the superstructure was much more susceptible to crippling damage than a comparable ship of an earlier era.

Propulsion Plant. The generally accepted need for speed markedly influenced modern designs, and all capital ships of the World War II era were designed for speeds of at least 27 knots. Such ships were distinguished by their great length and very fine lines, designed to attain higher speed by minimizing underwater hull resistance. The relatively fine hull form caused the machinery plant and main armament to be located farther aft than in World War I battleships. Wave-making resistance, created by the generation of waves as a ship moves through the water, depends on the hull form, speed-length ratio, and similar features.* Such resistance—as much as 40 percent of the total at high speeds—can be minimized by fine bow forms, well-designed stern shapes, and efficient longitudinal distribution of buoyancy. Frictional resistance, on the other hand, is a result of the relative motion of ship and water and varies in proportion to the wetted surface area of the hull. The benefits of efforts to reduce the speed-length ratio (i.e., increase length) to lessen wave-making resistance are somewhat offset because the increased length (greater surface area) results in more frictional resistance. Overall, the advantages of increased length offset the disadvantages of increased frictional resistance, hence the consistent adoption of long hull forms for high speed.

As speed requirements were increased, the demand for more powerful propulsion plants also enormously increased. This trend was particularly serious because of the effect of increased speed on the resistance of displacement hulls— power requirement varies approximately with the cube of the speed. For example, if 32,000 shaft horsepower drives a ship at 20 knots, some 62,500 shaft horsepower would be required at 25 knots, and 108,000 shaft horsepower for 30 knots—a 50 percent increase in speed requires more than a 235 percent increase in power.

Model basin tests. As a consequence, extensive model basin tests were made, seeking the best possible hull form to permit the highest speed with the least possible power. The tests generally indicated that a block coefficient† of 0.55 to 0.60 should be combined with a speed-length ratio of at most 1.0 in order to minimize power requirements for high-speed operations. At speed-length ratios greater than 1.0, the majority of the total resistance can be attributed to wave, form, and eddy resistance. The greater lengths and finer hull forms resulted in substantial fuel and power savings. Large bulbous bow structures became common, reducing wave-making resistance and the power required for high-speed operations. Although the details of the various hull forms varied, in all cases the hull geometry was the result of careful consideration of all factors involved.

Model-basin tests in the United States, Great Britain, and Japan evaluated various forms of twin keels from the standpoint of propulsion efficiency, hull resistance, and

*Ratio of the ship speed in knots to the square root of the waterline length in feet—an index of the effective speed of the ship in terms of resistance. The lower this ratio, the less the resistance per unit area of hull wetted surface. This is one of the prime reasons that high-speed ships are usually relatively long.

†The ratio of the actual hull displacement volume to the product of the waterline length, maximum waterline beam, and maximum draft. It is another index of the suitability of the hull form to high-speed operations. The lower this ratio (a rectangular form has a value of 1.00), the finer the hull form, but the less available volume for any given set of dimensions.

shielding the inner shaft. Interestingly, while American tests on twin-skeg models demonstrated such outstanding improvements in propulsive efficiency that all new American battleships were given such a stern form, Japanese and British tests demonstrated an *increase* in the required power. Notwithstanding such differences, and despite the complexity of the design problems, model tests were of considerable value to the development of hull forms.

Main propulsion machinery. Although improved hull forms somewhat alleviated the problems confronting the marine engineers, the remaining requirement—compact and powerful main-propulsion plants—was a major design challenge. In general, these problems were met by lighter-weight propulsive equipment, higher steam temperatures and increased pressures, improved boiler design, improved mechanical reliability, and electric arc welding.

The United States and German navies were especially bold in the adoption of steam turbine plants with relatively high pressures and temperatures. All other factors being equal, a steam turbine installation designed for extreme operating conditions can be more compact and efficient than a traditional, conservative design. As might be expected, there are complicated metallurgical problems involved in the adoption of more severe steam conditions, and German ships in particular were plagued by operational casualties caused by equipment failures. Many navies settled for more traditional designs, sacrificing speed for improved operational reliability and durability.

Diesel engines were proposed for the German battleships of the *Prinzregent Luitpold* class prior to World War I, but they were never installed. During the interwar years, the completion of the *Deutschland*-class armored ships with diesel propulsion was a significant achievement. Nevertheless, this type of propulsion plant, although proposed for the *Scharnhorst* and *Gneisenau,* could not be installed because of the inability to develop a successful engine of the necessary horsepower. The development of a successful diesel engine ultimately resulted in the "H"-and "O"-class designs having diesel propulsion plants proposed for them. There were no capital ships with diesel propulsion completed during the World War II era. Diesel engines were economical in terms of fuel consumption, but those engines available at the time were relatively heavy for their power. As a result, battleships with diesel installations had greater displacements for their power and were somewhat slower than those with turbine drive, even though they were capable of considerably better endurance.

Electrical installations. The increased electrical requirements of modern capital ships demanded powerful generator installations, particularly if the armament (main battery) were electrically powered, and the auxiliary machinery were all electric. Normally, electric plants were separated into several compartments in an effort to improve combat effectiveness after damage. Although there was no consistent trend toward either steam turbine or diesel drive, and there were variations between different classes of ships in the same navy, all navies with the exception of the U.S., French, and German used direct current for almost all shipboard purposes. The U.S. Navy had pioneered with alternating current, and all American ships featured alternating current installations. There are numerous advantages in using AC as compared to DC power—weight, efficiency, compactness, and reliability. The Italian and Japanese navies used more steam-powered auxiliaries, and thus their electric plants were much smaller. The large American capital ships of the *Iowa* and *Montana* classes had a maximum generating capacity of approximately 10,000 kilowatts.

German capital ships featured 100 percent reserve electric power, a practice that was consistent with their emphasis on damage control.

Ship Construction. In an era of treaty displacement limitations and emphasis on high speed, the efficient use of structural weights was of paramount importance. Careful analysis of structural design arrangements and construction techniques became of particular consequence and resulted in considerable progress. In addition, research into steel properties made it possible for many of the seafaring nations to develop higher-strength steels than had been employed during the First World War.

The conventional riveted construction of steel ships, requiring the overlapping of plates and additional material to ensure structural strength in areas of heavy loading, involved excess weight, and riveted joints were known to be weak points in the overall structure. Welded joints compared in strength to the base metal.

Welding. Unfortunately, welding technology was insufficiently developed when modern capital ships were designed and built. There was also a scarcity of skilled welders. There was, however, a consistent trend towards the greater use of welded joints in nonvital structural members.

As a consequence of the enormous advantages of structural welding, considerable experimental and developmental work was devoted to improvements in welding techniques and material. In Germany, great emphasis was placed on electric arc welding, and each of the three armored ships featured greater use of it as each was completed. The weight-saving aspects of welded plate and shapes were quickly realized by German shipyards, and the major yards were equipped with the necessary generators and equipment in the late twenties and early thirties. This facilitated the construction of the *Bismarck*- and *Scharnhorst*-class ships later in the decade. In Italy and Japan, on the other hand, there was great reliance on riveted structure for vital armor systems and major structural members, as the integrity of welded members under explosive loadings was suspect. Some welding was used, but this was limited to nonvital structure. In the long run, the trend in ship construction was decidedly in favor of welding.

As bulkheads in an underwater defense system absorb most energy through elastic deformation, the efficiency of joints between the individual plates becomes very important. If a joint is properly welded, its efficiency is 100 percent, that is, the weld metal in the joint is as strong as the original metal. Welding was still an experimental process in the 1930s and proper joints were difficult to achieve. Riveted joints were often used, even though they were less efficient in strength and weight. Welded structure tended to suffer other problems. As a welded plate is essentially one mass of metal, once a tear begins, it propagates rapidly and across the welded joint. In riveted construction, however, the rivet holes and the rivet seams act as crack arresters—thereby limiting the spread of tears. The Japanese and Germans had problems with some of their welded ships, and in the case of the Japanese the welded structure of the *Mogami*-class cruisers was replaced with riveted structure that featured different plate material and thickness.

Framing systems. Two basic structural systems were used in capital-ship construction. The transverse framing system consisted of closely spaced continuous frames, with widely spaced deep longitudinal members connecting adjacent frames. Such a system is relatively rigid and heavy. The longitudinal framing system is characterized by widely spaced transverse frames joined by numerous closely spaced shallow longitudinal members. This system is more flexible and considerably lighter for a given designed stress level. Normally, transverse framing arrangements were necessary amidships to help support the massive turret structures as well as the deck and side armor systems, while longitudinal systems at the extremities helped to conserve vital structural weight.

Political Factors. Domestic and international political considerations often played major roles in capital-ship design and construction. The enormous expense and industrial effort required to build even one capital ship often made the construction of such ships a matter of considerable domestic controversy. Construction was often delayed owing to political considerations, and at times such concerns led to the cancellation of major projects. Although these conditions were less important factors with the Axis powers, as they were essentially dictatorships, economic problems in Spain forced the cancellation of a battleship construction project.

The most obvious international factors were the treaties establishing displacement and armament limitations. Of somewhat lesser consequence was any one government's concern as to the reaction of another to the construction of new capital ships—for example, the German decision prompted by Adolf Hitler to retain the 283-mm guns in the *Scharnhorst* and *Gneisenau* in an effort to allay concern in the British government regarding German naval rearmament. Meanwhile, the Royal Navy arbitrarily limited the displacement of the *Lion*-class ships to 40,000 tons (despite the Treaty limit of 45,000 tons) standard in the (vain) hope that other European naval powers would follow the example of design moderation, thereby enhancing the relative combative value of the smaller *King George V* class. This did not happen, and when international tensions increased in 1938-39, Hitler finally wanted a battlefleet to protect his conquests from a belligerent United Kingdom.

All navies, of course, can be strongly influenced by a powerful chief of state. Classic examples cited in Volume 1 were the influence of President Franklin Roosevelt on the construction of the *Alaska*-class battlecruisers by the U.S. Navy and Prime Minister Winston Churchill in Volume 2 on the development of modern British battleships. The most active advocate of battleships among the Axis powers was Adolf Hitler, whose preoccupation with heavy guns in massive capital ships was a major factor in the German design studies that resulted in the "H"-class (1941-1944) designs, progressively larger battleship design schemes that eventually became such behemoths that they could not have been berthed in any existing German harbor. These German ships never reached the contractual state. During their conceptual stage of design, Hitler advocated that the H-class ships be armed with 800-mm guns. Such large guns were not beyond the capability of German technology of the period since the German Army ordered and built an 800-mm gun called "Gustav" for its siege artillery. Its characteristics were:

Gun	31.5"/40 (800 mm)
Shell weight (lbs)	15,653 (7,100 kg)
Muzzle velocity (fps)	2,362 (720 mps)
Maximum range (yds)	41,560 (38,000 m)
Maximum elevation	53°

Hitler's suggestions to Admiral Erich Raeder led the latter to appoint Admiral Werner Fuchs to show Hitler how enormous a ship would have to be to carry eight 800-mm guns. After being convinced of the massive ship size and the long construction time necessary to construct such guns and ships, Hitler finally relented and agreed to a 406-mm gun for H-39.

The Seydlitz *provided a classic example of the toughness of German capital ships, both battleships and battlecruisers. This 29-knot battlecruiser absorbed 23 shell hits and one devastating torpedo hit, which caused the ship to take on some 5,300 tons of water, yet the ship survived the Battle of Jutland. The* Seydlitz *had to steam astern to reach safety. (Foto Drüppel)*

The Admiral Scheer *was the second of the armored-ship or "pocket battleship" types. Slightly larger than her predecessor, she, too, saw extensive World War II service and was destroyed by British bombers in Kiel in April 1945. (Foto Drüppel)*

In Japan, there was a clash between the battleship and aircraft carrier advocates. Japanese naval operations in the interwar years had shown that aircraft and aircraft carriers were concepts that would have future potential and might possibly revolutionize naval warfare. There was much opposition to the construction of the four battleships of the *Yamato* class within the navy, but the battleship admirals were in charge. The construction proceeded and new fleet carrier construction lagged despite warnings from Admiral Isoroku Yamamato, who feared that even the giant *Yamato* class could fall victim to concerted air attack. His prophetic argument came true during 1944-45 when the *Musashi* and then the *Yamato* were sunk by massive air attacks originating from United States carrier task forces. Warnings of the battleship demise came earlier, however, and during the Battle of Midway four Japanese carriers were sunk not by battleships, but by carrier-based aircraft. This development brought about the conversion of the incomplete battleship *Shinano* to an aircraft carrier.

In Italy, Premier Benito Mussolini did not have the passionate interest of either Adolf Hitler or Winston Churchill in battleships. The development of the *Vittorio Veneto*-class battleships, however, came when his government realized that any Italian economic or political expansion in the Mediterranean would be opposed by a Franco-British alliance.

CHAPTER TWO

The Yamato Class

The design and construction of the battleships *Yamato* and *Musashi* and the aircraft carrier *Shinano* (a converted *Yamato*-class battleship) was one of the most noteworthy naval construction accomplishments of World War II. The careers of the two battleships marked the end of the development of that most expensive, complex machine of war, the modern battleship, which in 35 years had progressed from the 17,900-ton *Dreadnought*, armed with 305-mm guns, to these 64,000-ton giants armed with 460-mm guns.

In 1920–21, after the completion of the *Mutsu* and *Nagato*, Japan discarded the 14 remaining battleships and battlecruisers projected in her "8-8 fleet" program, as stipulated by the Washington Treaty. The *Kaga* and *Akagi* were converted into aircraft carriers. Several ships on launching ways were scrapped, while the *Tosa* was expended in ordnance trials that later influenced the design of the *Yamato*. Four projected battleships of the "Number 13" class, ancestors of the *Yamato* and *Musashi*, were never begun. Scheduled for completion in 1927, they were designed to displace 47,500 tons (normal) and carry eight 460-mm guns. Two decades passed before Japan completed another capital ship.

In early 1930, the General Staff (Gunreibu) wanted studies made for battleships that would replace the *Kongos*. The naval holiday was to expire on 31 December 1931, at which time Japan would replace these ships. It was decided that the displacement and armament would be limited as specified by the Washington Treaty, but that these battleships would be of an ocean-going type with greater endurance than that in earlier Japanese battleships or battlecruisers. Several designs were made, one by Admiral Y. Hiraga, who was then director of the Naval Technical Research Laboratory and had been responsible for the warships in the "8-8 fleet" program. Another design was advanced by constructor Captain K. Fujimoto, then the chief designer in the Naval Technical Bureau. He had been responsible for the design of the *Mogami*-class cruisers, the widely acclaimed *Fubuki*-class destroyer, and the torpedo boats of the *Tomozuru* class, one of which capsized in extremely heavy weather and unusually severe waves due to excessive superstructure and ordnance on such a small displacement. Table 2-1 summarizes the two proposed designs.

The Hiraga design was quite novel for its time because it featured a highly compact superstructure and main battery with superfiring triple turrets over twin turrets. This concentration reduced the length of the armored citadel (0.45 of waterline length, 0.65

A Yamato*-class battleship maneuvers violently during the Battle of Leyte Gulf, 24 October 1944. Japanese doctrine encouraged evasive maneuvering during air attacks. The U.S. Navy, on the other hand, concluded that the advantages of steaming on a steady course were greater, since radical maneuvering, such as practiced by the Japanese, tremendously reduced the effectiveness of antiaircraft gunnery. In the judgment of the U.S. Navy, this disadvantage more than offset the advantage of it being more difficult for aircraft to hit a maneuvering target. (James C. Fahey Collection)*

TABLE 2-1
Hiraga and Fujimoto Battleship Designs

	Hiraga	*Fujimoto*
Standard displacement	35,000 tons (35,562 mt)	35,000 tons (35,562 mt)
Trial displacement	38,630 tons (39,250 mt)	38,581 tons (39,200 mt)
Waterline length	760′ (231.6 m)	761.2′ (232.0 m)
Beam	105′ (32.0 m)	105′ (32.0 m)
Draft (trials)	30′ (9.14 m)	29.53′ (9.0 m)
Armament		
16.1″ (410 mm)	ten	ten
6.0″ (152 mm)	sixteen (8 in casemates) (8 in four twin turrets)	twelve (6 twin turrets)
4.7″ (120 mm)	eight (four twin turrets)	eight (four twin turrets)
aircraft	two planes/2 catapults	two planes/1 catapult
Speed	26.3 knots	25.9 knots
Shaft horsepower	78,906 (80,000 mhp)	72,002 (73,000 mhp)

in *Nagato* and *Mutsu*) and permitted a heavier main battery. The use of a superfiring triple over a twin was unusual, but the combination of restricted citadel length and fineness of hull forced such an arrangement. The use of casemates and twin mounts was also unusual at a time when most navies were discarding the casemate gun due to its limited arcs of train and elevation. The speed of 26 knots was modest, but it was 4–5 knots more than that of the U.S. *Colorado* and British *Nelson* classes. It would appear that speeds greater than about 26 knots were not possible for a 35,000-ton design with ten 406-mm guns.

The Naval Technical Bureau design was equally bold, with the secondary armament forward and aft of the main battery behind special blast shields and in special turrets. Even though the main armament and superstructure were not as concentrated as in the Hiraga design, the main armor box was only slightly longer (0.475 of the waterline length).

These designs were discussed and debated within the navy. However, the London Naval Treaty of 1930 extended the naval building holiday until 31 December 1936. Therefore, plans for construction of these designs had to be shelved. In 1934, however, the Japanese government, angered by sanctions imposed by the League of Nations over the Manchurian Affair, withdrew from that organization and renounced her naval treaty obligations. Preparations for the 35,000-ton battleships were dropped, and the General Staff requested a design with 460-mm guns (officially designated 406 mm), thick armor, good underwater protection and subdivision, high speed (30 knots, as the U.S. Navy would probably build 25–26-knot battleships), and size limited only by harbor and shipyard facilities. The specific requirements were:

Main armament	460 mm or greater
Secondary armament	12 155 mm tripled or 8 203 mm paired (no designs with 203 mm guns were developed)
Speed	more than 30 knots
Endurance	8,000 miles at 18 knots
Armor protection	To withstand 460-mm gunfire at ranges between 20,000 and 35,000 meters.
Torpedo protection	To withstand 300-kilogram TNT charges in contact with the side plating (standard established by designers).

The battleship Yamato *is shown here during the Battle of Leyte Gulf on 25 October 1944. Note the cluster of antiaircraft weapons amidships. These ships were armed with 25-mm, 127-mm, and 155-mm guns for AA defense. Even the 460-mm guns were equipped with fuzed shells for antiaircraft use during the latter part of the Second World War. The ship was struck by two bombs forward of Number 1 turret.*

Japanese naval authorities, well aware that their navy would be outnumbered in operations against the U.S. Navy, attempted to offset this inferiority by producing ships that were individually superior to their American opponents. To assist in their design efforts, they evolved a series of five estimates of the characteristics of U.S. battleships built to "no-Treaty" limitations, but restricted to a beam of 32 meters by the Panama Canal. Some of these designs, made in August 1935, were accurate predictions. Their estimate of the U.S. Navy's "1938 Type" assumed a 35,000-ton standard displacement, 33 knots, nine 406-mm guns, and a waterline length of 268 meters. The U.S. Navy's *Iowa*-class battleships, as completed, displaced 48,425 tons standard, had a speed of 33 knots, nine 406-mm/50-caliber guns, and a waterline length of 262 meters. The Japanese then proceeded to design ships that could not be matched by warships restricted in beam by the Panama Canal.*

Vice Admiral Keiji Fukuda (constructor) was in charge of the design effort and had the assistance of Japan's best naval constructors and naval architects. Chief design advisor was Yuzuru Hiraga (retired), who had designed the *Nagato* and *Mutsu* and other aborted ships in the 8-8 fleet program.

A total of 24 design schemes were prepared over a period of two years. Alternatives to the specifications were considered throughout the course of the preliminary design, involving changes in armament, protection, speed, and endurance. Early preliminary designs, completed from March until August 1935, are summarized in table 2-2. These designs were the foundations for the more intensive effort that came later.

---

*The battlecruisers of the B-65 type, planned as answers to the U.S. *Alaska* class but never built, are discussed briefly on pages 86–88.

TABLE 2-2*

Preliminary Design Studies—Yamato Class[a]

Design	*A-140*	*A-140A*	*A-140B$_2$*
Date	*10 March 1935*	*1 April 1935*	*1 April 1935*
Displacement			
Trials	68,402 (69,500)	66,926 (68,000)	61,021 (62,000)
Dimensions			
Waterline length	964.57′ (294 m)	908.79′ (277 m)	810.37′ (247 m)
Maximum beam	135.17′ (41.2 m)	132.55′ (40.4 m)	132.55′ (40.4 m)
Draft	34.12′ (10.4 m)	33.79′ (10.3 m)	33.79′ (10.3 m)
Armament			
18.1″ (460 mm)/45	nine	nine	eight
16.1″ (410 mm)	—	—	—
Arrangement[b]	3-3-3-A-O	3-3-3-A-O	2-2-A-2-2
6.1″ (155 mm)/60	twelve (tripled)	twelve (tripled)	twelve (tripled)
5″ (127 mm)/40	twelve (twin)	twelve (twin)	twelve (twin)
1″ (25 mm)/60	twenty-four (twin)	twenty-four (twin)	twenty-four (twin)
Armor			
Gun standard[c]	18.1″ (460 mm)	18.1″ (460 mm)	18.1″ (460 mm)
Immunity zone[d]			
inner limit	21,872 yds (20,000 m)	21,872 yds (20,000 m)	21,872 yds (20,000 m)
outer limit	32,808 yds (30,000 m)	32,808 yds (20,000 m)	32,808 yds (30,000 m)
Propulsion			
Diesel shp	—	67,070 (68,000 mhp)	138,874 (140,000)
Turbine shp	197,264 (200,000 mhp)	130,194 (132,000 mhp)	—
Total shp	197,264 (200,000 mhp)	197,264 (200,000 mhp)	138,874 (140,000)
Maximum speed	31 knots	30 knots	27.5 knots
Endurance	8,000 miles @ 18 knots	9,200 miles @ 18 knots	9,200 miles @ 18 knots

*See footnotes on page 51.

The first design in table 2-2 was A-140, completed on 10 March 1935. It featured a main battery of three triple turrets forward of the superstructure, as did 16 of the 24 designs. The armament was very concentrated amidships in order to keep the heavy deck and side armor to a minimum. This also permitted the magazines to be located in a wider part of the ship and ensured that the underwater protection for these compartments was the best possible. Providing the proper volume for the magazines of 460-mm guns was an important preliminary design consideration. Many studies were made regarding the turret machinery, as well. Another problem was the blast from the 460-mm and 155-mm guns. This would make it necessary to concentrate the guns so that the antiaircraft armament and aircraft and boat stowage facilities would be outside the danger zone. The geared-turbine power plant was capable of a maximum speed of 31 knots. The ship was too large in terms of displacement and length, however, and would have required extensive changes to harbors, naval dockyards, and shipbuilding yards (strengthening of building docks or launching ways). The *Yamato,* as built, approached the displacement of A-140, but with considerably reduced length and speed.

Design A-140-B$_2$, completed on 1 April 1935, was the only one to rely solely on diesel propulsion. Japanese marine engineers were impressed with the fuel economy of the diesel engine. Although a greater price was paid in the standard displacement—the

A-140-G *25 May 1935*	*A-140-G_1-A* *30 July 1935*	*A-140-I* *30 July 1935*	*A-140-J_0* *31 July 1935*
64,842 (65,883)	60,627 (61,600)	64,022 (65,050)	51,178 (52,000)
895.67′ (273 m)	805.44′ (245.5 m)	879.26′ (268 m)	793.96′ (242 m)
123.69′ (37.7 m)	127.62′ (38.9 m)	127.62′ (38.9 m)	118.77′ (36.2 m)
34.12′ (10.4 m)	34.12′ (10.4 m)	34.12′ (10.4 m)	33.14′ (10.1 m)
nine	nine	ten	—
—	—	—	nine
3-3-3-A-O	3-3-3-A-O (*Nelson*)[e]	2-3-A-3-2	3-3-3-A-O (*Nelson*)[e]
twelve (tripled)	twelve (tripled)	eight (twin)	nine (tripled)
twelve (twin)	twelve (twin)	sixteen (twin)	twelve (twin)
twenty-four (twin)	twenty-four (twin)	twenty-four (twin)	twenty-four (twin)
18.1″ (460 mm)	18.1″ (460 mm)	18.1″ (460 mm)	16.1″ (410 mm)
21,872 yds (20,000 m)	21,872 yds (20,000 m)	21,872 yds (20,000 m)	19,685 yds (18,000 m)
32,808 yds (30,000 m)	29,527 yds (27,000 m)	29,527 yds (27,000 m)	29,527 yds (27,000 m)
69,042 (70,000 mhp)	54,248 (55,000 mhp)	69,042 (70,000 mhp)	59,179 (60,000 mhp)
69,042 (70,000 mhp)	59,179 (60,000 mhp)	72,001 (73,000 mph)	59,179 (60,000 mhp)
138,084 (140,000 mhp)	113,427 (115,000 mhp)	141,043 (143,000 mhp)	118,358 (120,000 mhp)
28 knots	26 knots	28.0 knots	27.5 knots
19,000 miles @ 18 knots	6,600 miles @ 16 knots	7,200 @ 16 knots	7,200 miles @ 16 knots

engine weight was greater than a steam power plant—the amount of fuel to be carried to achieve the same endurance was less. Another alternative (A-140-A) introduced the diesel-steam plant combination, a concept first developed for the German battleship *Prinzregent Luitpold* (diesels never installed). This arrangement was to form the basis of the propulsion plant that was to be used during the rest of the preliminary design. The horsepower of the machinery would vary as the design continued, but this was due to the variations in ship speed and endurance, as well as to variations due in the engine designs, which were in progress at the time of the preliminary design of the *Yamato* class.

Four design studies (A-140-J_0, A-140-J_2, A-140-G_2-A_1, and A-140-J_3) were made to evaluate a 410-mm armament for these ships. The displacement of design A-140-J_0 was relatively modest, with protection required against 410-mm instead of 460-mm gunfire, thereby reducing the required armor thicknesses. The maximum speed was 27.5 knots. Also, the Japanese considered the effect of lesser armor protection (immunity zone predicated on 410-mm gun) as a means of reducing displacement and gaining speed. It was thought that the armor protection would be sufficient against an American 406-mm gun.

The slowest speed considered was 24 knots. Design A-140-K, completed in August

TABLE 2-2 *(Continued)*

Design	*A-140-J$_2$*	*A-140-K*	*A-140-K*
Date	*31 July 1935*	*1 August 1935*	*5 August 1935*
Displacement			
Trials	53,177 (54,030)	49,268 (50,059)	51,080 (51,900)
Dimensions			
Waterline length	836.61′ (255 m)	725.06′ (221 m)	771.0′ (235 m)
Maximum beam	126.31′ (38.5 m)	118.11′ (36.0 m)	121.39′ (37.0 m)
Draft	33.45′ (10.2 m)	33.14′ (10.1 m)	33.79′ (10.3 m)
Armament			
18.1″ (460 mm)/45	—	eight	eight
16.1″ (410 mm)	nine	—	—
Arrangement[b]	3-3-3-A-O (Nelson)[e]	2-3-3-A-O	2-3-3-A-O
6.1″ (155 mm)/60	nine (tripled)	nine (tripled)	nine (tripled)
5″ (127 mm)/40	twelve (twin)	twelve (twin)	twelve (twin)
1″ (25 mm)/60	twenty-four (twin)	twenty-four (twin)	twenty-four (twin)
Armor			
Gun standard[c]	16.1″ (410 mm)	16.1″ (410 mm)	16.1″ (410 mm)
Immunity zone[d]			
inner limit	19,685 yds (18,000 m)	21,872 yds (20,000 m)	21,872 yds (20,000 m)
outer limit	29,527 yds (27,000 m)	29,527 yds (27,000 m)	29,527 yds (27,000 m)
Propulsion			
Diesel shp	69,042 (70,000 mhp)	39,453 (40,000 mhp)	44,384 (45,000 mhp)
Turbine shp	64,111 (65,000 mhp)	39,453 (40,000 mhp)	49,316 (50,000 mhp)
Total shp	133,153 (135,000 mhp)	78,906 (80,000 mhp)	93,700 (95,000 mhp)
Maximum speed	29 knots	24.0 knots	26 knots
Endurance	6,000 miles @ 18 knots	6,600 miles @ 16 knots	6,600 miles @ 16 knots

1935, featured eight 460-mm guns forward in one twin and two triple turrets. The armor requirement was reduced—410-mm guns were used as the standard for protection, despite the main armament of 460-mm guns. Modest speed requirement gave this design the least displacement of any studied—50,059 metric tons in the trials condition. This was the absolute minimum battleship that could mount 460-mm guns and achieve superiority over a 35,000-ton American battleship with slightly greater speed.

Design scheme A-140-J$_3$ is of considerable interest because of its similarity in armament, protection, and speed to the U.S. Navy's *Montana* design of a few years later (see table 2-3).

Both designs featured a main battery evenly distributed forward and aft. Armor protection was designed to withstand 16-inch gunfire at normal battle ranges. The designed endurance of the *Montana* was to have been twice that of the Japanese design—15,000 nautical miles at 15 knots versus 7,200 miles at 16 knots. The Japanese, with their design concept of individual ship superiority, discarded the 410-mm gun as a main battery weapon for their future designs.

Table 2-4 summarizes designs A-140-F$_3$ and A-140-F$_4$, which were completed on 5 October 1935, and the final preliminary design of 20 July 1936. By October 1935, the primary characteristics of the *Yamato*-class battleship had been well defined: a main

A-140-K 5 August 1935	*A-140-K* 5 August 1935	*A-140-G_1-A* 10 August 1935	*A-140-J_0* 10 August 1935
53,049 (53,900)	52,753 (53,600)	58,560 (59,500)	52,163 (53,000)
807.08′ (246.0 m)	777.56′ (237.0 m)	800.52′ (244 m)	800.52′ (244 m)
121.39′ (37.0 m)	121.39′ (37.0 m)	127.62′ (38.9 m)	120.08′ (36.6 m)
33.79′ (10.3 m)	33.79′ (10.3 m)	34.12′ (10.4 m)	33.46 (10.2 m)
eight	eight	eight	—
—	—	—	nine
2-3-3-A-O	2-3-3-A-O	2-3-3-A-O	3-3-3-A-O
nine (tripled)	nine (tripled)	twelve (tripled)	nine (tripled)
twelve (twin)	twelve (twin)	twelve (twin)	twelve (twin)
twenty-four (twin)	twenty-four (twin)	twenty-four (twin)	twenty-four (twin)
16.1″ (410 mm)	16.1″ (410 mm)	18.1″ (460 mm)	16.1″ (406 mm)
21,872 yds (20,000 m)	21,872 yds (20,000 m)	21,872 yds (20,000 m)	19,685 yds (18,000 m)
29,527 yds (27,000 m)	29,527 yds (27,000 m)	29,527 yds (27,000 m)	29,527 yds (27,000 m)
64,111 (65,000 mhp)	49,316 (50,000 mhp)	54,248 (55,000 mhp)	64,111 (65,000 mhp)
64,111 (65,000 mhp)	49,316 (50,000 mhp)	54,248 (55,000 mph)	64,111 (65,000 mhp)
128,222 (130,000 mhp)	98,632 (100,000 mhp)	108,496 (110,000 mph)	128,222 (130,000 mhp)
28.0 knots	26.0 knots	26 knots	28.0 knots
6,600 miles @ 16 knots	7,200 @ 16 knots	6,600 miles @ 16 knots	7,200 miles @ 16 knots

battery of nine 460-mm/45-caliber guns in triple turrets, two forward and one aft; maximum speed 27 knots; armor protection designed to withstand 460-mm shells. The only material variation between the nearly identical designs involved the endurance, and this led to small differences in displacement and length. This was due to added fuel requirements for the larger endurance.

After the two October 1935 designs had been reviewed, it was decided to accept an endurance of 7,200 nautical miles, which was rather modest for Pacific operations. However, a larger endurance would have involved more fuel and a ship of greater size. The Japanese then began the final design for the *Yamato*, which would yield the characteristics necessary for construction and specifications for the building yards. The detailed plans of Design A-140-F_5 were completed on 20 July 1936 and accepted by the Japanese Navy. The characteristics are summarized in table 2-4. (Design A-140-F_5)

The Japanese Navy had a powerful new two-cycle, double-acting diesel engine, known as the Number-11 Type, in service in the *Taigei*-class submarine tenders. The Number-13 Type diesels projected for the *Yamato* class were actually developments of the Number-11 Type. However, the latter diesels developed some problems caused by a fundamental design defect, which made them very unreliable in service, and a major repair and maintenance effort was required to keep the tenders operational. This forced

TABLE 2-2 *(Continued)*

Design	*A-140-G_1-A*	*A-140-G_1-A*	*A-140-F*
Date	*12 August 1935*	*14 August 1935*	*14 August 1935*
Displacement			
Trials	59,052 (60,000)	59,987 (60,950)	59,397 (60,350)
Dimensions			
Waterline length	803.80′ (245.0 m)	816.93′ (249.0 m)	810.37′ (249 m)
Maximum beam	127.62′ (38.9 m)	123.62′ (38.9 m)	127.62′ (38.9 m)
Draft	34.12′ (10.4 m)	34.12′ (10.4 m)	34.12′ (10.4 m)
Armament			
18.1″ (460 mm)/45	eight	eight	eight
16.1″ (410 mm)	—	—	—
Arrangement[b]	2-3-3-A-O	2-3-3-A-O	2-3-A-3
6.1″ (155 mm)/60	nine (tripled)	nine (tripled)	twelve (tripled)
5″ (127 mm)/40	twelve (twin)	twelve (twin)	twelve (twin)
1″ (25 mm)/60	twenty-four (twin)	twenty-four (twin)	twenty-four (twin)
Armor			
Gun standard[c]	18.1″ (460 mm)	18.1″ (460 mm)	18.1″ (460 mm)
Immunity zone[d]			
inner limit	21,872 yds (20,000 m)	21,872 yds (20,000 m)	21,872 yds (20,000 m)
outer limit	29,527 yds (27,000 m)	29,527 yds (27,000 m)	29,527 yds (27,000 m)
Propulsion			
Diesel shp	54,248 (55,000 mhp)	64,111 (65,000 mhp)	64,111 (65,000 mhp)
Turbine shp	54,248 (55,000 mhp)	64,111 (65,000 mhp)	64,111 (65,000 mhp)
Total shp	108,496 (110,000 mhp)	108,496 (110,000 mhp)	128,222 (130,000 mhp)
Maximum speed	26.0 knots	26.0 knots	27.0 knots
Endurance	7,200 miles @ 16 knots	7,200 miles @ 16 knots	7,200 miles @ 16 knots

the Japanese to review the propulsion plant design of the *Yamato* class, because the 200-mm armor plates over the engine rooms made engine changes difficult; excellent reliability was mandatory. Therefore, the designers decided to use an all-steam turbine plant instead.

The final plans were completed in March 1937, featuring the all-steam propulsion plant. The characteristics are tabulated in table 2-5.

The first battleship of the Third Fleet Replenishment Program of 1937, the *Yamato*, was officially scheduled to be laid down at the Kure Navy Yard in November 1937, to be followed by the *Musashi* at Mitsubishi's Nagasaki yards in March 1938. Two other units of the *Yamato* class were also planned. One would be built at the Kure Yard after the *Yamato*, while the third would be built at the Yokosuka Navy Yard. The yards chosen for the construction of these ships were carefully selected. In some cases, approaches to them had to be dredged and the building facilities modified to handle the construction.

The Japanese were well aware that they could not hope to keep the construction of the *Yamato*-class ships a secret, but they did hope to keep the Allies ignorant of their extraordinary characteristics. At the Kure Navy Yard where the *Yamato* was built,

A-140-G_0-A	A-140-G_2-A	A-140-J_3
14 August 1935	*30 August 1935*	*30 August 1935*
64,416 (65,450)	62,448 (63,450)	57,477 (58,400)
879.26′ (268.0 m)	859.58′ (262.0 m)	826.77′ (252.0 m)
127.62′ (38.9 m)	127.62′ (38.9 m)	127.62′ (38.9 m)
34.12′ (10.4 m)	34.12′ (10.4 m)	34.12′ (10.4 m)
nine	nine	twelve
—	—	—
3-3-3-A-O (*Nelson*)[e]	3-3-3-A-O (*Nelson*)[e]	3-3-A-3-3
twelve (tripled)	twelve (tripled)	nine (tripled)
twelve (twin)	twelve (twin)	twelve (twin)
twenty-four (twin)	twenty-four (twin)	twenty-four (twin)
18.1″ (460 mm)	18.1″ (460 mm)	16.1″ (410 mm)
21,872 yds (20,000 m)	21,872 yds (20,000 m)	21,872 yds (20,000 m)
32,808 yds (30,000 m)	29,527 yds (27,000 m)	32,808 yds (30,000 m)
69,042 (70,000 mhp)	69,042 (70,000 mhp)	69,042 (70,000 mhp)
73,974 (75,000 mhp)	72,001 (73,000 mhp)	64,111 (65,000 mhp)
143,016 (145,000 mhp)	141,043 (143,000 mhp)	133,153 (135,000 mhp)
28 knots	28 knots	28 knots
7,200 miles @ 16 knots	7,200 miles @ 16 knots	7,200 miles @ 16 knots

[a]Values in parentheses are in equivalent metric units.
[b]3-3-3-A-O indicates that the main battery was concentrated forward in three triple turrets.
[c]The caliber gun for which the immunity zone was determined.
[d]The theoretical range band in which the armored citadel is immune to the shellfire of the enemy gun. The inner limit is determined by resistance of the side armor, while the outer limit is determined by horizontal protection.
[e]The *Nelson* reference means the same main-battery arrangement as British *Nelson* and *Rodney*, but with the third turret turned aft instead of forward.

about one-fourth of the dock was roofed over to prevent observation from a nearby hill. The slipway for the *Musashi* was covered by a 408-metric-ton sisal rope curtain. The amount of sisal rope needed was so great that its supply in Japan was materially affected, causing shortages in rope for the fishing industry. These efforts to maintain secrecy succeeded completely. By 1942, the U.S. Navy was aware of the existence of new battleships, and a rough sketch of the *Yamato* profile was captured in the southwest Pacific during the late spring of 1943. However, even by June 1945, after both ships had been sunk, Americans still believed that they carried 406-mm guns and displaced about 45,000 tons in the standard condition.

Before construction could commence, considerable preparations were necessary. At Kure, the construction dock was made one meter deeper to accommodate the *Yamato*

TABLE 2-3

Comparison of A-140-J3 with USS Montana

	A-140-J3	*Montana*
Displacement	57,478 tons (58,400 mt) (trials)	60,500 tons (61,471 mt) (standard)
Waterline length	826.77′ (252.0 m)	890.0′ (271.27 m)
Maximum beam	127.63′ (38.9 m)	121.17′ (36.93 m)
Main battery	twelve 16.1″ (410 mm)	twelve 16.0″ (406 mm)
Diesel shp	69,042 (70,000 mhp)	—
Steam shp	64,111 (65,000 mhp)	172,000 (174,386 mhp)
Total shp	133,153 (135,000 mhp)	172,000 (174,386 mhp)
Maximum speed	28 knots	28 knots

TABLE 2-4

Final Preliminary Design—Yamato Class[a]

Design	*A-140-F3*	*A-140-F4*	*A-140-F5*
Date	*5 Oct 1935*	*5 Oct 1935*	*20 July 1936*
Displacement			
Trials	60,036 (61,000)	61,557 (62,545)	64,170 (65,200)
Standard	57,776 (58,703)	58,260 (59,195)	62,315 (63,315)
Dimensions			
Waterline length	807.00′ (246 m)	813.65′ (248.0 m)	830.05′ (253.0 m)
Maximum beam	127.62′ (38.9 m)	127.62′ (38.9 m)	127.62′ (38.9 m)
Draft	34.12′ (10.4 m)	34.12′ (10.4 m)	34.12′ (10.4 m)
Armament			
18.1″ (460 mm)/45	nine	nine	nine
Arrangement[b]	3-3-A-3	3-3-A-3	3-3-A-3
6.1″ (155 mm)/60	twelve (tripled)	twelve (tripled)	twelve (tripled)
5″ (127 mm)/40	twelve (twin)	twelve (twin)	twelve (twin)
1″ (25 mm)/60	twelve (twin)	twelve (twin)	twelve (twin)
0.5″ (13 mm)	twelve (tripled)	twelve (tripled)	twelve (tripled)
Armor			
Gun standard[c]	18.1″ (460 mm)	18.1″ (460 mm)	18.1″ (460 mm)
Immunity zone[d]			
inner limit	21,872 yds (20,000 m)	21,872 yds (20,000 m)	21,872 yds (20,000 m)
outer limit	32,808 yds (30,000 m)	32,808 yds (30,000 m)	32,808 yds (30,000 m)
Machinery			
Diesel shp	59,179 (60,000 mhp)	59,179 (60,000 mhp)	59,179 (60,000 mhp)
Turbine shp	73,974 (75,000 mhp)	73,974 (75,000 mhp)	73,974 (75,000 mhp)
Total shp	133,153 (135,000 mhp)	133,153 (135,000 mhp)	133,153 (135,000 mhp)
Maximum speed	27 knots	27 knots	27 knots
Endurance @ speed	4,900 miles @ 16 knots	7,200 miles @ 16 knots	7,200 miles @ 16 knots

[a]Values in parenthesis are in equivalent metric units.
[b]3-3-3-A-O indicates that the main battery was concentrated forward in three triple turrets.
[c]The caliber gun for which the immunity zone was determined.
[d]The theoretical range band in which the armored citadel is immune to the shellfire of the enemy gun. The inner limit is determined by resistance of the side armor, while the outer limit is determined by the horizontal protection.

TABLE 2-5

Characteristics of the Yamato Class

Standard displacement	64,000 tons (65,027 mt)
Trial displacement	67,123 tons (68,200 mt)
Full-load displacement	69,988 tons (71,110 mt)
Waterline length	839.90′ (256.0 m)
Beam (design waterline)	121.06′ (36.90 m)
Draft (trials)	34.12′ (10.40 m)
Depth	61.23′ (18.667 m)
Armament	nine 18.1″ (460 mm)/45-caliber guns in three triple turrets twelve 6.1″ (155 mm)/55-caliber guns in four triple turrets twelve 5.0″ (127 mm)/40-caliber guns in six twin mounts twenty-four 1″ (25 mm)/60-caliber guns in twin mounts eight 0.5″ (13 mm) guns in quadruple mounts
Aircraft	six floatplanes
Speed	27 knots
Shaft horsepower	147,948 (150,000 mhp)
Endurance (6,300 tons of fuel)	7,200 miles at 16 knots
Protection	Main belt (410 mm on 16 mm) inclined at 20° Armor deck (200 mm) (230 mm on slopes)

and her future sister ship (No. 111). The gantry crane capacity was increased to 100 metric tons. At Nagasaki (where Mitsubishi Heavy Industries Co., Ltd, was the only other facility capable of building such a ship, even with expansion), the slipway was strengthened to take the *Musashi*, and the yard workshops were expanded to almost 240,000 square meters to handle the volume of material required during the construction. Floating cranes of 350- and 150-metric-ton capacity were built for heavy lifts. At Yokosuka, a new dry dock large enough to take *Yamato*-class ships was built, and the *Shinano*, third of the class, was eventually built there. A similar dock was built at Sasebo.

New facilities were needed, and in 1940 work commenced on a new navy yard at Uga, Kyushu, on the west side of Bungo Straits, to contain a building dock capable of handling battleships, two smaller docks, and berths for battleships, carriers, and submarines. All work on this new facility stopped in 1943. A repair dock capable of handling the *Yamato* was begun at Kitan Strait in Wakayama Prefecture as a new facility of the Kawasaki Company. This dock was half-completed when work stopped in 1943.

In addition to the tremendous efforts devoted to improving and building new industrial facilities, the Japanese designed and built a special cargo ship, the *Kashino*, to ferry guns and turrets from the Kure Navy Yard to other yards involved in the construction of the *Yamato*'s sister ships. This ship had a waterline length of 135 meters, a beam of 18.8 meters, and a displacement of 13,000 metric tons, of which 5,800 metric tons was the cargo deadweight.

Yamato—Operational History. The *Yamato* was the first Japanese capital ship completed in more than two decades. Her keel was laid at the Kure Navy Yard on 4 November 1937, she was launched on 8 August 1940, and commissioned on 16 December 1941, several months ahead of schedule. Construction had been accelerated early in 1941 when the imminence of war in the Pacific became apparent. It is a tribute to the designers and builders that the accelerated construction resulted in no problems or operational difficulties. Her trials in October 1941 were a great success, and a speed of 27.4 knots was realized during the full-power trials. However, due to the secrecy surrounding the construction of these ships, the launch and commissioning ceremonies were not as lavish as would normally have been the case.

Midway operation. On 12 February 1942 the *Yamato* became the flagship of the Combined Fleet, under Admiral Isoroku Yamamoto. On 27 May 1942 she left Hiroshima Bay for the Midway campaign. This was the first offensive action for Japanese battleships — the victories of the first few months were won by aircraft carriers. Midway was a disaster; carriers *Soryu, Hiryu, Kaga,* and *Akagi* were sunk by U.S. carrier aircraft. The battleship force took no active part, as the fleet disposition that so fatally weakened the vastly superior Japanese forces left them too far away to help support the carrier force.

Guadalcanal campaign. Photographs made on 4 July 1942 by a U.S. B-17 over Guadalcanal revealed that the Japanese were building an air strip there. Admiral Yamamoto had chosen Guadalcanal as a base for offensive operations in the South Pacific after the defeat at Midway. The U.S. reaction was remarkably prompt — Guadalcanal was invaded on 8 August.

The *Yamato,* still flagship of the Combined Fleet, arrived at Truk on 29 August 1942. From then until 8 May 1943, she was under way for only one day. Admiral Yamamoto transferred his flag to the *Musashi* on 11 February 1943. Neither ship took part in the Solomons campaign due to the lack of special shore bombardment ammunition. The dangers of uncharted waters there were another major factor in the Japanese reluctance to commit the new battleships and there was a lack of fuel.

The *Yamato* returned to Kure on 14 May 1943 and moved to the western Inland Sea on 21 July. Subsequently, the two virtually useless wing 155-mm triple turrets were removed and replaced by 25-mm machine gun mounts. Two Type-22 surface-search radars were added. On 16 August the ship again sailed for Truk, and during the next few months was relatively inactive. U.S. carrier raids on Wake Island on 5–8 October, followed by reconnaissance reports that few U.S. ships were at Pearl Harbor, convinced the Japanese that an all-out assault on Wake was impending. On 17 October the Japanese fleet sortied from Truk to Eniwetok to meet that threat, but after a week with no contacts, returned to Truk on 26 October.

Torpedo damage. In December of 1943 the Japanese decided to send heavy troop reinforcements to Kavieng and the Admiralty Islands. As transports were scarce, some troops were even carried by the *Yamato.* En route from Yokosuka to Truk on the morning of 25 December 1943, she was torpedoed by the U.S. submarine *Skate.* She was then about 180 miles from Truk, and arrived there that day.

The *Skate* scored one hit out of a spread of four torpedoes. The detonation ripped a hole on the starboard side aft, which extended downwards some 5 meters from the top of the blister and longitudinally some 25 meters between frames 151 and 173. The upper turret magazines flooded through a small hole punched in the longitudinal

The Yamato *is shown in her final phase of fitting out at Kure on 20 September 1941. The aircraft carrier on the right is the* Hosho. *The auxiliary in the background is the stores ship* Mamiya.

bulkhead, the hole being caused by the failure of the armor belt joint. Flooding amounted to more than 3,000 tons, far more than had been anticipated in the design of the side protective system.

On 16 January 1944 the ship reached Kure for repairs. A sloping plate was fitted at a 45-degree angle across the lower corner of the upper void compartment between the two longitudinal inboard bulkheads. This modification, proposed to run the full length of the citadel, was installed only in the *Yamato* in the region actually affected by torpedo damage. The measure, in any event, was hopelessly inadequate. A recommendation to use 5,000 tons of steel to reduce the volume of compartments beyond the citadel and so increase resistance to flooding was rejected because the extra weight would have increased displacement and draft beyond acceptable limits.

During February of 1944 six twin 127-mm gun mounts were added in place of the two 152-mm gun turrets removed earlier. No blast shields were available, so the new mounts were given the shields originally mounted on the previously installed 127-mm guns. These shields were never replaced. Subsequent to the damage repairs, 12 triple 25-mm machine gun mounts were added. The *Yamato* left Kure on 21 April 1944, stopped at Manila to hold "open house" in an effort to impress Filipinos with the might of the Imperial Japanese Navy, and reached Tawi Tawi on 1 May.

Battle of the Philippine Sea. A strong Japanese force left Tawi Tawi on 10 June in response to U.S. landings on Biak on 27 May. The Japanese hoped to reinforce the defending garrison on the island, and assembled a force at Batjan that included the *Yamato* and *Musashi*, a dozen cruisers, and other ships. The reinforcement run to Biak was scheduled for 15 June.

On 11 and 12 June, U.S. carrier raids on Guam and Saipan indicated that the United States planned to strike the Marianas next, and the Japanese, hoping for a decisive naval engagement, cancelled the Biak operation. The *Yamato*, with other fleet units, joined the First Mobile Fleet on 16 June. (On 1 March the Japanese, recognizing that carriers had supplanted battleships, had reorganized their fleet, and the Mobile Fleet resulted, with surface ships under tactical command of a carrier admiral, as the U.S. Navy had done since July 1942.) The Mobile Fleet had impressive strength: nine carriers, five battleships, and enough heavy and light cruisers and destroyers to bring the total to nearly sixty ships. But the U.S. forces for the impending battle were numerically greater in every respect. Of vastly greater importance, American air groups were combat veterans while most Japanese pilots had limited training, ranging from two to six months. In the resulting Battle of the Philippine Sea, 19–21 June 1944, Japanese air strength was rapidly reduced from a total of 430 carrier aircraft and 43 floatplanes to 35 carrier aircraft and 12 floatplanes.

The Japanese lost a total of 426 aircraft of all types and the carriers *Hiyo, Taiho*, and *Shokaku*, a blow from which their navy never recovered. In the same period the U.S. Navy lost 130 aircraft.

The *Yamato* returned to the Inland Sea on 24 June. The *Yamato* and *Musashi* each had five additional 25-mm triple mounts installed. In July the Japanese greatly increased the strength of the 25-mm machine gun batteries on nearly a hundred ships. The American use of bow and stern machine gun mounts had been noted with considerable interest and approval, but some captains refused to mount machine guns anywhere other than amidships.

At Kure the Japanese studied methods of improving damage control capabilities. As a result, several "emergency buoyancy-keeping procedures" were instituted, which included the removal of all possible combustible materials, such as furniture, linoleum, and bedding. After that, men actually slept on planks provided for damage control purposes. Inflammable paints were replaced by a light gray silicon-based paint, which gave the interiors of ships a drab and dreary appearance. More portable pumps and fire-fighting equipment were supplied. These measures were successful in limiting fires and explosions such as those which had contributed to the loss of carriers at Midway.

Battle for Leyte Gulf. The *Yamato* and *Musashi* left the Inland Sea on 8 July 1944 and arrived at Lingga Roads, off Singapore, on 16 July. By mid-October, a formidable aggregation of Japanese naval power had assembled. The Japanese tanker and merchant fleet had been so reduced by U.S. submarines that it was incapable of supplying a large number of warships anywhere else.

When U.S. forces began the long-expected invasion of the Philippines, the Japanese intended to concentrate their widely separated forces on the vulnerable invasion shipping. By mid-1944 defeat seemed inevitable, and the best they could hope for was a crushing attack on the invasion forces. If successful, perhaps a more favorable peace might be obtained.

Unfortunately for the Japanese, despite their correct evaluation of future U.S. plans, they did not know the specific time and place of the forthcoming invasion. This was fatal, for the resulting inevitable time lag prevented them from attacking during the most vulnerable phase of the amphibious assault. The Japanese plan was ingenious and complicated: the Main Body, a carrier force with few aircraft embarked, was to sortie from the Japanese Home Islands to decoy forces covering the U.S. landings to the north, away from the invasion beaches. In the meantime the First Striking Force—Force A—was to sortie from Lingga so as to approach the invasion beaches from the south.

Although the decoy force succeeded in its sacrificial mission, the powerful Force A was only partially successful in engaging the U.S. amphibious forces, as U.S. submarines sank the heavy cruisers *Atago* and *Maya* on 23 October 1944 in the opening action of the Battle for Leyte Gulf. The following day, the *Musashi* took numerous bomb and torpedo hits and went down. On 25 October, off Samar, Force A encountered one of three groups of U.S. escort carriers, which was all that kept them from the invasion shipping off Leyte. The screening destroyers and destroyer escorts began a series of suicidal torpedo attacks that convinced the Japanese that they had engaged a group of fleet carriers, and the *Yamato* turned away to avoid the torpedo attacks. As a result, the Japanese lost understanding of the tactical situation and broke off contact when the U.S. carriers were almost trapped by the faster Japanese cruisers and battleships.

The U.S. lost escort carriers *Gambier Bay* and *St. Lo*, destroyers *Hoel* and *Johnston*, and destroyer escort *Samuel B. Roberts.* The Japanese lost heavy cruisers *Chokai, Chikuma*, and *Suzuya* to air attack. Force A lost the light cruiser *Noshiro* to air attack, and a destroyer to gunfire.

The Main Body decoy force successfully lured major U.S. fleet units away from the invasion shipping that was the target of the Japanese attacks, although the carriers *Zuikaku, Zuiho*, and *Chitose*, one light cruiser, and two destroyers were sunk. This would have been a small price to pay for the victory which narrowly eluded the Japanese when Force A turned away from the invasion beaches at Leyte.

Force C, in Surigao Strait during the night of 24-25 October, ran into a trap set by Rear Admiral J.B. Oldendorf. The Japanese force ran a gauntlet of motor torpedo boat and destroyer torpedo attacks, finally to be confronted by an overwhelming U.S. force of six old battleships, four heavy and four light cruisers.* In the classic naval engagement known as the Battle of Surigao Strait, the battleships *Yamashiro* and *Fuso*, heavy cruiser *Mogami*, and three destroyers were sunk. Only one destroyer escaped. The rest of Force D, the Japanese northern force, three cruisers and four destroyers, were several hours away from this melee, but U.S. air attacks the next day sank one light cruiser and two destroyers of this force as they fled.

The *Yamato* and *Musashi* had sortied with Force A, which also included the battleship *Nagato* and the heavy cruisers *Atago, Takao, Chokai, Myoko, Haguro*, and *Maya.* At 0634 on 23 October 1944, four torpedoes from the U.S. submarine *Darter* hit the flagship *Atago*, and two overshoots hit the *Takao.* As the *Atago* was sinking, Admiral Kurita was forced to transfer his flag. The *Takao* fell out of formation and was attended to by several destroyers. At 0656 the U.S. submarine *Dace* torpedoed the heavy cruiser *Maya*, which sank shortly thereafter. The loss of these heavy cruisers to Force A would have serious repercussions in antiaircraft fire later that day when the force was attacked by U.S. carrier aircraft. The *Maya* was strategically located in the starboard column of the Force A formation ahead of the *Yamato* and *Musashi* to provide antiaircraft fire. Off the island of Samar (Battle of the Sibuyan Sea), U.S. carrier planes flew 259 sorties against Force A. The *Musashi* bore the brunt of these attacks, while the

---

*These were the *West Virginia, Mississippi, Maryland, Tennessee, California*, and *Pennsylvania.*

Yamato was hit by two bombs and sustained a damaging near miss. Important in these assaults on Force A was the loss of the three heavy cruisers and the punishing strafing attacks that took heavy tolls of the light antiaircraft personnel on all Japanese ships, substantially reducing their volume of antiaircraft fire. The *Nagato* was hit by two bombs, *Haruna* sustained five near misses, and the heavy cruiser *Myoko* was torpedoed in the stern and forced to withdraw.

Bomb damage. The first bomb to hit the Yamato exploded on the port side near frame 70 at the side shell plating and tore a 4- by 5-meter hole near the waterline that caused flooding amounting to 370 metric tons. A second bomb struck on the forecastle deck, four meters from the port chain locker, and penetrated all decks before detonating and causing flooding estimated at 3,000 metric tons. The anchor windlass compartment, chain lockers, and nearby storage spaces were flooded. A near-miss bomb on the port quarter punched 24 small holes in the shell plating. Nine men were killed or wounded.

The Yamato *is turning sharply at high speed to starboard with a bomb detonating in the water on her port side. Note the fire in way of the after 155-mm turret. The* Yamato *was sunk during this engagement on 7 April 1945. (James C. Fahey Collection)*

In this photo taken on 7 April 1945, the Yamato *has been severely damaged, with a list to port and a serious fire raging in the after 155-mm turret.*

During the battle off Samar on 25 October, the *Yamato,* for the only time in her career, fired on enemy surface ships—U.S. escort destroyers screening escort carriers. These American destroyers made torpedo attacks early in the engagement on the battleship and forced the *Yamato* to turn away from the fleeing American carriers to parallel the torpedo tracks. The *Yamato* was not hit, but her evasive action took her from the battle area, and as a result, she took no further effective part in the engagement. Because of the torpedo attack, the Japanese force was no longer a cohesive unit. During that engagement, the *Yamato* was damaged by a stray Japanese shell that punched a 0.6-meter hole in the overhead of the crew's galley. This was the result of Japanese ships firing at American ships caught between the two Japanese columns. One near-miss bomb on the starboard quarter punched a 0.5-meter hole in the shell plating, causing the flooding of a reefer space. A second near miss to port at frame 80 did minor damage, as did a third to starboard at frame 130. Bomb fragments and machine gun fire from strafing aircraft caused other minor damage and killed or wounded 17 men.

Admiral Kurita decided to regroup and assess the situation. He had four badly damaged cruisers, one of which later sank. After hesitating about two hours and at the threshold of his objective, Admiral Kurita turned north and abandoned the mission. The next day, as his force retired, U.S. carrier aircraft again damaged the *Yamato.* One bomb penetrated the upper deck at frame 63 and exploded on the deck below, slightly to the starboard of the centerline. Near misses caused 29 small punctures in the starboard outer shell plating. The attacks killed or wounded 21 men.

The air attacks of 24–26 October caused only minor damage but resulted in a surprising amount of flooding, particularly from the bomb hit on the forecastle. There were 33 fatalities. The ship reached Kure for repairs on 11 November 1944.

Bomb damage (in Japan). On 3 January 1945 the *Yamato* was moved to the Western Inland Sea and was made flagship of a drastically reduced fleet. All the carriers had been sunk or damaged, and of the five battleships only the *Yamato* was in service. In March the U.S. Navy's Task Force 58 raided the Japanese Home Islands, and on 19 March air strikes damaged 16 Japanese warships, including the *Yamato,* which took a single bomb hit on the forward bridge tower. The battleship was in Hiroshima Bay at the time of the attack and maneuvered very slowly in open waters near the Nasami Channel. Several near misses sent severe shock responses through the hull. This confrontation with American aircraft gave the Japanese renewed confidence that the *Yamato* could adequately defend herself against air attack.

Okinawa sortie (a kamikaze mission). Japan was in a desperate way in late March 1945. Her navy had been reduced to a fraction of its prewar strength, the army was fighting a losing battle in the Okinawa campaign, and suicide attacks were in vogue to save the empire. The Combined Fleet decided to use the *Yamato,* the light cruiser *Yahagi,* and eight destroyers in a combined action with kamikaze aircraft attacks on American ships off Okinawa to help relieve the pressure on the Japanese garrison there. When the ships and aircraft struck, the army was to take the offensive. It was a desperate plan drawn up by desperate men who were out of contact with reality. Some opposition was voiced against the plan, but the orders were never countermanded. In the end the ships were fueled to 60 percent of capacity instead of the 40 percent specified in their operational orders. At 1520 on 6 April the *Yamato* and her consorts—the "Surface Special Attack Forçe"—sortied from Tokuyama on a last offensive effort; they were to proceed to Bungo Strait, approach Okinawa from the northwest, and attack U.S. invasion shipping on 8 April. Upon completion of her mission, the *Yamato* was to be beached on the sands of Okinawa and become a fortress. The plan was known to the U.S. Navy, which had intercepted and decrypted the radio orders for this mission.

The Japanese ships refueled to 60 percent capacity, and carried a full load of ammunition that also included specially fuzed antiaircraft shells for the 460-mm guns. All combustibles, including ship's boats, were sent ashore. The futility of the mission was evident from the fact that they planned to attack a fleet strong in aircraft carriers, journey over 350 miles without air cover of their own, and enter battle with inadequately trained antiaircraft crews for the 25-mm machine guns against American aircraft. The Surface Special Attack Force was headed for almost certain destruction.

With the destroyers in the lead and the *Yahagi* on her port bow, the *Yamato* worked up to a speed of 22 knots. One-third of the crews were at battle stations due to the known presence of U.S. submarines in Bungo Strait. Once clear of the Kyushu coast, the formation began zigzagging. Two American submarines, the *Hackleback* and the *Threadfin,* reported the Japanese fleet around 2000, thus confirming the decoded combat intelligence. Both submarines could not press home torpedo attacks because of the speed and radical maneuvering of the Japanese ships. *Yamato's* radar detected the presence of these two submarines.

Shortly after midnight, Admiral R.A. Spruance, Commander U. S. Fifth Fleet, who was in command of the Okinawa Campaign, directed Admiral M.A. Mitscher, who commanded the U.S. fast carrier task force, to leave the Japanese force to the Fifth Fleet battleships. Mitscher had already drawn up an attack plan for his carrier aircraft to attack the Japanese. Spruance's orders were later countermanded, but Admiral Morton Deyo continued to plan for a surface engagement. Early during the afternoon of 7 April, a U.S. Navy force of six old battleships, seven cruisers, and 21 destroyers left the Okinawa area to engage the *Yamato* group. The outcome of the battle would likely have been a defeat for the Japanese, but probably at a heavy cost to the American surface fleet due to the *Yamato*'s more powerful guns, better armor protection, and greater speed. In any event, an attack by carrier aircraft would eliminate any doubt that air power had outclassed the battleship.

By sunrise on 7 April, the *Yamato* was in a condition of complete closure and ready for action. At 0823 that morning a scout plane from the *Essex* spotted the *Yamato* and her consorts, and later two PBM flying boats made contact. The PBMs shadowed the Japanese formation for the next five hours. At 1017 the *Yamato* fired at them, using her special 460-mm antiaircraft shells, but the two aircraft continued to keep in contact out of gunnery range and to send out position reports to direct incoming air strikes.

Three U.S. Navy carrier groups launched a total of 386 aircraft against the Japanese force, of which 227 planes, about one-third of them torpedo bombers, actually attacked.

Fifteen U.S. carriers launched aircraft, but planes from five carriers were lost in bad weather and failed to reach the target in time. At approximately 1000, the *Yamato* made an uncertain radar contact on aircraft. All ships went to complete closure and shortly thereafter all hands were at their action stations. At 1107 a tracking and covering group of 16 U.S. fighter planes (Hellcats) arrived to suppress any possible Japanese air support. The *Yamato* group had limited air cover—twenty reconnaissance planes flew ahead of the Japanese fleet shortly after dawn. They were intercepted and more than half were shot down by U.S. carrier fighters. No other Japanese aircraft appeared thereafter. The Japanese had decided to launch all their fighters against the carrier task forces and amphibious shipping instead of covering the Surface Special Attack Force.

At 1220 *Yamato*'s radar detected three large groups of American carrier planes approaching the task force, and the Japanese increased speed to 24 knots, with the screening destroyers commencing their circling tactics, the usual Japanese procedure against air attack. The first planes attacked at 1232, and their attacks lasted 10 minutes. During this attack the cruiser *Yahagi* increased speed to 35 knots and turned away from *Yamato* to draw some of the planes away from the battleship. Although the maneuver succeeded, there were enough aircraft to attack the *Yamato* and bracket her with near misses. At 1241 two 454-kilogram bombs penetrated the *Yamato*'s flying and upper decks on the starboard side. One exploded in the vicinity of frame 157, demolishing two 25-mm machine gun mounts as well as blowing a large hole some 6–7 meters in diameter in the weather deck. The other bomb also exploded on the starboard side near frame 159 on the main deck and ripped open a large hole in the superstructure and deck. The aft radar room was demolished. This bomb hit also obliterated the aftermost 127-mm gun mount on the starboard side. Five minutes later two more bombs struck this region of the ship. The first hit slightly to port of the centerline, just forward of the aft 155-mm centerline mount. It passed through the after secondary battery control station, destroyed the rangefinder and all equipment in the space, and penetrated several decks before exploding in several storerooms on the starboard side of the ship. Severe fires broke out and would not be extinguished. Fire-fighting efforts were disorganized and ineffective, partially because these bomb hits had killed the entire aft damage-control team. The other bomb exploded on the roof of the aft 155-mm turret and sent splinters into its magazine, which then caught fire. This blaze also raged out of control. The turret was gutted, and there was only one survivor. At this time *Yamato* was at her maximum speed.

Destruction of Yamato. About 1245 a torpedo struck the port bow near frame 8. It hit with devastating impact, and any unsecured equipment was violently shaken. Information on the flooding from this hit is indefinite, as is generally the case for all torpedo damage occurring beyond the armored citadel. Many of the men in these spaces were casualties from strafing attacks or were later trapped below deck when the ship capsized. Apparently the flooding and damage had only minor effects.

Near the end of the first attack the *Yamato* was struck by two torpedoes on the port side. One of these struck outboard of number 8 fireroom and the other outboard of the engine room. Both spaces reported very slow flooding. There is also the possibility that a third torpedo struck well aft, although no flooding was reported in the magazines for the after main turret. It is possible that this torpedo struck low enough so that the hull curvature muffled the torpedo plume, similar to the hit that occurred in the *Prince of Wales.* In any event, such a hit would account for the slow flooding of the auxiliary steering room. For this reason this hit is considered probable.

After the first attack, the *Yamato* was listing 5–6 degrees to port. Counterflooding of outer starboard voids reduced this to an approximately one-degree port list. There was a

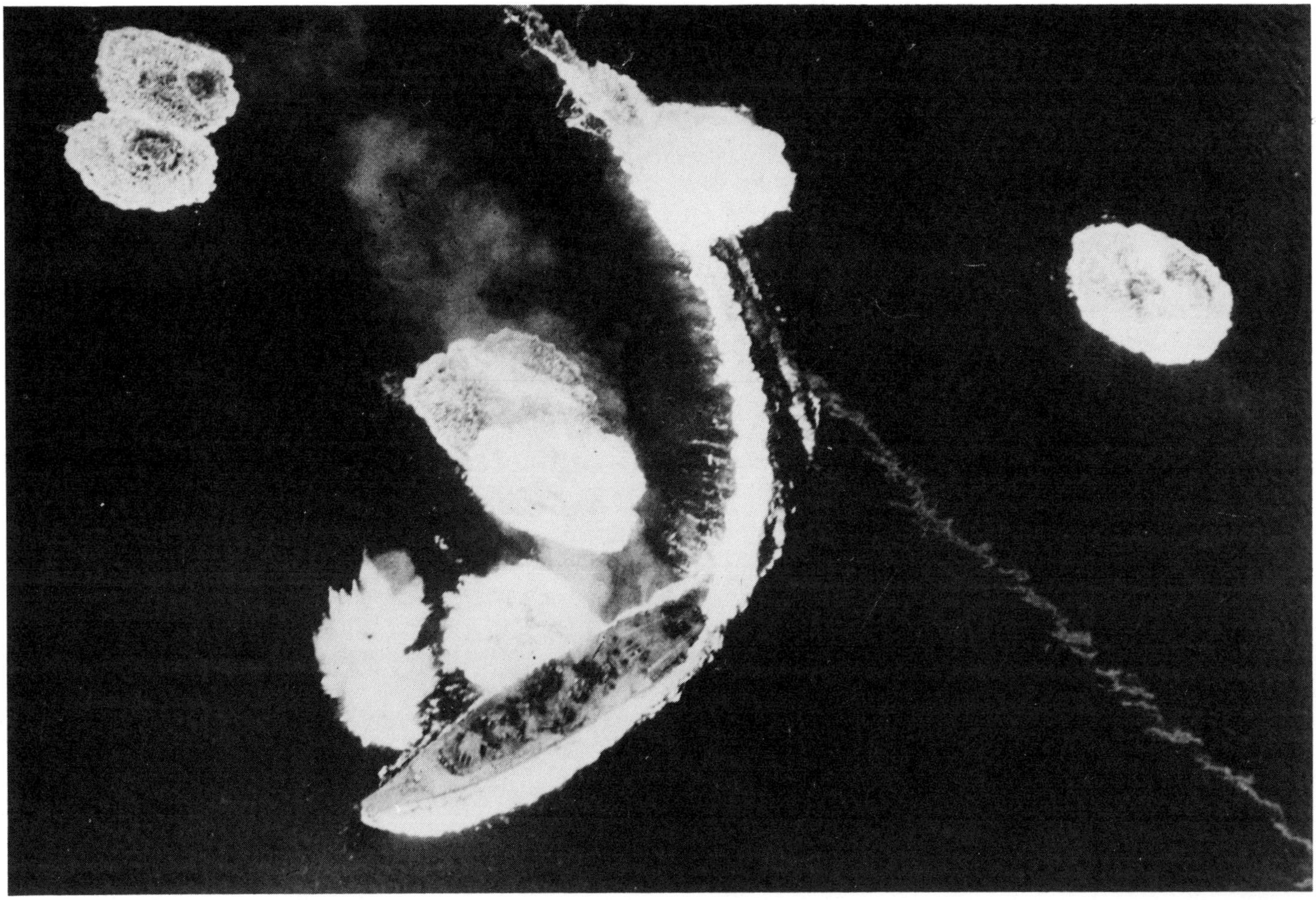

The Yamato *was attacked at her base in Kure Bay on 19 March 1945 by carrier-based aircraft. Note the tight pattern of the near-miss bombs. One bomb hit the battleship in her superstructure.*

small reduction in speed, although flooding in the port outboard engine room was under control. Number 8 fireroom was out of action. In addition, strafing of the ship had reduced the volume of 25-mm machine gun fire, as there were a number of crew casualties.

A second attack developed around 1259. This was a well-coordinated attack of dive bombers and torpedo bombers. While Japanese attention was drawn to the dive bombers, a number of torpedo bombers began their approaches from several points of the compass. The *Yamato* was in a turn to starboard when these planes dropped their torpedoes, and there was no way that she could avoid being hit. Three, possibly four, torpedoes struck to port, and a single torpedo hit the starboard side. It appears that the three hits to port were closely spaced, with hits occurring near frames 123, 131, and 142. A fourth torpedo could have struck farther aft on the port side near frame 148, but this could not be confirmed. One of the torpedoes exploded in way of number 8 fireroom, partly flooded by a previous hit, and accelerated the flooding there. Another detonated near fireroom 12 and caused some flooding in this space. The third, the most destructive of the three port hits, exploded outboard near the bulkhead separating the port outboard engine room and the hydraulic machinery room. The side shell and internal structure had already been weakened by a previous hit, and this hit accelerated the flooding of the outboard engine room. It is also possible that flooding occurred in the next set of spaces just inboard of these. A probable fourth torpedo hit to port around frame 148 would explain the very rapid flooding of this region of the ship.

One torpedo hit to starboard near frame 125 outboard of fireroom 7 prolonged the life of *Yamato*, as the rapid flooding of this space complemented the counterflooding measures. All starboard voids were now flooded and this, combined with inboard flooding from the starboard torpedo hit, reduced the port list about 10 degrees from an initial list of 15–18 degrees to port.

The *Yamato* was in serious condition after this second attack because the port list prevented the main turrets from firing. Prior to this their antiaircraft fire had not been very effective in breaking up the air attacks, and finally the chief gunnery officer had ordered their fuzes to be set at one second. This minimum setting would have caused the shells to explode some 1,000 meters from the ship — a last desperate measure of defense against torpedo bombers. The guns were fired, but the 460-mm shell fire could not break up the torpedo attacks that were so damaging during this second attack.

Although the ship was not sinking, the flooding and port list reduced the speed to 18 knots. Around 1340, with the ship listing heavily to port, the executive officer, Captain Jiro Nomura, informed Rear Admiral Kosaku Ariga, captain of the *Yamato*, that all possible voids and tanks on the starboard side had been flooded in an effort to right the ship. Any further reductions in the list would involve the drastic measure of counter-flooding the outboard firerooms and engine room on the starboard side. Such action would seriously reduce the reserve buoyancy and further cut the speed of the *Yamato*, which would make her very difficult to maneuver. The captain chose not to do this counterflooding at that time.

The third and final attack commenced around 1342. There were multiple bomb hits and four more devastating torpedo hits. Only four bomb hits can be readily identified; three occurred on the port side amidships and one forward at the bow. The three amidships bomb hits ripped large holes in the main deck and blew several of the 25-mm gun mounts along the port deck edge overboard. There were also a number of near misses in this attack, and these were sufficiently close to the hull to spring rivets or deflect shell plating or other structure inboard. This was particularly severe in way of the torpedo protection system because large struts, which had inadequate backing

The Yamato *is nearly dead in the water and down by the bow in this view, as escorting destroyers circle in a futile attempt to shield the battleship from further air attacks, 7 April 1945.*

This is one of the most famous photographs taken of the Yamato *and shows the battleship during her full power trials in October 1941. Note the bow spray from the speed of 27+ knots (Imperial War Museum)*

structure, were driven inboard and punctured the torpedo bulkhead. The holes increased the area of flooding. The bomb hits also destroyed power cabling to the machine guns, silencing many of these.

A bomb struck the forecastle, destroyed the anchor windlass, and severed the port anchor chain, sending the port anchor to the seabed below.

It was the torpedo hits in this last assault that doomed the *Yamato.* In fact, these hits almost brought the giant battleship to a halt. Three torpedoes struck the port side near frames 133, 153, and 211. The first two caused the flooding of the number 10 fireroom and increased the rate of flooding into the port inner engine room. The third torpedo exploded aft in way of the steering gear room, causing extensive flooding. Within a few minutes after the hit, the steering room crew was reporting heavy flooding. Before their message could be completed, voice communication was cut off. All hands in this space perished at their post. The rudder was jammed, and with the auxiliary steering room flooded, all maneuverability was lost. The critical damage control central and starboard pump room were also flooded.

During this final attack the port list began to increase rapidly. Captain Nomura ordered the outboard intact firerooms (numbers 5 and 9) and the starboard hydraulic machinery room to be flooded. This seems to have been done promptly, but had little effect other than to prevent the ship from listing further. Soon, however, the list to port began to increase, and the captain ordered the starboard engine room also flooded. It should be noted that many of the starboard voids could not yet be flooded completely by the pumping system, since the large lists made it difficult to fill these spaces.

Near the conclusion of the third attack, while the battleship was beginning her heavy list to port, one more torpedo struck the starboard side. The depth setting of this torpedo was 5 to 5.5 meters, which meant that it hit the vessel's underbottom somewhere in the vicinity of the starboard outer engine room. It appears that the decision to flood this space and the torpedo hit occurred very close together. The flooding of this space, fire room 3, the hydraulic pump room, and the refrigeration machinery room on the starboard side entombed their crews at their stations.

The speed of the *Yamato* was now reduced to 10 knots on a single shaft. Only the starboard inner engine room was still in operation. The *Yamato* was still in a sluggish turn to starboard. Earlier, the Japanese had thought that a sharp turn to starboard would aid in counterflooding the starboard voids, and the port torpedo hit in the vicinity of the steering-gear room left the rudder in a hard-over position. The *Yamato*, with both rudders out of action, began to turn in a wide circle. The port list grew to 22 degrees when the order to abandon ship was given. The order was given very late, and many of the crew below decks were trapped.

On *Yamato*'s bridge, the executive officer noticed that the red lights on the warning board were all lit, indicating that temperatures in the magazines aft were at the danger level. Warning alarms began their ominous buzzing, and the officer at the control panel screamed at Nomura, "Can we flood the magazines?" It was not possible to do so, for the pumping stations had been immobilized by the extensive flooding.

Shortly before 1420 all power was lost and the bridge tower was precariously close to sea level. Some of the damaged 25-mm mounts and their weakened foundations on the port side slid into the sea. At 1423 the ship capsized so rapidly that many men were trapped or sucked into the hull by the undertow. As she went over to 120 degrees one of the after magazines exploded. A giant smoke cloud rose over the site of the sinking. Out of a complement of 3,332 men, only 23 officers and 246 enlisted men were saved.

Table 2-6 summarizes the bomb and torpedo hits that the *Yamato* sustained in her last action. It should be noted that the frame locations are approximate. There is no question that this ship was sunk by torpedoes; the bomb hits and strafing only reduced her ability to defend herself against these attacks. See also figure 2-1.

TABLE 2-6

Damage to the Yamato (7 April 1945)

Time	*Torpedo Hits (frame and side)*	*Bomb Hits (frame and side)*
1245	8, port	150, starboard
		150, starboard
		157, port
		159, port
1259	123, port	
	150, port	none
	192, port (probable)	
1342	123, port	18, port
	125, starboard	amidships, port
	131, port	amidships, port
	142, port	amidships, port
	148, port (probable)	
1359–1402	133, port	
	153, port	none
	211, port	
	150, starboard	

Totals: 13 torpedo hits (11 certain, 2 probable) 8 bomb hits (confirmed)
Near-miss bomb hits are unknown, but a great number occurred during first and third attack. It is believed that these caused some damage to the hull envelope and contributed to the flooding.

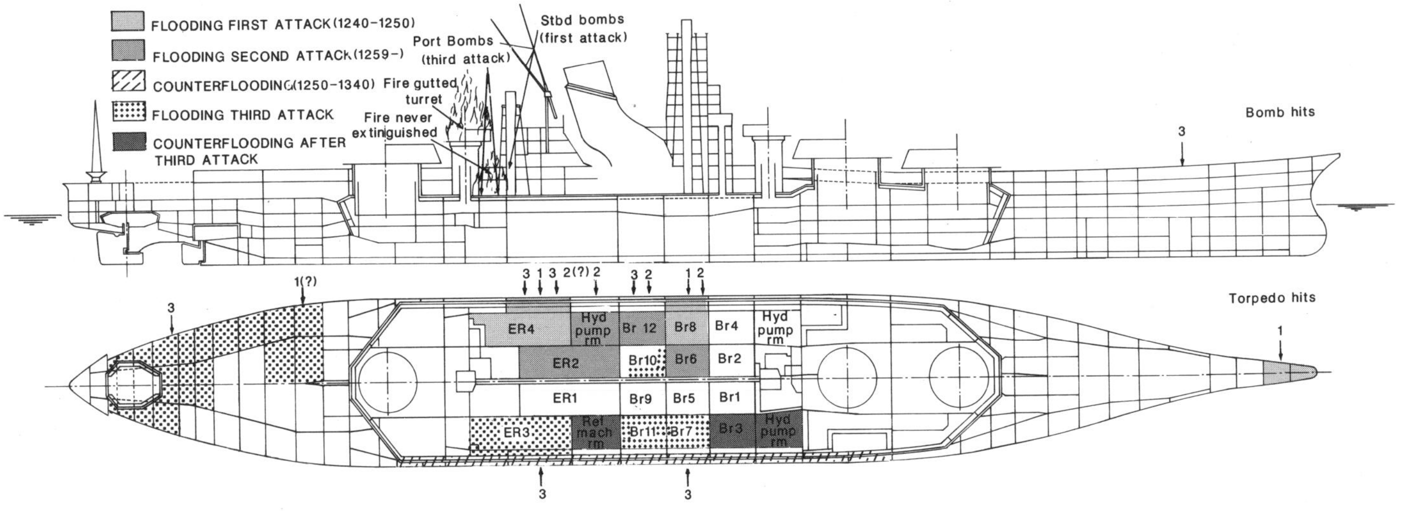

Figure 2-1. Yamato *loss—Approximate location of bomb and torpedo damage, 7 April 1945.*

The *Yamato* and *Yahagi* sank after massive damage, and four of the eight destroyers were also sunk. The naval minister, when informing Emperor Hirohito about the loss of the Special Surface Attack Force, commented that the entire operation was ill conceived. That this was true is evident from the loss of the *Prince of Wales* and *Repulse* in the South China Sea by Japanese forces on 10 December 1941. These ships had had no aircover, and the Japanese tried a similar operation at a time when air power had truly become dominant on the oceans.

Musashi—Operational Career. The *Musashi* was laid down at the Mitsubishi Yards at Nagasaki on 29 March 1938. She was launched on 1 November 1940, fitted out at Sasebo, and commissioned there on 5 August 1942. Her operational career closely paralleled that of the *Yamato*; she was torpedoed by an American submarine, first saw combat during the Battle of the Philippine Sea, and was sunk by U.S. carrier aircraft. Because of the enormous amount of fuel she required, her operations were to be limited during the course of the war.

The *Musashi* arrived at Truk on 15 January 1943, relieving the *Yamato* as flagship of the Combined Fleet on 11 February. The *Musashi* was to remain the flagship of Admiral Isoroku Yamamoto until his death. On 18 April 1943, Admiral Yamamoto and some of his staff were killed when their aircraft was intercepted and shot down by an American squadron of P-38s. The remains of the Japanese commander-in-chief were taken to Truk, and the *Musashi* returned them to Japan for a state funeral in June 1943. Following the U.S. invasion of the Aleutian island of Attu, some thought was given to using the *Musashi* and other ships for operations in Alaskan waters. Although the *Musashi* had arrived in Tokyo Bay on 21 May 1943, it was decided not to risk her in the operation. Japanese forces on Kiska in the Aleutians were evacuated on 29 July 1943. Admiral Mineichi Koga succeeded Admiral Yamamoto and made the *Musashi* his flagship.

The *Musashi* returned to Truk on 5 August and remained there for several months. In November 1943 she conducted a sweep eastward towards the Marshall Islands. The ship returned to Truk and remained there until a B-24 reconnaissance bomber made a photographic mission over Truk on 5 February 1944. This warned the Japanese of an impending air attack, and most of the combatant ships based there were moved to Palau. The *Musashi* was ordered back to Japan on 10 February 1944. Truk Atoll was raided by U.S. carrier aircraft on 17–18 February, in one of the U.S. Navy's most successful carrier operations of World War II. The Japanese lost the cruisers *Naka* and *Katori*, three destroyers, plus some three dozen other naval and merchant craft, for a total of 137,091 tons. In addition, from 250 to 275 Japanese aircraft were damaged or destroyed.

Torpedo damage. The *Musashi*, still the flagship of Admiral Koga, departed from Japan in early March 1944 for Palau. Late on the afternoon of 29 March 1944, as the *Musashi* approached Palau, she was attacked by the U.S. submarine *Tunny*, with one hit out of a spread of six torpedoes. The detonation, far foward on the port side near frame 27, caused fires and flooded the forward anchor windlass room and numerous other compartments with some 4,000 metric tons of seawater. This revealed a major design flaw—the vulnerability of the unprotected bow and stern areas. Seven men were killed.

The torpedo damage was extensive—272 kilograms of torpex, equivalent to about 550 kilograms of TNT, detonated some 6.1 meters below the waterline at a point devoid of underwater protection. Extensive flooding could have had serious consequences in the event of further damage.

After some makeshift repairs at Palau, *Musashi* returned to Japan. Admiral Koga departed *Musashi* and was later killed in a plane crash. After a high-speed run from Palau, the *Musashi* reached Kure on 3 April for permanent repairs. On 11 May, with five carriers and screening destroyers, she sailed for Tawi Tawi, where she arrived on 16 May. She took an inconsequential part in the Battle of the Philippine Sea and returned to the Inland Sea on 24 June.

Battle for Leyte Gulf. The war of attrition that U.S. submarines were waging against Japanese merchant shipping forced the Japanese to redeploy their battlefleet closer to anticipated American thrusts. The *Musashi* left the Inland Sea for the last time on 8 July 1944. She reached Lingga Roads on 16 July, and after several months left on 18 October as part of Force A, en route to the Battle of the Sibuyan Sea. The ship was fully loaded, with a draft of nearly 11 meters. En route, she burned about 1,000 tons of fuel and transferred another 800 tons to escorting destroyers. On the morning of 24 October, her draft was 10.5 meters.

That morning the crew went to general quarters at 0600. The crew was exceptionally well trained by Japanese standards, and morale was high. They had been drilled daily in damage control and counterflooding procedures.

Destruction of the Musashi. At 0810 the Japanese fleet of five battleships, ten cruisers, and 15 destroyers was sighted by aircraft from Task Group 38.2. The first strike of 21 fighters, 12 dive bombers, and 12 torpedo bombers was launched by the *Intrepid* and *Cabot* at 0910. The aircraft attacked at 1025; the *Musashi* was making 22 knots.

The attack lasted about five minutes. The ship was bracketed by near-miss bombs at frame 25 to starboard and frame 20 to port, which caused slow flooding of two forward peak tanks. Two near-miss bombs bracketed the ship at frame 145, causing minor flooding. One 227-kilogram bomb detonated harmlessly on the roof of the forward main battery turret.

Two minutes later a torpedo exploded at frame 130 on the starboard side in way of number 9 boiler room. This torpedo hit occurred in the region of the ship where structure had been already weakened by a near-miss bomb. Extensive damage was sustained, and leaks soon developed around rivets in the holding bulkheads bounding boiler rooms 5 and 9. This indicated the failure of the side protection system to resist heavy explosive charges. Quick flooding in the voids outboard of number 9 boiler room caused an initial list of 5.5 degrees, which was reduced to about a degree by counter-flooding on the port side. The peak tank flooding produced a bow trim of about 2 meters. Shock damage disabled some instruments in the main battery plotting room. Despite the damage, the ship maintained her speed of 22 knots and her position in the formation.

Another attack began at 1138 and lasted approximately seven minutes. Severe and crippling damage was inflicted by torpedoes and bombs. Three torpedoes struck the port side at frames 82, 102, and 140. The latter caused the only appreciable damage within the citadel, rapidly flooding the port hydraulic machinery room and causing minor leakage into the port outboard engine room. The flooding caused a port list, later corrected, of approximately five degrees. The extent of this damage indicates that the Japanese had used the voids in the area to help counter the flooding sustained in the previous attack, thereby reducing the effectiveness of the side-protection system.

The *Musashi* took two bomb hits and five near misses in that attack. The near misses, one to port near frame 70 and the others to starboard near frame 140, caused no material damage. A bomb hit at frame 15 to port destroyed the crew toilet and washroom and curled up the port bow deck plates. The second bomb penetrated three decks, detonated at frame 138 to port in a berthing compartment over the port inboard engine room, and started fires in the engine room and boiler rooms 11 and 12. Fragments ruptured the main steam lines, forcing men to evacuate the engine room, with control of this space transferred to the starboard engine room. The steam line to the whistle was severed at this time. Steam continued to escape from this broken line until the ship sank much later on. The sound of escaping steam, the roar of aircraft engines, and plunging bombs must have raised havoc with the nerves of those aboard *Musashi* at this time. A small fire at the entrace to boiler room 12 caused the upper watertight bulkhead to collapse, jamming the hatch. The fires were soon extinguished, but the port inboard engine room was never remanned.

During the 30 minutes before the next air assault, the port list was reduced to about a degree and the bow trim was reduced to one meter. At this time the most serious damage was the loss of the port inboard engine room, which was compensated for by increasing the power on the remaining three shafts. This enabled the *Musashi* to remain in formation and under a protective umbrella of antiaircraft guns.

At 1217 a third attack developed. A torpedo hit the starboard side and ripped into the shell at frame 60, forward of the side protection system. Several large storerooms flooded, and the fathometer room was damaged. The forward sickbay completely filled with carbon monoxide gas. Several near-misses disabled the aircraft crane on the fantail.

The flooding forward was extremely serious, for it further reduced residual stability, and the large compartments that flooded caused a trim by the bow of two meters. There was also a slight starboard list. Combat effectiveness was only partially diminished, primarily by the loss of the port inner engine room, and the ship was still in fighting condition, despite five torpedo hits, three bomb hits, and a dozen near misses.

A fourth series of attacks began at 1253, when four torpedoes and four bomb hits caused severe damage. Two torpedo hits, to port and starboard near frame 70, were particularly damaging. The first exploded in way of dry provision storerooms, flooding

A rare photograph of the Musashi, *shown here at 0800 on 22 October 1944, leaving Brunei for the Battle of Leyte Gulf. (Courtesy of T. Shiraishi)*

almost all of them. The second detonated slightly aft of frame 70, causing more flooding. The entire middle deck was flooded forward of the armored citadel, and the ship was down some 4 meters by the bow, with water at the upper deck level. The consequent reduction in metacentric height was a matter of serious concern for the damage control officer. There was little he could do, and the internal structural damage sustained in the bow region was very severe. At this time two more torpedoes struck to starboard, farther aft. One hit around frame 110, causing no damage within the citadel, but the other at frame 138 caused the instantaneous flooding of the reefer machinery room. Damage control teams had previously rigged some shoring to the reefer space piping alley manhole cover. This simply fell away. Three holes were punched in the side bulkhead of number 2 damage control center, and this contributed to further flooding. The rapid flooding in this region of the ship was probably the result of earlier near-miss bomb damage, as a bomb detonation sufficiently close to the side shell plating can have effects similar to that of a mine or torpedo, and there had been a near miss at framc 145 to starboard during thc 1025 attack.

The four bomb hits caused minor damage. One, at frame 45 to port, penetrated three decks before exploding in a crew living compartment. A second at frame 65 to port, somewhat forward of the citadel, penetrated two decks and exploded in a berthing compartment. Neither caused any flooding or fires. A third, slightly forward of frame 70 to port near the sloping citadel armored bulkhead, also penetrated two decks, detonated, and caused some fragmentation damage and warping of decks. The last bomb struck the exposed portion of the flying deck outboard of the funnel at frame 135 starboard, exploded, disabled several 25-mm machine gun mounts, and wrecked the chief steward's office.

At the conclusion of that attack the *Musashi* was down some 4 meters by the bow with a 2-degree starboard list. Port voids were flooded, and the small starboard list was removed. The situation regarding reserve buoyancy was becoming critical, as much water had been taken aboard *Musashi* by this time. Many of her counterflooding voids had been used, while others were open to the sea from bomb or torpedo damage. It was a grim picture for the damage control officer.

The extreme trim by the bow was then the most serious problem. It reduced the maximum speed to 16 knots, for the ship was directionally unstable at speeds above this. Reluctantly, the captain of the *Musashi* had to order his ship to slow and fall behind the formation. Another matter of grave concern was the reduction in antiaircraft firepower. Only a fourth of the 25-mm machine guns were operational, as bomb damage and strafing had taken a terrible toll. This was particularly serious, as the reduced speed forced the ship to drop out of formation, losing the antiaircraft support of other ships. With the ship's antiaircraft battery now materially reduced, the gunnery officer requested the captain, Rear Admiral Toshihira Inoguchi, to let him use the main battery guns with their special shells. These were a time-delayed type called "sanshiki-dan." It was not until the fifth attack that the request was finally granted. The captain's reluctance to use these guns was probably due to the blast effects, and many of the 25-mm machine gun mounts had no blast shields. However, the critical situation of the *Musashi* must have convinced him that the 460-mm guns should be used.

A fifth attack at 1315 resulted in no damage, as the *Musashi* limped northward at 16 knots. *Musashi* opened fire on her attackers with a full nine-gun 460-mm salvo. After this first salvo, Turret I was out of action, because one of its guns had a projectile jammed in one of the barrels. A second salvo of six 460-mm shells was fired. Neither salvo deterred the American aircraft that swooped in to drop their bombs and torpedoes. Three engine rooms and nine boiler rooms were still operational, and more than 80 percent of the armored citadel was still intact. After that attack the Japanese had an invaluable respite for damage control measures. The worsening bow trim had slowed the ship to 12 knots, but she was not in danger of sinking, despite nine torpedo hits, seven bomb hits, and a dozen near misses.

Almost all the outboard voids on both sides had been flooded. No attempt was made to remove some of this counterflooding water, although the voids had connections to steam eductors which, with a capacity of 200 metric tons per hour, could have removed at least part of it. This oversight cost the Japanese an excellent opportunity to regain some lost buoyancy.

The final attack, at 1445, was made by some 75 carrier aircraft. The fierce engagement lasted only a few minutes, but 11 torpedo hits, 10 bomb hits, and six near-miss bombs left the ship almost totally helpless, with her decks awash. An unusual quiet fell over the ship as her speed was now reduced to only six knots. The only sound other than

The Japanese battle force (Central Group) sorties from Brunei, 22 October 1944. The first six ships in the column to the left are heavy cruisers, followed by the battleships Yamato, Musashi, *and* Nagato. *Note the great beam of the* Yamato *and* Musashi *as compared to the other ships. (Courtesy of T. Shiraishi)*

The Musashi *is shown sinking late in the afternoon of 24 October 1944. Compare her freeboard at the bow with the similar view taken on 22 October. The poor quality of these photographs was a result of the exposed film being retained in the camera a number of years before being developed. (Courtesy of T. Shiraishi)*

the moan of the wounded or an occasional order was the escaping steam from the damaged whistle pipe.

One bomb hit the antiaircraft defense control tower, near frame 120, detonated in number 1 bridge house, sheared off the starboard control center, and damaged the bridge and operations room. The captain was seriously wounded by this explosion, which also killed or wounded 78 men.

Three bombs struck on the upper deck, at frames 105, 115, and 120, to port. They damaged 25-mm machine gun mounts, the communications command post, number 1 radio receiving station, the telecom room, and air intakes and exhaust uptakcs to boiler rooms 4 and 8.

Two bombs exploded on the starboard side at frame 115 near the captain's elevator, damaged more 25-mm machine gun mounts, and caused the forward bulkhead at the entrance to fireroom 7 to collapse. A bomb hit near frame 127, port side, destroyed a head and severely damaged the after side of the forward tower immediately below the flag deck.

A bomb hit at frame 62, to port, exploded in a berthing compartment, and damaged the sick bay area. This hit added to the damage already sustained during earlier attacks. Another exploded on the roof of number 2 turret, leaving a soup-bowl-sized gouge a millimeter deep in the armor plate. Another bomb exploded on the starboard side at frame 75 and wrecked administrative offices and the officers' day room.

Six near-miss bombs probably caused more damage to the shell plating. Four of these were observed to starboard near frames 130–140, while two more were noted to port.

At that time the combined heavy damage to the hull structure resulted in a longitudinal structural failure in shear, which ran from frames 70 to 95 on the starboard side of the flying deck some 2 meters from the outboard side.

Although the bomb hits caused numerous casualties and much damage to the superstructure, the *Musashi* was finally sunk by the added 11 torpedo hits. Three of these came on the port side at frames 40, 60, and 75, causing considerable flooding forward, including the instantaneous flooding of the number 4 bilge-pump room. Another hit to port at frame 125 flooded fireroom 8 through loosened rivets. There was also some minor leakage into fireroom 12. Three hits to port near frame 140 caused heavy damage; one broke through the side-protection system, and the others penetrated the port side 25-mm machine gun magazines, forward of the main machinery spaces, without detonating the ammunition. These magazines flooded, however, further aggravating the bow trim problem.

A torpedo hit at frame 145 to port resulted in the penetration of the port outboard engine room bulkhead over a length of some 10 meters. The bulkhead had been near collapse, bulging after a torpedo hit at frame 140 in the second attack. The port outboard engine room flooded rapidly, and the port inboard engine room began to flood, forcing the ship to steam on two shafts.

The after 460-mm magazine began to flood following a torpedo hit on the port side at frame 165. Damage control parties isolated the flooding. The number 6 antiaircraft gun-powder magazine, the after compass room, and the port evaporator room began flooding as a result of this damage.

The effects of all the hits on the port side were only very slightly offset by two hits on the starboard side at frames 80 and 105. No damage was reported in the citadel as a result of this damage, but the trim by the bow increased.

When the attack ended, the *Musashi* had a 10-degree port list and was down some 8 meters by the bow, with the port side of the forecastle submerged back to the forward main battery turret. The list was reduced to 6 degrees by counterflooding, but increased again when the bow submerged deeper into the water. Finally, in a desperate measure to reduce the bow trim, living quarters and storerooms on the starboard side aft were flooded. When the list reached 10 degrees, the three boiler rooms and the outboard starboard engine room were flooded, but to no avail. The flooding of these spaces at this time seriously reduced the reserve buoyancy and caused the ship to settle even deeper into the water without any positive effect on the trim or list. The increase in draft meant that the bow settled further into the water reducing the intact waterplane even more. The ship was close to sinking. Furthermore, the port list lessened the chances of free-flooding compartments on the starboard side, and the available pump capacity was inadequate to cope with such lists within the time frame necessary. At this point attempts to right the ship by counterflooding measures were abandoned.

Admiral Kurita did not attempt to negotiate the narrow San Bernardino Straits in daylight with the heavy aircraft attacks his fleet was being subjected to. Just before 1600 Admiral Kurita's ships, now distant from the heavily damaged *Musashi*, began to turn back, and by 1715 his ships were cautiously filing in column between the islands of Masbate and Burias.

Meanwhile, the death throes of *Musashi* continued. The ship lost its electric power and shortly after 1715 Admiral Kurita was informed by an emergency signal lamp from the crippled battleship that she was taking on an excessive amount of water. To this signal Kurita replied, "*Musashi*, go forward or backward at top speed and ground on the nearest island and become a land battery." Rear Admiral Inoguchi tried to comply with his admiral's order, but the listing *Musashi*, with her bow under water, would not respond to the helm. The ship continued to move slowly in a circle until she foundered.

Progressive flooding continued unchecked in the bow, despite efforts of damage control parties to establish flooding boundaries. By 1915, with a 12-degree list to port, preparations were made to abandon ship. With the extreme flooding and bow trim, the entire bow was submerged back to the forward main turret. Within 15 minutes the list grew to 30 degrees; the ship was obviously beginning to capsize as she plunged by the bow. Abandon ship was ordered around 1920, and destroyers moved in to pick up the crew. At approximately 1935 on 24 October 1944, the *Musashi* capsized with a sharp lurch to port. The executive officer swam off from one of the upper levels of the forward command tower. He saw two of the propellers slowly turning as the *Musashi* slid below the water. The chief engineer climbed over the starboard side amidships as the *Musashi* lurched to port. He walked and then scrambled around the vessel's girth, fighting the swift rolling action. He managed to climb over the bilge keel, but was then thrown into the water. He swam away. Some survivors also testified that they heard two large explosions deep within the ship after she had plunged under the surface. Out of her complement of 2,399, a total of 1,023 officers and men were lost.

Table 2-7 summarizes the bomb and torpedo hits, but specific details, especially on the final ordeal of the ship, are of uncertain accuracy. Eight American carriers launched 259 sorties against the ships of Force A on 24 October, with the *Musashi* absorbing the brunt of these attacks. Despite the enormous number of antiaircraft machine guns, the special 460-mm flak shell, and the dual-purpose guns that were defending the Japanese force, only 18 American planes were lost. The attacking planes sighted only four Japanese aircraft, all of which were shot down.

In the final analysis, the loss of the ship was due to the inability of the damage control teams to establish flooding boundaries. In fairness to them, however, it should

TABLE 2-7
Damage to Musashi—24 October 1944

Time	*Torpedo Hits (frame no. and side)*	*Bomb Hits (frame no. and side)*	*Near-Miss Bombs (frame no. and side)*
1025	none	No. 1 Turret	20, port
			25, starboard
			145, port
			145, starboard
1027	130, starboard	none	none
1138	85, port	15, port	70, port
	102, port	138, port	50, starboard
	140, port		50, starboard
			50, starboard
			50, starboard
1217	60, starboard	none	180, starboard
			180, starboard
			238, starboard
1253	70, port	45, port	none
	70, starboard	65, port	
	110, starboard	70, port	
	138, starboard	135, starboard	
1445	40, port	starboard AAFC Station	135, starboard
	60, port	62, port	135, starboard
	75, port	75, starboard	135, starboard
	80, starboard	No. 2 turret roof	135, starboard
	105, starboard	105, port	135, starboard
	125, port	115, port	135, starboard
	140, port	115, starboard	
	140, port	115, starboard	
	140, port	120, port	
	145, port	127, port	
	165, port	127, port	
Totals	20	17	18

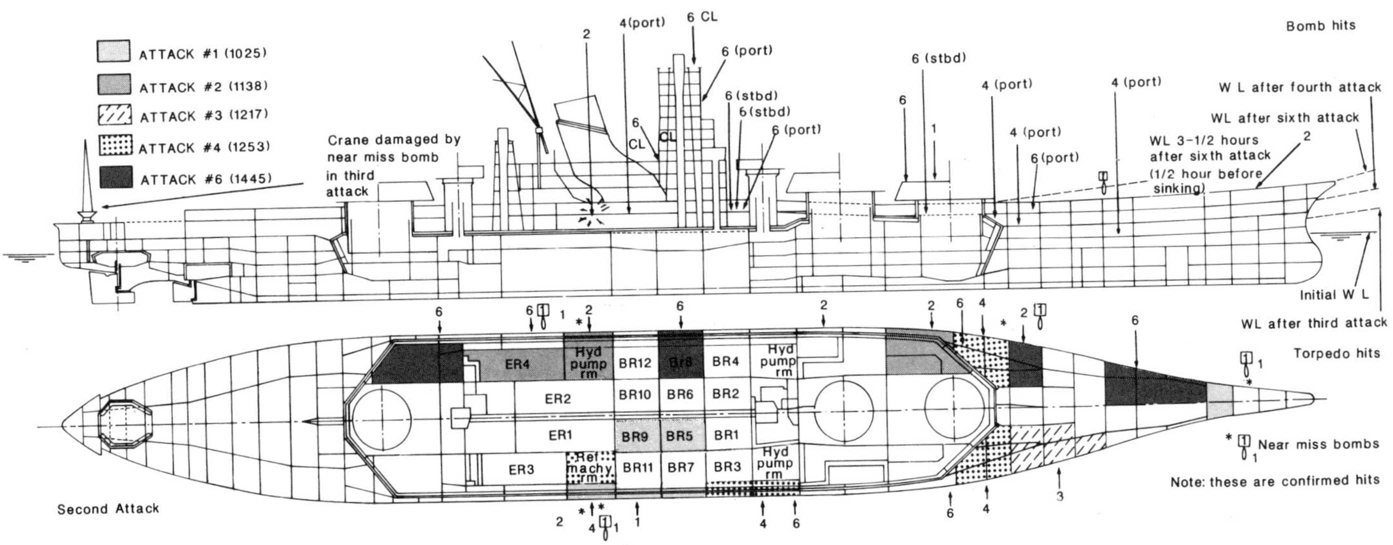

Figure 2-2. Musashi *loss—Approximate location of bomb and torpedo damage, 24 October 1944.*

be noted that they were faced with an insuperable situation. The counterflooding system did not have the pumping capacity to handle the massive flooding that ensued. For example, the storerooms and crew living spaces that were ordered counterflooded on the starboard side aft had no sea valves that would permit these spaces to be flooded from the sea. There were not enough fire pumps remaining in operation to flood these spaces from the firemain, so the attempt was abandoned. The partial flooding created a large free surface that severely reduced the ship's already impaired stability. The damage sustained by the *Musashi* was so great and the casualties to personnel so enormous that it was difficult to maintain a sustained effort in damage control or antiaircraft defense. The damage to *Musashi* was also so extensive that the authors believe that the ship would have been a total loss to the war effort even if she had survived her ordeal and returned to Japan. The structural failure in shear and the underwater damage to both sides of the hull would have required the total reconstruction of the hull, particularly in the bow region.

The sinking of the *Musashi* was an object lesson to U.S. experts. The number of bomb hits and torpedoes required to sink the ship amazed U.S. designers. A careful analysis of her last few hours resulted in a memorandum and briefing by a damage survey team from the U.S. Navy's Bureau of Ships to the effect that torpedo attacks against large warships should be concentrated on one side to maximize the tendency to capsize. Torpedo hits on both sides of capital ships tend to have the effect of counterflooding, thereby offsetting the effects of flooding. This advice was instrumental in the sinking of the *Yamato* several months later.

Shinano—Construction and Conversion. The keel of the third ship of the *Yamato* class—originally known as Warship No. 110 and eventually named *Shinano*—was laid down at the Yokosuka Navy Yard on 4 May 1940. The ship was built in Dry Dock 6, which was also under construction at the time. A new hull-plate workshop was to be built at Yokosuka during the same period. These developments slowed the construction of the *Shinano* compared to that of the *Musashi* and *Yamato.*

Warship No. 110 was a slightly modified version of the first two ships of the class. Proof firings of the 460-mm gun had indicated that the armor thicknesses of the first two ships exceeded design requirements, so the armor was modified as follows:

	Warship No. 110	*Yamato*
Main side belt	15.75″ (400 mm)	16.14″ (410 mm)
Armor deck	7.48″ (190 mm)	7.87″ (200 mm)
Turret face plates	24.80″ (630 mm)	25.59″ (650 mm)
Turret roofs	9.84″ (250 mm)	10.63″ (270 mm)
Turret sides	9.06″ (230 mm)	9.84″ (250 mm)
Turret rear plates	6.69″ (170 mm)	7.48″ (190 mm)

The weight savings permitted the extension of the triple-bottom protection and allowed better protection for fire-control and sky-lookout positions on the bridge tower. Staff accommodations were to be improved. Although the major variation from the original *Yamato* plan involved protection, there was an important modification to the armament. Instead of the 127-mm dual-purpose guns of the first two ships, Warship No. 110 was to be armed with the powerful new 100-mm/65-caliber, Type 98 gun that was ballistically superior to the 127-mm gun, although the older gun fired a heavier projectile:

Gun	5″/40 (127 mm)	3.9″/65 (100 mm)
Shell weight (lb.)	50.7 (23 kg)	28.7 (13 kg)
Muzzle velocity (fps)	2,379 (725 mps)	3,313 (1,010 mps)
Maximum range (yds)	16,185 (14,800 m)	21,325 (19,500 m)
AA ceiling (ft)	30,840 (9,400 m)	42,651 (13,000 m)

The new gun also had a vastly superior rate of fire—from 25–28 rounds per minute instead of 14 rounds per minute.

Although the antiaircraft armament was superior to that of the *Yamato* and *Musashi*, the superiority was strictly of secondary importance, and the Japanese were reluctant to spend immense sums of money, manpower, and materials for a ship not scheduled to complete before 1945. These considerations became of paramount importance in December 1941 after the successful attack on Pearl Harbor, and the sinking of the British *Prince of Wales* and *Repulse* dramatically emphasized the growing importance of air power and the declining role of the battleship. In addition, Admiral Isoroku Yamamoto, commander-in-chief of the Combined Fleet, long an opponent of battleship construction, held that

> Military people always carry history around with them in the shape of old campaigns. They carry obsolete weapons like swords and it is a long time before they realize they have become purely ornamental. These battleships will be as useful to Japan in modern warfare as a samurai sword.*

The decision in December 1941 to stop work left two alternatives open if the large dry dock was to be freed for other use—either scrap the ship or continue work on it until it could be floated out of the dock. The Japanese decided to continue construction, but the work force was reduced, and it was expected that a year would pass before the hull could be launched.

---

*John Deans Potter, *Yamamoto* (New York: Viking Press, 1966).

The Shinano *is shown undergoing trials in the Urage channel area of Tokyo around 1500 on 11 November 1944. The carrier was undergoing her steering trials and was steaming at 60 percent of capacity. She was turning from port to starboard seen in this photograph. Note the crew and shipyard workmen on the flight deck. The giant aircraft carrier sank late on the morning of 29 November 1944—the victim of four torpedo hits, a poorly trained crew, the errors of judgment of her captain, and her incomplete condition. (Courtesy Shizuo Fukui)*

The Japanese Yamato*-class ships were the largest battleships ever built. They mounted nine 460-mm guns in three triple turrets, each of which weighed more than many World War II destroyers. This is the* Yamato *in late 1941. Note the undulating sheer line, adopted in order to save weight. (Imperial War Museum)*

This port quarter view of the Yamato *shows the airplane and boat stowage facilities in the stern. Because of the severe problem with the blast pressures from the 460-mm guns, special precautions had to be taken to stow the boats and aircraft out of the firing arcs of the big guns. With the boat stowage in this stern location, the superstructure was freed for more antiaircraft guns. (Imperial War Museum)*

A large plume of smoke marks the destruction of the Yamato on 7 April 1945. A bathyscape, Pisces II, explored the wreck of the Yamato on 2 August 1985. The ship was found to be lying on the seabed at a depth of 340 meters. The enormous explosion shown here tore the ship in two. The forward section, about 170 meters long, was found 80 meters away from the stern section.

Conversion to an aircraft carrier. In June 1942, after the loss of four fleet carriers in the Battle of Midway, the Japanese decided to convert Warship No. 110 to an aircraft carrier to be named *Shinano.* The hull was only 45 percent complete, which permitted great latitude in the conversion. Structural work was essentially complete up to the lower deck and major machinery components were in place. The lower side belt armor was installed over the entire length of the armored citadel, and the main side belt and armor deck were in place over the main battery magazines. The forward barbettes were almost complete.

Battle experience had demonstrated that there was a need for heavy armor protection for aircraft carriers, especially against air attack. The *Shinano,* in addition to her conventional role of aircraft carrier, was also designed to serve as a heavily protected mobile air base for planes not assigned to the ship. Particular attention was given to damage control, and large gasoline tanks and ammunition magazines were provided to serve the dual mission.

Japanese carriers normally armed and refueled aircraft on the hangar deck. This was a major factor in the losses of the four carriers at Midway, as bombs started fires on the hangar decks, and the burning aircraft and ordnance could not be jettisoned. There were no sprinkling systems or fire curtains as in American carriers. Once the conflagration had started, there was little the Japanese firefighters could do to contain the holocaust. In the *Shinano,* by arming and refuelling planes on the armored flight deck, the Japanese expected to avoid such disastrous fires and explosions. Much of the hangar deck was left open, to facilitate jettisoning aircraft and ordnance if necessary. In addition, the armored flight deck would offer some protection against bombs, and the experiences of British carriers in World War II indicated that such a deck was a wise investment.

The conversion involved the deletion of the turrets, gun mounts, and superstructure characteristic of the battleships. The undulating weather deck was somewhat modified by the addition of shell plating forward to produce a more nearly orthodox profile. This was necessary to store planes on the hangar deck.

The plans for the conversion were accepted in September of 1942. Completed as an aircraft carrier, the *Shinano* had the characteristics summarized in table 2-8.

Aircraft. The assigned air group for the *Shinano* was very small for a ship of her dimensions and displacement because of the emphasis on armor protection and her dual role as a base for unassigned aircraft. She was to carry the following aircraft:

20 fighters (two in a dismantled condition)
"modified Zeke," 17-Shi, "Reppu"
20 bombers (two in a dismantled condition)
"Grace," B7A1, "Ryusei"
7 scouts (one in a dismantled condition)
"Myrt," 6N1, "Sai-un"

There were centerline elevators, forward and aft, but no catapults. The arresting gear had 15 wires with five arrayed forward, so that if the stern portion of the flight deck was damaged and could not land aircraft, the ship could steam downwind and recover aircraft over the bow. Actually, fate would decree that the ship would never carry a single aircraft.

Antiaircraft armament. The antiaircraft armament was unprecedentedly heavy for a Japanese aircraft carrier, comparing favorably with contemporary U.S. Navy carriers. Both the 127-mm guns and the 25-mm machine guns were considered by the Japanese to be reliable antiaircraft weapons. The 25-mm machine gun armament, however, was not as effective as the 20-mm and 40-mm armament used on the *Essex*-class carriers.

TABLE 2-8

Characteristics of Shinano

Standard displacement	62,000 tons (62,995 mt)
Trial displacement	66,984 tons (68,059 mt)
Full-load displacement	70,755 tons (71,890 mt)
Length on waterline	839.90′ (256.0 m)
Waterline beam	119.08′ (36.3 m)
Draft (trial displacement)	33.83′ (10.312 m)
Armament	sixteen 5″ (127-mm)/40 caliber guns — paired 105 1″ (25-mm)/60 caliber guns — tripled
Aircraft	forty-seven
Protection	6.30″ (160 mm) — main side belt, amidships 15.75″ (400 mm) — main side belt, citadel end 7.87″ (200 mm) — lower side belt tapered to 2.95″ (75 mm) 2.95″ (75 mm) — flight deck 0.47″ (12 mm) — hangar deck 0.98″ (25 mm) — upper deck, maximum 3.94″ (100 mm) — armor deck, amidships 7.48″ (190 mm) — armor deck, citadel ends 9.06″ (230 mm) — armor deck slopes
Shaft horsepower	147,948 (150,000 mhp)
Speed	27 knots
Endurance	10,000 miles at 18 knots

Armor protection. Armor protection was remarkably heavy for an aircraft carrier. Several standards were established: it was to protect against both gunfire and bombing attacks. It was intended that the citadel be able to withstand 203-mm shellfire. Since the ship was partially completed as a battleship, the extremities of the citadel were capable of withstanding 460-mm shellfire. The armored flight deck was designed to resist a 500-kilogram bomb dropped by a dive bomber. The total horizontal protective system, including the 100-mm armor deck, was intended to withstand an 800-kilogram bomb dropped from an altitude of 4,000 meters.

The Japanese were especially concerned about the vulnerability of the gasoline tanks. As originally designed, these tanks were shielded by armor capable of resisting 155-mm shellfire. After the *Taiho* sank, an investigation revealed that the structural arrangement of the fuel tanks was faulty. Torpedo hits in the vicinity of the tanks could permit gasoline vapors to leak beyond the cofferdam. An obvious solution to this problem was to ballast the gasoline tanks with salt water to prevent the build-up of gasoline vapors. Japanese pilots objected to the use of salt water, since the salt water could contaminate the fuel system if the tanks were improperly cleaned, creating problems in their aircraft engines. Moreover, ballasting solved only part of the problem, and it was felt necessary to isolate the gasoline tanks by a layer of concrete in the void spaces around such tanks. For these reasons, the cofferdams around the fuel tanks were filled with 2,400 metric tons of cement in an effort to minimize the chance of splinters rupturing the gasoline tanks. This was an expedient measure at best, but little else could be done in view of the then-advanced state of construction.

Since the *Taiho* had been destroyed by a gasoline vapor explosion, it was thought prudent to install large ventilation fans on the hangar deck to exhaust gasoline fumes in the event of damage to the gasoline system. Portable canvas wind scoops also were provided for rigging over elevator openings to force air into the hangar bays. Large capacity fog-foam systems were also installed to assist in fire fighting.

Underwater side protection. The underwater side-protection system followed the basic design of that of the *Yamato* and *Musashi*, with the exception of the reduced main side belt armor thickness amidships. This reduced armor thickness, in addition to the weak joint between the upper and lower armor belt, resulted in a side-protective system of inferior performance. The impressive resistance of the *Musashi* to torpedo damage on 24 October 1944 engendered great confidence in the underwater protection of the *Shinano*, which proved to be unwarranted.

Propulsion plant. The *Shinano* had exactly the same propulsion plant as the two battleships. The ship's accelerated construction, however, resulted in only eight boilers being operational when the *Shinano* was commissioned. Due to the incomplete state of the engineering plant, no dock trials were conducted, and actual power-plant performance was never determined. Trials were conducted in Tokyo Bay to test the completed propulsion plant. The endurance of the carrier was expected to be considerably greater than that of the battleships.

Structural design. Flight-deck design was based on that of the *Taiho*, the Japanese equivalent of the U.S. *Essex* class. The structure consisted of two decks separated by transverse and longitudinal girders with a depth of 0.84 meters. The flight deck—the strength deck—was armored over the hangar bays. As a result, aircraft landing loads were not controlling design factors. The entire structure, including supporting flight-deck bents, was designed as a rigid entity. There were no stanchions or columns in the hangar area, which had a clear deck height of 5 meters.

Flight deck. The flight deck was supported by transverse bents, located on every frame of the vessel. Every other bent was a heavy one—the heavy ones, however, were only 460-mm in depth. Longitudinals, relatively widely spaced, were to run across the transverses to assist in the support of the flight deck. With the transverse framing method used by the Japanese for flight-deck support, there was no necessity for deep transverse girders, and consequently no gallery deck was installed on this carrier.

Stack structure. One of the distinguishing features of Japanese aircraft carriers was the funnel arrangement. Downsweeping uptakes below the flight deck were used in many Japanese carrier designs to keep the boiler flue gases from interfering with flight operations. In the *Shinano*, the stack structural design of the *Taiho* was used, where the uptakes became an integral part of the island structure and were angled outboard some 26 degrees.

Construction. After the Battle of the Philippine Sea, construction of the ship was greatly accelerated, with completion desired for October 1944 instead of April 1945 as originally planned. The speed-up was ordered because of the fear that the United States would begin long-range bombing of Japan later in the year using B-29 bombers based in the Marianas. The builders complied to the best of their ability, but at the expense of good workmanship and the proper detailed fitting-out of the ship.

Completion was delayed by an accident during the launching on 8 October 1944. The ship was launched by flooding the dry dock in which she was built. The caisson at the end of the dock was not ballasted with water before the dry dock was flooded, as it should have been, and when the water level in the dock neared that in the harbor, the buoyant caisson suddenly floated, letting in a great surge of water that carried the carrier against the forward end of the building dock. So great were the impacts that the bow structure was damaged, necessitating dry-docking again to repair the underwater damage. Repairs were completed by 26 October.

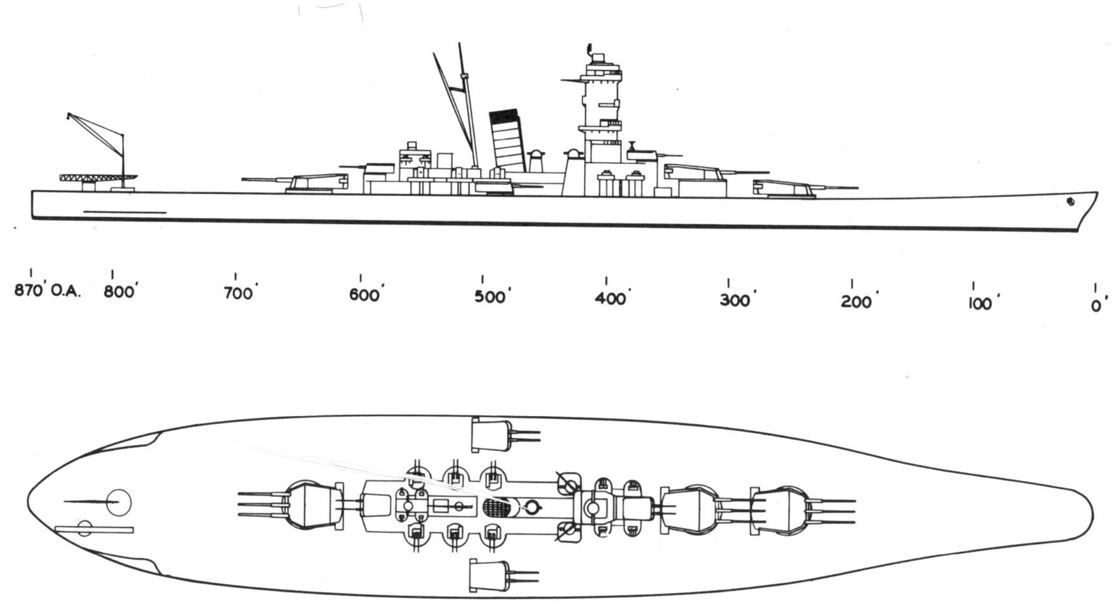

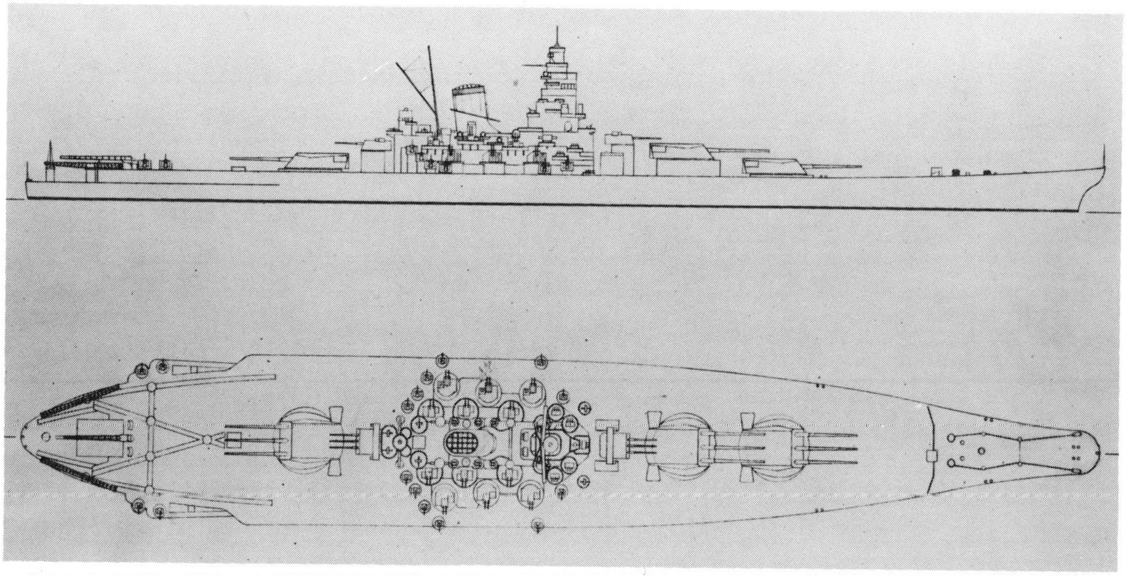

Little was known by the U.S. Navy of the characteristics of the Yamato *class until after World War II. The top view shows a summer 1944 sketch of the design. The second sketch, prepared after the Battle for Leyte Gulf in October, is a better likeness of the ships as there were a large number of photographs taken of the two ships during the battle. Even at this late date, March 1945, the U.S. Navy still believed that these ships were armed with 406-mm guns (James C. Fahey Collection)*

The *Shinano* was commissioned at Yokosuka on 19 November 1944. The ship was substantially complete, but several major items, including four boilers, were still unfinished, limiting the maximum speed to about 20 knots. The final completion compartment air tests had not been conducted, and the watertight integrity was unsatisfactory. Many holes in the bulkheads and decks, for electrical cables, ventilation ducts, and pipes, had not been sealed. Firemains and drainage systems were incomplete because pumps had not been delivered. The usual Japanese care in joining structure by welding and riveting was lacking, which could have had serious consequences if the ship had remained in service for several years. Japanese investigators later found that workmanship on the carrier had not been up to previous standards because of the pressure to deliver the ship as soon as possible.

Loss of Shinano. After the *Shinano* was commissioned, ten days were spent loading limited quantities of stores and ammunition. The crew, of which 70 to 75 percent had no previous sea duty, had reported on board 1 October 1944. Due to the threat of air raids, the Japanese decided to move the ship to the Inland Sea, despite the inexperienced crew and the ship's inadequate watertight integrity. It was decided to make a night passage of the Inland Sea to reduce the risks. At 1800 on 28 November the largest aircraft carrier ever built in Japan left Yokosuka for Kure to pick up her air complement and conduct training exercises while fitting out was completed.

The carrier steamed at 20 knots on six boilers and two shafts. Watertight doors on the armored deck, and many hatches below that deck, were left open for convenient access to the machinery spaces. Some manholes in the double and triple bottom were not closed. Around 2350 speed was reduced to 18 knots due to an overheated shaft bearing.

At 2048 that night the U.S. submarine *Archerfish*, patrolling the entrance to Tokyo Bay, detected the carrier and her escorts. The submarine tracked the group for several hours at her best surface speed, but hopes of reaching a firing position gradually dwindled. At 0300 on 29 November 1944 the *Shinano*'s zigzag course put her on a turn directly towards the submarine. Some eighteen minutes later, the *Archerfish* fired six torpedoes.

Four torpedoes hit the ship on the starboard side and breached the carrier's underwater defense system, about 4.27 meters below the waterline, causing rapid flooding that soon produced a starboard list of ten degrees. See figure 2-3. The dispersion and depth of the hits were decisive factors in the loss of the ship. One torpedo struck near frame 104 adjacent to the bulkhead separating some oil tanks from the air-compressor room. Both spaces were flooded. The antiaircraft magazines beneath the air-compressor rooms flooded through open hatches. No. 2 Damage Control Center began to flood.

A second torpedo exploded in way of boiler room 1 at frame 117. This space flooded so rapidly that there were no survivors. Fireroom 2 began to flood slowly through leaks in the too-rigid longitudinal bulkhead. Serious leakage also occurred in the outer bulkheads in way of the heavy H-beams tying the dividing bulkhead to the holding bulkead. Because there were no adequate backing structures, the torpedo explosion simply caused these beams to punch holes in the plating. Boiler room 5, abaft boiler room 1, also began to flood, indicating that the torpedo had hit near the separating bulkhead.

A third torpedo hit near the stuffing box compartment between frames 160 and 162. The starboard outer engine room began to flood rapidly through the damaged stuffing gland at frame 160 and less rapidly through the holding bulkhead at the after end of the engine room. All hands escaped from those areas.

The last torpedo detonated at frame 194 over the storerooms outboard of the empty after gasoline tanks. The refrigerated storerooms on the first platform directly over the gasoline tanks were flooded, as were the tanks. The deck above was badly ruptured and many men in the compartment there were killed.

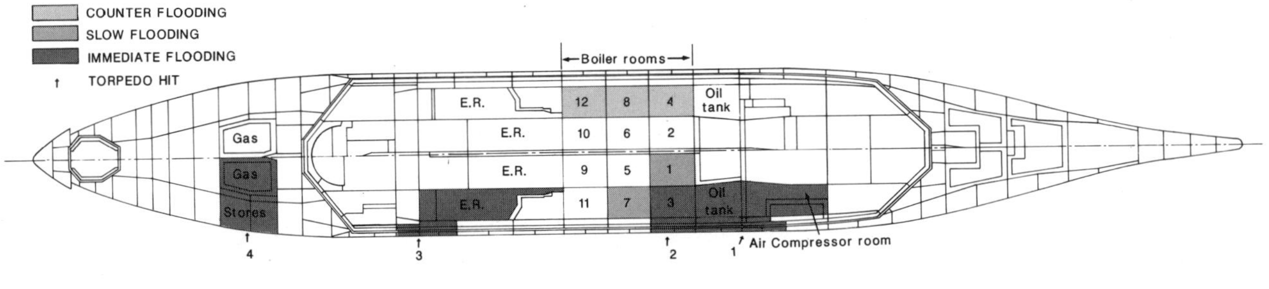

Figure 2-3. Loss of the Shinano*—Location of the torpedo hits, 29 November 1944.*

The Japanese battleships Nagato *(right),* Yamato *(center), and* Musashi *(left) at Brunei, Borneo, just before the Battle for Leyte Gulf, October 1944. (Courtesy Mr. K. Hando)*

Progressive flooding began as magazines within the after part of the armored citadel on the first and third platforms were inundated. The crew was overconfident in the carrier's ability to withstand damage, and as a result few effective efforts were made initially to save the ship. The port outboard voids were flooded in an attempt to right the ship, but the list was only temporarily checked at about 12 degrees to starboard. The flooding continued unchecked and gradually began to affect the stability as the ship continued to steam at 18 knots. Counterflooding measures and efforts to establish flooding boundaries included flooding the three outboard port boiler rooms. This desperate measure was too late. The lack of a crew skilled in damage-control procedures was a major factor in the loss of the ship as no effective measures were taken to establish flooding boundaries. Under normal circumstances, efforts to establish flooding boundaries would have been started soon after the torpedo hits occurred. In addition, the captain of the *Shinano* was guilty of poor judgment. After damage was sustained, the ship's speed was maintained, and this permitted large dynamic pressures to develop in areas open to the sea, causing structural damage and consequent progressive flooding. By the time rigorous measures were begun, the situation was beyond control.

Some flooding aft could have been controlled had the ship been in a better condition of readiness. Since the drainage system was incomplete, some portable pumps of limited capacity were put to use, and a few bucket brigades were formed. Flooding might have been better controlled by portable pumps, but few men knew how to use

The submarine Archerfish *(SS-311) torpedoed the aircraft carrier* Shinano, *the largest warship ever sunk by a submarine, early on the morning of 29 November, 1944. The ship sank about seven hours after being hit by four torpedoes. This view of the Archerfish was taken near the Golden Gate on 30 May 1945. (Courtesy Captain Joseph F. Enright, U.S. Navy, Ret.)*

them, and the men soon drifted away when they became fearful of being trapped in a sinking ship.

All discipline was lost as idle men began gathering on the hangar deck to await rescue. Civilian technicians, wearing what looked like naval uniforms, refused to obey orders to go below and were responsible for much of the confusion.

By dawn men were going overboard without orders. At 0430 the starboard list was increasing, and the ship was making very slow speed. The engineering officer ordered the three outboard port firerooms flooded in a last desperate attempt to right the ship. The list was temporarily checked, but increased again due to continued progressive flooding. By 0600, all boiler feed water was exhausted and power was lost. Progressive flooding had filled the three outboard firerooms to starboard and adjacent voids up to sea level. Flooding continued unabated in the after part of the ship. At 0800 the crew began transferring to destroyers alongside while the ship continued to list to starboard. Abandon ship was ordered at 1018 and shortly afterward the ship capsized to starboard. The *Shinano* rolled over bottom up and then sank stern first. The captain chose to go down with his ship; 25 percent of the crew was saved.

Warship Number 111—Construction. The fourth ship of the *Yamato* class, Warship No. 111, was laid down at the Kure Navy Yard on 7 November 1940. The ship, never named, was to be similar to the *Shinano* as she was conceived as a battleship. Construction began after the *Yamato* was launched. In December 1941, when the Japanese evaluated their capital-ship construction program, the hull was about 30 percent complete. Work was stopped and all the structure above the double bottom was dismantled and scrapped. Four large submarines were built on the double bottom. Completion of the nine 460-mm guns for this ship was canceled.

The Fifth Fleet Replenishment Program of 1942.

Five more capital ships were projected by this program—three battleships and two battlecruisers. These ships are described in the text that follows, in the order shown:

Battleships

- Warship No. 797 — An improved version of Warship No. 110
- Warship No. 798 — Battleship Design A-150
- Warship No. 799 — Battleship Design A-150

Battlecruisers

- Warship No. 795 — Battlecruiser Design B-65
- Warship No. 796 — Battlecruiser Design B-65

Although some developmental work was accomplished in support of the construction of these five ships, none of them were destined to be built.

Warship No. 797. This fifth member of the *Yamato* class, known as an "Improved No. 110," was never laid down. This ship incorporated the design modifications of the Warship No. 110 sub-type of the *Yamato* class. In addition, the two wing 155-mm triple turrets were removed, permitting an antiaircraft battery of 24 100-mm guns (authors' estimate) in an arrangement similar to that eventually given to the *Yamato.* This would have been an outstanding armament, far superior to that of the earlier ships.

Super Yamato. The last two ships projected in the 1942 program, Warships No. 798 and 799, were to have been battleships of a new design, armed with 510-mm/45-caliber guns. This was consistent with the Japanese policy of individual ship superiority, since the Japanese expected that the United States would learn of the actual armament of the *Yamato*-class ships when they became operational.

In 1920–1921 the Japanese had constructed a 480-mm gun, so they did not hesitate to plan the larger gun for this battleship class, which was designated Design A-150. However, neither ship was ordered, and all plans were destroyed in the confusion following the end of the war. The following general description is based on the best information available.

Initial design investigations for a ship with either eight guns in quadruple turrets or nine guns in triple turrets, and a speed of 30 knots, were based on the knowledge of the 27-knot maximum speed of the U.S. *North Carolina*-class battleships. These studies resulted in estimated trials displacements of some *90,000* metric tons! It was decided that such a ship was too large and too expensive and that the displacement would have to be similar to that of the *Yamato* class.

Design work commenced after such work on the *Yamato*-class ships was completed, probably in 1938–39. After various revisions and experiments, including caisson tests, the design was essentially completed early in 1941. At this time, war seemed imminent, and there was such a demand for aircraft carriers, cruisers, and smaller ships that work on the battleships was halted. The results of the Battle of Midway ensured that design work on battleships would never be resumed.

The last version of Design A-150 featured 510-mm guns in three twin turrets and a secondary battery of many 100-mm/65-caliber antiaircraft guns, which would have made these the most powerful battleships in history.

At least one of the giant 510-mm guns was actually under construction at the Kure Navy Yard early in 1941, and detailed tests for the new turrets were made that year. The armor-piercing shell would have weighed about 1,900 kilograms.* The powder bags—

---

*Authors' estimate, based on a plot of shell weight versus caliber for Japanese naval guns.

eight for each round—would have been too heavy for manual handling, and automatic transfer and loading devices were under development.

Japanese steel mills were incapable of handling the massive armor belt required, and the Japanese envisioned the use of double strakes of armor plates for the main side belt. This would have been unfortunate, for the effective thickness of doubled armor plates was appreciably less than that of a single plate of the same total thickness.

Warship No. 798 was to have been built at Yokosuka in the *Shinano* dock; Warship No. 799 at Kure in the *Yamato* dock, after the fourth ship of the class had cleared the dock. The ships were scheduled to have been laid down in late 1941 or early 1942 and launched about three years later, with a total building period of some five years.

Battlecruiser Design B-65. These battlecruisers were intended as replies to the *Alaska*-class ships known to be under construction in the United States. Design B-65 would have resulted in interim-sized surface ships, with displacements between that of heavy cruisers and the *Yamato*-class ships—fast, powerful, but with only moderate protection. The preliminary design of the two ships of the class, identified merely as Warships No. 795 and 796, satisfied every basic characteristic of the battlecruiser type.

Plans for these ships progressed well beyond the conceptual design stage. The underwater protection system was tested by full-scale experiments in 1940, while firing trials for the new 310-mm/50-caliber guns were conducted in 1941.

By September 1940, the preliminary design scheme, known as Design B-65, had been completed. The scheme, for which the principal characteristics follow, was never included on the agenda for technical conferences, because the Chinese conflict, as well as the steadily worsening diplomatic situation, focused attention on more urgent projects. The planning staffs of the Imperial Japanese Navy were overwhelmed by the work generated as Japan moved towards war.

Details of Design B-65 are summarized in table 2-9.

In exterior appearance, the ships would have resembled the *Yamato*, with a distinct clipper bow, flush-deck construction, and a generally similar superstructure. The

TABLE 2-9
Battlecruiser Characteristics

Standard displacement	31,400 tons (31,905 mt)
Full-load displacement	34,447 tons (35,000 mt)
Length overall	807.89′ (246.2 m)
Waterline length	787.40′ (240.0 m)
Beam (design waterline)	89.24′ (27.2 m)
Draft (design)	28.87′ (8.8 m)
Armament	nine 12.2″ (310 mm)/50-caliber, tripled sixteen 3.9″ (100 mm)/65, paired twelve 25-mm machine guns, paired four 13-mm machine guns, paired eight 24″ (610-mm) quadruple torpedo tubes, quadrupled
Shaft horsepower	167,674 (170,000 mhp)
Maximum speed	34 knots
Endurance	8,000 miles at 18 knots
Main side belt	7.48″ (190 mm), sloped at 20°
Decks	4.92″ (125 mm)

antiaircraft guns were to be concentrated on the main deck amidships, four twin mounts on either side.

Towards the end of 1941, when some of the basic characteristics of the *Alaska*-class design became known, it was suggested that Design B-65 be modified to feature six 356-mm guns in three twin turrets, with increased protection to resist 356-mm shellfire. These proposals were abandoned because of the added displacement and reduced performance that would have resulted.

Armament. The following detailed analysis is based in its entirety on the B-65 preliminary design, completed in September 1940, which included guns with these characteristics:

Gun	310 mm	100 mm	25 mm
Shell weight	1,265 lbs (561 kg)	28.67 lbs (13 kg)	0.55 lbs (0.2 kg)
Muzzle velocity	. . .	3281 fps (1000 mps)	2,953 fps (900 mps)
Maximum range	36,000 yds (32,920 m)	21,325 yds (19,500 m)	7,327 yds (6,700 m)
Maximum elevation	45°	90°	80°

The 100-mm gun (Type 98) was a superb antiaircraft weapon with a ceiling of 13,000 meters; its only fault was a short service life of approximately 350 rounds per barrel. It had a rate of fire of about 25–28 rounds per minute, a training rate of 12 degrees per second, and an elevating rate of 16 degrees per second.

The 25-mm machine gun, in service on almost all Japanese warships, had a ceiling of about 5,000 meters and a maximum rate of fire of approximately 220 rounds per minute. The Japanese considered it an excellent weapon with no operational or maintenance problems.

Undoubtedly, the machine-gun armament of these battlecruisers would have been tremendously increased had they been completed and placed in service.

The ships were to carry deck-mounted torpedo tubes. The Japanese naval constructors did not like underwater torpedo tubes because they caused problems in protection.

The ships were also to carry three aircraft, with one catapult and a hangar located between the single raked funnel and the after superstructure.

Protection. The armor protection was intended to provide an immunity zone of from 20,000 to 30,000 meters when exposed to 310-mm shellfire. The deck armor was calculated to be proof against a 1,000-kilogram bomb dropped from a dive bomber.

Very little is known of the details of the armor protection, other than the 190-mm thickness of the main side-belt armor, which was inclined some 20 degrees from the vertical. This inclination increased the equivalent vertical plate resistance of the side-armor system to about that of 241-mm plates. The deck armor was 125-mm.

The side-protection system was designed to resist 400-kilogram TNT charges detonated in contact with the side plating, as well as a 310-mm shell with an underwater trajectory. Full-scale explosion tests were conducted on a mock-up of the side-protection system at the Yokosuka Navy Yard.

Propulsion plant. Very high speed was a fundamental design requirement, with the maximum speed set at 33 knots. Eight Kampon boilers and four sets of geared turbines would develop a normal maximum of 170,000 metric horsepower. Endurance at 18 knots was 8,000 nautical miles. The machinery arrangement conformed to contemporary Japanese design practice, with individual compartments for each boiler and for each set of geared turbines.

By 1939, when the United States Navy was designing the 33-knot *Iowa*-class battleships, the battlecruiser concept was hopelessly obsolete. The development of fast-battleship designs, with battlecruiser speed but battleship protection and offensive power, rendered such "white elephant" schemes as the Design B-65 relatively useless.

The design, considered within the framework of a battlecruiser concept, was a good one. Offensive power was considerable, speed was adequate, and side protection had good designed resistance. It is probable, considering the deficiencies revealed in the side protection of the *Yamato* and *Musashi*, that the system would not have performed as well as expected. The armor protection was incapable of resisting fire from capital ships at normal battle ranges.

In fairness to the designers of the B-65 type, it must be admitted that the prime deficiency of the design was the obsolescence of the concept. Within the confines of the battlecruiser specification given the designers, a very good warship design was developed.

Armament.

The principle characteristics of the armament of the *Yamato* class are summarized in table 2-10.

TABLE 2-10
Gun Characteristics

Gun	18.1"/45 (460 mm)	6.1"/60 (155 mm)
Shell weight (lbs)	3,219 (1,460 kg)	123 (55.87 kg)
Muzzle velocity (fps)	2,559 (780 mps)	3,117 (950 mps)
Maximum range (yds)	45,275 (41,400 m)	29,965 (27,400 m)
Maximum elevation	45°	55°
Gun	5"/40 (127 mm)	1"/60 (25 mm)
Shell weight (lbs)	50.7 (23 kg)	0.55 (0.25 kg)
Muzzle velocity (fps)	2,379 (725 mps)	2,953 (900 mps)
Maximum range (yds)	16,185 (14,800 m)	7,439 (6,800 m)
Elevation	90°	80°

Main battery. The main armament consisted of the most powerful naval guns ever placed in service aboard a battleship; these guns are the most interesting aspect of the *Yamato*-class design. The only larger weapons, other than a few very heavy mortars, were the 80-cm guns used by the Germans during the siege of Sevastopol. The Japanese designated the guns as "Type 94 40 cm" (Model 1939 40 cm) to keep the actual caliber a secret. Even at the end of World War II, the U.S. Navy believed that the *Yamato*-class ships carried 406-mm guns. The 460-mm gun was superior to the 410-mm gun briefly contemplated by the Japanese as an alternative, as the comparison in table 2-11 shows. It is rather apparent that once the Japanese reviewed these data, they would choose the larger gun. These ships would be fighting at longer ranges than European battleships, whose battleground was mainly in waters where there was less visibility.

It was felt that the greater accuracy and penetration of the 460-mm gun would better suit the Japanese need to fight an opponent who had more ships.

TABLE 2-11

Comparison of Japanese 460-mm and 410-mm Guns

Gun	18.1"/45 (460 mm)	16.1"/45 (410 mm)
Shell weight (lbs)	3,219 (1,460 kg)	2,249 (1,020 kg)
Muzzle velocity (fps)	2,559 (780 mps)	2,559 (780 mps)
Maximum range (yds)	45,275 (41,400 m)	41,995 (38,400 m)
Maximum elevation	45°	43°
Penetration: vertical (side) armor plates		
@ 21,872 yards (20,000 m)	22.3" (566 mm)	10.6" (269 mm)
@ 32,808 yards (30,000 m)	16.4" (417 mm)	8.0" (203 mm)
horizontal (deck) armor plates		
@ 21,872 yards (20,000 m)	6.6" (168 mm)	—
@ 32,808 yards (30,000 m)	9.1" (231 mm)	—

Despite the great power of the new 460-mm guns, calculations using empirical equations developed by the U.S. Navy indicate that they were at best mediocre for their caliber, possessing only slightly better armor-piercing capabilities than the 406-mm/50-caliber Mark VII guns developed for the *Iowa*- and *Montana*-class battleships. Admittedly, the values expressed are of only general accuracy, but the relative merits of the guns are believed to be as indicated in table 2-12

Each gun weighed about 165 metric tons and each turret about 2,774 metric tons, more than the displacement of most destroyers of the period. The maximum training rate was 2 degrees per second, the maximum elevating rate was 10 degrees per second, and the maximum elevation was 45 degrees. The guns were loaded at 3 degrees and could be depressed to 5 degrees. The designed rate of fire was 1.5 rounds per minute at the maximum elevation; it was possible for a better rate of fire at lower elevations. The guns were wire-wound and radially expanded with a screw-type breech. The guns' design was similar to that used in the British 381-mm/42 (Mark I) gun as well as to the 410-mm gun used in the *Nagato* and *Mutsu*. This meant that the Japanese emphasized durability and reliability in their gun design at the expense of ballistic performance. A total of 27 guns were produced, including nine barrels for the *Shinano*, completed as an

TABLE 2-12

Comparison of Japanese 460-mm/45 and American 406-mm/50 Guns

Yamato 460-mm gun (18.1"/45)	
Shell weight	3,219 lbs. (1,460 kg)
Muzzle velocity	2,559 fps (780 mps)
Armor penetration	34.06" @ muzzle (865 mm @ muzzle)
	19.48" @ 21,872 yards (495 mm @ 20,000 m)
	14.22" @ 32,808 yards (361 mm @ 30,000 m)
Iowa 16"/50 gun (406 mm)	
Shell weight	2,700 lbs (1,225 kg)
Muzzle velocity	2,500 fps (762 mps)
Armor penetration	32.62" @ muzzle (829 mm @ muzzle)
	20.04" @ 20,000 yards (509 mm @ 18,289 m)
	17.36" @ 25,000 yards (441 mm @ 22,860 m)
	14.97" @ 30,000 yards (380 mm @ 27,432 m)
	12.97" @ 35,000 yards (330 mm @ 32,004 m)

aircraft carrier. The latter were captured intact by American occupation forces, and two of the guns were taken to the Dahlgren Proving Grounds for testing.

The design of the very heavy turrets, turret rollers, and barbettes was a formidable problem. The rotating structure was arrived at by scaling up the design that had been used for the *Nagato.* There were serious apprehensions regarding the turret roller bearings, which were supporting unprecedented weights, but they performed well, although the guns were used sparingly over the three years that the *Yamato* and *Musashi* were in service. Therefore, inherent defects, if there were any, may not have had an opportunity to develop. The biggest problems encountered in service were:

- Blast from the guns, particularly in the bridge area.
- The large amount of lubricating oil used by the cordite hoist racks and winches and by the training gears. The latter had shown heavy pitting.
- Noise made by the "coaster gear," and difficulty experienced in hoisting shells when the ship was rolling more than 5 degrees.

These problems seemed small in comparison with those experienced in the *King George V* and *Prince of Wales.* The 356-mm gun turrets on the British ships used very complicated safety interlocks, whereas the Japanese used very few. Instead, the Japanese depended upon well-drilled crews rather than mechanical devices to prevent accidents.

The shell and handling rooms were located above the magazines, which were located around both the upper and lower cordite handling rooms. The lower magazine supplied the center gun only, while the upper supplied the two outer guns by a cage-and-wire-type hoist operated by a hydraulic winch. The cage was hoisted directly from the handling room to the gun house. A total of 180 shells (60 rounds per gun) were stowed in the turret's rotating structure, with the remainder in the shell rooms from which they were supplied to the lower shell-handling rooms only. An original push-and-pull-type geared mechanism was designed by the Japanese to move the heavy shells around these compartments. "Pusher" type hoists, capable of being loaded in either of the shell-handling rooms, raised the shells to the gun house. The shells were stowed and moved to the gun house vertically and were tilted to the horizontal in the gun house. The movement of the shells from the shell rooms was extremely slow; the Japanese considered the 180 shells in the rotating structure of each turret sufficient for a surface engagement. Separate rammers were used for the shells and cordite, the latter being loaded in six equal charges by a single stroke.

Considerable research was devoted to the problem of reducing the size of salvo spreads. Great care was taken in the design and manufacture of the turrets to minimize recoil in all elevating and training gears and all other sources of wide spreads. The normal spreads when firing five- and four-gun salvos at maximum range were about 457 to 549 meters. Broadsides produced larger spreads. The reduction in salvo spreads was further enhanced by the Type 98 Firing Device, which prevented the simultaneous firing of adjacent guns of the main turrets to prevent interference between two shells during their trajectories. This equipment permitted the *Yamato* to fire very tight salvos during her encounter with the U.S. escort carriers and destroyers in the Battle of Leyte Gulf.

During the action at Leyte Gulf, the Japanese first used their 460-mm guns against aircraft, mainly torpedo bombers. The shell used was of a special incendiary shrapnel type. Some 10 to 40 percent of the projectiles stowed aboard ship after 1943 were of this type. There was no rapid method of changing the type of shell to be fired by these guns other than loading some hoists with one type and the remainder with another. Fuzes

This sketch of the Shinano *was drawn in 1950 by Mr. Shizuo Yano, one of the participants in the design of the ship from prior to the laying of her keel. (Courtesy Shizuo Fukui)*

This sketch of the Shinano *was prepared by Mr. Shizuo Fukui in 1952. Note the characteristic lines of the flying deck of the battleship hull form. The funnel was canted outboard at an angle of 26 degrees from the vertical. (Courtesy Shizuo Fukui)*

were set in the shell-handling rooms and fuze caps were used to prevent damage to the fuze before loading.

The designers of the 460-mm gun mounts were pleased with their design efforts, as these guns performed well in service. The Japanese fully expected that a new turret with so many innovations would have complaints from the operating personnel. This did not happen. There was only one serious accident during the 3–4 years of service, and this led to a fatality. A man was killed in the space between the shell and shell-handling rooms, by being cut in two by a shell bogie. The Japanese concluded that his loss of life was due to his own carelessness.

Four 15-meter rangefinders were used in the *Yamato*-class battleship, one being mounted atop the main control tower, the other three in the main-battery turrets. A 10-meter rangefinder was mounted on the aft control tower. The 15-meter rangefinder set on the forward tower was of a stereoscopic type, the largest set of its kind ever installed on a battleship. Japanese optics were excellent, equal to those of the Germans.

Secondary battery. The secondary battery of twelve 155-mm/60-caliber guns in four triple turrets was intended for use against surface targets, for which their long range admirably suited them. The 155-mm gun was a very effective weapon against surface

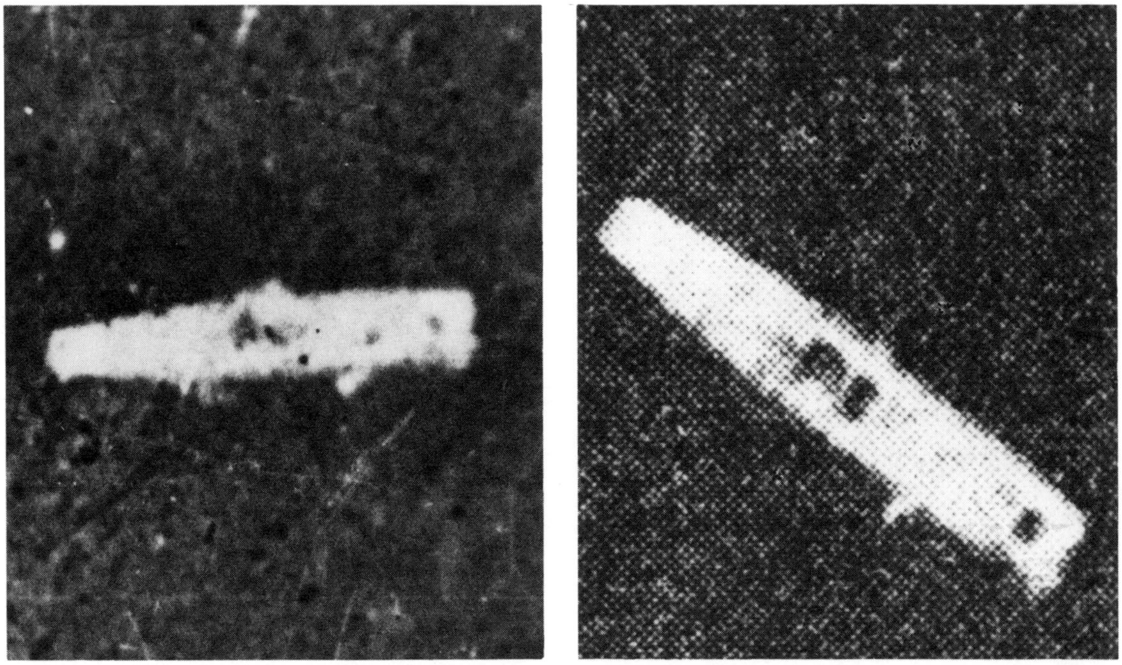

These two views of the Shinano *were taken by an American B-29 reconnaissance bomber on November 1, 1944 high over the naval base at Yokosuka. The aircraft was seen by the Japanese, and precipitated the decision to move the* Shinano *to the Inland Sea, which led to her loss. (Courtesy William Somerville)*

targets, but did not perform efficiently as an antiaircraft gun due to its slow rate of fire—5 rounds per minute, modest maximum elevation of 55 degrees, and slow training rate of 5 degrees per second. The triple turrets were made available by the conversion of four *Mogami*-class cruisers, which were given twin turrets with 203-mm guns in their place. The triple turrets were installed in a diamond arrangement, turrets in superfiring positions over the main battery fore and aft, and the others in wing positions, one on a side. This was a very effective arrangement of these heavy guns; they had arcs of fire that permitted nine guns to bear against most targets. In the autumn of 1943 the wing turrets were removed to provide space for additional 127-mm antiaircraft guns.

The primary antiaircraft weapons were twelve 127-mm/40-caliber dual-purpose guns arranged in three twin mounts on each side. These were efficient weapons, with a rate of fire of 14 rounds per minute and mount elevating and training rates of 16 degrees per second. Their only drawback was a short range of 14,900 meters and a limited ceiling of 9,400 meters. Due to the blast pressures from the 460-mm guns, these guns had to be enclosed. The *Yamato* was fitted with twelve more 127-mm guns in twin mounts in 1943.

The secondary battery had a similar fire-control system to the main battery, with 4.5-meter rangefinders on the port and starboard base of the superstructure.

Antiaircraft battery. The original antiaircraft defense was provided by twenty-four 25-mm machine guns in triple mounts, supplemented by 13-mm machine guns in two twin mounts on the forward bridge tower. The 25-mm machine gun, widely used throughout the Japanese Navy, was considered an excellent weapon. Its design was based on the French Hotchkiss machine gun, and its effective range for antiaircraft fire was 3,000 meters, with a rate of fire of about 220 rounds per minute. Each mount required a crew of nine men—pointer, trainer, sightsetter, and two loaders per gun.

Blast shields protected the triple mounts from main-battery blast effects. The blast problem created a unique situation, however, in the arrangement of the antiaircraft battery. Since it was not feasible to locate the aircraft or boats amidships, as in some modern battleship designs, because of the blast effects of the 460-mm guns, a concentrated antiaircraft battery with excellent hemispherical coverage was possible, unimpeded with boat clutter.

By 1945, the Japanese 25-mm antiaircraft machine gun was far too light and too short-ranged a weapon to be effective in engaging heavier and faster American aircraft. The Imperial Japanese Navy was the only major navy not to employ antiaircraft machine guns of over 25-mm bore.

Two twin 13-mm machine gun mounts on the forward bridge tower provided close-in protection against dive bombers. They had a muzzle velocity of 800 meters per second, an effective range of 1,524 meters, and fired about 450 rounds per minute.

The antiaircraft armament of the completed ships was virtually unchanged from that of the contract design. During the war, the antiaircraft batteries were systematically improved. Late in 1943 the 155-mm wing turrets on both ships were removed to make room for three twin 127-mm gun mounts on each side. Material shortages prevented installing the mounts on the *Musashi* (triple 25-mm mounts were substituted instead), but tremendous improvements were made in the antiaircraft armament of both ships, as indicated by this tabulation for the *Yamato*:

		5" (127 mm)/40	*1" (25 mm)/60*	*13 mm*
December	1941	12	24	4
Fall	1943	12	36	4
February	1944	24	36	4
May	1944	24	98	4
July	1944	24	113	4
April	1945	24	150	4

In 1944 single, free-swinging 25-mm mounts were added in open areas and on turret roofs to increase antiaircraft fire. Unfortunately, much of the added firepower was wasted because material shortages prevented the installation of additional fire-control directors. As it was, a Type-95 short-range high-angle director was used to control these guns. This was a simple course-and-speed system and was not up to American or British standards, although the guns could be fired in remote control. Furthermore, blast shields were not provided for those machine gun mounts added after 1943—a serious problem when the 460-mm guns were used in antiaircraft defense. The radically modified armament emphasized the new role of antiaircraft defense that battleships assumed during World War II.

Blast from the main-battery guns was a serious design problem, and during the course of the careers of the *Yamato* and *Musashi*, gun blast pressures created problems in the forward control tower. Firing three guns in any one turret was calculated to produce blast pressures of 7 kilograms per square centimeter (about 100 psi) 15 meters from the muzzles. Blast pressure of only 0.28 kilograms per square centimeter could destroy boats on board ship, while 1.16 kilograms per square centimeter could temporarily knock a man out. Consequently, all antiaircraft weapons exposed to main-battery blast required blast shields.

Aircraft. When the design of the *Yamato* class began, aircraft spotting, reconnaissance, and limited offensive missions were considered normal functions for a battleship floatplane. The ships had a designed aircraft complement of seven aircraft, but rarely did they carry more than four. The two-seat Mitsubishi F1M2 (Pete) observation floatplane was standard equipment carried by the *Yamato* and *Musashi*. Each ship was equipped with two catapults, a large aircraft crane, and an elevator. Temporary stowage of aircraft on the open deck aft was possible by the provision of rails, although special below-deck storage facilities were provided in the stern for all ship's boats and aircraft to protect them from gun blast.

Armor Protection.

For the armored citadel, very heavy armor was required to provide an immunity zone of 20,000 to 30,000 meters against 460-mm shellfire; this armor contributed materially to the performance of the side-protection system, which was designed to withstand five torpedo hits on either side. Such requirements made the *Yamato*-class battleships the most massively armored ships ever built.

Throughout the length of the citadel, the main side belt was comprised of 410-mm face-hardened armor (VH) plates inclined outboard at 20 degrees, providing protection equivalent to that of 584-mm vertical plates.* Over the main-battery magazines a lower side belt inclined some 25 degrees from the vertical, and was tapered from 270 mm to 160 mm at its intersection with the armor protection provided under the magazines. This belt was not face-hardened armor steel, but a new noncemented armor plate using copper instead of nickel, since the latter was in short supply. Over the propulsion plant, the lower side-belt inclination was 14 degrees; thicknesses varied from 200 mm to 100 mm at the bottom shell of the ship. The greater thickness over the magazines was dictated by the reduced extent of the side protection system necessitated by the hull form. This heavy lower side-belt armor was installed as a result of the 1924 gunnery experiments on the *Tosa* and the Japanese development of a very successful Type 91 projectile that had the capability of long underwater trajectories, with the potential for armor penetration well below the waterline. The development of this type of projectile for 152-mm and larger guns and the threat of mine explosions below the ship prompted Japanese designers to give much attention to underbottom protection. Relatively heavy armor protection was given to the underside of the main-battery magazines by inclining the armor inboard about 6 meters below the design waterline to a flat portion that formed the top of a triple bottom. Coupled with heavy deck armor and protection under the magazines, ranging from 50 mm to 80 mm, the armor system of the *Yamato* class was the thickest armor protection ever mounted on a warship.

Deck armor. Deck armor was designed to resist long-range 460-mm shells striking at an angle of 55 to 65 degrees. This dictated that the armored citadel have thick horizontal protection to act in conjunction with the heavy side-belt armor. Deck-armor thicknesses are tabulated in table 2-13.

Partial protection on the weather deck forward and aft was provided by plates ranging in thickness from 50 mm to 35 mm.

The main armor deck was concentrated on the third deck except in way of B and X turrets. Forward, the armor was elevated, while around the aft turret, it dipped down to meet the after transverse armored bulkhead. This unconventional disposition of armor plate was designed to enhance the protection, to limit the weight of B barbette, which

*In firing tests after World War II at the Dahlgren Proving Grounds two 406-mm (Mk 8, Mod 6) projectiles were fired normal to 660-mm face plates intended for the *Shinano*. One projectile failed to penetrate, but the second, fired with a higher muzzle velocity, went right through. No damage was done to this projectile. It appears that Japanese VH face-hardened armor was approximately 10 percent inferior to American type A armor.

TABLE 2-13
Deck Armor Thickness

	Amidships—Centerline	
Main deck	0.47″	(12 mm)
Second deck	0.39″	(10 mm)
Third deck	7.87″	(200 mm)
Splinter protection	0.35″	(9 mm)
Fourth deck	0.35″	(9 mm)
	9.45″	(240 mm)

	Amidships—Outboard	
Main deck	0.79″ + 0.71″	(20 + 18 mm)
Second deck	0.98″	(25 mm)
Third deck slopes	9.06″	(230 mm)*
	11.54″	(293 mm)

*Sloped 7 degrees from horizontal.

was 560-mm thick, for stability and weight considerations, and to provide three barbette rings of uniform height. This uniformity would ease production requirements.

The deck structure that supported the horizontal armor plates was designed to withstand stresses imposed by longitudinal bending; nonlocal loads were provided for in the structural design of the strength deck. The calculations for the deck areas that might be subjected to gun blast pressures were based on experimental firing trials. The weather deck forward might have been subjected to severe loadings by heavy seas, but the Japanese were content to design the structure on the merits of earlier designs.

In way of the uptakes, the armor thickness was increased to 380 mm. This was a homogeneous plate with many holes drilled in it, which made this armor equivalent to 140 mm of armor-deck protection, less than the surrounding 200-mm deck armor.

Turret armor. The main-battery turrets were the most heavily protected parts of the ship. Even though heavy armor plate was used, the Japanese sloped some of the armor to enhance the tendency of shells to ricochet. The turret armor protection was massive: 650-mm face plates, 250-mm sides, 270-mm roofs, and 190-mm backwalls. Barbette protection was equally heavy, with 560-mm plates on the beam; thicknesses fore and aft varied from 400 mm to 380 mm depending on the protection offered by other structure.

The cruiser-type secondary turrets were protected against strafing or fragmentation damage by 25-mm armor. The very tall barbettes were originally equipped with only 50-mm-thick armor plate, but this was later increased by the use of doubler plates to a total thickness of 75 mm. Flash-repellent hatches leading to the powder magazines slightly lessened the vulnerability of these turrets. All battleship designers realized the practical impossibility of providing truly adequate armor for gunnery installations other than the main armament.

Conning tower. The conning tower protection was of a conventional design, with sides protected by 500-mm plates, the roof by 200-mm armor, and the floor by 75-mm plating. The Japanese considered such protection essential to continue ship operations in battle.

Transverse armored bulkheads. Forward and after main transverse armored bulkheads were sharply inclined from the vertical, the upper segment sloping down and away from the citadel while the lower segment inclined in the opposite direction. Forward, the initial slopes were 340-mm thick, with the lower sections varied from centerline values of 300 mm and 270 mm at the lower segment to 350 mm and 330 mm off the centerline. The inboard profile in the gatefold at the end of this chapter shows the complicated shape of the armored bulkheads. The reason for such a complicated armor arrangement was to ensure protection against a diving 460-mm shell and to provide an efficient weight-saving scheme of armor.

Steering-gear room armor. The steering-gear room was heavily protected. The main steering engine room had inclined side armor 350-mm thick; the forward bulkhead was 360-mm thick and the after bulkhead 350-mm thick. Protection was also given to the auxiliary steering position by 250-mm inclined side armor, a 300-mm forward bulkhead, and a 250-mm after bulkhead.

Three major types of armor plate were used in the *Yamato, Musashi*, and *Shinano*. All were newly developed and designed to surpass previously used grades of armor. They had the following properties:

- *Vickers Hardened (VH)* — A face-hardened steel armor widely used on the main-battery turrets and in the main side belt. Generally, it was used for thicknesses in excess of 200 mm.
- *Molybdenum Non-Cemented (MNC)* — A nickel-chromium-molybdenum quality armor used in armor-deck plating of 75 mm and greater thicknesses. Ballistic tests conducted at the proving grounds at Kamegabuki proved this armor plate to be superior to the homogeneous Vickers steel plates by a factor of 10 to 15 percent.
- *Copper Included Non-Cemented (CNC, CNC_1, CNC_2)* — Intended for thin armor-deck protection, primarily in areas where splinter protection was desired. The exact nature of the armor desired was obtained by varying the chromium and nickel content. Considered superior to the New Vickers Hardened Non-Cemented (NVNC) plates used on the *Nagato*. The presence of the nickel permitted the steel to be rolled and not develop brittle fracture properties. Characterized by very good energy-absorption performance, a criterion of crucial importance in armor plates.

Side-protection system. The underwater side-protection system was derived using empirical methods, relying heavily on model experiments as well as on the *Tosa* experiments. The Japanese naval constructors had noticed the penetrating power of the shell with an underwater trajectory during their analysis of the Battle of Jutland and their experiments with the *Tosa*. It was decided that equal consideration had to be given to this type of attack and to that of the torpedo. In the design of the *Mogami*-class cruisers, this principle was first used, that is, to provide equal emphasis to resisting the underwater explosion of both the torpedo and penetration by the diving shell. To accomplish this, the sloping side armor was continued to the keel, albeit with reduced thicknesses. The space between the shell plating and the tapering armor plate below the design waterline was to be kept void.

The underwater protection of the *Yamato*-class battleships was based upon this principle, because the main-battery guns were considered to be the primary weapon in a clash of battlefleets. This system also was checked by analytical and small-scale tests of its ability to withstand torpedo attack. Variables dealt with included the thickness of the internal armor bulkhead, its distance from the shell plating, the liquid-loading of the longitudinal compartments, and the designed resistance. The Japanese appear to have been unfamiliar with the elastic deformation theory that was employed in the design of American, French, and British battleships. A series of ballistic tests were made in January 1935 on 1/3-scale model caissons to develop a system that would adequately resist the diving shell and yet provide a good underwater protection system against torpedoes.

The threat of the diving shell was enhanced, in Japanese eyes, by the development of the Type-91 armor-piercing projectile, which was designed to enter the water outboard of its target, maintain a stable underwater trajectory, and penetrate the hull below the main side belt. Since the shell also had a breakaway nose cap and an extremely long fuze delay, armor protection against such shells had to be provided to the keel. The Japanese also thought that a thick lower belt would compensate for the absence of the liquid-void scheme used in foreign battleship designs.

The 1935 explosion tests showed that a number of sharp-edged fragments hit the surface of the torpedo bulkhead in a 65° cone-shaped area with its apex at the point where the detonation occurred at the shell. The torpedo bulkhead was ruptured by these notches and penetrations, but it did not split completely. It was estimated that the bulkhead would be more effective if the fragments could be prevented from striking the bulkhead as splinters. The most effective way of achieving this would be to have the void space backed by a liquid layer of a required minimum depth immediately outboard of the torpedo bulkhead. This was not practical because the outboard spaces were to be kept void and used for counterflooding. The only other solution was to strengthen the tapered lower armor belt from an 80-mm to 25-mm thickness to a thickness of 200 mm to 75 mm sloped at 25 degrees. Even with the increased armor thicknesses, there was a large disparity in armor thickness between the upper and lower belts at their juncture. This created a problem with regard to the design of an adequate connection. An important consideration in the solution of this dilemma was the ability of the Japanese steelmakers to produce the required shapes and rivets. The joint design finally selected is shown in figure 2-4. Several key naval constructors involved in the *Yamato* design objected to this joint design and suggested a delay in the ship's construction was justified to permit the improvement of this design.

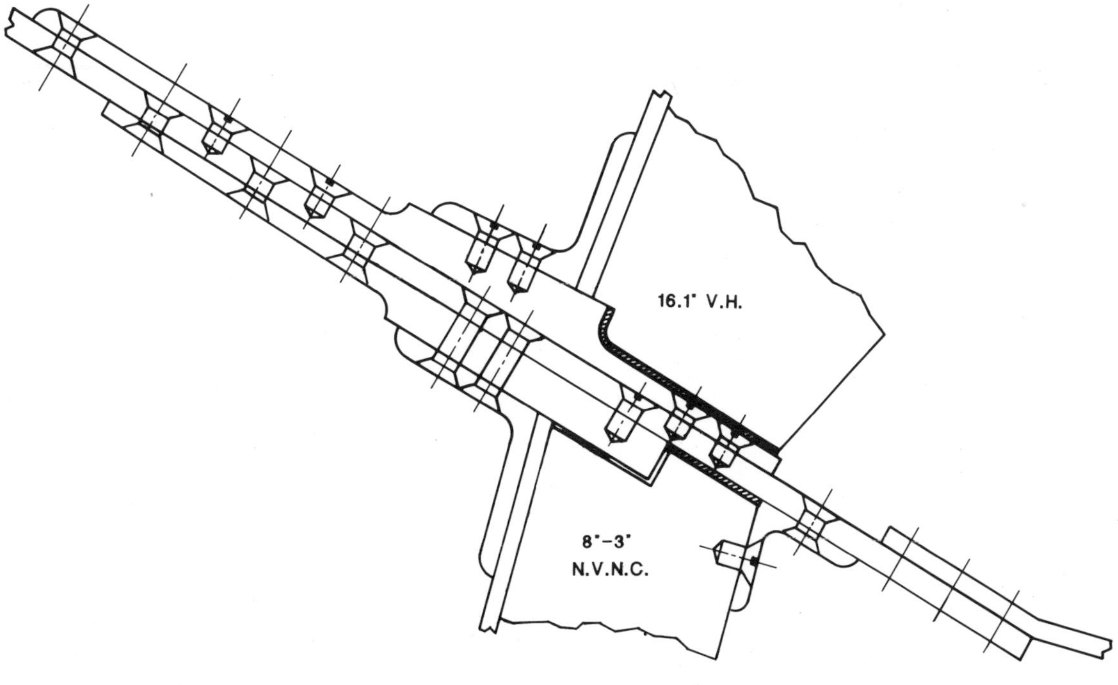

Figure 2-4. Armor-belt connection—Yamato *class.*

The Yamato *is depicted in this Richard Allison painting as she sortied from Tokuyama the evening of 6 April 1945, en route to Okinawa as flagship of the "Surface Special Attack Force." The light cruiser* Yahagi *is shown broad on the battleship's port bow. The battleship, the light cruiser, and four of the eight accompanying destroyers were sunk by concentrated attacks from American carrier-based aircraft the next day.*

This Richard Allison painting depicts two 1930 Japanese design studies of treaty battleship replacements for the Kongo-class battlecruisers/fast battleships. The Hiraga design, to the left, featured an unusually compact superstructure with a main battery of ten 410-mm guns, with triple turrets superfiring over twins. This arrangement was forced by the limited citadel length and the fine hull lines. The Fujimoto design provided for a similar armament, but the secondary battery of twelve 152-mm guns in twin turrets was unusually positioned, nearer the bow and stern than the main-battery turrets, with special shields to protect them from the main armament's blast. Both these design studies provided for maximum speeds of about 26 knots.

The unusual canted funnel characteristic of a number of Japanese aircraft carrier classes is apparent in this Richard Allison painting of the Shinano. *This aircraft carrier was torpedoed and sunk by the American submarine* Archerfish *on 29 November 1944.* Shinano, *the largest warship ever to be sunk by a submarine, never launched or recovered any aircraft—the ship was en route to Kure to pick up her air complement and begin training exercises when she was torpedoed.*

The joint between the upper and lower belt armor systems relied upon the shearing strength of rivets, with very little transverse continuous structure inboard of the joint. As a result, the initial thrust of the torpedo explosion would place the tap and three-ply rivets in shear. This design weakness permitted the main armor belt to rotate under explosive loadings, driving the supporting structure at the joint through the longitudinal bulkhead inboard of the armor belts. The defective joint diminished ship resistance to torpedo charges and also made it possible for the bounding bulkhead of the vitals to be penetrated by part of the defective joint supporting structure. The weakness of the joint was first demonstrated in December 1943 when the *Yamato* was struck by a torpedo outboard of X turret. The joint between the upper and lower belt failed and permitted some 25 meters of armor belt to simply cave in. In moving inboard, failed sections of the armor belt caused punctures in the bounding bulkheads of X turret magazines, resulting in some 3,000 tons of flooding. Subsequent damage to the *Yamato, Musashi*, and *Shinano* showed repeatedly that U.S. torpedoes armed with 272-kilogram torpex warheads (roughly equivalent to 550 kilograms of TNT) had no difficulty in rupturing these battleships' underwater protection systems. The amount of damage and extent of flooding was a function of the depth at which the torpedo struck. A hit at the level of the upper belt would not be as damaging as one below the level of the joint.

In the case of the *Yamato* class, the armor plates for underwater protection against 460-mm shells with underwater trajectories were sufficiently thick that a liquid layer was felt unnecessary. As a consequence, the side-protection system was exclusively composed of a series of void compartments. Fuel oil was stored in tanks in the double bottom under the propulsion plant, or in deep tanks beyond the citadel—a unique departure from the normal design practice that utilized the need for liquid loading of side-protection systems by using many compartments within the protective system as fuel tanks.

The side-protection system featured a maximum of four void compartments, the two inboard of the side armor being designated as flooding compartments. The exclusive use of void compartments reduced the chances of progressive flooding by lessening the need for interconnecting piping systems, as would have been required for fuel tanks. The system was riveted construction exclusively, as the steel used was a high-strength type for which no acceptable welding techniques had been developed.

Compartmentation within the side-protection system varied radically, as the lower side-belt inclination differed over the machinery spaces and magazines. The best indication of the intended system is the tabulation of breadths and bulkhead thicknesses in table 2-14.

The underwater protection system had these basic design requirements:

- 1 hit — The ship is to remain in battle, with temporary repairs and counterflooding complete within five minutes.
- 2 hits — The ship will be able to return to battle condition after 30 minutes.
- 3 hits — The ship might be ready for battle within 60 minutes.
- 4 hits — All lists should be limited to a 5-degree maximum after completion of damage-control procedures. (Above 5-degree shell handling was not possible.)

Underbottom protection. In the design of the *Yamato*-class battleships, a unique feature of the armor protection was the 50–80-mm armor plate under the main magazines. The Japanese were quite concerned that a mine explosion below the keel could cause serious damage to the magazines, so they decided to use armor protection to prevent fragments from entering the vulnerable powder magazines. The armor thicknesses were based upon a 200-kilogram charge in contact with the ship's bottom. A triple bottom was used under the magazines, while a simple double bottom was used in

TABLE 2-14
Characteristics of Side-Protection System

Location	Breadth at 17.06′ (5.2 m) Below Design Waterline (Half Draft)	Total Bulkhead Thickness	Torpedo Bulkhead Thickness
Turret A	—	—	3.98″ (100 mm)
Turret B	8.53′ (2.60 m)	—	3.98″ (100 mm)
Amidships	16.40′ (5.00 m)	5.78″ (141 mm)	3.98″ (100 mm)
Turret X	13.69′ (4.16 m)	—	3.98″ (100 mm)

way of the propulsion plant. As in most battleship designs, the Japanese believed that compartmentation of the machinery spaces would be the best protective feature to deal with underbottom damage.

The *Yamato* class was well protected due to the thick armor used, but the enormous thicknesses were compromised by somewhat inferior armor quality, poor connections, and inefficient armor distribution. The *Yamato* battleships were an example of an "all-or-nothing" concept of armor distribution. Their ratio of armor box length to waterline length compared to other battleships shows this: *Yamato*, 53.5; *Nelson*, 54.7; *Nagato*, 63.1; *Fuso*, 65.1; *Bismarck*, 67.0; *Richelieu*, 54.2; *Iowa*, 53.9; Montana, 59.1. The relative reduction in citadel length permitted heavy protection, but exposed a greater proportion of the hull to torpedo damage. Design specifications required that the ship remain stable with all areas beyond the citadel flooded, but this required that the citadel remain intact. The ships might have been able to overcome the effects of the extreme concentration of the protection, if not for the defective armor joint.

Stability and subdivision. The stability criteria for the *Yamato*-class battleships were developed with the greatest of care. The importance of adequate ship stability and the result of neglecting it were emphasized when the torpedoboat *Tomozuru* capsized during heavy weather on 12 March 1934. This resulted in added attention to weight control, topside weight reduction, and adequate stability in the *Yamato* design.

The following stability criteria were established:

- With all unprotected spaces flooded, the volume of the armor box above the damaged condition waterline should still be 20 percent of the total volume of the armored citadel.
- With all voids within the side-protection systems on both sides flooded, the ship was to have positive transverse stability and not capsize. With either the bow or stern portions flooded, the vessel must still have positive longitudinal stability so that it would not plunge by either bow or stern. In such conditions, the ship must float and have a range of stability of 22 degrees.
- Since the *Yamato* was to withstand 5 torpedo hits on one side, the stability after damage necessitated an initial metacentric height of from 3.05 to 3.66 meters, with a righting arm of at least 2.33 meters. All lists had to be limited to 17 degrees in the damaged condition, since the weather deck would be awash at any greater angles.
- The period of roll should be between 16 and 18 seconds, but not less than 15 seconds, so that a stable gun platform could be provided even in fairly rough weather.
- With the outboard engine and corresponding firerooms flooded, the ship should still have positive transverse stability and not capsize.

Battlecruiser Design B-65, completed in September of 1940, is shown in this Richard Allison painting. With a main battery of nine 310-mm guns and a full-load displacement of 35,000 metric tons, the general characteristics of this design were very similar to that of the American Alaska-class large cruisers.

Battleship Design A-150, as shown in this Richard Allison painting, would have been generally similar to the Yamato, but with a main battery of six 510-mm guns and a dual-purpose antiaircraft battery with the high-performance 100-mm gun. The authors have estimated that the armor-piercing shell for these 510-mm guns would have weighed some 1,905 kilograms.

These requirements were difficult to satisfy, particularly due to the weaknesses in the armor-belt joint design. More than 18 degrees of list could be corrected by counter-flooding and transferring fuel oil. This approach to damage-control had the following capabilities:

Degree of list	*Counterflooding procedure*
9.8	Flood 64 quick-flooding compartments on the undamaged side.
13.8	Flood an additional 39 normal flooding compartments.
18.3	Transfer fuel oil from damaged to undamaged side, after flooding 103 compartments on undamaged side.

The stability investigation was thorough, but the defective armor-belt joint reduced the resistance to underwater damage.

Much attention was devoted to increasing the number of watertight compartments, as in German designs, with the following results:

	Below armor deck	*Above armor deck*	*Total*
Fuso (1915)	574	163	737
Nagato (1920)	865	224	1,089
Yamato (1941)	1,065	82	1,147

The small number of watertight compartments above the armor deck in the *Yamato* was due to the greater relative height of the armor deck. The relatively small increase in total number of such compartments in the *Yamato* as compared to the *Nagato*, a ship of about half the displacement, shows the greater average volume of spaces on board the *Yamato*. Bow and stern compartments on both ships were excessively large, a serious weakness in resistance to battle damage.

The Japanese managed at least to approach closely all their stability criteria. At the designed trial displacement of 68,200 metric tons, the metacentric height was 2.60 meters, the range of stability was 70.8 degrees, and the rolling period was 17.5 seconds.

These ships had massive, effective protection against shellfire and underwater attack. The armor was the most massive ever given a warship; their side protection and overall stability permitted these ships to absorb numerous torpedo hits before sinking. The unusual side-protection system, devoid of liquid load, proved effective, albeit weakened by the defective armor joint.

Radar. The electronics equipment used on the *Yamato* and *Musashi* was the latest, most modern available in Japan, and eventually included several types of radars, passive hydrophone arrays, and infrared signaling gear. The first Japanese radars were developed by the army for air search. Radar development began in April 1941, and initial efforts were concentrated on a three-meter wavelength. By December 1941, some thirty sets were available. Japanese radar research and development was greatly assisted after the fall of Singapore where the army captured five British fire-directing radars. The equipment recovered was brought back to Japan and aided in Japanese development of search and gunnery radar. The Japanese had hoped to recover some radar equipment from the *Prince of Wales*, which had been sunk in the South China Sea, but the water was too deep. The Japanese Army and Navy had different priorities assigned to radar development. The army was concerned with air search, and the first radar sets used by the navy were the 12 Go Dentan installed aboard the *Ise* and *Hyuga*. The navy was not satisfied with this equipment, and in August 1942 a very important modification was made that involved separate transmitting and receiving arrays. The *Musashi* received this equipment (Type 21-Kai 1) in September 1942, and during the fall of 1942 the *Yamato* was fitted with the new Type 21-Kai 2 air-and-surface warning radar, with antennas mounted on both sides of the main-battery optical rangefinder on the forward bridge tower. The *Musashi* equipment had these characteristics:

Target	*Accurate Detection Ranges*
Seaplane	21,872 yards (20,000 m)
Seaplane formation	32,808 yards (30,000 m)
Torpedo bombers	43,745 yards (40,000 m)
Battleship	38,277 yards (35,000 m)
Surfaced submarine	10,936 yards (10,000 m)
	Bearing Errors
At maximum range	Less than 5 degrees
At close range	Less than 10 degrees
	Ranging Errors
	Less than 328 yards (300 m) for ranges less than 27,340 yards (25,000 m).
	Less than 547 yards (500 m) for ranges less than 43,745 yards (40,000 m).

Few difficulties were experienced in the maintenance and operation of the Type 21 equipment.

Japanese analysis of the Battle off Savo Island (October 10-11, 1942) indicated that American ships had very effective radar-controlled gunfire. The Japanese reliance on optics, searchlights, and illuminating shells was no longer sufficient for night engagements. There was an urgent need for radar-controlled gunfire — initial efforts were concentrated on perfecting the Type 21 radar for surface fire control. This development was initiated in October 1942, and finally the *Yamato* was equipped with a Type 21-Kai 3 radar in September 1943. This radar featured a large mattress-type antenna with the power increased to 25-30 kw.

Early in 1943 each ship received two Type 22 surface-search radars, one on each side of the bridge tower. The Type 22, using a wavelength of 10 centimeters instead of the 150 centimeters of the Type 21, was capable of much improved range discrimination. Effective detection range on a battleship target was 22,000 meters for the *Musashi* radar, improved to 24,625 meters for the *Yamato*. Operational difficulties were not considered serious.

In 1944 both ships received Type 13 search radars. Two units were mounted on each ship, on the raked mast abaft the funnel. Performance was similar to the Type 21, with a few maintenance problems and no operational difficulties. The Type 13 radar did not replace the Type 21, but was fitted as an additional surface-search radar, although serving as a very reliable air-search radar. This set worked on a two-meter wavelength at 10-kw power. The radar antenna consisted of ladder-type broadside arrays of four steps with two elements, one meter in length.

The Type O passive listening sonar equipment installed early in 1944 consisted of circular arrays of 15 hydrophone units on either side of the bow, giving limited detection capability when the ship was making low speed or stopped.

Each ship carried some 40 radio receivers and 17 transmitters, with operating frequencies ranging from low frequency (LF) to very high frequency (VHF). Most of the equipment operated at LF, MF, and HF.

The Navy Type II infrared communications equipment was developed in 1942 for large warships. Speed of transmission was about half that of conventional light systems. Operation was considered quite satisfactory, with an effective range of about 15,000 meters. The system, not stabilized in level and cross-level, had an angle of view of about 10 degrees. A pair of 20-power binoculars mounted coaxially with the transmitting equipment assisted in search procedure and keeping contact.

Propulsion Plant. The desire for as high a maximum speed as possible was complicated by the relatively low length-to-beam ratio of 6:6:1. The citadel length was fixed by the requirements of machinery and gunnery installations, and the low length-to-beam ratio was selected in order to minimize the relative length of the unprotected portions of the hull: the greater the proportionate length of the citadel, the better the resistance to battle damage. The relatively short hull form contributed directly to the adoption of the bulbous bow.

Hull length was inadequate for the desired speed of 27 knots when the design was first completed. This design featured a V-shaped bow and cruiser stern. Extensive model-basin tests led to the adoption of a bulbous bow and a semitransom stern form. The large bulbous bow, with 13.3 percent of the underwater cross-sectional area amidships, reduced hull resistance by about 8.2 percent at maximum speed. Refinements to the shaft brackets and bilge keels also contributed to the eventual service speed of 27 knots. The modified stern form improved the turning characteristics. For a speed of 27 knots, the following power savings were realized:

bulbous bow	13,261 shp	(13,445 mhp)
shaft bracket	1,874 shp	(1,900 mhp)
bilge keel	469 shp	(475 mhp)

The stern form was not a true transom stern, as conventionally known. The waterlines at the stern were rounded instead of being sharply knuckled as in the United States Navy and Royal Navy. The stern form was considered to be of little importance to the propulsive qualities, as the Japanese planned to have no transom immersion in normal loading conditions. This design might have contributed to difficulties in attaining high speeds at the heavier displacments, if the stern was immersed, thereby adding materially to the hull resistance. In addition to improving turning qualities, the design provided added volume aft for an airplane hangar and below-decks boat stowage.

A study of the speed-power curves for the *Yamato*-class battleships reveals that they very nearly attained the maximum possible speed for their relatively short, broad hulls. The rated maximum speed of 27 knots was easily exceeded during trial runs; the *Yamato* made 28.05 knots in June, 1942. The actual performance on trials indicated a potential endurance about 50 percent greater than the rated endurance of 7,200 nautical miles at a speed of 16 knots. Trial conditions, of course, are indications of optimum performance for ships rather than conditions normally obtained in service.

Boilers. Despite the need for a conservative machinery design to ensure good reliability, the *Yamato* and *Musashi* had the most powerful installations of any Japanese battleships. The conventional steam-turbine power plant consisted of 12 Mark RO bureau-type boilers with four sets of geared turbines. Conservative design practices dictated

The fore tower of the Musashi *is shown in this dramatic view, taken from the main deck on the port side, near the Number Two main-battery turret. The fore 155-mm gun turret and its optical rangefinder hood are partially obscured by the massive protective hood for the 15-meter rangefinder for the 460-mm turret. This picture was taken early in 1942. (Courtesy Shizuo Fukui)*

relatively low steam pressure and temperature, 25 kilograms per square centimeter and 325 degrees centigrade. The tube banks were 19 rows deep and the boilers equipped with two-pass four-loop superheaters. This resulted in a peculiar necking down of the boilers between boiler banks above the furnace and the superheaters. In service, these boilers were reliable, but they had poor circulation and experienced burner problems.

The Japanese, confident in the underwater side protection given the ships, adopted longitudinal subdivision of the machinery spaces in order to isolate each boiler and geared-turbine set in individual compartments. The 12 boiler rooms, each with superheater and air preheater, were concentrated four abreast, forward of the engine rooms, which likewise were arranged four abreast. A maximum of nine burners could be used on each boiler at any given time. The propulsion machinery was not at all unique, but the arrangement in four rows was noteworthy. The three boilers in any longitudinal row were intended to provide steam to the turbine set located directly aft. Such an arrangement, considered most desirable from the standpoint of damage control and protection, could only be used in ships of considerable beam.

Turbines. The geared turbines were designed for 10 percent overload. The rated normal maximum for each shaft was 37,500 metric horsepower. Each engine room was equipped with high- and low-pressure turbines and cruising turbines, coupled by reduction gears to the propeller shafts.

The effects of the conservative design practice in the machinery are evident in a comparison with the U.S. *South Dakota*-class ships:

	South Dakota	*Yamato*
Normal maximum shp	130,000 (131,803 mhp)	147,498 (150,000 mhp)
Machinery weight (tons)	3,580 (3,647 mt)	5,216 (5,300 mt)
Pounds per shp	61.69 (27.60 kg/mhp)	79.0 (35.33 kg/mhp)

Electric plant. The electrical capacity of 4,800 kilowatts was provided by four turbogenerators and four diesel generators, each rated at 600 kilowatts, 225 volts. The low electric load for ships of this size reflected Japanese philosophy on the design of auxiliary machinery. There were a number of steam-driven auxiliaries, while electric power for hotel services was severely limited.

Rudders. The *Yamato*-class battleships had two rudders, main and auxiliary, located on the longitudinal centerline, with the auxiliary rudder 15 meters forward of the main rudder. Such separation was intended to prevent loss of steering control in the event of damage to either rudder. The Japanese considered the arrangement justified by the loss of the *Bismarck* due to rudder damage. A retractable bow rudder was proposed, but model tests indicated that such a rudder was unsatisfactory. Stern rudder effectiveness is augmented by proximity to the propellers; a bow rudder would have only the movement of the ship to help develop the turning moment. As an alternative to the bow rudder, the area of the auxiliary rudder was increased.

The trials of the *Yamato* revealed that the auxiliary rudder alone could not reduce the turning momentum of the ship sufficiently to bring it to a steady course. A better arrangement would have been to locate the rudders abreast of each other in the propeller races, which would have increased their effectiveness. The Japanese obtained better damage resistance at the expense of reduced rudder effectiveness. Even so, the *Yamato* and *Musashi* were relatively maneuverable for their size and displacement. At 26 knots, the tactical diameter was 640 meters, using maximum rudder angle. This produced a heel of 9 degrees.

Hull Characteristics. Despite the great displacement of the *Yamato*-class ships, the Japanese conscientiously attempted to minimize structural weight wherever possible. Part of the armor system served structural purposes as well. In particular, the lower side-belt armor was used as a strength member in the hull structure, an innovation Vice Admiral Hiraga first used with medium armor plating. In addition, the Japanese used a high-strength steel, similar to the British D1 steel, in the hull and superstructure. No armor was used for splinter protection.

Welding. Arc welding in Japan in the 1930s was not widely accepted as it was after World War II. It was still a new technique and did not have the theory developed, as was the case with riveting. However, welding offered great weight reduction; as early as 1932, the minelayer *Yaeyama* had been extensively welded. The submarine tender *Taigei*, completed in March 1934, was the first Japanese naval vessel with a completely welded hull. After two destroyers broke up during heavy weather in September 1935, an investigation led to the decision not to employ welding in longitudinal strength members. This finding was reinforced by experiences with the cruiser *Mogami*, which sustained cracks in the welded portion of her stern. This was attributed to very light steel plate, and the welded structure was replaced with riveted construction.

Despite these problems and limitations, much welding was done on the *Yamato*-class ships. A total of 7,507,536 welding rods were used during the construction of the *Yamato*, for a total weld bead length of 463,786 meters. The largest welded block was 11 meters high and weighed 79 tons.

The primary longitudinal strength members were made of Ducol high-tensile steel, while mild steel was used on members of secondary structural importance. The centerline longitudinal bulkhead was built in duplicate in order to provide added strength to support the heavy armor deck. The resultant small watertight centerline compartment was used for the protection of important primary electrical cableways running the length of the ship.

Sheer line. The unorthodox sheer line of the *Yamato*-class battleships were the result of strictly utilitarian considerations. Bow freeboard was determined for reasons of seaworthiness, the freeboard amidships was fixed by the criteria for stability, while the stern freeboard was reduced to the minimum practicable. This undulating line, which

The Yamato *and* Musashi *are shown early in 1943 at anchor at the major Japanese naval base at Truk, in the Caroline Islands in the Central Pacific. It is probable that the ship to the right is the* Yamato, *which was relieved as flagship of the Combined Fleet on 11 February 1943. The* Musashi *had arrived at Truk on 15 January 1943. (Courtesy Shizuo Fukui)*

The great beam of the Musashi *is apparent in this striking view, which emphasizes the undulating profile of the battleship's main deck, termed the flying deck by the Japanese. Note the relatively uncluttered deck in this early 1942 view. The Japanese were much more reluctant than the Americans to accept blast-exposed antiaircraft positions for their machine-gun batteries. Even very late in the war, there were only a few added machine guns on the forecastles of Japanese capital ships. (Courtesy Shizuo Fukui)*

had a straight run for the length of the citadel, was characteristic of Japanese warship design since the completion of the heavy cruiser *Furutuka* in 1925. The Japanese felt this arrangement saved structural weight without seriously diminishing the overall hull girder strength.

Construction time. Months of construction time were saved by procuring some of the lighter armor under the appropriations for the reconstruction of the old battleship *Hiei* and the cruisers *Tone* and *Chikuma.* This also permitted the construction of the *Yamato* to begin in November 1937.

Habitability. The living accommodations of the *Yamato* class were the most advanced in the Japanese Navy. Special color studies were made to improve the appearance of the spaces. The officers had staterooms, and the petty officers were separated from the rest of the crew. However, the Japanese habitability standards for their accommodation were still below those of the western navies. The following is a comparison of the *Yamato* with other ships in the Imperial Japanese Navy:

Ship		*Square meters/man*
Yamato	(battleship)	3.2
Nagato	(battleship)	2.6
Ise	(battleship)	2.2
Myoko	(cruiser)	1.54
Sendai	(cruiser)	1.3

In every respect, Japanese sailors were proud to serve aboard these ships, and the *Yamato* soon earned the nickname of the "Palace."

Even though these ships had luxurious accommodations by Japanese standards, compared with ships of other navies they were very cramped. Deck heights varied from 1.9 to 2.3 meters, and with the cabling and ducting in the overhead, these clearances were further reduced.

Like the Royal Navy, the crew berthed and ate in the same space. The officers had better accommodations and slightly better privacy.

Access in Japanese battleships was difficult due to steeper than normal access ladders, and the hatches and watertight doors were small. These were not quick-acting types and this made it difficult to achieve watertightness in short periods of time. A great amount of drill was required to increase crew efficiency. This was lacking on the *Shinano*, however, and helped contribute to her loss.

The complement of the *Yamato* at the time of her loss had grown to over 3,300 officers and men. Some new officers who were assigned to the ship a few days before her final mission were put ashore. The largest contributor to this manning increase was the triple 25-mm gun mount, which required nine men to man it. In addition, there was the magazine complement that had to be increased as the number of guns increased, as well as the storekeepers, cooks, etc., necessary to keep these men fed and comfortable. Conditions aboard the *Yamato* were very bad, as new men had to sleep in hammocks in passageways and in any other space they could find. Also contributing to this overcrowding was the fact that the *Yamato* was the only capital ship in service, and she was the flagship of the Second Fleet.

Summary. The Japanese originally planned seven new battleships, five of the *Yamato*-class and two of the more powerful *798*-class. Only the *Yamato* and *Musashi* were completed, and they represent the end of the battleship era, for they were the most powerful battleships ever built. Although they were capable of defeating any single potential adversary, several fundamental design defects somewhat diminished their combat effectiveness.

The massive side-protection system performed moderately well in service, but improved design techniques could have greatly increased resistance to torpedo damage. Unquestionably, the defective joint joining the upper and lower side belt within the underwater protection system diminished its potential to resist damage. The system relied on a series of void compartments supplemented by heavy armor plate to prevent fragmentation damage inboard of the system. Other navies normally used liquid-loaded compartments to help minimize such damage, which also increased resistance to underwater detonations. While battleships of other nations stored a considerable amount of fuel within the side-protection system, the *Yamato* and *Musashi* used tanks

The Musashi, *shown here early in 1942, and the* Yamato *were the most powerful battleships ever built. The spacious forecastle area was characteristic of these enormous warships.*

This photograph shows the Yamato *under attack in her approach to the Philippine Islands in October 1944. Note the heavy bow wave from high-speed steaming. The ship was beginning to make a starboard turn when this photo was taken. (James C. Fahey Collection)*

within the underbottom structure and beyond the armored citadel exclusively. The system was good, but could have been much more effective for its weight.

Great Japanese reliance on counterflooding was not borne out by experience. Before the London Treaty, they managed to obtain a U.S. Navy damage-control manual. Shortly before the war, their standards of damage control were slightly inferior to those of the U.S. Navy; this disparity grew during the war. With a list of 16 degrees, outboard voids on the undamaged sides of these ships could be filled only to 55 percent of capacity by free-flooding from the sea, and adequate pumping capacity was not provided to offset this fault. Pumps of higher capacity, and a better system of counterflooding, would have assisted in counteracting the effects of torpedo hits, but no realistic system could have coped with the damage these ships received.

The machinery plant was extensively subdivided longitudinally, thereby creating the possibility of large listing moments resulting from the effects of asymmetrical flooding. Given a very effective side-protection system, which the Japanese were confident these ships had, such subdivision is merited, since it is one of the few means of keeping within tolerable limits the flooding resulting from underbottom explosions. However, the side-protection system proved ineffective against U.S. torpedoes. The considerable list of about 10 degrees quickly taken by the *Shinano* after four torpedo hits was in part due to the flooding of major engineering spaces.

A design detail shortcoming was revealed in the sinking of the *Shinano* when heavy H-beams, used to strengthen longitudinal bulkheads, created "hard spots" that could result in tearing the bulkheads by punctures. The massive deflections in the steel plate set up by the torpedo explosions would find the areas around these H-beams rather immobile.

The loss of the *Yamato and Musashi* revealed the vulnerability of the bow to torpedo damage; the bow simply did not have adequate resistance to extensive flooding. To overcome such a weakness in design illustrates a most difficult and complex problem, and this weakness of the *Yamato*-class battleship was by no means an exception to the rule. A difficult compromise is required between conveniently large compartments, required by the users, and sufficiently numerous watertight compartments, required for damage control. The compromise is not always the best solution, and in the case of these ships, the compartment arrangements led to large flooding and excessive trims.

Emperor Hirohito visited aboard the Musashi *at Yokosuka in June of 1943. The emperor is seated in front, in the center of the row. Interesting details are shown of the triple 25-mm antiaircraft machine-gun mount in the left background, and the twin 5-inch/40-caliber dual-purpose gun mounts in the right background.*

While underwater protection was repeatedly tested by torpedoes, the massive armor protection was tested only by bombs. Those bombs that struck the *Musashi* and *Yamato* did superficial damage, but did not damage the vitals of either ship. Other than the virtually inescapable vulnerability of secondary and antiaircraft guns, the primary deficiency in the armor protection was the inadequate superstructure protection. This, too, is most difficult to obtain, as weight and stability considerations preclude truly adequate protection in the superstructure. The massive armor protection of the *Yamato*-class ships was partially reduced in effectiveness by the somewhat inferior quality of Japanese heavy armor plate and by the defects in the belt armor joint design.

Japanese radar was markedly inferior to U.S. equipment. Even as late as October 1944, U.S. smoke screens during the battle off Samar decreased the effectiveness of Japanese gunnery. This was a serious weakness, for by that time the U.S. and Royal navies used radar fire control extensively.

The *Yamato, Musashi*, and *Shinano* had a very inefficient propulsion plant, due to the Japanese failure to develop an adequate diesel engine that was mechanically reliable and the subsequent adoption of a steam turbine plant of a very conservative, relatively low-power design. As a result their fuel rate was high. This was one of the reasons why they were not used more than they were. During the Guadalcanal Campaign, the

The confined waters in which the Musashi *was operating during the Battle of Leyte Gulf are apparent here. The* Musashi *was destroyed on 24 October 1944 in the Sibuyan Sea by an unprecedented number of bombs and torpedoes. (The picture has been cropped at the top.)*

Japanese were reluctant to commit these ships to any of the battles due to the uncharted waters and the fact that the naval staff believed that battleships should not be used in shore bombardment. A major reason why the *Yamato* and *Musashi* were not used at Guadalcanal was the fuel situation. The fleet's daily consumption of fuel was greater than 10,000 tons per day in the summer of 1942. The fuel reserves at Kure had dwindled to 65,000 tons. Thus, it was imperative to conserve fuel, and the two large battleships were not used for most of 1942. A more efficient propulsion plant would have made their use in action more attractive.

Despite the shortcomings in the propulsion plant, the design of the *Yamato* and *Musashi* must be considered successful. The ships were very well protected, and their main battery was without question the most powerful ever mounted on any battleship. The mixed-caliber secondary battery proved somewhat anachronistic. As was the case in all navies, the antiaircraft batteries were progressively increased throughout the war, but the inferior performance of the Japanese batteries and the lack of radar reduced the effectiveness of the massive machine gun battery. The best testimony to the design of these ships was their impressive resistance to battle damage which, coupled with their great offensive power, certainly helped make them the most formidable battleships ever built.

The decision to design and build these giant battleships was contested within the Japanese Navy. Admiral Yamamoto was strongly opposed to the plan to construct such large ships. He believed that the amount of money and resources required for these ships would have been put to better use in the development of naval aviation. No matter how large a battleship, it would never be unsinkable. The potential of aircraft was such, however, that Admiral Yamamoto envisioned naval battles being fought between opposing fleets without any visual contact between the opposing ships. The final decision to build the *Yamato* class rested with those who believed that a superbattleship would overwhelm any conceivable opponent. To accomplish this, the Japanese did design a battleship that was difficult to sink, and every eventuality was considered. The *Yamato, Musashi,* and *Shinano* were enormous warships, and with their sinking, the era of the modern battleship and Japan's ambitions for an empire died.

Name & hull number	*Yamato* (1)	*Musashi* (2)	*Shinano* (110)
Builder	Kure Navy Yard	Mitsubishi, Nagasaki	Yokosuka Navy Yard
Laid down	4 November 1937	29 March 1938	4 May 1940
Launched	8 August 1940	1 November 1940	8 October 1944*
Completed	16 December 1941	5 August 1942	19 November 1944*
Operational	27 May 1942	January 1943	. . .
Disposition	Sunk on 7 April 1945, East China Sea	Sunk on 24 October 1944 in Sibuyan Sea, 13°7′North, 122°32′East	Sunk on 29 November 1944 in Inland Sea

Name & hull number	Hull 111	Hull 797
Builder	Kure Navy Yard	. . .
Laid down	7 November 1940	. . .
Launched	. . .	. . .
Completed	. . .	. . .
Operational	. . .	. . .
Disposition	Cancelled in December 1941 with hull 30% complete; later scrapped	. . .

Displacement

Yamato—December 1941
59,163 tons (60,113 mt) Light ship
64,000 tons (65,027 mt) Standard
67,123 tons (68,200 mt) Trials
69,988 tons (71,110 mt) Full load

Yamato—July 1944
65,000 tons (66,043 mt) Standard
68,009 tons (69,100 mt) Trials
71,658 tons (72,808 mt) Full load

Musashi—October 1944
68,595 tons (69,594 mt) Trials

Shinano (as an aircraft carrier) — November 1944
58,150 tons (59,083 mt) Light ship
62,000 tons (62,995 mt) Standard
66,984 tons (68,059 mt) Trials
70,755 tons (71,890 mt) Full load

*As an aircraft carrier

Frames *(Yamato)*

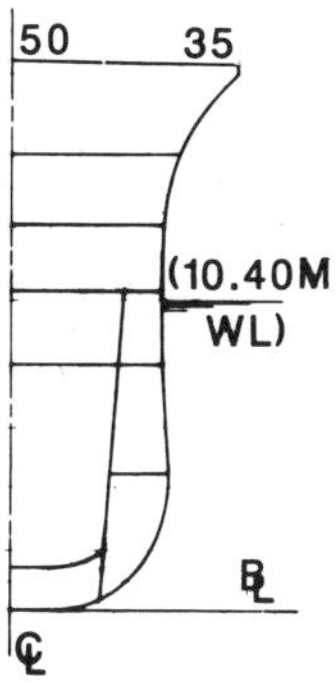

Frame 30

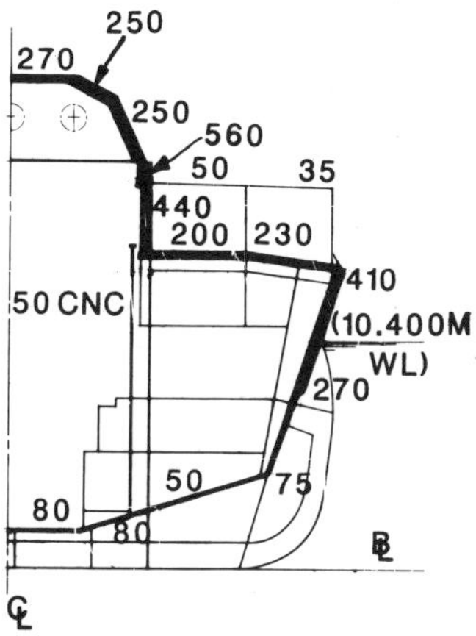

Frame 77

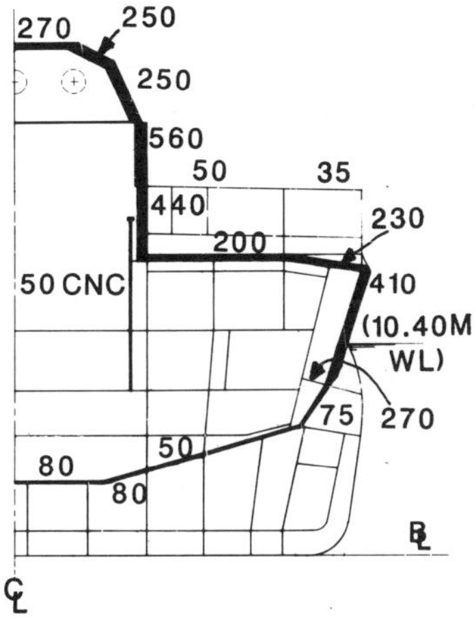

Frame 96

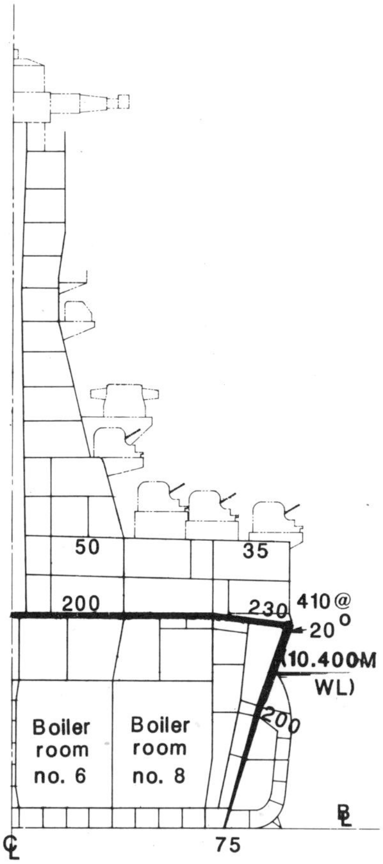

Frame 124

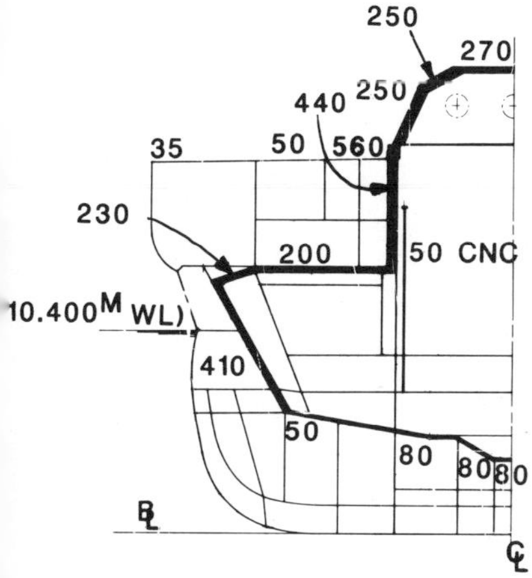

Frame 176

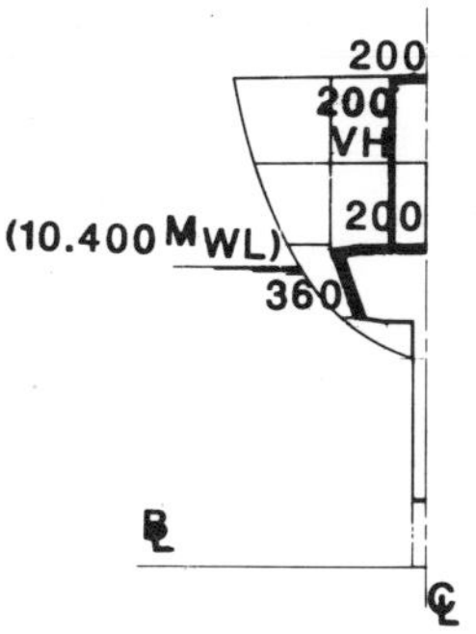

Section at A. P.

Midship Schematic *(Yamato)*

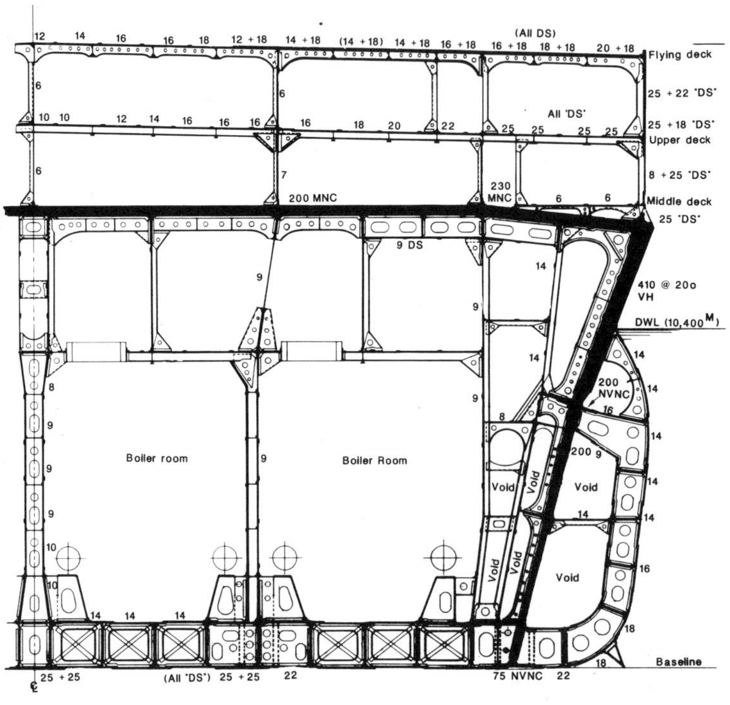

Dimensions

Yamato and *Musashi*

862′ 10″ (263.00 m) Length overall
839′ 11″ (256.00 m) Waterline length
800′ 6″ (244.00 m) Length between perpendiculars
127′ 8″ (38.90 m) Maximum beam
121′ 1″ (36.90 m) Waterline beam
30′ 4″ (9.25 m) Mean draft @ 59,163 tons (60,113 mt)
34′ 3″ (10.40 m) Mean draft @ 67,123 tons (68,200 mt)
34′ 9″ (10.58 m) Mean draft @ 68,009 tons (69,100 mt)
35′ 5″ (10.80 m) Mean draft @ 69,988 tons (71,110 mt)

Shinano

872′ 8″ (266.00 m) Length overall
839′ 11″ (256.00 m) Waterline length
800′ 6″ (244.00 m) Length between perpendiculars
839′ 11″ (256.00 m) Flight-deck length
129′ 3″ (39.40 m) Flight-deck width
127′ 8″ (38.90 m) Maximum beam
119′ 1″ (36.30 m) Waterline beam
28′ 11″ (8.875 m) Mean draft @ 58,150 tons (59,083 mt)
33′ 10″ (10.31 m) Mean draft @ 66,984 tons (68,060 mt)
35′ 4½″ (10.81 m) Mean draft @ 70,755 tons (71,890 mt)
34′ 10″ (10.62 m) Maximum draft @ 58,150 tons (59,083 mt)
35′ 5″ (10.80 m) Maximum draft @ 66,894 tons (68,060 mt)
38′ 3½″ (11.67 m) Maximum draft @ 70,755 tons (71,890 mt)

Machinery Schematic *(Yamato)*

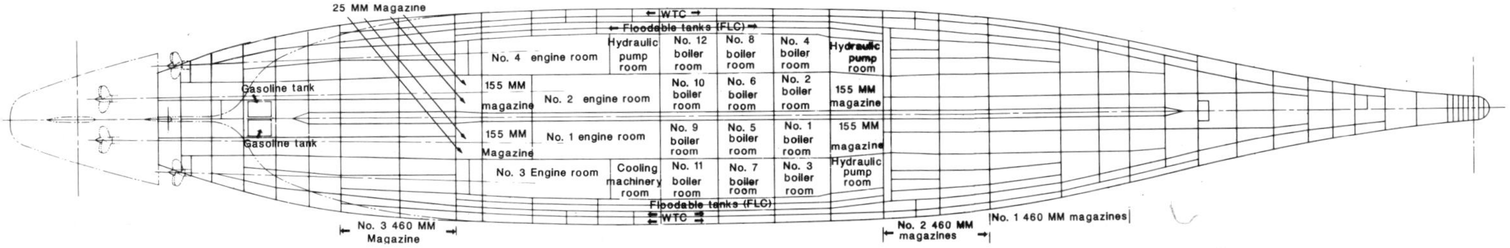

Hull Characteristics at D.W.L.

Yamato and *Musashi*

Displacement	68,009 tons (69,100 mt)
Mean draft	34′ 1″ (10.40 m)
Hull depth amidships	62′ 1″ (18.915 m)
Freeboard at bow	32′ 10″ (10.00 m)
Freeboard at stern	21′ 0″ (6.40 m)
Block coefficient	0.596
Prismatic coefficient	0.612
Midship section coefficient	1.121
Actual metacentric height	*Yamato* — 1941
	9.84′ (3.00 m) @ 59,163 tons (60,113 mt)
	8.53′ (2.60 m) @ 67,123 tons (68,200 mt)
	9.81′ (2.99 m) @ 69,987 tons (71,110 mt)

Shinano

Displacement	66,984 tons (68,059 mt)
Mean draft	32′ 10″ (10.31 m)
Hull depth amidships	81′ 4½″ (24.81 m) — to flight deck
	Shinano—1944
Actual metacentric height	11.22′ (3.42 m) @ 58,150 tons (59,083 mt)
	11.31′ (3.45 m) @ 66,984 tons (68,059 mt)
	12.65′ (3.86 m) @ 70,755 tons (71,890 mt)

Armament

Yamato and *Musashi*

Nine 18.1″/45-caliber guns (Model 1939) (460 mm)
Twelve 6.1″/60-caliber guns (Model 1933) (155 mm)
Twelve 5″/40-caliber guns (Model 1930) (127 mm)

This data is for the *Yamato* and *Musashi* as completed; refer to data following for variations throughout operational careers.

Seven aircraft; two catapults

Body Plan *(Yamato)*

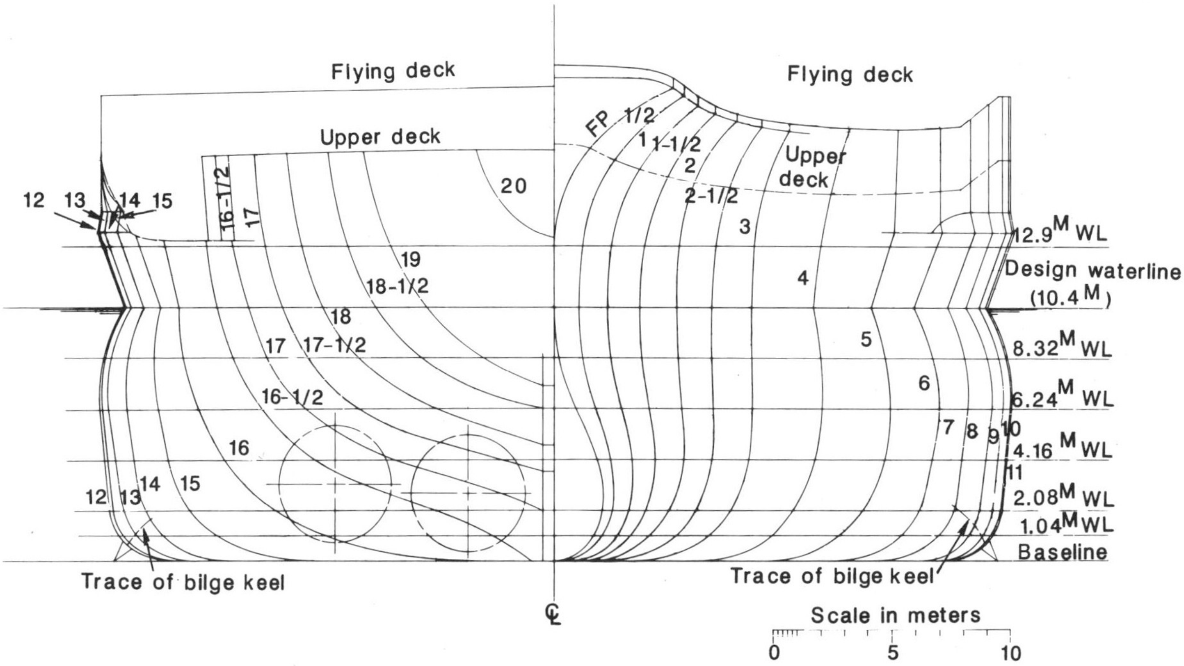

	6.1″/60 (155 mm)	5″/40 (127 mm)	1″/60 (25 mm)	0.51″ (13 mm)
Yamato				
December 1941	12	12	24	4
Fall 1943	6	12	36	4
February 1944	6	24	36	4
May 1944	6	24	98	4
July 1944	6	24	113	4
April 1945	6	24	150	4
Musashi				
August 1942	12	12	24	4
Fall 1943	6	12	36	4
April 1944	6	12	54	4
May 1944	6	12	116	4
July 1944	6	12	130	4

Shinano

Sixteen 5″/40-caliber guns (Model 1930) (127 mm)
One hundred five 1″/60-caliber guns (25 mm) (triple mounts)
About forty 1″/60-caliber guns (25 mm) (single mount)
Twelve twin 4.7″ (120 mm) rocket launchers
Forty-seven aircraft
- 20 fighters (including 2 dismantled)
- 20 bombers (including 2 dismantled)
- 7 scouts (including 1 dismantled)

Outboard Profile (April 194

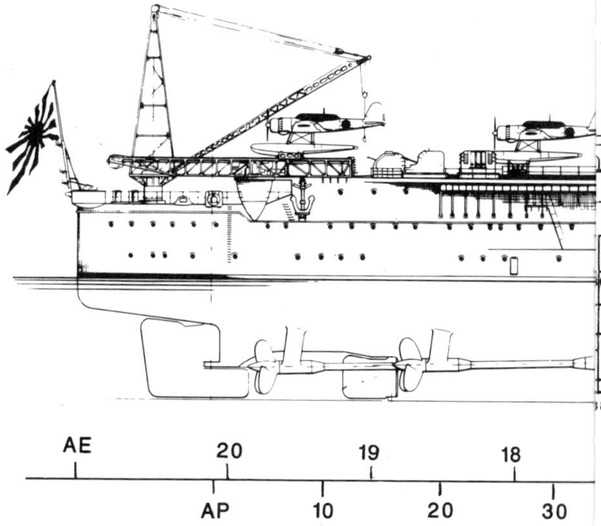

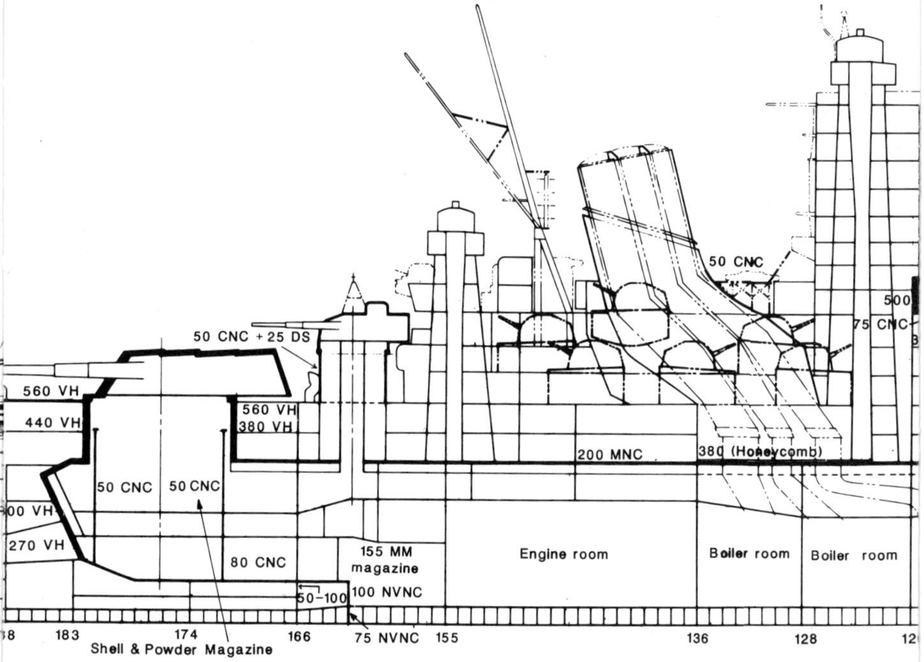

Topside View (April 1945)

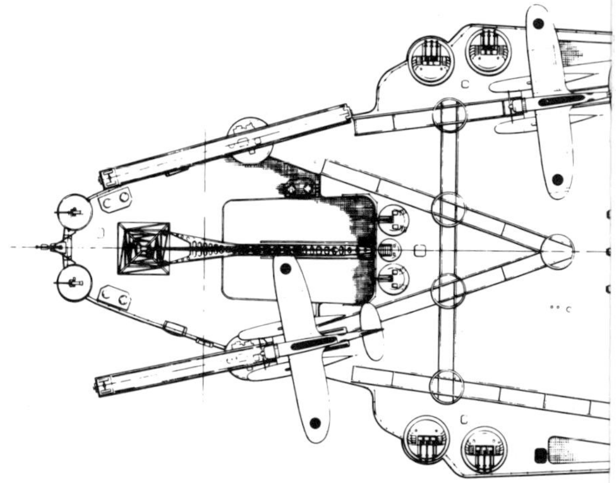

Machinery Schematic *(Shinano)*

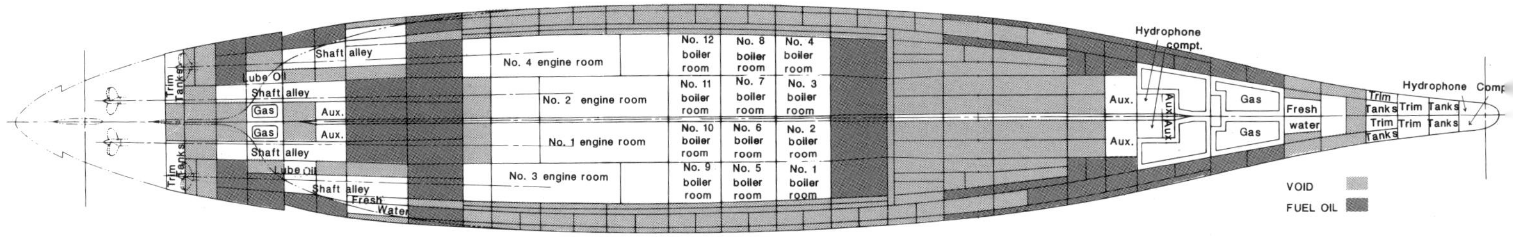

Armor Protection (refer to plans for arrangement details)

Yamato and *Musashi*

Immunity zone — From 21,872 to 32,808 yards (20,000 to 30,000 m) — Citadel versus Japanese 18.1″/45 (460 mm) firing a 3,219-pound (1,460-kg.) projectile.

Amidships

Belt armor — 16.14″ on 0.63″ (410 mm on 16 mm) inclined at 20 degrees

Lower belt armor — 10.63″ tapered to 2.95″ (270 mm to 75 mm) inclined at 20 degrees over magazines

7.87″ tapered to 2.95″ (200 mm to 75 mm) inclined at 20 degrees over machinery

Deck armor

	(centerline)	(outboard)	
main deck	0.47″ (12 mm)	0.79″ + 0.71″	(20 mm + 18 mm)
second deck	0.39″ (10 mm)	0.98″	(25 mm)
third deck	7.87″ (200 mm)	9.06″	(230 mm) — Slopes
splinter deck	0.35″ (9 mm)	—	—
fourth deck	0.35″ (9 mm)	—	—
total	9.43″ (240 mm)	11.54″	(293 mm)

Turret armor	
face plates	25.59″ (650 mm)
side plates	9.84″ (250 mm)
back plates	7.48″ (190 mm)
roof plates	10.63″ (270 mm)
Barbette armor	
centerline forward	17.32″ (440 mm)
sides	21.65″ (560 mm)
centerline aft	14.96″ (380 mm)
Secondary gun armor	
turrets	1.97″ (50 mm)
barbettes	1.97″ + 0.98″ (50 mm + 25 mm)
Conning tower armor	
centerline forward	19.69″ (500 mm)
sides	19.69″ (500 mm)
centerline aft	19.69″ (500 mm)
roof plates	7.87″ (200 mm)
communications tube	11.81″ (300 mm)

Shinano (No. 110—as battleship)/No. 111/No. 797

Armor was generally to be as the first two ships, except for the following modifications:

Main side belt	15.75″ (400 mm) sloped at 20 degrees
Third deck	7.48″ (190 mm)
Turret armor	
face plates	24.80″ (630 mm)
side plates	9.06″ (230 mm)
back plates	6.69″ (170 mm)
roof plates	9.84″ (250 mm)

Body Plan *(Shinano)*

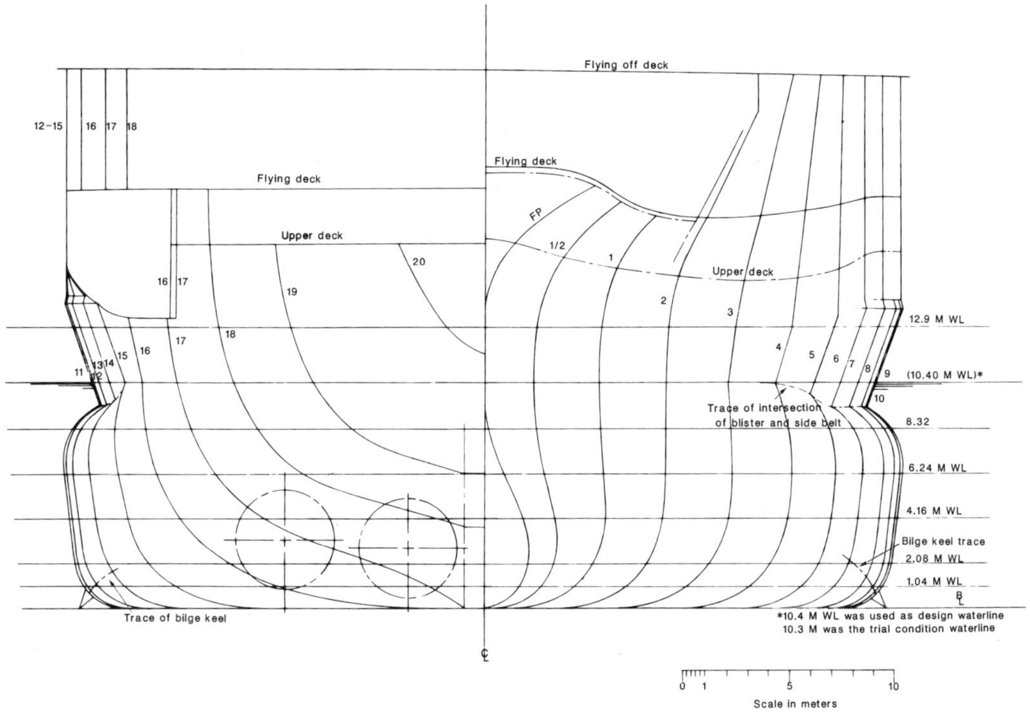

Shinano (as an aircraft carrier)

Immunity zone	Designed to withstand 8″/50 (203 mm) projectile	
Amidships		
Belt armor	6.30″ on 0.63″ (160 mm on 16 mm) inclined at 20 degrees	
Lower belt armor	7.87″ tapered to 2.95″ (200 mm to 75 mm)	
Deck armor (Centerline)		
flight deck	2.95″ + 0.79″	(75 mm + 20 mm)
hangar deck	0.47″ + 0.71″	(12 mm + 18 mm)
upper (second) deck	0.98″	(25 mm)
armor (third) deck	3.94″	(100 mm)
total	9.84″	(250 mm)
Deck armor (Outboard)		
flight deck	2.95″ + 0.79″	(75 mm + 20 mm)
hangar deck	0.47″ + 0.71″	(12 mm + 18 mm)
upper (second deck	0.98″	(25 mm)
armor (third) deck	9.06″	(230 mm) Slopes
total	14.96″	(380 mm)
Gasoline tanks		
decks	2.76″ to 1.18″ over 0.98″	(70 mm to 35 mm over 25 mm)
sides	0.98″ over 0.98″	(25 mm + 25 mm)
longitudinal bhd	0.98″ over 0.98″	(25 mm + 25 mm)
transverse bhd	0.98″ over 0.98″	(25 mm + 25 mm)
cement	2,362 tons (2,400 mt) in cofferdams surrounding tanks	

Underwater Protection	
Designed resistance	882 pounds TNT (400 kg)
Side-protective system depth	23.46′ (7.15 m) @ half-draft amidships
S.P.S. design loading	all void
Total plating thickness	2.44″ (62 mm) plus lower side belt
Bottom protective system depth	4.59′ (1.4 m)
Designed loading	liquid
Total plating thickness	1.54″ (39 mm)

Tank Capacities

Yamato	
Fuel oil	6,201 tons (6,300 mt)
Reserve feed water	313 tons (318 mt)
Lubricating oil	89 tons (91 mt)
Light oil	71 tons (72 mt)
Shinano (as an aircraft carrier)	
Fuel oil	8,763 tons (8,904 mt)
Reserve feed water	295 tons (300 mt)
Lubricating oil	217 tons (220 mt)
Light oil	591 tons (600 mt)

Machinery

Boilers: twelve Kampon Mark RO Bureau-type boilers with superheaters and air preheaters
pressure: 356 psi (25 kg/cm^2)
temperature: 618° F. (326 C.)

Turbines	four sets geared turbines	
Shaft horsepower	overload	162,743 (165,000 mhp)
	normal maximum	147,498 (150,000 mhp)
	service normal (100%)	150,992 (153,553 mhp)
	(80%)	118,642 (120,655 mhp)
	(60%)	88,577 (90,080 mhp)
	(40%)	60,405 (61,430 mhp)
	(20%)	18,285 (18,595 mhp)
Maximum speed	*Shinano*	27 knots (as aircraft carrier)
	Yamato	
	at 100% power	27.46 knots
	at 80%	26.60 knots
	at 60%	25.60 knots
	at 40%	23.20 knots
	at 20%	16.47 knots
Nominal endurance	*Yamato*	7,200 miles @ 16 knots
	Shinano	10,000 miles @ 18 knots (as aircraft carrier)
Generators	four ship's service turbogenerators (600 kw) four ship's service diesel generators (600 kw) total ship's service capacity — 4,800 kw @ 225 volts, DC (all ships)	
Propellers	four three-bladed propellers, 19.69′ (6.0 m) diameter	
Rudders	one main rudder on centerline one auxiliary rudder on centerline	

Outboard Profile

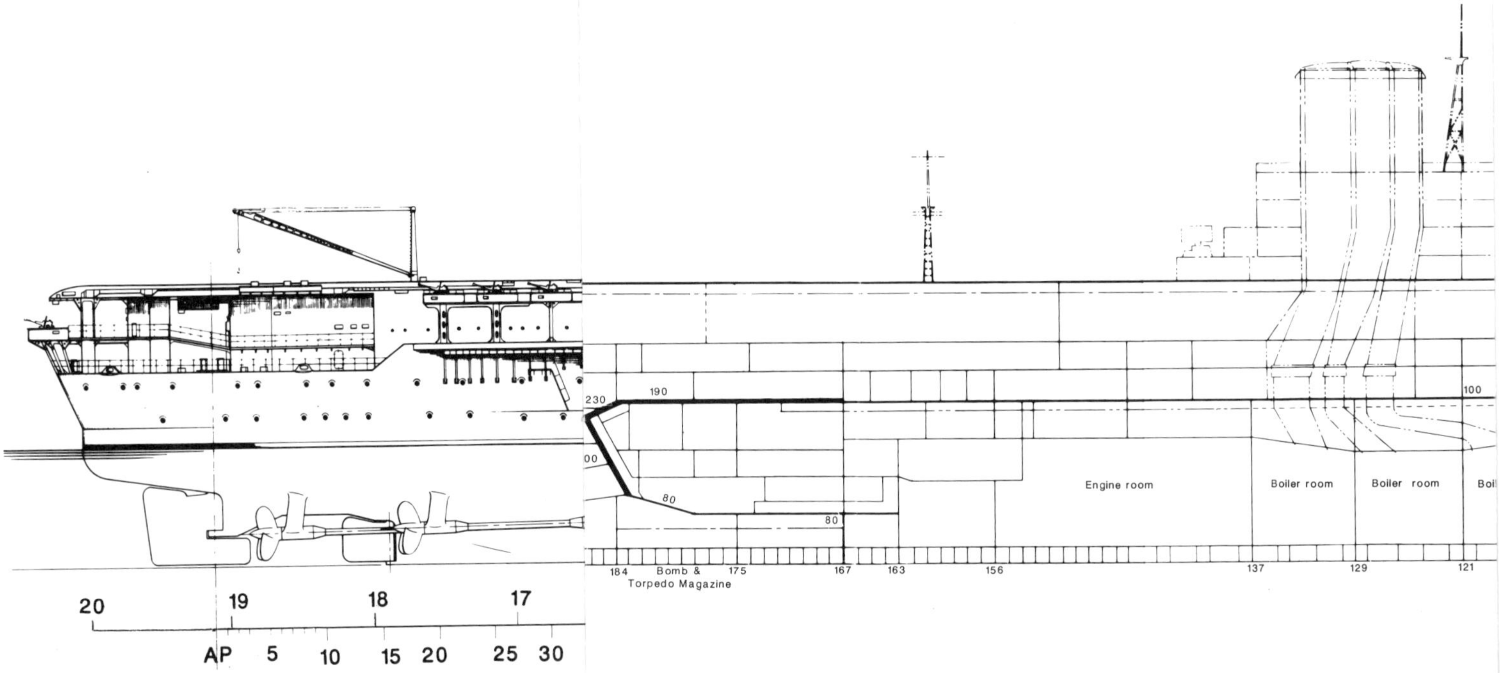

Topside View (October 194

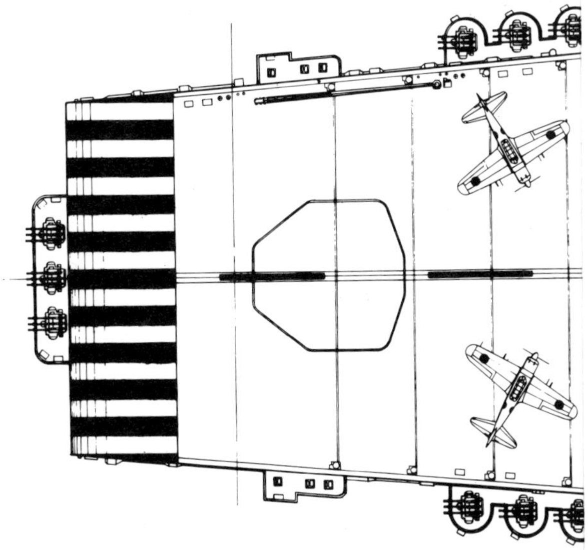

CHAPTER THREE

The Scharnhorst Class

After World War I, the Versailles Treaty had restricted Germany from building any warship larger than a 10,000-ton cruiser, and the German fleet of the interwar period was reduced to a token force (1,500 officers and 15,000 enlisted men). The German Naval Staff wanted to replace some of the older predreadnoughts that were the basis of the post–World War I German fleet, but intense debates arose in the Reichstag of the young Weimar Republic over whether to even use the tonnage allotted to them by the Versailles Treaty. These debates over the use, type of deployment, and size of armament for new warships in the 1920s created intense friction between political factions who did not want the return of a "Tirpitz" fleet of prestige or even a limited fleet-replacement program. Only after years of deliberation was the navy finally permitted to build the first armored ships, but the Versailles Treaty restrictions had to be observed. By the early 1930s, however, large naval programs advocated by the French and Soviet navies soon caused the Germans to consider warships larger than the permitted armored ships.

The Construction Office of the German Navy—like its foreign counterparts—followed warship developments abroad. This meant that studies were made at times to evaluate various types of warships being proposed by the major navies. In 1928-29 this led to several small capital-ship designs, which were based upon battleship sizes being advocated by the Royal Navy. The main objective of these studies was to develop a ship that could mount a 305-mm gun and achieve 17,500–25,000 tons of standard displacement. Such a ship would be a match for the large number of heavy cruisers being built by many other navies. During 1928 a battlecruiser design was prepared with the characteristics summarized in table 3-1.

This design featured very light protection and minimum underwater protection against torpedoes because of the emphasis on speed and gun power. The design had certain imbalances, especially when compared to the 26,200-ton battlecruisers of the *Derfflinger* type, which also had eight 305-mm guns. The reason for this was the use of four main-battery turrets and the need for considerable hull volume for a 34-knot propulsion plant. This design was not completed because German naval constructors recognized that to retain a 30+ knot speed, it would probably be necessary to reduce the number of 305-mm gun positions by using triple or quadruple turrets or to accept a smaller main-battery caliber. Even so, it was felt that more displacement would be required to produce a balanced design with the necessary characteristics.

This striking view of the battlecruiser *Scharnhorst* shows much of her armament. The main armament of 283 mm, however, was politically motivated and a severe handicap since it was the smallest caliber used in any capital ship designed, projected, or built in World War II. In this view the secondary battery of 150-mm and 105-mm guns also can be seen. This was a standard arrangement for all German battleships and battlecruisers. (Drüppel)

TABLE 3-1

17,500-ton Battlecruiser Design

Standard displacement	17,500 tons (17,786 mt)
Design displacement	19,192 tons (19,505 mt)
Full-load displacement	21,000 tons (21,343 mt)
Waterline length	675.9 feet (206 m)
Waterline beam	82.0 feet (25 m)
Depth at side	43.6 feet (13.3 m)
Draft — design	25.6 feet (7.8 m)
Armament	Eight 12″ (305 mm) in four twin turrets Eight 5.9″ (150 mm) in eight single mounts Four 3.46″ (88 mm) antiaircraft guns
Protection	3.94″ (100 mm) side belt 1.18″ (30 mm) deck outboard 0.79″ (20 mm) deck inboard
Shaft horsepower	157,811 (160,000 mhp)
Speed	34 knots

When Adolf Hitler assumed power in Germany in 1933, he made it clear to Admiral Erich Raeder, the fleet commander-in-chief, that he had no intention of pursuing a naval policy like that of Admiral von Tirpitz in the years before World War I; that is, he would not build a navy to challenge British supremacy of the seas. He did feel, however, that it was important to counter the naval construction program underway in France. Thus, he would permit a fourth and fifth armored ship to be built, but only with increased protection, a displacement limited to 19,000 tons standard, and two triple 283-mm turrets.

The scope of German naval planning in 1933 was limited to the defense of Baltic Sea commerce against Polish forces that were possibly acting under French protection, or the defense of German sea lanes in a commerce war against France. The Russian Navy was still too weak to be much of a factor in German plans, even though there were hints of an extensive Soviet naval construction program. By 1935, the German naval staff planned a limited fleet expansion to 421,000 tons (standard), with 184,000 tons reserved for battleship construction. It was believed that this figure would be surpassed, as the Washington Treaty would expire at the end of 1936, at which time increased naval construction by the United States, Great Britain, Japan, France, and Italy would give Germany justification for further naval expansion.

In 1934, the French Navy announced the construction of a second battlecruiser—the *Strasbourg.* In February 1934, the German Navy decided to pursue the design of a battlecruiser displacing 26,000 tons and armed with three triple turrets of 283-mm guns. Before the construction of these ships was announced, Hitler wanted to reassure the British in the matter. He also wanted to have sufficient warships to deal with the French Navy, so he proposed that a treaty be concluded with Great Britain that would give the Royal Navy a three-to-one superiority in sea power. Such a pact would also legally remove the restrictions of the Versailles Treaty. The Anglo-German Naval Treaty, which permitted Germany to build a fleet, was concluded on 18 June 1935 in London. This treaty was a tacit admission that German naval policy would not be directed against Great Britain. The Germans, however, were also bound to observe any future naval treaties that might be concluded.

The *Gneisenau* (upper) and *Scharnhorst* (lower) are shown here after their reconstruction and the addition of "Atlantic bows." The distinctive differences between the two ships can also be seen. The *Gneisenau* had her mainmast attached to the stack, while the *Scharnhorst* had hers mounted forward of the after rangefinder. The *Scharnhorst* had three anchors, with one at the stem. (Albrecht Schnarke Collection)

Design characteristics. In 1933 Admiral Raeder had advocated augmenting the defensive characteristics of armored ships D and E. He also suggested that the inclusion of a third turret be considered. These could not be accomplished with the displacement of the *Admiral Graf Spee*, which was the largest of the armored ships. After some discussion, Hitler permitted an improvement in their defensive qualities only—the armor and compartmentation—and accepted the corresponding increase in displacement. A limit of 19,000 tons was stipulated on the standard displacement, which would not support a third turret and the increases in protection sought. The German Navy favored the use of 380-mm guns in three twin turrets. Hitler, however, barred any increase in gun caliber from 283-mm to 380-mm, as this would be more difficult to justify politically. Furthermore, the increase in protection and a proposed third turret would have meant a much larger ship. Finally, in February 1934 Hitler permitted the addition of a third turret, but then he would only permit a preliminary design for a ship of 26,000 tons armed with nine 283-mm guns. Thus, the design concept of the *Scharnhorst* and *Gneisenau* was born. These ships were not outgrowths of World War I battlecruiser developments, but were descendants and expansions of the armored ships. Even the secondary armament was dictated by what was left over from the cancellation of the fourth and fifth armored ships. As eight single 150-mm mounts were available, the new ships had mixed 150-mm battery turrets and single mounts with open shields. These ships also had heavy armor protection, but lacked the traditional upper citadel belt that characterized many of the German battlecruiser and battleship designs of World War I.

Orders were placed with Deutsche Werke, Kiel, and the Wilhelmshaven Navy Yard in February 1934, but without contract drawings. It was not possible to commence construction until 14 months later. In the months prior to the signing of the Anglo-German Naval Treaty, the construction drawings and specifications were completed for a ship that would carry three triple turrets, using some of the equipment originally ordered for the fourth and fifth armored ships.

Even though the German Navy regarded the *Scharnhorst* and *Gneisenau* as battleships, they were, in a sense, battlecruisers with heavy protection, high speed, and medium-caliber guns, a development of the *Deutschland*-class armored ships and a transitional type with a design constrained by technical limitations and political considerations. The 1934 design requirements are tabulated in table 3-2.

Since the *Scharnhorst*-class battlecruisers were being built to rival the French *Dunkerque* ships, utmost consideration was given to firepower and protection.

The new-model 283-mm main-battery gun, by Krupp, had great power and range for its modest caliber, somewhat more than had been provided the similar guns of the

TABLE 3-2

1934 Design Requirements for a 26,000-ton Battlecruiser

Standard displacement	26,000 tons (26,425 mt)
Armament	11.1″ (283-mm) Model 1934 guns 5.9″ (150-mm) guns 4.1″ (105-mm) guns
Maximum speed	30 knots
Endurance	10,000 miles at 14 knots
Protection	Armor to withstand at least the shell fire of the *Dunkerque* guns (330-mm) and the best possible protection against bombs and torpedoes.

armored ships. The problem that confronted German naval constructors in providing adequate armor protection can be better understood when the sizes of the French and German armor-piercing shells are compared. The *Dunkerque*'s 330-mm shell weighed 572 kilograms, while the 283-mm German shell would weigh only 330 kilograms. Nevertheless, the smaller guns were retained because they had a very rapid rate of fire, much of the *Dunkerque*'s hull was believed to be unprotected, and destruction of equipment in those areas could possibly disable her. The overriding consideration, however, was the fact that a smaller main-battey caliber was politically dictated.

During the preliminary design, it was thought that an armament of nine 283-mm guns and protection against the French 330-mm gun at close ranges in the North Sea could be achieved on a 26,000-ton displacement. In fact, it was recognized in 1936 from shipyard weight estimates while the ships were under construction that the displacement was going to be much larger than had first been estimated. This large increase threatened the stability, seakeeping, and protection of the ships, since the armor deck would now be below the design waterline instead of above it. The loss of freeboard was also serious, because this meant that the range of stability would be reduced. Later, ship operations showed that the increase in draft created serious deck wetness in heavy head seas, necessitating further alterations to the sheer line and bow flare.

The problem of increased displacement was taken up by the Construction Office. Since changes to the basic design characteristics were non-negotiable, the office concluded that the only way to gain back the lost freeboard would be to increase the hull beam. This was not possible with the ships already on the building ways. Adding blisters was an undesirable option, since it would adversely affect speed and make the ships even larger. Further changes, however, were more diligently investigated before being approved. The increase from 26,000 to 30,000 tons in the standard displacement was accepted. When construction commenced, the ships had the characteristics summarized in table 3-3.

TABLE 3-3

1935 Characteristics of the Scharnhorst Class

Standard displacement	31,053 tons (31,552 mt) — *Scharnhorst* 31,132 tons (31,632 mt) — *Gneisenau*
Full-load displacement	37,224 tons (37,822 mt) — *Scharnhorst* 37,303 tons (37,902 mt) — *Gneisenau*
Waterline length	741.5 feet (226.0 m)
Waterline beam	98.43 feet (30.0 m)
Draft - design	28.44 feet (8.69 m)
Armament	Nine 11.1″/54.5 tripled (283 mm) Twelve 5.9″/55 paired and single (150 mm) Fourteen 4.1″/65 paired (105 mm) Sixteen 1.40″/83 paired (37 mm)
Maximum speed	30 knots
Shaft horsepower	157,811 (160,000 mhp) — Maximum 123,290 (125,000 mhp) — Normal
Endurance	10,000 miles at 14 knots
Protection	13.78″ (320 mm) — main side belt 1.97″ (50 mm) — upper deck 4.13″ (105 mm) — lower armor deck

The design of these ships was concluded in 1935, by which time the new 283-mm gun had been successfully tested at the Naval Proving Grounds. The most controversial element in the design of these ships, however, was to be the type of propulsion plant. To ensure early delivery and to develop the necessary horsepower and speed, it was decided early in the design process to install a high-temperature, high-pressure, geared-turbine drive. The diesel power plant of the armored ships could not propel these larger ships at the speeds sought, and there were no diesel engines of sufficient size available. Admiral Raeder decided to take the risk and go to the steam turbine with the new high-pressure, high-temperature boiler. The new type of power plant had been tested in shipyards, and the engineering officers who were to run the plants on the ships at sea participated in the tests and were favorably impressed. To wait for the necessary diesel engine to be developed and tested seemed unwise under such circumstances, even though it was recognized that the ships' endurance would be less with steam propulsion.

Gneisenau—Operational History. The keel of the *Gneisenau* was laid at the Deutsche Werke Shipyard in Kiel on 6 May 1935. Construction proceeded with little or no difficulty, and the ship was launched on 8 December 1936. The *Gneisenau* sustained some grounding and stern damage after her launch when her chain drags broke. The battlecruiser drifted to the shore opposite the launch area, and it took several hours to free her. The *Gneisenau* was commissioned on 21 May 1938.

As she was operational before the *Scharnhorst*, her first year was spent largely in trials and training. Cruises to the North Sea and the Baltic soon indicated that the ship would have trouble in rough seas due to insufficient bow freeboard. In her first yard availability period during the winter of 1938–1939 her bow was modified by adding more flare and sheer. The anchors remained in hawsepipes.

This early view of the *Gneisenau* shows her original profile—the straight stem, no radar on top of the forward rangefinder, two catapults (with one on top of turret Caesar), no funnel cap, and anchors mounted on the shell plating. (Blohm & Voss)

These three views of a *Scharnhorst*-class battlecruiser at various times in their careers show changes made in the ships. The top view is the original profile of the *Scharnhorst* before her 1939 reconstruction. The middle view shows the *Gneisenau*'s modified bow after a refit in 1939. A clipper stem was added and the bow freeboard increased. The lower view shows the *Gneisenau*'s "Atlantic bow" with the anchors now stowed on deck and the bow freeboard further increased. Both battlecruisers had severe problems with seakeeping throughout their careers and were very wet ships forward.

In June 1939 Admiral Raeder sent the *Gneisenau* into the Atlantic for a trial trip, principally for target practice, on Hitler's assurance that the political situation was well in hand. The *Gneisenau* carried practice ammunition and had almost no battle ammunition. Admiral Raeder later commented:

> In any but the most peaceful times, no warship ever leaves home waters without full war-readiness magazines, and if I had apprehensions of war, I would have never permitted her to sail unprepared. Neither the commander of the Battleships Division nor the commander-in-chief of the fleet expressed to me any concern on the matter.

The *Gneisenau* was ready for action when the war began in September 1939. On 4 September, 14 Wellington bombers of the Royal Air Force attacked the *Gneisenau* and *Scharnhorst* in Brunsbüttelkoog. No damage was sustained, and the Luftwaffe managed to shoot down two of the attacking aircraft.

On 8 October the *Gneisenau* sailed with the light cruiser *Köln* and nine destroyers to create a diversion and compel the dispersion of Anglo-French naval forces searching for the *Deutschland* and *Admiral Graf Spee*. After the fleet had been spotted by British reconnaissance aircraft, the ships went no farther than the Isle of Utsire off the southern coast of Norway before returning to Germany through the Skaggerak and Kattegat. The operation raised concern in the British Admiralty, and the Home Fleet tried to intercept. The Luftwaffe attacked the British ships without success.

Sinking of HMS Rawalpindi. On 21 November the *Gneisenau, Scharnhorst*, cruisers *Leipzig* and *Köln*, and three destroyers left Wilhelmshaven for a sweep against British patrols between Iceland and the Faroes. This operation was intended to divert French and British naval strength from the South Atlantic where the *Admiral Graf Spee* was being pursued. Once the larger ships cleared the minefields, the cruisers and destroyers returned to port. The battlecruisers, with the *Gneisenau* in the lead and the *Scharnhorst* slightly to stern and some 20,000 meters to starboard, headed into a fierce gale at 27 knots; there was minor structural damage to both ships, and some water entered the forward turrets and magazines. Speed was reduced to 12 knots while lookouts scanned the horizon for British ships. At 1607 on 23 November the armed British merchant cruiser *Rawalpindi* was sighted by lookouts on the *Scharnhorst*.

The *Rawalpindi* was a former P&O liner whose seven old 152-mm guns and lack of armor protection made her no match for such opponents. At 1703 the *Scharnhorst* opened fire and was joined by the *Gneisenau* some eight minutes later. At 1706 a salvo of 283-mm shells ripped into the bridge of the auxiliary cruiser, killing the captain and most of his officers. There was much structural damage and a number of fires broke out. By 1716 the two battlecruisers had reduced the British ship to a burning and sinking hulk. Admiral Marschall, on board the *Gneisenau*, ordered the *Scharnhorst* to pick up survivors. This continued until the British cruiser *Newcastle* appeared. The *Scharnhorst* laid down a smokescreen, and the battlecruisers quickly left the scene. They steamed north towards the Arctic Circle and waited for bad weather before returning to Wilhelmshaven on 27 November through the Bergen-Shetland Straits. The return trip was made in severe weather, and both ships sustained further storm damage. The forward turrets took in water and there was minor flooding in the crew's quarters forward, the result of shell plating damaged from slamming into heavy seas. Due to insufficient freeboard, both vessels shipped water over their bows for most of one day; the heavy seas inundated the bridge decks, necessitating handling of the ships from their conning towers. In the meantime, an Anglo-French naval force, consisting of the *Hood* and *Dunkerque* and the Home Fleet (*Nelson* and *Rodney*), were in unsuccessful pursuit.

In a seven-week overhaul at the Naval Dockyard, Kiel, the *Gneisenau*'s bow was once again reconfigured to provide additional flare and sheer. On 15 January 1940 she resumed training, but thick ice in the Kiel Canal blocked transit to the North Sea until 4 February.

On 18 February, the *Gneisenau, Scharnhorst, Admiral Hipper*, and two destroyers left Wilhelmshaven to attack convoy shipping between Norway and Great Britain. The force was under the command of Admiral Marschall and reached the Shetland-Norway passage. He returned his force to Wilhelmshaven when no shipping was sighted.

Invasion of Norway. On 7 April 1940, the *Gneisenau, Scharnhorst, Admiral Hipper*, and 14 destroyers headed for Norwegian waters at 24 knots. Norway was to be assaulted by a combination of sea and land forces from Narvik in the north to Oslo in the south. The destroyers and *Admiral Hipper* carried assault troops for Trondheim and Narvik, while the two battlecruisers were to provide the covering force. Steaming north towards their destinations, the ships were attacked by British bombers. No hits were made, but the position of the German fleet was reported. At 2015 on 7 April the Home Fleet put to sea on the premise that the large German Fleet would attack vital British shipping in the North Atlantic.

About 0800 on 8 April the destroyer *Bernd Von Arnim* emerged from a squall to discover the British destroyer *Glowworm*, which was trailing a British force consisting of the battlecruiser *Renown* and light cruiser *Birmingham* that were some distance west of Trondheim. A running battle, in which the *Admiral Hipper* and four destroyers took part, ensued some 100 miles west of Trondheim. The *Scharnhorst* and *Gneisenau* manned battle stations, but did not enter the action. The *Glowworm* was sunk at 0924 after she rammed the *Admiral Hipper*, which was not seriously damaged and was thereafter dispatched to Trondheim with four destroyers. At 2100 the Narvik landing force, aboard ten destroyers, was detached from the main battle group. Meanwhile, the *Scharnhorst* and *Gneisenau* took position west of West Fjord. High winds and seas made movements of the ships difficult, so speed was reduced to only 7 knots.

Action with Renown. At 0430, 9 April, the *Gneisenau* made radar contact with the *Renown*, and battle stations were sounded on both German ships. Squalls and low clouds limited the visibility until 0505, when the *Renown* opened fire on the *Gneisenau* with 381-mm guns at a range of 11,800 meters. Three minutes later the Germans returned the fire. The *Renown* shifted fire to the *Scharnhorst* at 0513. Gunfire was sporadic until 0600; then, after a lull, both sides resumed ineffective fire at 0620, continuing until 0715.

Within five minutes after the opening salvo, the *Gneisenau* made two hits on the *Renown*, with the aid of radar. One 283-mm shell went through the main leg of the British battlecruiser's foremast and passed overboard without detonating. The second struck aft of "Y" turret on the starboard side plating, between the upper and main decks above the steering-gear room, passed across the ship and out the port side. Almost simultaneously, the *Gneisenau* was hit by two 381-mm shells. The first went through her director tower, severed many electric cables, and passed out of the ship without detonating. Debris killed one officer and five men. The optical rangefinders for the forward 150-mm guns were destroyed, and fire control was shifted aft. The second shell damaged and silenced the after turret. Shortly after these hits, the *Gneisenau* ceased fire and increased speed. It had been determined that the *Renown* was escorted by eight destroyers, and there were fears that these destroyers could torpedo the disabled ship if either one of the battlecruisers were damaged and lost speed. The *Gneisenau* fired sixty

This is the *Gneisenau* during a fleet review, probably in August 1938. Note the crew lining the rail, an admiral's flag flying from the foremast, and an HE-114 seaplane on the main catapult. Another catapult was fitted on turret Caesar but was later removed.

283-mm and eight 150-mm shells during the action; two men were killed and another eight were injured.

The German ships withdrew at high speed in the heavy seas, taking much water over their bows. This water cascaded down the main deck and entered their forward turrets, which had been trained to port and abaft the beam. It caused a short circuit in the ammunition hoist of the forward turret. Although action was broken off, the *Renown* continued pursuit with eight destroyers. Gun flashes from the destroyers evidently led the Germans to believe that they were being pursued by a strong force.

The Germans were able to outdistance the British battlegroup and steamed westward into the Arctic Ocean, where they remained for several days before beginning their return to Germany. Both ships reached Wilhelmshaven on 12 April in company with the *Admiral Hipper*, which had joined up with the battlecruisers at 0830 on 11 April off Trondheim.

Mine damage. After the battle damage was repaired, the *Gneisenau* went to Bremerhaven for dry-docking from 26-29 April. On 5 May, while steaming at 22 knots in the North Sea off the Elbe estuary en route to the Baltic, she exploded a magnetic mine about 21 meters off the port quarter and 24 meters below the keel. The side-armor belt was pushed inboard near the after turret, and there were numerous tears and cracks in the shell plating. The port shaft alley, several storerooms, and a void space adjacent to the shell were flooded. The ship took a ½-degree port list with a slight stern trim.

Shock damage was particularly severe. The starboard low-pressure turbine bearing and several auxiliary units, including condensate pumps and searchlight transformers, were out of operation due to the failure of their foundations under the severe shock loadings. The after rangefinders and the optics for the target indicator were also damaged. Almost all automatic cut-off switches to the main and auxiliary machinery were damaged and put out of service.

The rotating structure of the after turret was lifted by the whiplash action of the hull and appeared to be jammed, but strenuous efforts by the crew put it back into operation within an hour. The severe shock caused by the mine explosion was best demonstrated by the tensile failure of hold-down bolts for the 150-mm turrets, which made them inoperative. Only the four single 150-mm mounts in the secondary battery remained operational. The shock loads were more severe than usual because the mine detonated very near to the hull in shallow water. Steering control was lost for 18 minutes due to a power failure. Repairs to the ship, made in a floating dock at Kiel, took from 6 to 21 May, after which the *Gneisenau* left for a shakedown cruise in the eastern Baltic. On 27 May she returned to Kiel ready for action.

Norway operations. The *Gneisenau* and *Scharnhorst* sailed on 4 June 1940, in company with the *Admiral Hipper* and four destroyers, to attack Allied naval forces and transports near Harstad, Norway. On 7 June the German ships joined with the tanker *Dithmarschen* so that the heavy cruiser and destroyers could be refuelled. At 0555 on 8 June, the *Admiral Hipper* located and sank a British escort trawler. The tanker *Oil Pioneer* was also set ablaze by the 105-mm guns of the *Gneisenau*, and a torpedo from a German destroyer sank the tanker around 0800. Reconnaissance planes were launched from the *Admiral Hipper* and *Gneisenau* to locate convoys. A cruiser and merchant ship were spotted to the south of the German fleet and a passenger ship and hospital ship to the north. The *Admiral Hipper* and destroyers were dispatched to intercept the passenger ship, the 19,500-ton *Orama*. This ship was sunk, and her distress signals were successfully jammed. The hospital ship, *Atlantis*, was not attacked.

After these actions, Admiral Marschall decided to detach the destroyers, which were low on fuel, and the *Admiral Hipper*, and sent them to refuel in Trondheim. Meanwhile, he would proceed with the two battlecruisers to the Harstad area.

At 1645 on 8 June, the aircraft carrier *Glorious* was sighted heading south at a range of 50,000 meters. The German battlecruisers were then heading north, with the carrier on the starboard beam and closer to the *Scharnhorst*, which opened fire first. The carrier was accompanied by the destroyers *Ardent* and *Acasta*, which attempted to lay a smokescreen when the Germans were sighted. The two battlecruisers altered course and quickly closed range. Although the *Glorious* was screened by smoke, the German

This stern quarter view of the *Gneisenau* in 1938 clearly shows the aircraft launch facility. A Heinkel-114 sesquiplane floatplane is stowed on the catapult over the hangar, while an Arado-95 float biplane is on the catapult on turret Caesar.

ships used their radar to assist the optical rangefinding to make hits that crippled the British carrier and later contributed to her sinking. At 1726 the battlecruisers closed range to 25,600 meters and both fired salvos. When the range dropped to 22,900 meters, they also fired on the destroyers with 150-mm guns, and then with 105-mm guns. Starting at 1738, three hits were made on the *Glorious* in rapid succession at a range of 23,000 meters. One shell demolished her bridge, another hit the flight deck, destroying some planes preparing for takeoff. At 1752 the *Glorious* was burning fiercely, but still moving at high speed.

The *Gneisenau* shifted fire to the *Acasta*, and the *Scharnhorst* engaged the *Ardent*. The *Acasta* dodged shells long enough to fire four torpedoes at 1830 before the *Gneisenau* was able to hit her. The *Gneisenau* evaded the torpedoes, but as the *Acasta* sank, one of her torpedoes hit the *Scharnhorst*'s stern, sharply reducing her speed. Around 1900 the *Glorious* capsized and sank. With the *Scharnhorst* seriously damaged, both German ships headed for Trondheim as quickly as possible.

On 8 June Admiral Marschall put to sea with the *Gneisenau, Admiral Hipper*, and four destroyers, but on 10 June returned to Trondheim when it became certain that the convoys were heavily guarded.

Torpedo damage. On 20 June the *Gneisenau, Admiral Hipper*, and four destroyers, under the command of Admiral Günther Lütjens, made a voyage into the Norwegian Sea towards Iceland. The German Naval Staff intended sending the larger ships as a feint breakout towards the Iceland-Faroes Passage. This would provide a diversion that would permit the *Scharnhorst* to return to Germany for repairs. Some 40 miles northwest of Halten with poor weather hampering visibility, the British submarine *Clyde* hit the *Gneisenau* with one torpedo that exploded on the starboard side, 16 meters from the stem and very close to the termination of the forward splinter belt. The splinter belt armor was unaffected, but the bow was severely damaged, to the point of its possible loss. The detonation, 3 meters below the waterline, affected shell plating and framing on both sides of the ship. The hole on the starboard side measured 15 meters across and 6 to 10 meters along the girth. A smaller hole was also made on the port side. A large crack developed in the shell plating at the after end of each hole. Significant flooding occurred in watertight compartments I and II, and speed was reduced due to the possibility of losing a good portion of the bow structure. To prevent this from happening, flat bars were welded across the cracks on both sides. A severe bow trim was corrected by counterflooding stern voids. The ship's fighting ability and her propulsion plant were not affected, and she returned to Trondheim without further mishap. A position report by the *Clyde* brought about an air attack by Coastal Command aircraft, in which none of the ships was damaged.

In Trondheim, workmen from the repair ship *Huascaran* completed welding steel bars over the torn shell plating below the waterline to reinforce temporary repairs. En route to Kiel on 25–27 July, she was escorted by the *Admiral Hipper, Nürnberg,* four destroyers, and six torpedo boats. Units of the British Home Fleet (*Renown, Repulse,* cruisers, and destroyers), were dispatched to intercept, but failed to make contact. The submarine *Swordfish* sank one of the torpedo boats, but the German task force was able to arrive at Kiel on 28 July without any further casualties.

The *Gneisenau* spent five months in Howaldt Werft in Kiel while her damaged bow was repaired. This was the yard's first job on a major combatant ship, and the workmen were more than generous in pleasing the crew. For example, they simply left out specified stanchions in the 'midships petty officers' quarters on the battery deck because they bothered the crew in that space. This was a serious omission that would have later consequences.

Atlantic operations. On 28 December 1940 the *Gneisenau* sailed with the *Scharnhorst,* under the command of Admiral Günther Lütjens, for operations against British shipping in the North Atlantic. A heavy gale in the North Sea forced both ships to reduce speed. The *Gneisenau* sustained heavy damage in the space where the stanchions had been left out of the petty officers' quarters. Shell frames and deck beams buckled under the severe impact loads, and there was some distortion in the main deck and forward

This starboard view of *Gneisenau* (post-summer 1939) shows the radar devices on the forward command tower and above the aft rangefinder. The stack has been lowered and a funnel cap added to keep the exhaust gases from the command tower.

shell plating. Various defects occurred in the fire-control system and the antiaircraft battery, caused by the ship's slamming in the heavy seas.

Although the *Scharnhorst* was undamaged, the ships were forced back—the *Scharnhorst* to Gotenhaven (now Polish Gdynia) and the *Gneisenau* to Kiel. Because of the extensive storm damage, changes in the *Gneisenau*'s bow structure were personally requested by her captain. Repairs were completed in record time by Marine Werft.

On 22 January 1941 the two ships, again under Admiral Lütjens, sailed from Kiel for the Atlantic. Their movement through the Kattegat and Skaggerak was detected, and the Home Fleet left Scapa Flow to cover the Iceland-Faroes passage. Later, the Germans were headed towards the position of the British force, but their radar detected large and small ships ahead, whereupon Admiral Lütjens ordered course reversed. The ships ahead were units of the Home Fleet (*Nelson, Rodney, Repulse*, eight cruisers, and 11 destroyers.) During the Germans' turning maneuver, the British light cruiser *Naiad* spotted the Germans about 5,000 meters abeam. She reported the sighting and maintained contact until the ships were lost in a squall. Admiral Sir John Tovey, commanding the Home Fleet, discontinued the search, assuming the *Naiad*'s sighting was erroneous.

After refueling east of Jan Mayen Island, the Germans headed south at high speed for the Denmark Strait and the Atlantic Ocean. Their operation in the Atlantic began at 0300 on 3 February, when they successfully eluded the last British cruiser patrol. To confuse British strategy and force a dispersion of their naval forces, the *Admiral Hipper* left Brest around 1 February for her second Atlantic operation. She was to operate against Gibraltar–Freetown convoys.

Early on 6 February the two battlecruisers refuelled from the tanker *Schlettstadt* south of Cape Farewell. The next day they began their search for shipping in such rough seas that a warrant officer on the *Gneisenau* suffered a fractured skull when he was hurled against a turret by the violent ship motions. The ships slowed in order to prevent any structural damage and to decrease the violent motions that were severely affecting the crew.

At 0835 on 8 February, the *Gneisenau* sighted the masts of an outbound convoy from Halifax, HX-106, and the two German battlecruisers prepared to attack the convoy. The *Scharnhorst* was sent ahead to attack from the north. At 0947, however, lookouts on the *Scharnhorst* spotted the old British battleship *Ramillies*, which was armed with eight 381-mm guns. When informed of the presence of the British battleship, Admiral Lütjens called off the attack, as his orders directed him not to engage enemy capital ships. The *Scharnhorst*, however, closed to 23,000 meters by 0950 in an attempt to lure the British ship away from the convoy so the *Gneisenau* could move in and attack it. Admiral Lütjens issued a direct order to the *Scharnhorst* to end this foray and join the flagship.

The two ships sailed northwest towards the Davis Straits, where they refueled from two tankers before heading south into the Atlantic Ocean. Heavy seas hampered the search for shipping. Ventilation cowls were damaged, preventing proper ventilation of the machinery spaces, and the antiaircraft batteries could not be properly manned. Speed was reduced to prevent the shipping of water and spray formation until the seas quieted down. On 22 February, about 500 miles east of Newfoundland, a convoy of empty cargo ships, heading west, was sighted. With the appearance of the German battlecruisers, these ships dispersed. The *Gneisenau* and *Scharnhorst* sank four of them and later sank a fifth ship for a total of 25,587 gross tons. The last ship sunk, the SS *Harlesdan*, transmitted a radio warning before her antenna was destroyed. This forced Admiral Lütjens to shift his ships to a new area of operations.

Admiral Lütjens next shifted operations southeastward in the North Atlantic to cover the Capetown-Gibraltar convoy route from a point north of the Cape Verde

This bow view of the *Gneisenau* was taken after her last bow modification. The number of anchors was reduced to two, and a chock was installed at the stem. This arrangement was another feature distinguishing the *Gneisenau* from the *Scharnhorst*. (Drüppel)

This 1941 view of the *Gneisenau* in the Atlantic shows some of the major modifications made to the ship since her commissioning. Note the removal of the catapult from turret Caesar, the new "Atlantic bow," the radar on top of the forward command tower, the enclosed secondary armament directors, and the deck-stowed anchor arrangement. (Drüppel)

The *Gneisenau* is shown engaging the British carrier *Glorious* off Norway on 8 June 1940. The deck-stowed anchor arrangement can also be seen in this photo.

Islands. Before reaching this position, he refueled in mid-Atlantic. On 8 March the British battleship *Malaya*, also armed with eight 381-mm guns, was sighted by lookouts on the *Scharnhorst*. Over the horizon were masts of at least twelve merchant ships. Admiral Lütjens ordered the *Scharnhorst* to leave the area and join the *Gneisenau*, some 40 miles away. In making this decision, Admiral Lütjens was probably guided by German experiences in the Battle of Jutland. *Ramillies* and *Malaya*, with speeds of 23–25 knots, were slower than the 30-knot battlecruisers, but high speed in battle is a fragile commodity. A single 381-mm shell hit could cancel out this advantage.

A dozen merchant ships from Convoy SL67, bound for Great Britain, were also sighted. To avoid action with the *Malaya* and other escorts, Admiral Lütjens decided to remain astern and guide U-boats to attack by radioing the convoy's position to Berlin. The German Naval Command then directed the U-124 and U-105 to the attack, and during the night of 8–9 March they sank a total of 28,488 tons. The Germans were sighted by a seaplane from a convoy escort, and that afternoon the *Malaya* closed to 24,000 meters, within range of the Germans' 283-mm guns, but again Admiral Lütjens considered a gunnery engagement contrary to his orders and decided to shift operations to the mid-Atlantic. While en route to a refueling position, the Germans sunk a solitary Greek steamer.

On 11–12 March the two battlecruisers refueled and replenished from the armed supply ships *Uckermark* (ex-*Altmark*) and *Ermland*. The same day all four ships turned north and formed a 120-mile-long scouting line, and on 15 March intercepted some tankers. The *Gneisenau* captured three ships, for a total of 20,139 tons, and sank a fourth. The German fleet continued north to hunt for ships on the Halifax convoy route.

Meanwhile, the *Rodney, Nelson,* and *King George V* were waiting in the North Atlantic for the German battlecruisers to head towards Germany or renew their attacks on the northern convoy routes.

Early on 15 March dispersed tankers from a convoy were sighted, and the *Gneisenau* sank seven ships totalling 26,693 tons, while the *Scharnhorst* sent six ships of 35,088 tons to the bottom. These ships were able to radio distress calls bringing the *Rodney* and *King George V* to the area. On 16 March the German ships continued action, sinking ten more ships. In her last action with convoy shipping, the *Gneisenau* opened fire on a small 1,800-tonner, the *Chilean Reefer,* which skillfully evaded her gunfire and accurately radioed her position, then turned on the battlecruiser with a single deck gun and also laid down a smokescreen. Such determined resistance convinced the Germans that she must be an auxiliary cruiser. Therefore, it was decided to finish her off with long-range gunfire—73 rounds of 283-mm shells—more than both ships had used on all other ships sunk during the entire cruise. This unusual action terminated when an unidentified large ship was picked up on the radarscope on the *Gneisenau.* The Germans made a high-speed escape with the help of a rain squall.

Admiral Lütjens headed for Brest, as both battlecruisers had definitely been sighted, and chances of further successes seemed small. The German Naval Command believed that their continued presence at sea would make it difficult for the *Admiral Hipper* and *Admiral Scheer,* at that time in the Atlantic, to return to Germany. Furthermore, the *Scharnhorst* was experiencing difficulties with her boilers. Late on 20 March, a Swordfish reconnaissance plane from the *Ark Royal* spotted the German ships about 700 miles from Brest. After being sighted, Admiral Lütjens headed the two ships north until the aircraft vanished over the horizon. Course was altered to the east and France. At this point the British carrier was only 160 miles away, but approaching darkness and bad weather made it impossible to launch a strike. On 21 March the RAF Coastal Command sighted the two ships less than 200 miles from Brest. On 22 March the battlecruisers reached the French port, despite their pursuit by the British Home Fleet. During the two-month cruise, the *Gneisenau* sank 22 ships of 115,662 tons, most of them in ballast. The actions of the *Scharnhorst* will be described later. The *Gneisenau* remained in the Brest Navy Yard for almost a year. The yard had been taken over and partially restored by workmen from the Wilhelmshaven Navy Yard. After the arrival of the two battlecruisers in Brest, the British stepped up their air raids on the port, using special 227-kilogram armor-piercing bombs. The nights of 30–31 March and 4–5 April saw particularly heavy air raids. During the latter raid a 227-kilogram bomb barely missed the ship and landed in 4 meters of water in the dry dock. It did not explode. The *Gneisenau* was moved from her dry dock and moored to a buoy in Brest Harbor.

Torpedo damage. Surrounded by torpedo nets and protected by the seawall, she was the target of an attack on 6 April by four Beaufort bombers equipped to launch torpedoes. Only one plane found the *Gneisenau* and attacked her. It was shot down, but it did manage to launch its torpedo. The torpedo ran straight and true and exploded 204 kilograms of TNT against the starboard side, just forward of the after turret. The ship listed two degrees to starboard and settled 1.2 meters by the stern as 3,050 metric tons of water flooded the damaged area. The explosion virtually demolished the shell plating over an area of some 210 square meters; the first longitudinal bulkhead, the armored torpedo bulkhead, and the transverse and light longitudinal bulkheads in the area were also destroyed. The inner bottom and the upper and middle platforms were deformed and damaged. The torpedo had struck a particularly vulnerable area where the outboard shaft passed through the lower portion of the side-protection system adjacent to the after turret.

The starboard shaft tunnel was caved in over a length of 36 meters. All the stuffing glands of the intact bulkheads in the stern were loosened, allowing extensive flooding in the shaft alleys, which made the centerline shaft inoperative. The starboard shaft bearings were severely damaged. Some auxiliary machinery and the main propulsion turbines in numbers 1 and 3 engine rooms were out of service due to the flooding. Some electrical equipment in the damaged areas had to be thoroughly overhauled because of salt-water damage.

The fuel and fresh-water tanks were 80 percent filled at the time of the attack. Fuel tanks in the explosion zone ruptured and a mixture of salt water and fuel oil poured into the central control station, the after gyro room, and the after gun-control station; some control equipment was put out of service. The magazines for the after main-battery turret were flooded by salt water only, which simplified the clean up. There was some shock damage to transformers, gun phones, optical gear, gyrocompasses, and other small electrical equipment.

As water spread through damaged stuffing glands, the crew vainly attempted to use nonporous material to keep them watertight. Water entered number 3 engine room through shaft stuffing glands, and two pumps were put into use there. A salvage tug pumped out salt water and assisted in controlling flooding in the four main compartments after watertight integrity had been restored.

Bomb damage. The *Gneisenau* was dry-docked for repairs, which took several months. During the night of 9–10 April, British bombers dropped approximately 25 tons of 227-kg semi-armor-piercing bombs on the Brest Navy Yard. The *Scharnhorst* escaped damage, but the *Gneisenau* sustained four hits on the starboard side of her forward superstructure, and near misses on the quay showered her with fragments and shell splinters. One bomb exploded on the lower armor deck, a dud hit the conning tower, the third exploded on the lower armor deck after piercing the superstructure plating and the upper and battery decks, and the fourth pierced the upper deck and buried itself in the debris on the battery deck that resulted from the third hit, but failed to explode. In the attack 72 men were killed and 90 others were injured (16 of these later died).

The main armored deck was slightly damaged. The outboard longitudinal bulkhead was bulged outward on the starboard side near turret Bruno, and its welded seams were torn open. The torn seam in the armored longitudinal bulkhead revealed a poor weld. Some barbette armor plates were splinter-damaged, others were warped by the concussion.

The antiaircraft control station and central automatic control station were partially flooded, but the propulsion plant and gun batteries were not affected.

The bomb explosion in the superstructure caused many fatalities and injuries, and damaged electrical circuits to equipment in the bow. Gyrocompasses, welding transformers, and forward living and medical spaces were all without electric power, and fire-control and interior communication systems were seriously affected. The bow section of the ship filled with blinding, toxic gas, hampering firefighting. The magazines for turret Bruno were flooded as a precautionary measure, but were pumped out after all fires were extinguished.

With the *Gneisenau* damaged, it was decided that in addition to the battle damage repairs and the overhaul of the *Scharnhorst*'s boilers, there would be further improvements to both ships, including the fire-control system and radar installations. Torpedo tubes would be installed. Additional 20-mm guns were installed—14 on the *Gneisenau* and 13 on the *Scharnhorst*. The action with the *Chilean Reefer* had convinced Admiral Lütjens that torpedoes were necessary to sink merchant ships, and six triple-mounted 533-mm torpedo tubes were installed forward of the after group of 150-mm guns, one

These three views of the *Gneisenau* were taken in Norway shortly after her being torpedoed by a British submarine on 20 June 1940. The bow was being held to the ship by the intact section between the main and second decks. However, due to her forward movement, some cracks developed in the intact portion of the damaged bow, necessitating flat bars to be welded over them so that the bow would not be lost. Four months were required to repair this damage. (Albrecht Schnarke Collection)

triple mounting, on either side of the ships. The catapult above the hangar was removed, and the entire hangar arrangement was changed. This meant that the repair period was extended, so that neither ship could participate in the scheduled April sortie of the *Bismarck* and *Prinz Eugen*. While her repairs were under way, the *Gneisenau* was subjected to periodic air raids by the Royal Air Force. A particularly heavy raid occurred on 24 July 1941, but no damage was done to the ship.

Operation Cerberus (the Channel Dash). When Germany and Italy declared war on the United States on 11 December 1941, there seemed to be very little chance for successful operations against merchant shipping in the Atlantic. Besides, the *Scharnhorst, Gneisenau,* and *Prinz Eugen* had had large crew turnovers, and the time needed for indoctrination and training would not permit an extensive Atlantic voyage. All three ships would have to refuel at sea if they were to head for Norway by way of the Denmark or Iceland-Faroes Straits, but almost all German supply ships in the Atlantic had been sunk after the loss of the *Bismarck.** Moreover, Hitler would no longer risk his large ships in Atlantic operations, since he was convinced that they were vital to the defense of Norway. With radar and aircraft reconnaissance becoming more effective, the navy proposed to Hitler that the three ships return to Germany through the English Channel—a most daring and risky venture. The German Naval Command foresaw the northern waters as a vital theater, especially with the entry of the United States into the war and the increase in convoys to Russia. Meanwhile, repairs and alterations to the *Gneisenau* were completed; however, British air raids on the French port were increasing in severity and accuracy. On 18 December the *Gneisenau* was slightly damaged by splinters from bomb hits on the dry dock. On 23 December she went alongside a jetty to test electronic gear, and by the end of the month she was ready for sea duty.

By the end of 1941 preliminary plans were being drawn up to return the *Scharnhorst, Gneisenau,* and *Prinz Eugen* to Germany for duty in Norway, which Hitler believed would be the "zone of destiny." The bomb damage sustained by the *Gneisenau* in Brest demonstrated that the German Luftwaffe could not provide the air cover that these ships required to maintain their operational status. This was demonstrated when at 2030 on 6 January 1942 a bomb fell and exploded between the ship's side and the dry dock wall. Splinter damage to the side shell around the waterline caused some flooding in the wing tanks of two main compartments. Damage to the ship's structure was light and was repaired in 10 days.

On 12 January 1942, the final decision to return the Brest squadron to Germany was made during a high-level conference in Hitler's headquarters with naval and air representatives present. Great secrecy was attached to the planning of Operation Cerberus, and some deception was planned to mislead the French populace on the destination of these ships. Toward the end of January, Vice Admiral Otto Ciliax was presented with the final detailed plan.

During the evening of 26 January, the *Gneisenau* got under way for engineering trials and gunnery practice that ended a few days later. She was ready for her voyage through the English Channel.

During the first few days of February, minesweeping of a passage through the English Channel commenced under the cover of darkness. The minesweeping operations were not detected by the British, but the passage of the destroyer squadron assigned to the *Tirpitz* was noted. The latter caused some concern that the Germans might be planning an Atlantic exercise with the Brest squadron.

Around 2300 on 11 February 1942, the Germans began one of the most spectacular and daring naval operations of World War II, Operation Cerberus (the Channel Dash). The departure of the *Gneisenau, Scharnhorst,* and *Prinz Eugen* was delayed about two hours by an air raid, but the ships entered the Channel just after midnight. At 27 knots they hugged the French coast and by 0630 had passed Cherbourg, where a flotilla of torpedo boats joined them. Luftwaffe fighter planes circled the ships at masthead height to avoid detection by British radar. Close cooperation between German ships and aircraft was ensured by General Adolf Galland, who assigned Luftwaffe officers aboard each large ship. Later, German planes jammed British radar with chaff. By 1300 the force had passed the cliffs of Dover without much serious opposition. At 1334 six Fairey Swordfish torpedo bombers with Spitfire cover attacked the formation. The heavy air cover distracted the Spitfires, and the slow torpedo bombers were left to other German

*Before the *Bismarck* was sunk, some secret codes had been removed from *U-100* before she sank. This code book enabled the Royal Navy to track down almost all the German supply ships at sea in May–June 1941.

The *Gneisenau* is shown at sea in 1941 during her Atlantic cruise with the *Scharnhorst*. Note the camouflage scheme. This was similar to that used in the *Bismarck*.

aircraft and the heavy and accurate antiaircraft fire from the formation. All six planes were destroyed. There were no torpedo hits.

When the *Scharnhorst* struck a mine and was forced to stop, the *Gneisenau* continued on with the *Prince Eugen* and five destroyers. At 1445 five twin-engine Whirlwind fighter-bombers attempted to attack, but were driven off by German fighters. During the next two hours several more air attacks were repulsed by fighters and antiaircraft fire. A total of 242 British aircraft were dispatched to attack the German ships, but only 39 could find them. Although no bomb hits were scored on any of the larger ships, two of the escorts sustained splinter damage and were forced to seek shelter.

At 1617 five British destroyers* made an unsuccessful torpedo attack at ranges from 3,700 down to 2,200 meters on the *Gneisenau* squadron. The *Gneisenau* engaged these ships and fired full 283-mm gun salvos at them; she was joined by the *Prinz Eugen*. One destroyer, the *Worcester*, closed to 2,200 meters to fire her torpedoes when she was heavily damaged by 203- and 283-mm shell hits. The destroyer was forced to stop and for a period of ten minutes presented her broadside to concentrated German shellfire. The *Gneisenau* had difficulty in observing her fall of shot as she was masked by her own smoke, caused by a strong following wind. The *Worcester* was heavily damaged, with

*This destroyer group was hastily assembled and consisted of two destroyers from the 16th Destroyer Flotilla in Harwich and four from the 21st Destroyer Flotilla in Sheerness. One destroyer developed bearing problems and had to be left behind. It should also be noted that the Home Fleet was moved from Scapa Flow to Hvalfjord, Iceland, to be in a better intercepting position if the German force in Brest attempted to reach Germany via the Atlantic Ocean. The *Victorious, King George V*, and *Rodney* were the major units at this time.

The *Gneisenau* is shown dry-docked in Brest after being torpedoed on 6 April 1941. Note that water is pouring from her damaged starboard side aft and the damaged crane to the left of the ship. Heavy air attacks on the Brest Navy Yard made it difficult for the Germans to keep repair schedules or keep necessary shipyard equipment functional.

most of her superstructure and masts shot away and several large holes in her hull from 283-mm shells, which in some cases simply passed right through the ship without exploding. The heavy overcast and rough seas prevented these British destroyers from attacking with decisive results. The *Scharnhorst*, under way again, trailed the main force by 23,000 meters, and at 1608 about 100 bombers (Hudsons and Beauforts) attacked the Germans without any success. At least five planes were shot down.

Mine damage. At 1806 one torpedo bomber flew through an intense barrage to attack the *Gneisenau*, but the torpedo was a surface runner and was easily avoided. As darkness set in, British Wellingtons bombed the formation for 30 minutes. The Germans fought them off and shot down several. The bombing attack failed to slow the German battlegroup, so the British aircraft began dropping mines in the Elbe Estuary and the approaches to the Kiel Canal.

A mine, dropped several days earlier, damaged the *Gneisenau* as she neared Terschelling, Holland, at 27 knots. The explosion at 1955, just forward of the after turret, produced small dents in the hull, tears in some welded seams, and cracks in the shell plating. The middle turbine lost power, and the captain ordered the others to be stopped. The starboard shaft alley was flooding and the shaft, slightly out of alignment, displaced a stuffing gland. Salt water began to trickle through; the list and trim that resulted from this flooding were insignificant. The leak was plugged with nonporous material, and pumps controlled the flooding. The shock of the explosion damaged navigation instruments, but the guns and propulsion plant were intact. Within 30 minutes the ship was under way, but at a slow speed. At 0350 the *Gneisenau*, with two destroyers, reached the Heligoland anchorage.

The bold and successful return of the three large ships was made possible by the cooperation of the Luftwaffe, which supplied 252 fighters for the operation. It was one of the few times in the war that the German Navy and Air Force worked so well together.

Bomb damage—Kiel. The *Gneisenau* departed from the Heligoland anchorage in company with the *Prinz Eugen* to head for the Kiel Canal. The entrance of the canal was blocked by severe ice, so that the two ships were forced to stop in Brunsbüttelkoog. In maneuvering into her berth, a strong current swept the *Gneisenau* around, causing her stern to strike a submerged wreck. A hole was punched in the hull below the waterline at a point just forward of the starboard shaft's intersection with the shell plating. This caused some flooding into the starboard shaft alley, and the maneuvering of the ship in the heavy ice also damaged the propeller blades, necessitating their replacement.

The *Gneisenau* finally steamed through the Kiel Canal and was repaired at the naval dockyard there. Work was completed by 26 February 1942 and the ship was scheduled to depart for Norway on 6 March. She had taken on ammunition and was ready for dock

The *Gneisenau*, above, and the *Scharnhorst* are shown in the Brest Navy Yard in December 1941 under heavy camouflage. Under constant surveillance and attack by the Royal Air Force, the Germans were looking for a way for these ships to leave Brest and be used in other theaters of war. Note that turret Bruno of the *Gneisenau* is trained to starboard while the turrets of the *Scharnhorst* are trained in.

trials when British bombers attacked the night of 26–27 February. One 454-kilogram general purpose bomb struck the upper deck at an angle of 20 degrees from the vertical between the right-hand and centerline guns of the forward turret, about 1.5 meters off the centerline. The bomb exploded in the opening of a small uncovered ventilator, and a red hot splinter ignited some powder charges in the magazine. The resulting explosion severely damaged the turret and magazine, and fire from the burning powder spread to a ready-service magazine. Pressure caused by the deflagration blew off the turret top. The whole revolving part of the turret substructure to the gun platform was lifted 0.5 meters, falling back onto the rotating track when the pressure created by the burning powder diminished. This was discovered in a subsequent investigation of the damage, when it was noted that a crew station that was free from obstructing the turret was damaged. Further inspection showed that the roller bearings were no longer inside the spur gears in the stationary part of the turret, but resting on them. The shells in the magazines did not explode, but fire broke out in adjacent fuel tanks. The ventilation system in this section of the ship was completely destroyed, the pneumatic armor hatch was ruptured, and the armor plates on the turret were ruined. Intense heat from fuel oil and 230 burning powder cartridges destroyed the structural properties of the shell and armor plating in the vicinity of turret Anton, and this necessitated the complete renewal of the side shell and armor in the damaged area. The turret was gutted by fire, and a devastating explosion was averted by partially flooding the shell magazine.

Outside of the turret and magazines the lower armor deck was deformed; the substructure and plating of the upper deck was forced outward; the longitudinal torpedo bulkhead had torn seams and a puffout; a weak connection of the torpedo bulkhead failed. All adjacent decks, platforms, and bulkheads were damaged, deformed, or ruptured. The forward part of the ship, over a length of 42 meters, was flooded by fire-fighting water.

The propulsion plant was not affected, but the anchor windlass, overboard discharge pumps, fire pumps, and other electrical equipment lost power due to the flooding. The electrical system was partially restored by the crew before the ship underwent further repairs. The fire pumps, which might have extinguished the magazine fire, were the most serious loss.

The explosion killed 112 men and injured 21 others. Repairs involved rebuilding turret Anton and the magazines as well as replacing large sections of shell plating, face-hardened armor plate, and bulkheads. This work, estimated to require as much as two years to accomplish, would mean the loss of the *Gneisenau* to the war effort, a severe blow to the German Navy.

Gneisenau Reconstruction. Since the damage was so extensive and repairs would take so long, a study was initiated to determine whether it would be practical to convert the main battery to six 380-mm guns. Such a conversion had been planned earlier, but put aside due to the war, and because of objections due to the resulting decrease in freeboard and increased bow trim. The turret exchange would have increased the draft and resulted in an unsatisfactory trim situation if the hull form forward remained unaltered. With the forward section of the ship so badly damaged, a new situation was created. Studies revealed that the armament change was feasible if the bow was lengthened to decrease the trim and draft. Detailed studies in 1942 on the feasibility of carrying six 380-mm/47-caliber guns in three turrets showed the following:

- Electric power requirements would be considerably greater.
- Barbettes could adequately support turrets for the heavier guns provided the substructure were reinforced.
- Special turrets that would fit the existing barbettes would have to be designed and then built.
- Ammunition-handling and fire control would have to be altered.
- The hull would have to be lengthened to provide additional buoyancy and a shift in the longitudinal center of buoyancy.

These would be major alterations. Higher electrical loads would require new cables. If the same reserve electrical power of 100 percent was to be maintained, the generating plant would have to be re-engineered to increase its output. Nevertheless, the changes were authorized. Krupp designed and built the turrets that would meet the requirements. The armor for one turret had been completed when the cancellation of the program was announced in February. The guns and turret machinery were delivered to the shipyard in Gydnia, but they were later used for coastal defense artillery (see photograph on page 190. Also see photograph in Appendix C for disposition of old turrets and 283-mm guns).

The preponderant addition of weight would have been forward, and the resulting bow trim would have been reduced by lengthening the bow section by 10 meters. The new bow would have had no bulbous form and resembled the foresections of the proposed O-class battlecruisers. The changed hull form and lengthened waterline alleviated the trim and draft problems that would otherwise have resulted from the proposed conversion. The longitudinal center of buoyancy would be shifted forward, reducing troublesome trims under full-load conditions. The increased hull length would provide additional volume, but no advantage was taken of this to improve the fuel capacity because of the trim problem. The new bow form was designed to ship less water than it had before.

The armament changes principally involved the main battery, but a study was made of the secondary battery. The Germans had recently developed a successful dual-purpose 127-mm gun, superior to the dual-purpose 105-mm weapons already installed. The changeover from 105- to 127-mm would involve considerable work and time, and as the army and Luftwaffe had priorities for the new 127-mm guns, the decision was made to retain the 150-mm/105-mm secondary armament. The Channel Dash had demonstrated that antiaircraft defense was an important factor in modern warship operations, so additional 20-mm guns were provided for.

The German Naval Staff approved the reconstruction of the *Gneisenau* and her conversion to mount 380-mm guns once the studies had concluded in early 1942. The necessary floating dry dock was to be moved to Gdynia as soon as ice conditions permitted, then the ship would be taken there for repairs. The reconstruction and rearmament plan provided for:

- The removal of turret Anton prior to sending the ship to Gdynia.
- The replacement of the old bow section with a new one of reinforced construction 10 meters longer.
- The replacement of the nine 283-mm guns in triple turrets with six 380-mm guns in twin turrets.

On 24 March 1942 Admiral Raeder visited the ship and told the crew she was to have a stronger armament and become a more powerful warship. On 4 April the ship left for Gdynia, accompanied by the icebreaker *Castor* and the predreadnought *Schlesien*, and made the trip in two days despite ice. At Gdynia the bow was cut off at frame 185.7, and some deck and side armor and portions of the torpedo bulkhead adjacent to turret Anton were removed. The other turrets were also dismantled.

The *Gneisenau* is shown at Gdynia (Gotenhafen) shortly after the end of the war in Europe. (Drüppel) After having her foreship heavily damaged in an air raid on 26 February 1941 at Kiel, some 30 meters of her bow were cut off and the ship towed to Gotenhafen to be converted to carry six 380-mm guns.

It was estimated that the gun change would require 75,000 man days and the new bow would require another 45,000, on the basis of a work force of 300 men, with a possible increase to 500. For general repairs, yard workmen assisted the crew on a limited basis. Necessary material was to be purchased from outside contractors. Krupp was to supply the new turrets, different from those in the *Bismarck* and *Tirpitz* in that they could rotate within the existing barbettes without obstructions. Guns and turrets were available, but interior fittings, turret accessories, and fire-control equipment had to be manufactured.

There was some concern about the overhang of the substituted turrets, especially that of turret Bruno, but this problem was resolved by ensuring that the turret overhang completely cleared the superstructure. The substructure for the turrets was checked by the Deutsche Werke Shipyard in Kiel to determine the need for additional structure to properly support a 380-mm turret. The shipyard found that it would be necessary to strengthen the supporting foundations for the 283-mm barbettes.

All was ready by the specified delivery dates. The ship was ready for the new turrets and bow section by early 1943, when Hitler, angered by the lack of success a German fleet had in attacking a convoy to Russia on 31 December 1942, decided to scrap all German capital ships and cruisers. This brought all work to a stop, and necessary material was diverted to other urgent work. Finally, the 283-mm guns were removed from the yard; guns for turret Anton went to Denmark, and turrets Bruno and Caesar and their guns went to Norway.

Final Disposition — Gneisenau. On 1 July 1942 most of the crew had been transferred from the ship, leaving only a small detachment to man antiaircraft guns and do routine maintenance. A skeleton force remained on board to learn about the new 380-mm guns until 2 February 1943 when all work officially ceased upon receipt of a directive by Admiral Dönitz to lay up the German battleships and cruisers. The *Gneisenau* would be disarmed and some of her equipment removed. For the remainder of the war, the ship remained a useless hulk in Gdynia. Finally, as the Russians advanced on the city, the *Gneisenau* was sunk on 23 March 1945 as a block ship. The Polish authorities ordered the hulk removed, and between 1947 and 1951 it was salvaged and scrapped.

It has been stated frequently, but incorrectly, that the *Scharnhorst* and *Gneisenau* were initially designed for 380-mm turrets. The diameters of the barbettes and roller-bearing tracks for the 283-mm triple turrets were nearly the same as those for the 380-mm twin turrets. The first investigation of the possibility of converting the main armament to the larger caliber was made in 1935, when it was determined that such a conversion was feasible.

The new *Gneisenau* would have differed considerably from her sister ship if the proposed work had been completed. With her main battery of six 380-mm guns, she would have been reminiscent of the British *Renown* and *Repulse*. A tripod mast was to have been installed between the hangar and the after conning tower and in fact had been prefabricated at Kiel, but because of time limitations in getting the ship to Norway, it was not fitted during the yard availability period before the damage occurred. The revised characteristics of the *Gneisenau* are tabulated in table 3-4.

Another important insight on the importance of the conversion of the main armament can be appreciated when the results of prewar tests on the target ship *Hannover* with 283-mm and 380-mm armor-piercing projectiles in the engine and boiler rooms

TABLE 3-4

Characteristics of Reconstructed Gneisenau (1942)

Standard displacement*	32,980 tons (33,510 mt)
Full-load displacement*	40,080 tons (40,720 mt)
Waterline length	774.3′ (236 m)
Draft (full load)	32.0′ (9.75 m)
Armament	Six 15″/47 twin (380 mm)
	Twelve 5.9″/55 paired single (150 mm)
	Fourteen 4.1″/65 paired (105 mm)
	Sixteen 1.4″/83 paired (37 mm)
	Twenty-eight 20-mm machine guns in quadruple and single mounts

*Authors' estimates

are analyzed. In the case of the engine room, the detonation of a 283-mm shell would have caused the engine room to be abandoned because of ruptures in the steam lines. The main machinery was damaged, but it could be repaired during a yard overhaul. In the case of the 380-mm projectile, the main engine was completely destroyed, requiring a new installation. Based on these tests and the serious bomb damage sustained during the February 1942 air raid on Kiel, it is easy to understand why the Germans decided to convert the *Gneisenau* to 380-mm guns at this time in her career.

The conversion would have increased the displacement by approximately 1,200 tons, but the longer hull would have diminished the draft, restored the freeboard sought during the preliminary designs, and brought the longitudinal center of buoyancy more in line with that of the center of gravity, thereby eliminating the bow trim.

Scharnhorst — Operational History. The *Scharnhorst* was destined to become one of the most famous German surface ships of World War II. Her keel was laid in the Wilhelmshaven Navy Yard on 16 May 1935; she was launched on 3 October 1936, and commissioned on 7 January 1939. Her launching was a spectacular affair and was attended by Adolf Hitler. Construction was delayed because the yard was also busy with ship repair and upkeep, and late delivery of vendor equipment, such as turbo-generators, held up some other work. Preliminary trials in mid-1939 indicated that adjustments and alterations were necessary in some systems and equipment, such as the new boilers, and that problems would exist because of the low freeboard and bow trim. In August 1939 an aircraft hangar was added and the bow was modified as the result of the trials. On 2 September 1939 the *Scharnhorst* made a brief trial run to check out such adjustments and added new features. It was found that some work had to be accomplished on the superheater tubes, as they were not functioning properly.

The ship was ready for combat operations in October and, as described earlier, sailed on 21 November with the *Gneisenau* for a sweep into the Iceland-Faroes passages. On that operation she sank the armed British merchant cruiser *Rawalpindi*. During this engagement, the *Scharnhorst* sustained a 152-mm shell hit on her stern. There were some casualties and minor splinter damage. Making skillful use of bad weather, the ships returned to Wilhelmshaven for repairs to structural damage caused by heavy seas and battle damage from her action with the British auxiliary cruiser. During this yard availability period, she also had her boilers overhauled.

Norwegian operations. In January 1940 the *Scharnhorst* was sent to the Baltic Sea for training and gunnery practice. Ice-bound there until February, she returned to Wilhelmshaven on 5 February for operations in the North Sea. On 7 April she left Wilhelmshaven with the *Gneisenau* to be part of the main covering force for the invasion of Narvik and Trondheim. Bad weather — low clouds and frequent rain and snow — made visibility very poor. At 0430 on 9 April the *Gneisenau* reported a radar contact, and both ships went to battle stations. About 0500 the navigator of the *Scharnhorst* spotted a flash of heavy gunfire in his sextant mirror. In the subsequent engagement with the *Renown*, the *Scharnhorst*'s radar malfunctioned, and she could not track the target. She came under fire from the *Renown* briefly at 0518, but repeated course changes allowed her to escape undamaged. By 0715 the German battlecruisers had outdistanced their pursuers. However, the *Scharnhorst*'s turret Anton was put out of action by heavy seas that cascaded over the bows and into the turret through the cartridge ejection scuttles, rangefinder gear, and the gun bloomers. The ammunition

The *Scharnhorst* is shown under construction in Wilhelmshaven in 1938. Note the pedestal for the catapult and the foundation for the after main-battery rangefinder. Also note the opening over the amidships machinery space. (Bilddienst)

These two prewar views of the *Scharnhorst* show the vessel before a major overhaul during the summer of 1939. The lower photo shows the *Scharnhorst* firing a salute for a visiting dignitary on 1 April 1939. Note the absence of gun bloomers on the turrets. These were added later after problems with boarding seas caused short circuits in the turrets' machinery. (Drüppel)

hoist motor was short-circuited by seawater. When the *Scharnhorst* increased speed to the maximum possible, the starboard turbine had to be stopped, which slowed the two ships to 25 knots. Both ships, after joining the *Admiral Hipper*, reached Wilhelmshaven on 12 April, and the *Scharnhorst* underwent an extensive overhaul of her guns and propulsion plant.

Again ready for action, the two battlecruisers left Wilhelmshaven on 4 June, in company with the *Admiral Hipper* and four destroyers. Their foray against merchant shipping and their action with the *Glorious* has been described earlier. In that action the *Scharnhorst* fired some 212 shells. During the battle with the three British ships, the destroyer *Acasta* crossed the bows of the *Scharnhorst* from port to starboard and launched a spread of four torpedoes at extreme range before reversing course. The *Scharnhorst* altered course to comb these torpedoes but then too quickly resumed her original course to continue firing with her 152-mm guns on the destroyer, now on her port side. At 1839 one torpedo hit the starboard side near turret Caesar — the most vulnerable area of the ship — at an angle of 15 degrees from the horizontal plane of the ship and 3 meters below the main side belt.

Torpedo damage. The damage was severe. The turret and magazines filled with smoke, gun crews were evacuated, and magazine flooding was ordered but canceled when no danger of fire was reported. The shell plating, which offered enough impact resistance to activate the warhead detonator, bore the brunt of the detonation, and a section 6 by 14 meters was destroyed. The explosion was deep enough so that a major portion of its energy was vented into the ship, where it tore the torpedo bulkhead from the armor deck and bent its top edge inboard 1.7 meters. That bulkhead was damaged from the level of the side armor to a platform above the shaft alley, for 10 meters. Two transverse bulkheads, the battery deck, and the first platform deck were damaged. The armor shelf and some adjacent structure were slightly damaged.

The torpedo struck at a point where the propeller shafts passed through the torpedo bulkhead, which had had to be knuckled to fit in place and accordingly had reduced ability to deflect elastically. Also, there was an inadequate connection of the torpedo bulkhead to the armor deck, and the structure was not continuous, so that stress flow could not transfer to adjoining structure. The bulkhead began to deflect elastically, as designed, but the upper end connection failed and permitted extensive flooding of inboard compartments. As a result of this damage, four of the 22 main watertight compartments had some flooding; 30 spaces in the area took on some 2,500 tons of water and 48 men perished. The *Scharnhorst* listed 3 degrees to starboard and was down 3 meters by the stern.

The propulsion plant was seriously affected by flooding and damage. The starboard shaft, which passed through the lower part of the underwater side-protection system abreast of turret Caesar, was destroyed, and the shaft alley began to flood immediately. A seaman was trapped there, and when another man opened a watertight door in a rescue attempt, the after engine room, which supplied power to the centerline shaft, began flooding so rapidly that it was impossible to properly secure the plant. One of the turbines, under maximum load, cooled so quickly that the housing came in contact with the turbine blades, and it had to be stopped. All steam connections were shut off in this space. With the starboard engine room also secured, the ship had only one shaft in operation.

Turret Caesar was out of action. Some compartments below the magazine were flooded, and electrical and other equipment in the magazine was damaged. Some cartridges and a few powder cases burned; many were damaged. The cartridge magazine contained 283 projectiles and cartridges combined, ready to be fed to the turret above;

other projectiles, without cartridges, were on a loading platform a few meters from the impact area.

The starboard after 150-mm twin turret was put out of action by flooding of subturret compartments and damage to the electrical system. The fire-control system for the after group of 105-mm guns was damaged.

The ship was limited to a maximum speed of 20 knots en route to Trondheim. Collision mats were rigged in an attempt to prevent further structural damage, but could not be secured, and the attempt was abandoned. The afternoon of 9 June both ships reached Trondheim, where the repair ship *Huaskaran* was moored. On 10 June an RAF Coastal Command plane spotted the *Scharnhorst*. On 11 June a dozen Hudson bombers dropped thirty-six 113-kilogram semi-armor-piercing bombs on the *Scharnhorst* from an altitude of 4,570 meters, but all missed.

The battleship *Nelson* and the aircraft carrier *Ark Royal* were then dispatched; the latter, when 170 miles from Trondheim on 13 June, launched 15 Skua bombers. They were intercepted by German fighters, and eight of them were shot down; the others attacked the *Scharnhorst*, but only one bomb hit was made and that one, although it penetrated the upper deck, failed to explode.

The propulsion turbine for the centerline shaft was repaired in 10 days. The damage to the starboard shaft was believed to be serious and could only be surveyed in a dry dock; since it was feared that the shaft had been severed by the explosion, the propeller was lashed to the hull. By 20 June two shafts were in operation and the ship sailed for Germany, under escort, making 24 knots. On 21 June RAF Coastal Command planes spotted them off the Isle of Utsire, and around 1500 six Swordfish torpedo planes attacked, but were easily repulsed by antiaircraft fire. At 1630 nine Beauforts attacked with 227-kilogram armor-piercing bombs, but were also driven off by antiaircraft fire and German fighters. In these attacks the ship expended 900 rounds of 105-mm and 3,600 rounds of machine-gun ammunition. When German interception of British radio messages revealed that much of the British Home Fleet was at sea, the *Scharnhorst* was ordered into Stavanger; some British ships had closed to within 35 miles of her position when that decision was made. On 22 June she left Stavanger for Kiel, where the next six months were spent in making repairs, after which she ran trials in the Baltic.

Atlantic operations. Early in December 1940, the *Gneisenau* and *Scharnhorst* went to Kiel to prepare for a breakout into the North Atlantic, but storm damage forced the *Gneisenau* to return to Kiel and the *Scharnhorst* to Gdynia. On 22 January 1941, as related earlier, both ships put to sea. Although sighted by the cruiser *Naiad*, they eluded British search and reached the Atlantic. A convoy was sighted on 8 February, but the appearance of the battleship *Ramillies* forced the Germans to call off their attack. They refueled south of Greenland on 16 February, and on 22 February sighted another convoy, out of which the *Scharnhorst* sank the 6,000-ton tanker *Lustrous*. The two ships replenished near the Azores and then headed for the Cape Verde Islands. On 8 March the *Scharnhorst* sighted the battleship *Malaya* and, two hours later, 12 merchantmen, but avoided action. Two days later they headed north after refuelling and replenishing from the *Uckermark* and *Ermland*. On the way to the replenishment area, the *Scharnhorst* sank the Greek cargo ship *Marathon*. After refueling on 12 March, all four ships turned north.

On 15 March a dispersed convoy was encountered, and the *Scharnhorst* sank two ships. Several days later she located the main body and sank seven more ships for a total of 27,277 tons. Both battlecruisers reached Brest on 22 March.

During that operation the *Scharnhorst* had constant trouble with defective boiler superheater tubes, which could have terminated the cruise many times, but hard work

This wartime view of the *Scharnhorst* in Kiel shows the new "Atlantic bow." Note the anchor now stowed at the stem and the large number of portholes. The latter had to be equipped with special watertight closures because of the large increase in draft. (Albrecht Schnarke Collection)

by the crew prevented any major breakdowns. The shipyard estimated that at least ten weeks would be required to re-tube the superheaters. Damage was of a purely technical nature — the metal was not adequate for the operating temperature. Repairs required for both ships made it certain that they could not participate in the *Bismarck* operation, which had been planned for April but was later postponed to May.

Boiler repairs to the *Scharnhorst* were completed in July, even though the ship was under frequent air attack, and she went to La Pallice on 21 July for trials, during which she made 30 knots with no difficulties. Before returning to Brest, the *Scharnhorst* was anchored at La Pallice. There, on 24 July, several squadrons of Handley Page Halifaxes bombed from altitudes of 3,000 to 3,700 meters. Five bombs hit the starboard side simultaneously in a nearly straight line parallel to the centerline. Two bombs were of the 227-kilogram high-explosive type, the others were 454-kilogram semi-armor-piercing type bombs.

Bomb damage. One of the 227-kilogram bombs hit abeam of the conning tower, just forward of the starboard 150-mm twin turret. It passed through the upper and middle decks before exploding on the armor deck, which remained intact. The first platform deck was torn, with significant bulging in the explosion area. The side-armor plating was deflected outboard about 200 mm, and a small hole was torn in it. Rivets that joined the armored torpedo bulkhead to the main deck were loosened enough to cause leakage.

Ammunition for the 150-mm guns, stowed about 3 meters from the center of the explosion, was not affected. Splinter damage was insignificant.

A 454-kilogram bomb hit the port side between the 100-mm and the 150-mm guns, 3.5 meters from the deck edge, and penetrated the upper deck, lower armor deck, and first platform before being deflected downward along the torpedo bulkhead and out through the double bottom without exploding. The bottom plating was holed and local flooding occurred. The wing tanks had their restraining walls holed by splinters. Number 4 generator room was flooded, several electrical installations were put out of action, and cables, damaged by splinters or flooding, disrupted operations in the battle, command, and fire-control stations, including those for the forward antiaircraft battery and turret Anton.

A second 454-kilogram bomb hit midway between the 150-mm and 105-mm guns, 2.6 meters from the deck edge; it, too, penetrated all decks and platforms before passing through the side shell below the armor belt without exploding. Five spaces on the starboard side over a length of 10 meters were flooded. Some lights were extinguished, water leaked into the magazines for the 150-mm single mounts, and the living spaces were damaged by splinters.

The third 454-kilogram bomb hit slightly abaft the after turret, 3 meters from the deck edge, tore through the upper deck, passed through the side plating, and buried itself in the sea bed, unexploded; it was later recovered. The shell plating was severely damaged, and 10 watertight spaces, including the starboard shaft alley, were flooded. Flooding also occurred in the magazines for turret Caesar, and the ammunition hoist was put out of service.

The other 227-kilogram bomb fell forward of the after turret, to starboard, 3 meters from the deck edge; it penetrated two decks and exploded on the main armor deck, where it made a small hole. Several frames were holed by splinters, and the connection at the top of the torpedo bulkhead was damaged. The penetrated decks bulged from the explosion and were holed by splinters. Some flooding occurred in the outboard spaces. Heating, potable, and plumbing piping under the battery and middle decks was damaged. The ammunition hoists for the 37-mm guns were put out of action, although the ammunition was not affected.

The ship took an 8-degree list to starboard, as most of the void tanks used for counterflooding were flooded. Damage would have been more extensive if all three 454-kilogram bombs had not been duds. Trim by the stern increased 3 meters due to 1,520 to 3,050 metric tons of water taken on board. The forward and after turrets were temporarily out of action, and half the antiaircraft battery was out. Several small fires broke out but were extinguished. Two men were killed and 15 others were injured.

Quick damage-control action corrected trim and list, and steam was raised in record time. Draft was increased by one meter, but the ship was able to get under way for Brest by 1930, at a speed of 25 knots. At dawn, a British patrol plane was spotted and shot down by an escorting destroyer. When the *Scharnhorst* reached Brest on 25 July, the only visible sign of damage was her excessive draft, which put her stern portholes awash. Four months were spent in repairing the damage. Changes were also made to the ship, which included a new radar aft, increased output for the forward radar to 100 kw, and triple 533-mm torpedo tubes between the after 105-mm mounts and 150-mm turrets on either beam.

Operation Cerberus (Channel Dash). Early in 1942 the *Scharnhorst* made preparations, described earlier, for Operation Cerberus. Leaving Brest under cover of darkness on 11 February, the ship saw no action until after she cleared the English Channel. Just east of Dover, she fired on one of a flight of six Swordfish torpedo planes at a range of

The *Scharnhorst* and *Gneisenau* are shown during training maneuvers in the Baltic. The *Gneisenau* is in the lead. (Drüppel)

approximately 1,000 meters. The plane came down about 100 meters off the port side. It had released its torpedo, which the ship evaded by a turn. At 1531 an air-dropped magnetic mine, dropped several days ahead of the operation, detonated 30 meters to port, abreast of turret Bruno and 38 meters below the surface. The electrical system failed, due to damage to circuit breakers, darkening all interior spaces for 20 minutes, and the turbines could not be stopped immediately because loss of power knocked out emergency switches to the boilers and turbines.

While the *Scharnhorst* was immobile, Vice Admiral Ciliax shifted his flag to the destroyer Z-29. During the transfer, the rising sea caused the destroyer to surge against the larger ship, tearing off part of her bridge wing and leaving it hanging on the battlecruiser's superstructure. The detonation of the mine had caused a large gash and deep dishing in the starboard shell plating abeam of turret Bruno. Some 1,220 metric tons of water flooded 30 watertight spaces within the limits of five main watertight compartments — a distance of 40 meters — and the ship listed one degree to port while taking a one-meter trim by the bow.

Shock damage was moderately severe. Turret Bruno was temporarily jammed, and its main electric motor was seriously damaged. The twin 150-mm turret and the single 150-mm mount, on the port side forward, were also jammed. Number 2 105-mm mount was damaged. Some transformers and fire-control equipment were destroyed. Foundations had not been designed for such shock loadings, as evidenced by the failure of bearings in the fuel oil pumps and turbo-generators, which forced the ship to a complete stop. All turbo-generators, except those in generator room 4, were out of commission

This port view of the *Scharnhorst* in 1941 shows a camouflage scheme designed to confuse enemy gunner estimates of her heading. (Drüppel)

due to bearing failures. The after gyrocompass was out of service for a short while, as were the direction finder and the echo sounder. As a result of the whiplash response to the non-contact underwater explosion, cracks, possibly from poor welds, developed in the keel and the bottom shell forward of turret Anton. A crack also developed at the deck hawse pipe casting on the starboard side, possibly due to a casting flaw that the shock response finally exposed.

The mine detonated at 1531. The first turbine was restarted at 1549, the second at 1555, and the third at 1601, allowing a speed of 27 knots on three shafts. Shortly before the ship was under way again, a twin-engined bomber dropped a stick of bombs about 90 meters off her port side. No damage was sustained. After she was under way, a dozen Bristol Beauforts staged a 10-minute attack that failed due to antiaircraft fire, Luftwaffe fighter support, and evasive maneuvering. Next, the ship evaded a torpedo launched by a torpedo bomber that attacked from the stern. Several more attacks were made before the action ceased; antiaircraft guns were red hot, one 20-mm gun barrel burst, and several others had jammed training gears.

Mine damage. By 1800 the *Scharnhorst* was off the coast of Holland. At 1916 high-flying aircraft dropped bombs astern. At 2234 a magnetic mine explosion to starboard in some 24 meters of water resulted in a brief loss of steering control. The gyrocompass and lighting systems were out for about two minutes. All turbines had to be stopped. Two were jammed; the turbines for the starboard shaft had only minor difficulties.

The starboard shell plating was dished and cracked near the after 105-mm guns. Ten spaces in four main watertight compartments flooded with an additional 300 metric tons of water. Steam-pipe bracing in the starboard engine room failed, and due to concern over possible leaks, the boilers for that turbine were shut down. The foundation bolts for the outboard shaft bearings failed under combined shear and tensile loads, making it necessary to steam on only the centerline shaft at 10 knots, until partial power was restored on the starboard shaft and speed was increased to 14 knots. The port list resulting from the first explosion was corrected, but the draft was increased. The starboard turbine and centerline engine room were in operation again by 2311, but the port turbine would require shipyard repairs. There was additional damage to the electrical system, switches were inoperative for 30 minutes due to damage to the automatic equipment, generator room 5 was out of action, and generator room 2 had difficulty in providing a steady electric current.

The main battery turret machinery sustained some light damage from a shock response; all mounts and turrets had some superficial damage to their rotating parts. Three 150-mm turrets were seriously jammed; the 105-mm mounts were partially damaged.

Shortly after 2300 the *Scharnhorst* was under way again with partial power on the starboard shaft. By 0800 on 13 February she was delayed by ice in the Jade bight, where Admiral Ciliax once again shifted his flag back aboard. By noon she reached Wilhelmshaven, whereas the *Prinz Eugen* and *Gneisenau* went on to Brunsbüttelkoog. The *Scharnhorst* was dry-docked in the Wilhelmshaven Navy Yard for inspection of underwater damage caused by the mine detonations. It was found, however, that this was not serious enough to keep her in Wilhelmshaven for an extended period of time. Since Wilhelmshaven was closer to British air bases, this would mean a greater chance of air attack and possible bomb damage than if the ship was at Kiel. Therefore, the ship remained in Wilhelmshaven only two days before going to Kiel for permanent repairs to the mine damage.

It was quite unusual for Germany to have two capital ships under repair at the same time in the same shipyard, as only one battleship (the *Tirpitz*) and two battlecruisers were operational. A contributory factor was the extreme pressure exercised by Grand Admiral Erich Raeder, who stressed the necessity for speed in completing the repairs and getting the two ships to Norway. All Construction Office personnel concerned with the repairs were sent to Kiel to expedite the work. The German Coastal Command had stated that its antiaircraft defense in the area around Kiel was more than adequate to defend the ships. Nonetheless, the *Gneisenau* was seriously damaged by a single armor-piercing bomb dropped during an intensive raid on 26 February 1942. The *Scharnhorst* escaped damage during that attack. Due to the menace of air attacks the floating dry dock, in which she was being repaired, was towed from the yard and anchored off Kiel until antiaircraft defenses were strengthened and more fighter protection was available. This delayed the repair work, but by July 1942 the ship was ready for trials. Although the trials proved successful, the ship was detained in Kiel for boiler repairs. Several of the boilers needed new boiler tubes. The guns, radar, and other equipment were considered to be satisfactory.

During early August the *Scharnhorst* was conducting exercises with a number of U-boats from Kiel when she collided with U-523. There was some minor damage that forced the battlecruiser to be dry-docked for repairs. By September the ship was again operational and began training exercises in the Baltic. In late October she steamed to Gotenhafen for a new rudder of a design based on lessons learned from the torpedoing of the *Prinz Eugen* and *Lützow*. There were some troubles with the boilers and turbines throughout 1942, which prevented her from being sent to Norway. By December 1942 only two shafts were operable, which cut speed to 25 knots. A thorough overhaul of the propulsion plant was considered mandatory. Early in January 1943, the propulsion plant was in working order, and trials were conducted satisfactorily. The *Scharnhorst* was pronounced ready for reassignment to Norway. On 7 January 1943 the *Scharnhorst, Prinz Eugen*, and five destroyers headed for Norway, but were forced back when heavy activity was reported at British airfields along the coast of Scotland and England.* On

*By this time Ultra was decoding German cipher traffic and had intercepted the order deploying the *Scharnhorst* to Norway. Sufficient warnings were given to the Coastal Command, and appropriate air measures were taken that forced the Germans to return the *Scharnhorst* to port twice. Bad weather made the third attempt possible.

While in Brest the two battlecruisers were frequently attacked by British aircraft. The *Scharnhorst* is shown here in late 1941 under heavy camouflage. The Germans also used an old French warship disguised as the *Scharnhorst* and *Gneisenau* to confuse British bomber and reconnaissance planes. (Albrecht Schnarke Collection)

23 January the *Scharnhorst* had to turn back again for the same reason. While preparing for the next sortie to Norway on 10 February, she ran aground while avoiding a collision with a U-boat. Repairs to a hole in her bow put her in dry dock from 24–26 February.

Norwegian operations. On 8 March 1943 the *Scharnhorst* left Gotenhafen and sailed for Norway in company with four destroyers. By 11 March they were in the Norwegian Sea where they encountered a severe storm off Bergen. The destroyers were forced to seek shelter. The *Scharnhorst* continued on, her speed reduced to 17 knots until the gale moderated, when the speed was brought up to 28 knots. Port side gun positions sustained some storm damage. At 1600 on 14 March the *Scharnhorst* anchored in Bogen Bay, opposite Narvik, and on 22 March she steamed in company with the *Tirpitz* and *Lützow* to Alten Fjord where the storm damage was repaired. In early April the *Scharnhorst* and *Tirpitz* conducted a training exercise in the Arctic Ocean and steamed towards Bear Island in company with nine destroyers. During the afternoon of 8 April a serious accidental explosion in the after auxiliary machinery space above the armor deck killed or injured 34 men. The magazines of turret Caesar were flooded as a precaution. With the assistance of a repair ship, the damage was repaired in 14 days.

For almost six months the ship was inactive except for brief training exercises. Fuel restrictions hampered any major fleet operations. In June 1943 a British and Norwegian force seized the German weather station at Spitzbergen. During the summer, the German Navy planned a countermove using the *Scharnhorst* and *Tirpitz*. The use of the two capital ships was considered superfluous for such an operation, but it would provide an opportunity for the two ships to work together and increase the morale of the crews, which was low due to their inactivity. On 8 September 1943, in company with the *Tirpitz* and 10 destroyers, the *Scharnhorst* bombarded land installations at Sveagruva, Spitzbergen, where she knocked out a battery of two 76-mm guns with one salvo.

The two ships shelled buildings, coal mines, fuel tanks, gun emplacements, and loading facilities. Some 1,000 troops were also landed, but the Norwegian garrison escaped into the mountains. By 1100 the, *Scharnhorst* had completed her mission.

Later that month, the *Tirpitz* was attacked in Kaa Fjord and severely damaged by British midget submarines. Two of these craft had been assigned to attack the *Scharnhorst*, which, absent from her anchorage, was moored off the island of Aaroy awaiting a practice shoot at a drogue towed by an aircraft. With the *Tirpitz* immobilized, the Arctic Task Force was limited to the *Scharnhorst* and five destroyers.

Scharnhorst's last battle. The German Army in Russia was in a desperate situation in the latter half of 1943. After the Allied convoys to Murmansk resumed, on 20 December Hitler approved sending the *Scharnhorst* against the next convoy. Admiral Erich Bey, in temporary command of thc task force, was advised on 22 December by Admiral Karl Dönitz to be ready to go to sea on 3-hour notice.

That same day a German aircraft sighted a convoy of 20 merchant ships and tankers, guarded by cruisers and destroyers, some 400 miles west of Tromsø, Norway. The convoy, spotted again two days later north of Norway, was definitely headed for the Soviet Union. At 0900 on Christmas Day a U-boat reported the convoy's position, and Admiral Dönitz ordered it intercepted. His orders to Admiral Bey included the following:

- The engagement will be broken off at your discretion. In principle, you should break off on the appearance of strong enemy forces.
- The tactical situation must be exploited with skill and daring.
- The fight must not end in a stalemate. The opportunity must be seized to attack in force. The superior firepower of the *Scharnhorst* affords the best chance of success, and hers must be the main contribution. The destroyers should be employed later.
- Inform all the crews accordingly. I have full confidence in your offensive spirit.

The orders were contradictory, as they exhorted Admiral Bey to attack in force, yet required him to disengage in the face of superior enemy forces. Admiral Bey planned to attack the convoy around 1000 on 26 December if the weather and visibility conditions, as well as information on enemy forces, were favorable. With six hours of twilight and only 45 minutes of full daylight, there was little time for action.

The force got under way about 1900 and cleared the Norwegian coast around 2300. Admiral Bey was in constant touch with Naval Group North and finally at 0319 the German Fleet Command conveyed the decision of the German Admiralty that in the event that heavy weather should hamper the destroyers' ability for action, the *Scharnhorst* should carry on the attack alone. The British were able to decode German Navy codes during this operation. At the same time Admiral Bey was reading his orders, the British Admirals Burnett and Fraser were also looking at their English translations. At 0703 on 26 December, the German battlegroup, some 40 miles southwest of Bear Island, changed course towards a point where they could intercept the convoy at twilight, which would occur around 1000. The destroyers scouted 10 miles to the southwest of the *Scharnhorst*. All ships were at battle stations from 0300 on. In the moderate gale, the destroyers began to labor heavily, and speed had to be reduced to 10 knots.

Shell damage (cruiser action). Convoy JWBB was screened by the cruisers *Sheffield*, *Belfast*, and *Norfolk*. Admiral Burnett positioned his force of three cruisers between the convoy and the possible approach course of the *Scharnhorst*. Meanwhile, Admiral

This is a view of the main deck on the starboard side showing the forward 150-mm twin turret and turret Bruno, trained aft. (Albrecht Schnarke Collection)

Fraser, who was in the *Duke of York*, had left Iceland on 23 December, accompanied by the cruiser *Jamaica* and 4 destroyers. Admiral Fraser's force was 270 miles west of North Cape when he received the Ultra intercept that the *Scharnhorst* was to intercept the convoy.* Admiral Fraser at once made preparations to cut off the *Scharnhorst*'s return to Norway. By 0925 his ships were 125 miles southwest of the German battlecruiser's position. Admiral Burnett was advised of the plans of Admiral Fraser and told of the Ultra intercept.

The close British covering forces were at battle stations when the *Belfast* made radar contact with the *Scharnhorst* at a range of 33,000 meters at 0840 on 26 December. At this point the German battlecruiser was approximately 60,000 meters from the convoy. All three British cruisers continued to track the approach of the German ship. In the meantime, the *Scharnhorst* was unaware of the presence of these British ships, since her radar had been secured by orders of her admiral to prevent possible detection. At 0921 the *Sheffield*'s lookouts sighted the *Scharnhorst* at a range of 11,000 meters; three minutes later the *Belfast* opened fire at a range of 8,600 meters with star shells. About 0925, the *Norfolk* opened fire, her shells falling about 500 meters from the *Scharnhorst*. The German battlecruiser responded with a salvo from turret Caesar, increased speed to 30 knots, and altered course to disengage the British cruisers. During this 20-minute encounter, the *Scharnhorst* took two 203-mm hits. One shell struck the upper deck

*British intelligence was able to decode the order (Ostfront — 1700 hours) for the *Scharnhorst* to sail. Admiral Fraser was advised that the battlecruiser was going to sea and he could make advanced preparations.

This 1938 view of the *Gneisenau* shows her 283-mm and 150-mm guns trained to port.

This is a starboard view of the main deck of the *Scharnhorst* from a position abeam of the after main-battery rangefinder. Note the after starboard 150-mm twin turret, the single-mount 150-mm gun, and the 105-mm dual-purpose gun on the deck above. A salvo has been fired from one of the forward turrets. (Drüppel)

These two views of the *Scharnhorst* show the problem that these ships had with seakeeping despite the changes in bow freeboard and deck-stowed anchors. (U.S. Navy and Drüppel)

This is a prewar view of a twin 105-mm gun mount at the breech end during firing practice. The absence of shields made the crew very vulnerable to aircraft strafing attacks. Also, four gun mounts were fed from a centrally positioned hoist—another vulnerable feature.

between the forward single 150-mm mount and the torpedo tubes. The unexploded projectile landed in a berthing space in watertight compartment IX and started a small fire that was quickly extinguished. Several minutes later, another shell struck the forward rangefinders and showered exposed antiaircraft gun crews with splinters. The forward radar antenna was destroyed, and splinters entered the radar receiving space, killing all personnel there. Radar search in the forward sector was sharply curtailed. The aft radar, much lower in the ship, had only a limited radar sweep forward, being blind over an arc of 60–80 degrees.

In a desperate attempt to disengage, the *Scharnhorst* changed course several times to evade the three British cruisers. At 0955 Admiral Bey radioed that he was in action with British cruisers, but minutes later he was able to successfully disengage. The British ships could only make 24 knots in the heavy seas. With a 4–6 knot speed advantage, the German ship was able to outdistance her pursuers. The *Belfast* and *Sheffield* were using flashless powder, while the *Norfolk* was not. Also, the British were using their gunnery radars, and these factors placed the German battlecruiser at a slight disadvantage, even though she had superiority in gun power and speed. Having succeeded in breaking off the gunnery duel, the *Scharnhorst* continued her search for the convoy. By 1200 the battlecruiser was north and east of the convoy, and five minutes later the *Belfast* reestablished radar contact. It was not until 1221, however, that the British cruisers could close the range. During this interval, the British ships were detected by the aft radar set on the *Scharnhorst* and by her lookouts. The British fired illuminating shells, but the *Scharnhorst* quickly opened fire from her forward turrets and changed course. As she turned, the after turret went into action. This turning maneuver prevented British destroyers from making a torpedo attack.

At 1223 the *Norfolk* was hit abreast of her after funnel. A few seconds later, another 283-mm shell struck the barbette of her "X" turret, which was instantly put out of action. The magazines for that turret were flooded as a precautionary measure. The first hit on the *Norfolk* was quite serious. The 283-mm shell pierced the starboard superstructure and exploded against the port side plating, where a large hole was ripped open. Splinters disabled the entire radar installation, making it impossible to continue accurate fire. A few minutes later the *Sheffield* was showered with large splinters. At 1241, as the situation began to look bleak for the British, the *Scharnhorst* altered course and increased speed. Admiral Bey thought by steering a course away from these cruisers, he might encounter the convoy. With the action now broken off and Admiral Fraser coming up from the southwest, the British cruisers kept out of range, but maintained radar contact while sending position reports.

The German destroyers had observed the star shells in the morning action, but were far from the *Scharnhorst*'s position. Admiral Bey, however, ordered them to head northeast and rejoin the *Scharnhorst*. At 1158, he had ordered the destroyers to proceed west once again and search for the elusive convoy. After this, there was to be no further tactical coordination between the destroyers and the *Scharnhorst*. Unknown to the German destroyers, they passed 15,000 meters to the south of the convoy around 1300. Finally, at 1343 Admiral Bey ordered them to cease their search and return to base. They returned to Kaa Fjord at 1000 on 27 December, and their absence during the final stages of the Battle of North Cape would prove to be fatal to the *Scharnhorst*.

The senior surviving crew member of the *Scharnhorst*, a petty officer named Goeddes, whose battle station was on the bridge, described the action with the cruisers as follows:

> Shortly after 1230, I and several others sighted three shadows ahead and reported it accordingly. The alarm had already been sounded as a result of a previous radar report. Before our guns could open fire, however, star shells were bursting over the *Scharnhorst*. The enemy's salvos were falling pretty close to the ship. The first salvos from our own 283-mm guns straddled the target. I myself observed that after three or four salvos, a large fire broke out on one of the cruisers near her after funnel, while another cruiser was burning fiercely fore and aft and was enveloped in thick smoke.
>
> After further salvos I saw that the third cruiser had been hit in the bows. For a moment, a huge tongue of flame shot up and then went out. From the dense smoke that enveloped her, I presumed that the ship was on fire. The enemy's gunfire then began to become irregular, and when we altered course, the enemy cruisers turned away and disappeared in the rain and snow squalls. During this action the enemy had been ahead and visible on both sides. Our turrets Anton and Bruno had been firing at these cruisers and were joined in only shortly by the two forward 150-mm turrets. I did not hear either by telephone or through any other source of any hit received during this phase of action by the *Scharnhorst*. While the enemy had been scarcely discernible during the first encounter, this time with the midday twilight, we could easily distinguish the cruisers' outlines. The range, too, was much shorter than it had been during the morning action.*

Action with the Duke of York. Around 1315 Admiral Bey decided to return to base, as no further action was anticipated. Since the crew on the battlecruiser had not eaten since morning, they were fed in relays. A battle readiness condition still existed. The aft radar set was turned off to prevent detection by enemy forces. Unknown to the Germans, the *Scharnhorst* would have to cross the course of the *Duke of York, Jamaica*, and four destroyers that were steaming towards the German ship's position, guided by

*Quoted from "Battleship *Scharnhorst*" by Vulliez and Mordal.

This is a new 20-mm quadruple mount aboard the German heavy cruiser *Prinz Eugen*. This was a very successful mounting capable of a high volume of fire and was used in all large German warships.

the three cruisers' radio reports. The *Scharnhorst* with her forward radar set destroyed and the aft set not in a good position to search the forward sector, was heading into a trap from which there would be no escape. At 1617, the *Duke of York* made radar contact with the German battlecruiser at a range of 42,500 meters and continued to monitor her progress until she was at a range close enough for accurate shellfire.

Gunnery engagement (afternoon action). At 1643 Admiral Fraser in the *Duke of York* ordered the *Belfast* to illuminate the *Scharnhorst* with star shells. When the *Belfast* fired those shells at 1647, to the surprise of the British they revealed that the *Scharnhorst* had her main battery trained in! At 1650 the *Duke of York* began salvo fire at a range of 11,000 meters. The *Jamaica* followed at 1652 at a range of 12,000 meters, straddling the *Scharnhorst* on her third broadside. Although the German ship seemed taken by surprise when the star shells illuminated her, she quickly returned the fire. The duel between the *Duke of York* and the German ship was very unequal because the 283-mm shells could not penetrate the vitals of the British battleship. At 1655, a 356-mm shell struck the *Scharnhorst*'s starboard side abreast of turret Anton. The turret jammed, with the training and elevating gears out of action. The magazines caught fire from incandescent splinters, which also penetrated the fireproof bulkhead into the magazines for turret Bruno. Both magazines were flooded; turret Bruno's magazines were drained so rapidly that there was little effect on its rate of fire. Magazine crews worked waist deep in freezing water while attempting to save undamaged ammunition and serve the guns. The ship maintained her speed despite the damage, which had silenced a third of her main battery.

Another shell hit damaged the ventilation trunk to turret Bruno, causing it to fill with choking and blinding fumes each time the breeches were opened.

Still another shell hit abreast of turret Caesar and penetrated the battery deck, making a hole 0.5 meters in diameter. This was later patched, but the spaces in which it detonated were flooded and not drained. The two aircraft were riddled, and splinters wiped out many of the crew in nearby exposed locations.

Around 1800 a shell hit the starboard side (figure 3-1), passed through the light upper citadel belt and battery deck, ricocheted at the lower armor deck, penetrated into the raised portion of the lower armor deck over the boilers, and exploded in number 1 boiler room. This was first thought to be a torpedo hit because of the destruction that it caused. Many steam mains were ruptured, as there were four boilers in the fireroom. Shell fragments tore into the double bottom, causing flooding up to the level of the floor plates. The ship slowed to a speed of 8 knots. Damage control measures were prompt and effective, but as the watertight doors and hatches were dogged down, 25 men were trapped in the boiler room. Steam pressure was soon raised and the chief engineer telephoned the bridge stating, "I can maintain 22 knots," to which Captain Hintze responded, "Bravo, keep it up!" The *Scharnhorst* opened the range from 15,500 to 20,000 meters, then straddled the *Duke of York* with several salvos. Shell splinters came aboard the *Duke of York*, and one struck the foremast, temporarily disabling the fire-control radar and gouging away part of one mast support.

The gun duel continued for almost 90 minutes, and the *Scharnhorst* absorbed considerable punishment. Her topsides had been holed by shell splinters, and some of the superstructure had been wrecked by direct hits from 152-mm, 203-mm, and 356-mm shells. Fires broke out, some followed by explosions. One fire broke out in the hangar, and the two floatplanes were destroyed. This was extinguished in 10 minutes, but an attempt to launch the plane on the catapult met with failure as the compressed air main had been destroyed. Almost all gun positions and the port torpedo tubes were destroyed or put out of action. The gun crews that survived were ordered to take cover while the fires were being fought. At 1730 the two forward 150-mm turrets were struck by 356-mm shells. The starboard turret was demolished, and the entire turret and magazine crews were killed. The port turret was jammed. Ten minutes later it was permanently disabled.

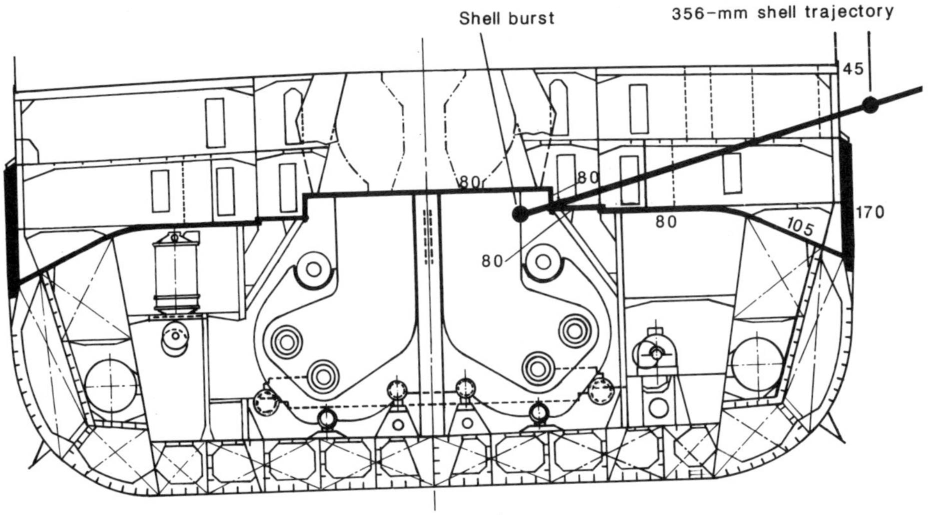

Figure 3-1. A 356-mm shell hit on the Scharnhorst.

The absence of aircraft carriers in the German Navy made it necessary for German battleships and battlecruisers to carry a large complement of aircraft. Two aircraft could be carried in the hangar with their wings folded, while another would be stowed on the catapult above the hangar. Two airplane/boat cranes abeam the hangar would retrieve the aircraft from the sea or hoist them to the catapult for launching. The other two aircraft are suspended from cranes, while the aircraft on the lower left is poised for a takeoff. Note the large number of men involved in the operation. (Drüppel)

The torpedo officer braved the heavy shellfire to reach the port torpedo tubes before they were put out of action. He managed to train these outboard and fire two of the torpedoes. The third became jammed in the tube. It appears from the fragmentary evidence available that this officer was killed by the shell splinters whizzing around or by a shell that probably detonated the jammed torpedo in the launcher.

A shell hit in the forecastle severed the anchor chain for the starboard anchor. The anchor and its remaining shot of chain fell into the sea. Another hit severed the forward anchor chain, and the anchor and chain fell into the sea.

At 1842 the *Duke of York* ceased fire, having fired 52 salvos for 31 straddles and at least 13 hits. These shell hits and those from the cruisers had caused a number of casualties aboard the *Scharnhorst*. By this time most of the 150-mm guns were out of action. The *Scharnhorst*, however, was still making good speed, and Admiral Fraser, afraid that she might escape, ordered destroyers to torpedo her. The German battlecruiser had no defense against such attacks, and the destroyers made their approach to close range without opposition. The *Stord* and *Scorpion* fired eight torpedoes around 1850. The *Scharnhorst* swung to starboard to comb the tracks, but took three confirmed hits. In turning, she also exposed her broadside to the *Savage* and *Saumarez*. The former fired eight torpedoes. The latter managed to close to 1,600 meters where she was taken under fire by some of the lighter guns on the starboard side of the *Scharnhorst*. Several

The *Scharnhorst* in the foreground and the *Tirpitz* in the background at anchor before the Spitzbergen raid in September 1943. Admiral Dönitz believed that the Spitzbergen operation would give the ship crews some offensive action as well as the opportunity to work together. (Bilddienst)

shells went through the destroyer's director and under the rangefinder, splinters pierced the side plates and superstructure, and the destroyer's speed was finally cut to 10 knots, but she did launch four torpedoes.

Torpedo damage. Definite information on torpedo hits during this phase of the battle is fragmentary. One torpedo exploded abreast of turret Bruno, which caused its elevating and training gears to jam. The main entrance hatch to the turret was jammed, and it took some time for the crew to escape. The magazine began to flood. Another torpedo struck outboard of a port boiler room, and some flooding occurred inboard of the torpedo bulkhead. A third struck aft on the port side where several compartments were already flooded, and the shaft was damaged. A fourth torpedo struck the bow on the port side. All these torpedoes were armed with 340-kilogram charges.

Considering the torpedo damage to the *Gneisenau* and *Scharnhorst* by surface and aerial attacks in earlier actions, it seems reasonable to assume that the *Scharnhorst* must have had substantial inboard flooding from these hits. Evidently, the torpedo hit in way of turret Bruno was devastating. Not only did it generate a severe shock response, but the inboard flooding indicates that the torpedo defense system was breached at this point in the hull. This system was quite vulnerable in this area, as it narrowed due to the fine hull lines. The 340-kilogram charges of these British torpedoes were also sufficient to defeat the torpedo protection outboard of the port boiler room. Damage there might have been more extensive if water had been present in the void layer of the torpedo defense system. The weak connection and the structural discontinuities probably contributed to the extensive flooding.

As a result of the torpedo damage, the *Scharnhorst* was slowed to 12 knots, although the chief engineer reported that she could still maintain 22 knots. The *Duke of York* was now able to close to 9,100 meters—point-blank range for her 356-mm guns—and the final assault against the *Scharnhorst* began. The German battlecruiser had two turrets jammed, and the third was low on ammunition. All available men (probably the crews of the 4.1-inch guns) were put to work transferring ammunition from turret Anton to turret Caesar, and several minutes later that turret resumed fire.

As the *Scharnhorst* took on more water, she slowed to 5 knots and her steering became erratic. The British had learned from the action with the *Bismarck*, and the same seemed to be true in the present engagement with the *Scharnhorst*, that it was impossible to sink a German capital ship by gunfire alone, so Admiral Fraser decided to sink her with torpedoes. The *Jamaica* fired three torpedoes at 1925, after having expended 22 broadsides. Two minutes later the *Belfast* fired three torpedoes, and ten minutes later she fired three more from a final range of 3,500 meters. No hits could be seen in the smoke and haze. This ended the cruiser action, but destroyers made two more attacks. The *Musketeer* closed to 900 meters and fired torpedoes at 1933, followed by the *Matchless* minutes later. The *Opportune* fired from a range of 2,300 meters, and lookouts clearly observed two hits on the starboard side between the mainmast and funnel, but their effect would have been diminished because the *Scharnhorst* was very deep in the water by this time, and the detonations probably took place against the main side belt. The *Virago* fired seven torpedoes at 1934 from a range of 2,000 meters. At the conclusion of the torpedo attacks by the destroyers, the *Scharnhorst* was almost dead in the water and shrouded by dense smoke and haze.

Scharnhorst sinks. Around 1900 the captain of the *Scharnhorst* ordered all confidential papers burned. As all other guns were silenced, he told the crew of number 4 150-mm turret, ". . . everything depends upon you."

The ship was listing to starboard and down by the bow. The only other gun still firing was a 20-mm machine gun atop turret Bruno. The 150-mm turret fired until the list jammed its ammunition hoist. By 1940 the ship was listing heavily, and the bow was deep in the water. All hatches and watertight doors had been secured to contain flooding and allow more time for the crew to escape. The torpedo damage, however, had eliminated all but a fraction of the reserve buoyancy. At 1945 the bow dipped under and the *Scharnhorst* then capsized to starboard. She sank with her propellers slowly turning.

The *Scharnhorst* in Norway in early 1943. This ship was sent to Norway to join the *Tirpitz* as a battle squadron to attack Russia-bound convoys. Suspension of the convoys to Russia in the spring and summer of 1943 left these two German capital ships idle. (Bilddienst)

After she disappeared sharp rumbles were heard underwater. Of the crew of 1,968 officers and enlisted men, only 36 enlisted men survived.

Admiral Fraser was deeply touched by the heroic action of the *Scharnhorst*'s crew. Later that evening he told the officers of the *Duke of York*:

> Gentlemen, the battle against the *Scharnhorst* has ended in victory for us. I hope that any of you who are ever called upon to lead a ship into action against an opponent many times superior, will command your ship as gallantly as the *Scharnhorst* was commanded today.

On her return voyage to Scapa Flow, when the *Duke of York* passed the position where the battlecruiser sank, Admiral Fraser ordered a wreath dropped. The *Scharnhorst* was the last German capital ship in an offensive action; her loss ended the threat of the German surface fleet and gravely affected the German position in Norway.

The loss of the *Scharnhorst* can be attributed to the lack of a destroyer escort and the superiority the Allies had attained in radar by 1944. After the war, Admiral Karl Dönitz wrote:

> . . . An operation by the battlecruiser *Scharnhorst* with a destroyer group in December, after a successfully concealed start, seemed to have good prospects of success in view of the enemy dispositions and weather conditions. It proved to be a failure, apparently through a misjudgment of the local situation, and the *Scharnhorst* was lost. . . .

Unfortunately, as we know now, the operation was not concealed from the start, as Ultra had breached the German codes. The British commanders of the two battlegroups were quite well informed of the *Scharnhorst*'s planned movements, and under such conditions they were able to plan their actions in advance of battle.

Armament. The principal characteristics for the guns of the *Gneisenau* and *Scharnhorst* are summarized in table 3-5.

TABLE 3-5
Gun Characteristics

Gun	11.1" (283 mm)/54.5	5.9" (150 mm)/55
Shell weight	727.5 lbs (330 kgs)	99.8 lbs (45.3 kgs)
Muzzle velocity	2,920 fps (890 mps)	2,871 fps (875 mps)
Maximum range	46,749 yds (42,500 m)	24,060 yds (22,000 m) 25,153 yds (23,000 m)*
Elevation — maximum	40°	35° — single 40° — twin*
Gun	4.1" (105 mm)/65	1.46" (37 mm)/83
Shell weight	33.3 lbs (15.1 kgs)	1.64 lbs (0.745 kgs)
Muzzle velocity	2,952 fps (900 mps)	3,281 fps (1,000 mps)
Maximum range	19,357 yds (17,700 m)	7,382 yds (6,750 m)
Elevation — maximum	85°	80°
Gun	0.79" (20 mm)/65	
Shell weight	0.291 lbs (0.132 kgs)	
Muzzle velocity	2,952 fps (900 mps)	
Maximum range	5,249 yds (4,800 m)	
Elevation — maximum	90°	

*These are figures for the twin-mounted guns.

Main battery. The 283-mm SKC-34 gun for the *Scharnhorst* and *Gneisenau* was an improved gun over that provided the *Deutschland*-class armored ships. The projectile for the newer guns was longer, but carried less explosive than that for the earlier SKC-28 model. Important here is that the extra weight of the new shell was concentrated mostly in the cap and windshield, which made it more potent in piercing armor plate. Table 3-6 is a comparison of the shell properties of the 283-mm guns of the *Scharnhorst*- and *Deutschland*-class ships.

TABLE 3-6

Comparison of the SKC-28 and SKC-34 283-mm Projectiles

Gun	*Type projectile*	*Length (calibers)*	*Total weight*	*Weight of cap*	*Weight of explosive*
SKC-28	High explosive	4.2	661.4 lbs (300 kg)	—	51.4 lbs (23.3)
	Common	4.2	661.4 lbs (300 kg)	—	37.3 lbs (16.9)
	Armor-piercing	3.7	661.4 lbs (300 kg)	70.4 lbs (32.0 kg)	17.3 lbs (7.8)
SKC-34	High-explosive	4.4	694.5 lbs (315 kg)	—	48.1 lbs (21.8)
	Common	4.4	727.5 lbs (330 kg)	—	35.3 lbs (16.0)
	Armor-piercing	4.4	727.5 lbs (330 kg)	98.4 lbs (44.7 kg)	14.6 lbs (6.6)

The Germans used three types of shells for their ships. The properties and intent of these were as follows:

- Armor-piercing—primary use against strongly armored targets. The projectile had a small amount of explosive and the base fuze was of the long-delay type.
- Common—these projectiles contained slightly more explosive and thus were capable of increased fragmentation damage. They were used only when the maximum armor of the target was such that it could be penetrated by these projectiles.
- High-explosive—these shells were used when fragment effect was desired, for example, against exposed personnel, antiaircraft positions, fire-control instruments, searchlights, etc. These shells were also to be used against lightly protected ships up to and including destroyers.

This philosophy on the use of projectiles and the type of ammunition carried in the magazines was carried through on all the main-battery guns of German capital ships of the World War II era.

Ballistics of the 283-mm guns made them effective against the new French battlecruiser *Dunkerque* at normal battle ranges, as shown in table 3-7.

TABLE 3-7

Armor Penetration Data - 11.1" (283 mm) Guns

Yards	*Side*	*Deck*
0	23.79" (604 mm)	
8,640 (7,900 m)	18.09" (460 mm)	0.76" (19 mm)
16,514 (15,100 m)	13.18" (335 mm)	1.63" (41 mm)
20,013 (18,288 m)	11.47" (291 mm)	1.87" (48 mm)

Due to displacement limitations, a weight-efficient distribution of the main armament was considered necessary for success in an encounter with a *Dunkerque*-class opponent. The three-triple-turret arrangement was finally chosen, because the turrets for the fourth and fifth armored ships were already designed and material procured, and more barrels could be arranged in triple turrets. The use of available material was necessary because German steel production had also to meet the needs of the army and air force. Although former German battleships and battlecruisers had always mounted four twin turrets for better fore-and-aft grouping of firepower, that preferred distribution had to be sacrificed due to weight and material-procurement considerations.

Ammunition-handling arrangements were the same as in the armored ships, with some modifications to handle the larger shell. Both classes of ships had the same Model 1928 turrets, which speeded the design and construction of the *Gneisenau* and *Scharnhorst*.

During the design period, Hitler completely rejected Admiral Raeder's proposal to increase main-battery caliber to 380 mm. However, after the Anglo-German Treaty had been signed and the French Navy had commenced the design of the *Richelieu*, Hitler authorized the increase. In *Mein Kampf* he was very critical of the armament in some German World War I capital ships. He now felt that these battlecruisers were also weakly armed, and contracts were renegotiated to provide for a later rearming with 380-mm/47-caliber Model 1934 guns. This change was scheduled for the winter of 1940–41.

In 1935–36, construction was well advanced, and the armament change would have seriously delayed building and delivery, as work would have had to halt while the redesign was accomplished. In addition, there would have been delays due to the development of special 380-mm turrets and mountings for these ships. Some bulkheads would have to be shifted, and small modifications would have been necessary in the magazine stowage. It was decided that the re-gunning would take place as soon as the turrets were ready. One major problem in the conversion was the effect on the draft and trim, which already exceeded design limits. The re-gunning would further aggravate this problem.

After several studies of the draft problem had been made, it was decided to increase the beam, at the same time moving the side armor belt outboard. New frame plans were reviewed, with special care to see that the curvature of the frames remained constant in the region of the armor belt, since reconstruction on cemented armor plate was impossible. The plans were set aside when war came, but were revived after the *Gneisenau* was badly damaged at Kiel on 26 February 1942.

Secondary battery. The secondary armament was chosen on the basis of two factors: eight single mounts were available from the equipment and material slated for the fourth and fifth armored ships, and a new twin turret for 150-mm guns had been developed by the Rheinmetall-Borsig concern. Therefore, these ships were to feature an unusual combination of four single- and eight twin-mounted guns. The 150-mm gun had been the standard caliber used on most German capital ships and cruisers since the turn of the century.

The four single-mounted guns, located abreast of the funnel, two on each side, were grouped together with a centrally located ammunition supply and had very light, 25-mm armor, sufficient only for splinter protection. Anything heavier might have compromised the operation of the mounts and would have meant more weight in a ship already weight critical. These guns had questionable usefulness in battle—they were soon silenced in the *Scharnhorst*'s engagement with British warships on 26 December 1943. Fire control presented a problem when the single- and twin-mounted guns had to work together. This was due to the differences in their rates of fire. Night action with star shells was equally difficult.

An early war photograph of the *Gneisenau* showing her post-conversion changes at Deutsche Werke, Kiel. This photo also shows the second modification to the funnel, where the inclined hood on the funnel was narrower and had a top plate without visible steam-pipe sockets. (Bilddienst)

The four twin turrets eliminated the relatively exposed operation of the 150-mm single mounts and gave gun crews more armor protection and a better ammunition supply. As the turrets were power-driven, it was possible to provide them with some protection, and their front plates had modest armor. With such protection and equipment, each turret had a rotating weight of approximately 126 tons.

Primary defense against destroyers and cruisers was to be the twelve 150-mm guns, which could elevate 40 degrees and be depressed 10 degrees. Rate of fire was six rounds per minute, good against surface ships. Rear Admiral Karl Witzell, chief of the Marine Waffenamt (Bureau of Ordnance), wrote the authors concerning the problem of choosing the 150-mm armament:

> We had to cope with an adversary possessing a great quantitative superiority in destroyers, and accordingly needed medium artillery that could be used rapidly and effectively against the largest destroyers and also the mass attacks of small-sized destroyers. Therefore, on the basis of thorough proving-ground testing and test shell fire on target ships, it was concluded that the 150-mm gun was necessary for very rapid action against patrol craft and also in action against convoy shipping. However, in antiaircraft defense the same 150-mm gun was determined to be ineffective because of its heavy construction and slow rate of fire. The American 127-mm and British 134-mm dual-purpose guns would not have satisfied our needs for the rapid and decisive action against a destroyer.

The question of a dual-purpose gun had been examined, but in 1935 the Germans believed it not possible to develop a really dependable rapid-firing gun with great mobility and ready conversion for antiaircraft defense. German ordnance experts believed that a battleship's secondary armament should be not less than 150 mm. Such a gun could not be an effective dual-purpose weapon, as shells had no fuze-setting provision for action against aircraft, and turret training and elevating rates were designed for surface action. The fuzes were set for each distance manually on the turret platform.

Antiaircraft battery. The main antiaircraft defense was furnished by fourteen 105-mm/65-caliber Model 1933 dual-purpose guns, located one deck above the 150-mm guns. Being higher, with better distribution around the superstructure, their arcs of fire were little impeded by top hamper. These guns could elevate 85 degrees and be depressed to 8 degrees, and had a rate of fire of 15 rounds per minute at a muzzle velocity of 900 meters per second. They were effective against aircraft at short ranges, but could not provide the long-range protection offered by larger guns in other navies.

The entire 105-mm mount weighed approximately 27 tons, due to two key construction factors. The guns had high elevation angles and a mechanically driven loading device with transport rollers. Their relatively large angles of depression, permitting use against surface targets at close range, made it necessary to keep the trunnions of the gun as near the breech end as possible. There was also the need to lessen the recoil forces, which was solved partially by making the barrels heavy—9.3 tons for each pair.

The gun cradles permitted three axes of movement. Training and elevation were remotely controlled; cross-level adjustments were made by a continuously operating hydraulic drive developed by the Pittler-Thoma concern. The third axis of motion was the result of earlier studies of high-elevation firing from a moving platform. By making guns independent of ship motion, their control drive would be less burdened and a more efficient loading made possible. The third axis of motion also made it possible for uninterrupted fire against aircraft, decreasing the effects of ship motions.

The mixed-battery arrangement was effective, although the few attacking planes the British used in the early phase of the war were relatively slow and not as effective as those developed by the Japanese and Americans. The Germans were also spared the problem of defense against the type of dive bombing used in the Pacific theater of operation. The British had not mastered that technique as effectively as the Germans, Japanese, and Americans.

Machine gun battery. Sixteen 37-mm/83-caliber machine guns were provided in eight twin mounts. These guns—not fully automatic, but with a rate of fire of 30 to 40 rounds per minute—were used for close-in protection against aircraft by every major combatant ship in the German Navy. Like the 105-mm guns, they had three axes of movement.

The number of 20-mm guns progressively increased on both battlecruisers during the war, with most of them added while the ships were in Brest. Eighteen 20-mm guns were added to the *Scharnhorst* and fourteen to the *Gneisenau.* These guns, in single or quadruple mounts, were installed wherever deck space permitted, as well as on the turret roofs and specially constructed platforms on the mainmast. In fact, several searchlights were removed to provide a mounting platform for some of these guns. The reconstructed *Gneisenau* of 1942 would have had additional 20-mm guns added during her projected conversion. Her battery of 20-mm guns would have included 24 quadruple and 8 single mounts.

Torpedo armament. After the *Gneisenau* and *Scharnhorst* returned from their Atlantic operation in late March 1941, Admiral Günther Lütjens, who commanded that operation, suggested in his report that torpedo tubes be mounted on these ships. He indicated that the sinking of merchant ships required a disproportionate amount of time and/or ammunition and that torpedoes offered a much quicker destruction, particularly at close ranges. The OKM (Oberkommando Kriegs Marine) concurred and authorized the placement of two triple 533-mm deck-mounted tubes installed on both ships while they were in Brest. No fire-control equipment was provided, and the mounts would have to be manned by antiaircraft gunners. An aiming apparatus was provided on the torpedo mount. Torpedo stowage was provided in nearby lockers. These mounts were unprotected and could have easily been put out of action by small-caliber gunfire or fragmentation attack through shell or bomb bursts. In the *Scharnhorst*'s final action one torpedo mount was put out of action by a near-miss shell.

Aircraft. With the lack of aircraft carriers in the prewar German Navy, there was a need for German capital ships to carry aircraft. Both battlecruisers carried aircraft for reconnaissance and spotting of salvos, although limited offensive measures were possible since an aircraft bomb magazine was provided. Both ships featured different hangar arrangements. When the *Gneisenau* first went into service in 1938, she was equipped with a small hangar; the *Scharnhorst*, on the other hand, had a much larger one. During a major overhaul in 1939, the hangar on the *Scharnhorst* was lengthened some 8 meters to provide stowage for three Arado-196 aircraft. The hangar roof was retractable, the two forward sliding doors being pushed over the after one. Two cranes lifted the planes out of the hangar and placed them on the catapult at the forward end of the hangar.

Another aircraft could be stowed on top of turret Caesar. A catapult was fitted to the turret, but during winter overhauls on both ships from December 1939 to February 1940, this catapult was removed as it did not prove to be successful.

The *Gneisenau*'s hangar was completely altered in Brest in 1941. A catapult was mounted inside the hangar, which was lengthened and broadened, and the side walls were fitted with large doors. Two other aircraft could be stowed below the catapult.

The aircraft of these ships were extensively used throughout their careers, particularly during the Atlantic cruise in early 1941. Carrying aircraft on board, however, involved risks, particularly with the aircraft fuel system. During the Battle of the North Cape, the *Scharnhorst*'s unprotected hangar sustained several severe hits, which started serious fires.

Protection. Traditionally, German capital ships had heavy armor protection, extensive subdivision, and good underwater protection. With the advances in long-range gunnery and the greater threat of aerial bombing, the Germans were forced to pay greater attention to horizontal protection. The armor-deck thicknesses of World War I capital ships would not suffice in 1934. They accepted the principle that neither vertical armor nor horizontal armor alone could prevent the heaviest projectiles from penetrating a given armor plate. It was also recognized that heavy guns had won the contest with armor plate and truly massive protection would not be possible. Therefore, it was decided to distribute the armor in such a manner that at critical ranges both the horizontal and vertical armor would work together to protect the vitals of a capital ship. This meant a risk of exposing a ship to long-range gunfire or bombing from very high altitudes. The Germans, however, believed that their ships would be engaged in close combat in the North Sea where visibility was limited.

German armor protection was not predicated on the immunity-zone concept that was used in other navies. Instead, complicated range and target-angle tables for the guns of likely French or British adversaries were worked out so that the German capital-ship captains could select battle ranges that minimized the chances of damaging hits. These tabulations were an aid, but not a guarantee, for some of the data were based on erroneous information. Nevertheless, the estimates given in the tables were accurate enough for them to form their general tactics. In general, however, German armor protection design practice was based upon short-range engagements, with the armor arranged so that no projectile could pass through the side belt and into the vitals without encountering some portion of the deck structure. The Germans did not consider an underwater diving shell to be as serious a problem as did the Japanese, American, and British navies. With the advent of accurate long-range gunnery, the German armor scheme was weak against more modern battleships at the longer battle ranges.

Main side belt. The vertical main side belt was 320-mm thick and tapered to 170 mm at the outboard slope of the lower armor deck. The main side belt was vertical and continued 1.7 meters below the design waterline before tapering. There the lower edge of the armor slope was joined to the main side belt, which then tapered from 320 mm to 170 mm over a length of one meter below the slope-belt connection. The protection provided was comparable to that of the *Bismarck*-class battleships, and the sides, plus armored slopes, were proof against a 1,016-kilogram 406-mm shell fired at any range over 11,000 meters. German naval constructors opposed the use of sloping internal armor belts in capital ships because they left the outboard areas unprotected. The main side belt had constant thickness throughout the armored citadel above the taper and extended 3.0 meters above the design waterline; from that point, a splinter belt 45-mm thick extended to the upper armor deck. The thin splinter belt was all that could be afforded in these battlecruisers. More armor was needed above the design waterline than below to protect the ship when she rolled, listed, or the draft increased due to damage.

The main side belt extended beyond the forward armored bulkhead to the bow, tapering from 320 mm to 70 mm from the armored bulkhead to a point 41 meters from the stem. The belt was carried sufficiently below and above the design waterline to provide splinter protection for the unprotected portions of the bow.

The steering gear and shafts were protected from the after armored bulkhead to the after bulkhead in the steering gear room by an armored slope. A splinter belt extended aft to a point 37 meters from the stern ending. The side armor in the stern region was 170 mm; 80-mm-thick horizontal protection was provided on the lower armor deck. It was considered essential that propellers and rudders be protected from plunging shell fire and bombs, but it was also recognized that they could never be completely protected, especially from torpedoes.

Deck armor. The Germans deployed their deck armor using the so-called turtle-deck armor system, with the deck armor connected to the lower edge of the main side belt by armor slopes, instead of to the upper edge as had been done in the Japanese *Yamato* class. German naval constructors were of the opinion that placing the armor deck higher in the ship would have caused shells or bombs to detonate high in the ship and also would not allow the deck to contribute anything to the protection if the main side belt were struck. It was decided, however, to provide a 50-mm upper armor deck to keep out general-purpose bombs. If a semi-armor-piercing projectile were to hit the ship, it would penetrate the 50-mm deck; thus, transverse and longitudinal splinter bulkheads were provided between the upper and lower armor decks to contain fragmentation damage. It was considered possible that heavy armor-piercing bombs could penetrate both decks, but they would require extreme altitudes or rocket assistance to accomplish this. This was the basis of the armor deck system in all German capital ships of the World War II era.

Since it was not possible to test armor-piercing bombs from extreme altitudes with high terminal velocities, proving-ground tests were made on vertical armor with large-caliber shells stabilized to simulate the effects of armor-piercing bombs. Results indicated that complete horizontal protection was not practical and that the armor-deck structure would be more satisfactory if it decreased the kinetic energy of the bomb and caused deformation of its casing, which would deform the fuze and disarm the bomb. Therefore, it was decided to use horizontal armor over two decks, with the heaviest thicknesses concentrated outboard on the lower armor deck, on the armored slopes. The two decks had a separation of 5.1 meters between them, with the battery deck of ordinary steel plate in between. It was expected that the 50-mm upper deck would initiate the fuze action and cause the projectile to explode on the lower armor deck. This arrangement conceded damage to the spaces between the two decks, but such occurrences would have a minimal effect on the ship's fighting efficiency. The various thicknesses are tabulated in table 3-8.

TABLE 3-8
Deck Armor Thicknesses

	Deck at Centerline		*Deck at Sides*	
	Magazines	Machinery	Magazines	Machinery
Upper Deck	1.97″ (50 mm)	1.97″ (50 mm)	1.97″ (50 mm)	1.97″ (50 mm)
Armor Deck	3.74″ (95 mm)	3.15″ (80 mm)	4.14″ (105 mm)	4.14″ (105 mm)
	5.71″ (145 mm)	5.12″ (130 mm)	6.11″ (155 mm)	6.11″ (155 mm)

The deck-armor arrangement for the lower armor deck closely followed that used in the design of the *Ersatz Yorck* class, where a single deck was positioned just above the design waterline for a greater portion of its breadth. This deck was then sloped down and outboard to the lower edge of the main side belt. The armor deck terminated one deck lower than in most battleships or battlecruisers of that period. The armor slope tremendously increased the protection of the vitals. In fact, barring an underwater diving hit, the vitals of these battlecruisers were well protected against any shell used by existing battleships at ranges where the belt and slope both had to be penetrated. At very long ranges, the deck armor was vulnerable to penetration. The slope also served as a buttress for the armor belt.

With a displacement exceeding 26,000 tons, it was not possible to provide a single armor deck one deck higher than the lower armor deck, because it would have meant extending the main side belt one additional deck height. In the planning of the *Scharnhorst* class the emphasis was on survival from shell attack and not bomb attack. In fact, it was not until 29 May 1937, when the armored ship *Deutschland* received two bomb hits in the harbor of Ibiza, Spain, that the danger of airplane bombing was taken more seriously. These two bomb hits caused much internal damage, killing 31 men and injuring scores of others. Later, the bombing of the *Scharnhorst* at La Pallice in July 1941 caused much concern in the Construction Office, for several of the bombs had passed through both armor decks, fortunately without exploding. Ordnance disposal teams were sent to the French port to retrieve the bombs for cxamination. Although these ships had a recognized deficiency in armor-deck protection, it was not possible to correct this due to the limitations on displacement and design draft.

The main armor deck also had some discontinuities. It was raised 0.7 meters above the boiler rooms over a distance of 9.62 meters to provide sufficient boiler clearance. This was not foreseen in the planning; the boilers were larger than originally estimated and could not be built to give dependable service within the original dimensions. Inboard, over the boiler rooms, the vertical portion of the armor deck was protected by 80-mm armor plate.

Part of the German fleet in Kiel Bay entrapped in ice during the extremely harsh winter of 1939/40. *Gneisenau,* with her catapult on turret Caesar, is in the foreground with the new heavy cruiser *Blücher* off her starboard bow and the *Admiral Hipper* off her starboard stern quarter.

Originally, the main armor deck was to be located 530 mm above the design waterline, but changes in the armor and other additions increased the displacement and draft until the deck was practically level with the design waterline. In the full-load condition, the armor deck was 730 mm below the waterline. This situation could not be corrected easily, as the only practical measure left was to change the dimensions by bulging the hull or increasing the beam. As it was, nothing was ever done to correct the overload situation until severe bow damage to the *Gneisenau* presented an opportunity partially to correct the condition by lengthening the ship some 10 meters during her planned conversion and reconstruction.

Conning tower. German naval constructors and line officers believed that there was a need for heavy armor protection for the main and secondary conning positions. Therefore, the conning towers of these two battlecruisers were heavily protected. An armored communications tube rose from the lower armor deck to the floor of the main conning position. The secondary conning tower aft had thick armor too, but less than that used in the main conning tower forward. The main-battery fire-control station forward was also protected by plate of 60-mm noncemented armor with 20-mm plate protecting its access.

Armored bulkheads. The *Scharnhorst*-class battlecruisers had armored bulkheads at one end of the armored citadel, and another at the after end of the steering gear room. The forward bulkhead extended from the upper platform deck to the upper deck. Below the armor deck the thickness was the same as that used in the steering gear room, while above that deck it was 150-mm thick from the battery deck to the armor deck and 70-mm thick from the battery deck to the upper deck. Armor thicker than 100 mm was of face-hardened (Class A type) plate.

Turret armor. An unusually great proportion of the turret protection was concentrated on the sides, with 350-mm armor. The roof was 150-mm thick, while the face plate was 360-mm thick. The extremely heavy armor was further supplemented by heavy barbette armor that tapered from a maximum of 350 mm on the exposed sides to 200 mm on the less-exposed centerline where the turret shielded it. The turrets were the most heavily protected portions of these ships.

The secondary armament of capital ships was invariably a vulnerable point in the armor system. The Germans provided relatively heavy armor protection for the 150-mm turrets, more than was provided in most Allied battleships of the same era. Such protection, however, was not adequate against direct hits by large-caliber shells. The 150-mm single mounts had only 25-mm armor plate protection against strafing and fragmentation damage. These were the real weak points of the armament of the *Scharnhorst*-class battlecruisers, as was shown during the action of the *Scharnhorst* with the *Duke of York*.

There was extensive splinter protection in these and other German capital ships due to the location of the main armor deck. Most of this splinter protection was concentrated in the citadel to contain shell fragments from shells exploding on the lower armor deck. There was 20-mm splinter protection around the uptakes as well as around such spaces as the night fire-control and aft main-battery fire-control stations.

Torpedo protection. The underwater protection system was designed to resist a 250-kilogram charge of TNT in contact with the ship's side at half design draft. This charge was slightly less than those carried by British ship-launched torpedoes, but greater than those contained in their aerial torpedoes. The structural arrangement and compartmentation of the side-protection system was developed from a series of full-scale underwater explosion tests conducted on several sections cut from the old predreadnought *Pruessen*. These tests proved that welded structure, although lighter in weight than comparable riveted construction, better withstood the forces of a 250-kilogram charge, and the resulting damage could be more easily repaired. Previously, welded connections of non-heat-treated armor steel had been successfully proven in gunfire tests at the naval proving grounds. These particular tests confirmed that welds using the Nichrotherm electrode developed by Krupp could withstand the stresses and bending expected for the torpedo bulkhead, which was to be constructed from non-heat-treated armor plate. The Construction Office, therefore, had decided to have the structure of the armored ship *Admiral Graf Spee* entirely welded.

The conclusions drawn from these tests were that the underwater protection system as planned was structurally sound against an explosive charge of at least 250 kilograms of TNT. It was decided, however, that the torpedo bulkhead should be riveted in view of the fact that poorly welded connections could not withstand explosive shock forces. Most of the faulty joints were attributed to bad welding, and this could have been detected only through X-ray examinations, which the Germans had not developed for shipyard testing. Also, it was determined that no connections should be made on the torpedo bulkheads to support cables or pipes, as they would hinder the elastic action of the plating and might damage the lines during the elastic deformation of the plating.

The side-protection system had light armor plate on the side shell. Below the main side belt an armor-type plate varying in thickness from 12 to 16 mm would detonate the explosive charge. There was always the problem that fragments from the shell plating might riddle the inboard bulkheads, if the torpedo charge was large enough. A large void behind the plate acted as an expansion chamber into which the explosive gases would be allowed to expand to reduce pressure and energy. Inboard of this void, a bunkering space absorbed the remaining energy by deflection or rupture of its 8-mm outer boundary, which was backed by longitudinal stiffeners and transverse frames. Some energy would be absorbed by elastic and plastic deformation of the 45-mm torpedo bulkhead. Where possible in the citadel, spaces outboard of the essential areas acted as flooding boundaries in case of leakage through the torpedo bulkhead.

The torpedo protection system resembled that of the armored ships except that the main side belt was farther outboard and made vertical on the ship's outer skin. The bulge was discarded in favor of a semi-bulkhead system. Major characteristics of the system are tabulated in table 3-9.

TABLE 3-9

Characteristics of the Side Protection System

Location	*Depth at ½ draft*	*Total bulkhead thickness*	*Torpedo bulkhead thickness*
Turret Anton	8.45′ (2.58 m)	2.09″ (53 mm)	1.77″ (45 mm)
Turret Bruno	10.10′ (3.35 m)	2.09″ (53 mm)	1.77″ (45 mm)
Amidships	14.76′ (4.50 m)	2.09″ (53 mm)	1.77″ (45 mm)
Turret Caesar	12.28′ (3.74 m)	2.09″ (53 mm)	1.77″ (45 mm)

The system was extremely effective amidships, but grew progressively weaker toward the ends of the citadel, where the hull narrowed; resistance there dropped to 200 kilograms of TNT. Structural design around the after turret was very complicated due to the fine stern shape and the passage of the outboard shafts through the underwater protection system. The shaft alleys were used as part of the underwater protection system.

Another serious weakness in the underwater protection of these ships was revealed every time they sustained torpedo damage. The torpedo bulkhead extended from the bottom shell to the underside of the slopes at an angle of approximately 10 degrees, where it was riveted in place by two angle bars. This area was always under stress from normal ship bending forces, which were disastrously intensified by an explosive loading. The rivets would take shear loads in excess of their designed stress levels and fail. In fact, a sharp impact load from a bomb or shell hit on the outboard slope was also a serious threat to the connection. The riveted joints that held the torpedo bulkhead to the armor deck in way of the after turret would always fail in an area of high stress concentration, mainly because of the abrupt ending of the bulkhead, and in part due to the discontinuity in the bulkhead around half draft. The bulkhead was curved inboard to gain more depth, but in doing so it decreased the structural effectiveness of the system.

The torpedo defense system had to be narrow, due to the needs of the propulsion plant. These conditions barred the use of the system employed in the *Baden* and *Bayern*. Later, the *Bismarck* and *Tirpitz* used the *Baden* system. The *Scharnhorst*-class ships were designed for 30 knots plus, with beams limited to 30 meters; only by increasing the beam and displacement could better protection be provided around the turrets. Accordingly, the system given these battlecruisers was not as efficient as had been desired.

Stability and subdivision. The principle of achieving a large degree of stability by means of comprehensive compartmentation, which had been established in World War I, was repeated in these ships. Because of their size, such stability was demanded in a higher degree than in the armored ships. The Germans thought that a capital ship should be divided into a number of watertight compartments, and the principles of the International Convention on Safety of Life at Sea (SOLAS) concerning passenger-ship design were extended to German warship design in that these ships were to be designed as two-compartment ships. This meant that any two adjacent compartments of any length and location could be flooded without the submergence of the watertight bulkhead deck. Each major watertight section, except those in the narrow extremities of the hull, was further subdivided into watertight spaces. The experiences of the *Lützow* and *Seydlitz* at the Battle of Jutland proved these principles to be an absolute necessity for survival. In the design of these two new battlecruisers, the details of the World War I mine damage to the *Bayern* were also instructive. As a result, the *Scharnhorst* and *Gneisenau* had a greater number of transverse and longitudinal bulkheads than did previous German capital ships. The propulsion plant was in several large compartments to provide separate power sources for each shaft; the navy consciously accepted a more complicated ship operation arising from the lack of accessibility. This applied especially to the increased number of watertight ducts through bulkheads and the use of watertight closures.

As the result of a thorough damage-stability analysis on both ships during their early design, they were subdivided into 21 main compartments by six transverse bulkheads that extended to the battery deck and fourteen others that extended to the upper deck forward. The torpedo bulkheads in the side-protection system had 15 wing compartments on each side. These bulkheads were located 10.56 meters from the centerline. The ships were designed to lose buoyancy in any three main watertight compartments without sinking.

The location of the transverse bulkheads was determined by the calculation of floodable length curves using procedures similar to those established by merchant ship classification societies for compensation of those type ships. It was understood, however, that such a calculation assumed symmetrical flooding. Therefore, these and other major German warships were purposely built with wide beams and large metacentric heights, so that expected asymmetrical flooding could be controlled within a certain time by counterflooding, which would correct trim and list resulting from damage. The length of damage for a wing compartment from a torpedo hit was estimated to be 30 meters or at least three compartments. Such flooding would be faster than that from a shell hit, which would be more easily controlled. A separate damage stability calculation was made for each ship to account for differences in arrangements and weights/centers of gravity.

These ships had only fair stability characteristics for their size, as is shown by the tabulation of metacentric heights and range of stability shown in table 3-10.

TABLE 3-10
Stability Characteristics

Condition	*Displacement*	*Metacentric Height*	*Range of Stability (Degrees)*
Light ship	31,045 tons (31,552 mt)	5.25′ (1.6 m)	58
Standard displacement	31,848 tons (32,368 mt)	6.56′ (2.0 m)	. . .
Design displacement	33,454 tons (34,000 mt)	7.38′ (2.25 m)	. . .
Full load	37,214 tons (37,822 mt)	7.70′ (2.5 m)	62

The relatively small range of stability of these ships when compared to foreign battleships or battlecruisers was due to the need to save weight of armor and achieve as small a target area as possible. The length-to-depth ratio of these battlecruisers was 16:4. Ships of the First World War had been constructed with even lower freeboard, but in view of the fact that intended operations had shifted from the Baltic and North Seas to the Atlantic Ocean, a greater freeboard had been chosen for the *Gneisenau* and *Scharnhorst.* Aside from the increases in displacement and dimensions, more consideration was given to the strength of the hull girder. Even if these ships' hull girders should be damaged, the ship should be able to continue in action. This point was considered extremely important from experiences gathered in naval battles of World War I. Therefore, portions of the hull above the waterline were provided with heavier structural thicknesses. This, coupled with the protective requirements, is why the center of gravity of these ships was rather high. The limited beam did not permit the relatively large metacentric heights that would be achieved in later battleship designs.

The *Scharnhorst*-class battlecruisers had a double bottom 1.7 meters deep. When design work commenced, the magnetic detonator for torpedoes and mines had not yet been developed, and it was assumed that these ships would enter and leave the Atlantic through mine-free routes, or be escorted. Damage to the *Bayern* during World War I had shown that only a deep double bottom could absorb the effects of an underbottom explosion, and this only in a ship with a very large displacement.

Vital spaces within the armored citadel were extensively subdivided by two longitudinal bulkheads 7.3 meters from the centerline, a practice commenced in the predreadnoughts. Such bulkheading preserved the separate watertight integrity of the engine rooms, boiler rooms, and magazines, and actually diminished the probability of large lists if the ship was severely damaged on one side, instead of increasing the probability, as might be expected. Single centerline bulkheads, although they diminish the mass of water entering the ship, allow a greater increase in the listing moment than two off-centerline bulkheads. The Germans were convinced that the three-shaft arrangement for the propulsion plant was more efficient in subdividing a ship than was the four-shaft scheme.

Radar. German research on shipborne radar commenced during the summer of 1933 and by October 1934 a set was installed on the yacht *Grille.* The range of this radar set was only 12 kilometers. Later on, successful studies of a 50-cm wavelength radar apparatus by Dr. Wilhelm Runge in 1935 had permitted the German Navy to award a contract to the GEMA firm in February 1936. GEMA would develop an 80-cm air- and surface-search set capable of tracking surface ships at 35 kilometers and aircraft at 48 kilometers. GEMA developed this radar set under protest, because it believed that a smaller-centimeter wavelength would be far superior. Tests were made at Swinemünde for land and marine use. The marine version was given the code name "Seetakt," while the land version was named "Freya." The light cruiser *Köningsberg* was equipped with an early version of the Seetakt radar. The pocket battleships *Admiral Graf Spee, Admiral Scheer,* and *Deutschland* (later *Lützow*) were equipped with a more advanced type.

The *Scharnhorst* and *Gneisenau* were equipped with two Seetakt radar sets, one on the forward gun director on top of the bridge-command tower and the other on the aft main-battery director. This radar set operated at 368 megacycles with an output of 14 kilowatts, which was later raised to 100 kilowatts. The forward set could rotate 360 degrees, but its sweep astern was not effective because of structural interferences. This led to the need for a second radar to be installed on the after conning tower. These radar sets operated on an 80-cm wavelength, and their design was similar to that of the Luftwaffe's Freya set.

The Germans had stopped the development of a shorter-centimeter radar set, despite the protests of the GEMA firm. There were many individuals responsible for this, the most influential being Hermann Göring. He and others did not direct the electronic industry properly, however; if Hermann Göring could have been shown how the smaller-centimeter wavelength radar performed, then he might have taken more positive steps to have it perfected. In addition, Admiral Witzell had a tenacious influence on the OKM liaison officer to GEMA, which proved to be unwise according to all communications on centimeter radar research from the naval proving grounds. Although Germany was technically more advanced in the development of radar at the time World War II commenced, it would be surpassed in radar technology by Great Britain by mid-1941.

Propulsion Plant. The *Scharnhorst*-class battlecruisers were designed in an era of great emphasis on speed, the reason for the large horsepower provided. When the ships were redesigned for more speed than their armored-ship predecessors (26 knots), they conformed to the axiom, "faster than a stronger opponent, stronger than a faster opponent."

German marine engineers had spent a long time developing and perfecting a successful diesel propulsion plant for the armored ships. Diesel engines offered economical fuel consumption and good endurance, allowing a ship to stay at sea for long periods without refueling. Such advantages were not possible in ships with a maximum speed in excess of 30 knots and limited propulsion-plant space. The development of the necessary high-speed diesel drive would have involved much time and expense and further delays in construction.

A steam-turbine main propulsion system seemed to be the only alternative, as the Construction Office opposed a mixed diesel and geared-turbine plant. Successful high-pressure, high-temperature steam plants were operating in the new East Asian Express steamers, the *Scharnhorst, Gneisenau,* and *Potsdam.* Land power plants had successfully used the principles of high pressure and high temperature in steam generation for several years; thus, it did not appear too difficult to attain the same conditions in a marine power plant. Marine engineers, however, disagreed as to the relative advantages and disadvantages of such a plant, with the strongest argument against it being the attendant danger. The whole matter could have been settled by systematic and rigorous tests, but time would not allow this. Admiral Raeder decided to intervene personally and decided to risk high-pressure, high-temperature steam in the new capital ships.

Boilers and turbines. The high-pressure, high-temperature installation was designed to operate at a pressure of 58 kilograms per square centimeter and 450 degrees centigrade, higher than any other battlecruiser or battleship to be built in this era with the exception of the *Bismarck* and *Tirpitz.* Twelve Bauer-Wagner boilers supplied steam to three turbines that developed a normal maximum of 150,000 metric horsepower and 160,000 mhp for a short period in the overload condition. Turbines for the *Scharnhorst* were produced by Brown Boveri; those for the *Gneisenau* were designed and built by Deschimag in Bremen.

The steam plant used in these battlecruisers was lighter and more compact than a comparable diesel plant of the same power. The fuel consumption, however, was

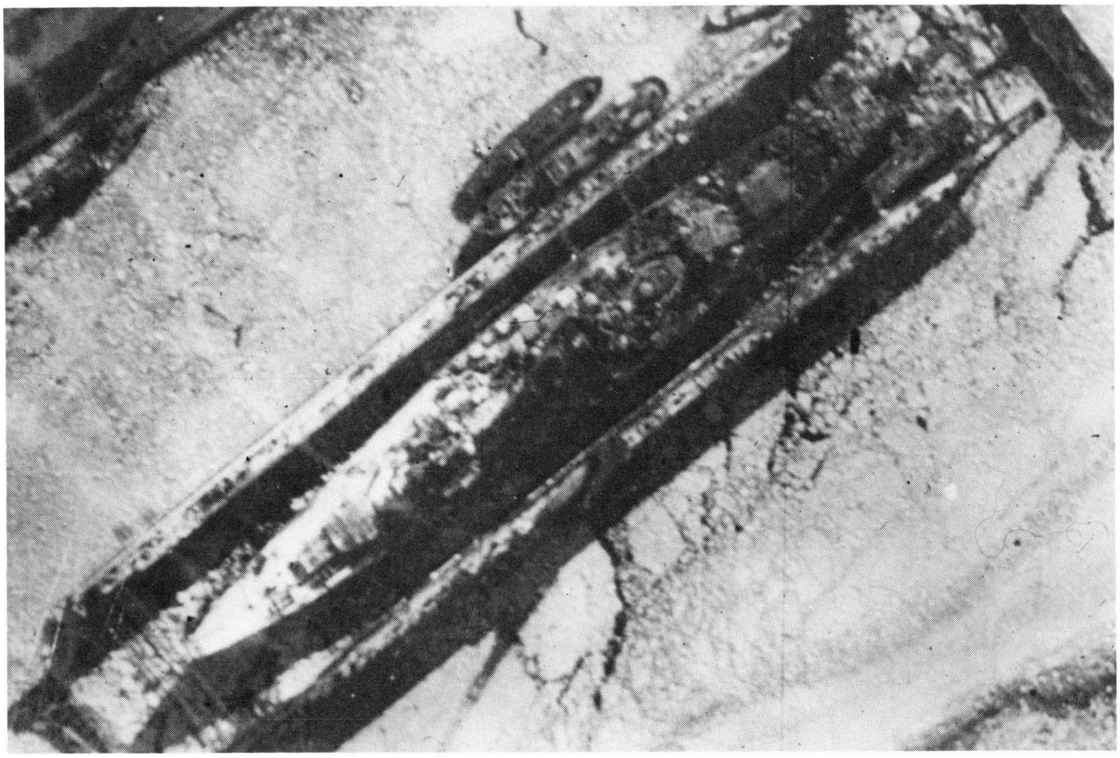

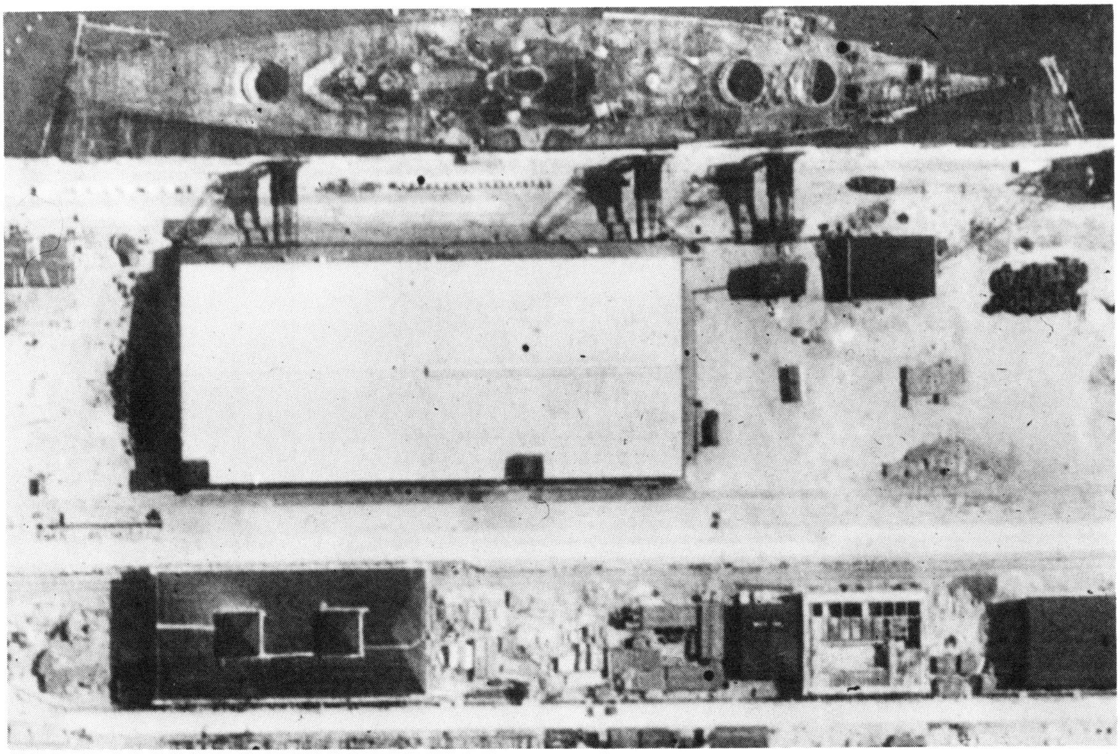

These four photographs illustrate the story of the *Gneisenau* after the Channel Dash of February 1942. After successfully reaching Germany, she was severely damaged in an air raid on the Deutsche Werke shipyard in Kiel on 26 February 1942. The upper photo shows the ship in dry dock in March 1942, where work was already in progress. A significant amount of armor plating had been removed from her bow area, which had been devastated by a magazine fire in turret Anton. In early 1942 the battlecruiser was moved to Gdynia (Gotenhafen), where the final work was to be completed. The second photo shows reconstruction well advanced. The 380-mm

turrets are alongside the *Gneisenau* while work is underway in the barbettes to prepare for their mounting. Some 30 meters of her bow have been removed down to the waterline. In the third photo, the *Gneisenau* is shown in November 1942 when all preparatory work for the new bow and turrets had been completed. The lower photo shows the ship in October 1943, where temporary sheds now cover the barbettes and an attempt has been made to camouflage the ships from Allied aerial reconnaissance. By this time, work on the hulk had been suspended. (Imperial War Museum)

higher, and therefore more fuel was required to attain the same endurance. Displacement limitations and the priority of heavy armor protection barred an increase in the fuel-oil capacity, so the large endurance had to be abandoned. The large fuel capacity and high consumption rate created some stability problems, as fuel-oil expenditure might adversely affect the ships' ability to survive underwater damage and subsequent flooding. As the Germans did not like to use their fuel-oil tanks for salt-water ballast, except the wing tanks in a side-protection system in an emergency, a means had to be found to lessen the effects of fuel consumption. It was decided to locate the wing tanks rather high in the hull so that as fuel was consumed in normal ship operation, the stability would be improved. This consideration led to the unorthodox location and structure of the fuel tanks in the side-protection system.

After its introduction in the *Admiral Scheer*, an oil-water separator was installed in all German ships so that oil in tanks used for salt-water ballast could be used in the propulsion plant. The inner and outer wing tanks were filled, at least partially, with fuel to increase the range of the ship. The low tanks in the forward section of the ship had to be emptied first to establish the proper trim. Stability of these ships was not always improved by fuel consumption, so it was necessary at times to use salt-water ballast.

Engineering spaces were well subdivided, with each turbine set in a separate engine room and four boilers in each of three firerooms. Each engine room had a high- and low-pressure turbine and cruising and astern turbines, coupled to a shaft through a single reduction gear. Maximum rated output for each turbine was 53,300 mhp; the main turbines were able to take a four percent overload.

Electric plant. As a result of the greatly increased use of electric power, particularly in the operation of the armament, each ship had a total generating capacity of 4,520 kw, including a 900 kw reserve in case of battle damage. The plant consisted of six 460-kw turbo-generators and two 150-kw diesel generators. The diesel generators provided the reserve electrical power and had a special electrical circuit. The various generators were located in five different spaces. The normal ship's-service electric power was 230-volt direct current, but a limited amount of 110-volt alternating current was also provided.

Hull Characteristics.

The *Gneisenau* and *Scharnhorst* had modest bulbous bows that reduced high-speed wave resistance, but the great hull length and fine form were even larger factors in reducing wave resistance. These ships were very different from their World War I predecessors and existing capital ships in foreign navies. The emphasis was on speed, and thus there was an absence of blisters, which made them easier and more economical to propel.

Seakeeping. A most noticeable structural feature of both battlecruisers as they were completed for service in 1938 was the flat sheer and steep straight stem. The anchors were stowed in a conventional manner. The freeboard, which was relatively low in comparison with other capital ships, had been diminished appreciably during construction by the addition of equipment, changes in scantlings, and heavier-than-anticipated weights for some vendor-ordered items. Further aggravating the lack of adequate freeboard was a 0.8-meter trim by the bow in the full-load condition, which could only be removed by using oil from the forward deep tanks during the first 24 hours of steaming. It was these first 24 hours that were the most critical in the ships' operation when water swept over the bow, causing problems with the operation of turret Anton and making handling of the ships difficult. In certain sea states, the lack of freeboard, the bow form, and the standard anchoring arrangement combined to cause spray formation and green water to wash over the forward half of the ships to such an extent that they had to be handled from the conning towers. The trials of the *Gneisenau* at high speed showed that the straight-stem bow form also pushed a large amount of water ahead of it, similar to the armored ships.

During the winter of 1938–39 the *Gneisenau* underwent a bow modification that increased the bow freeboard and broadened the flare. Waterline endings were also made finer. This improved the shiphandling somewhat, but there still was much spray formation, particularly in the areas of the anchors. During a major overhaul during the summer of 1939, the *Scharnhorst* was fitted with a new bow configuration (the "Atlantic" bow), which featured the stowage of anchors in recesses at the deck edge. This change added approximately five meters to the length of the battlecruiser; however, a major change in the anchoring arrangement was accomplished that was to provide a distinctive feature between these two battlecruisers for the rest of their careers. The *Scharnhorst* was provided with a bow anchor as well as two recessed side anchors. After her action with the *Rawalpindi*, the *Scharnhorst* had a breakwater installed on the forecastle and extra-heavy watertight scuttles placed on the upper deck to eliminate the flooding of lower-deck living spaces during heavy seas.

The *Gneisenau* required several bow modifications to alleviate the problem of insufficient freeboard. After the action with the *Rawalpindi*, the recessed anchor arrangement was finally adopted, but unlike the *Scharnhorst*, one anchor was eliminated, and a bow chock fitting was located at the stem. After the *Gneisenau* was severely damaged in a North Sea gale in December 1940, the foredeck was further strengthened and breakwaters added. However, the "Atlantic" bow did not completely solve the problem of deck wetness and spray formation, but did reduce it to tolerable limits. The seakeeping qualities of both ships remained bad throughout their careers. The real answer to the problem was to increase the freeboard by providing additional hull depth, but this would have resulted in an increase in the armor and target area. Therefore, some seagoing qualities were sacrificed.

Welding. These were the first extensively welded capital ships in any navy. Electric arc welding had been first used in the construction of the *Kirdoff*-class cargo ships at Wilhelmshaven after World War I. Later, the cruiser *Emden* had her inner bottom, bulkheads, and platform decks welded, which saved considerable weight. Next, a class of small destroyers was entirely welded. When the armored ships were built, all possible weight-saving measures were taken to restrict displacement. At first, all connections of beam to plate and the seams of plating, with the exception of the shell, were welded. The *Admiral Graf Spee*, in which the shell plating butts were welded, was the first German warship of great size with a totally welded hull.

Welding technology had greatly improved by the time the *Scharnhorst* and *Gneisenau* were being constructed, and their hulls were entirely welded, except for the connections of the torpedo bulkheads and deck structure for the lower armor deck. There was some faulty welding that was revealed when the *Gneisenau* was damaged in June 1940. Welding techniques at the time of this vessel's construction depended on electrodes that were not capable of the length that exists in present-day welded construction. In certain cases of torpedo or bomb damage, welded bulkhead joints failed. These were principally caused by poor electrodes, although there was some trace of poor workmanship. Nevertheless, German naval constructors unanimously agreed that welded construction was superior to riveted.

Hull form. The *Gneisenau* and *Scharnhorst* had a rather unusual hull form to facilitate their construction, repair, launching, and docking. Instead of a smooth curve fairing into a flat bottom or deadrise (slight slope of ship's bottom), the midbody had chine-type sections that gradually faired out towards the stem and stern. The straight sections greatly facilitated the construction of the ships, and if the ships were damaged, the repairs could be expedited. Such construction was limited in the afterbody where there was double curvature in the vicinity of the propellers and shafting.

Summary. The *Gneisenau* and *Scharnhorst* were the first capital ships built in the initial phases of the German fleet expansion. Their design was completed under many restrictions. A high-pressure, high-temperature steam plant was used instead of diesel propulsion, and endurance was limited. The Germans decided to construct these ships despite possible shortages in steel and equipment. They gambled, and both ships were ready for action in the first months of World War II. Their active careers and the amount of damage they endured offer an excellent opportunity to evaluate their strengths and weaknesses.

Their outstanding deficiency, politically motivated but technically unnecessary, was the 283-mm main battery. It was a personal disappointment to Grand Admiral Erich Raeder, and would later prove a decisive factor in the loss of the *Scharnhorst*, as such projectiles could not defeat the armor of a 35,000-ton battleship. The action with the *Duke of York* is mute testimony to this fact. If the conversion to 380-mm guns could have been accomplished and the delay in placing these vessels into service accepted, main-battery effectiveness would have been increased significantly. Nonetheless, the smaller guns performed well under the conditions for which they were designed. The triple-turret arrangement was the only one either contemplated or actually used on a German capital ship.

Although these battlecruisers lacked the heavy upper citadel belt of the *Bismarck*-class battleships, they did have heavy armor protection for their size and displacement. Certainly the 350-mm thickness of the main side belt was far in excess of the main-battery caliber of 283 mm. It did not follow the principle of corresponding protection used in most capital ships of that period. The principal weaknesses in the armor system were the horizontal protection and the lack of an upper citadel belt. The latter was serious, since it permitted large-caliber shells with flat trajectories to penetrate the lower armor deck. The combination of side armor, the armored slopes, and torpedo bulkhead provided excellent protection for the vitals. Moreover, the armored slope served as a structural buttress for the main side belt. The weaknesses of the armor-deck system were revealed in the bombing of the *Scharnhorst* at La Pallice and the *Gneisenau* at Kiel. Furthermore, the 356-mm shell hit in the boiler room of the *Scharnhorst* during her battle with the *Duke of York* underscored the need for a thicker upper citadel belt to reinforce the lower armor deck.

The *Scharnhorst*-class battlecruisers had insufficient protection against torpedoes, particularly in way of the after turret. This was caused by an inherent design factor due to the emphasis on speed, which brought about fine lines, and the displacement restrictions imposed on the design from its outset. This limited the amount of material that could be used on these ships. The structural design of the side-protection system was also faulty, as demonstrated by the damage inflicted by 204-kilogram and 340-kilogram torpedo warheads. The basic weakness was that the torpedo bulkhead had an abrupt termination at the lower armor deck and was further aggravated by faulty welding in the butt joints of the bulkheads. Overweight problems prohibited extending the torpedo bulkhead to the upper armor deck for structural continuity and a firm foundation for plastic deformation of the torpedo bulkhead. German naval constructors were aware of this shortcoming when they designed the ships. The weakness in the side-protection system was more apparent at the extremities of the armored citadel. The underwater protection was designed to resist 250 kilograms of TNT amidships, but only 200 kilograms of TNT at the extremities, which offered only marginal protection against British aerial torpedoes. A very serious flaw existed in the torpedo defense system where the propeller shafts passed through it. In both instances of torpedo damage in that area, the shafts were severely affected, as the explosive gases could expend their energy on the exposed sections.

The secondary armament was adequate for the type of action anticipated, but not for the action that developed during World War II. The combination of lightly protected single mounts and heavier-protected twin turrets was an unfortunate, but understandable, choice. Against cruisers, these mounts could be easily put out of action, as they were in the *Scharnhorst*'s last battle. Two twin turrets instead of the four single mounts would have been a better arrangement and would have given better splinter protection to the gun crews. The turrets should have been adapted for antiaircraft defense, as in the French *Richelieu*-class battleships. Such conversion was possible and should have been completed before the war began. The 105-mm guns lacked the range necessary to down aircraft before they could release their projectiles. The use of a mixed-caliber secondary battery involved extra weight and cluttered the topsides with many gun positions that could have been better consolidated into a single dual-purpose battery.

The slow British Fairey Swordfish torpedo bomber and imperfections in British bombing tactics failed to reveal the inadequacies of the 105-mm gun, but its deficiency in antiaircraft defense would have been clearly demonstrated in actions in the Pacific against aircraft with higher speeds and better performance. Both times the antiaircraft batteries of these ships met a severe test, they had air support and were not totally dependent on themselves.

Mine damage emphasized that the main and auxiliary machinery, electronics, ordnance, and electrical equipment, and the foundations to support such systems, were not adequately designed to resist shock loadings. Small auxiliary equipment in the propulsion plant was put out of action for extended periods by noncontact mine explosions that caused only moderate flooding and damage.

The introduction of high-pressure, high-temperature steam in the German Navy came too early and over the objections of the Construction Office. The simultaneous outfitting of several types of combatant ships with different boiler designs caused many setbacks and increased costs in maintenance, as well as the loss of the ships from service for extended periods of time. Defective superheater tubes were only one example; much worse and more time-consuming in terms of repairs were the weaknesses in the auxiliary equipment and fittings for these boilers, and the efforts to eliminate them.

The design of the *Scharnhorst*-class battlecruisers must be considered only satisfactory in view of the limitations faced by the designers. The ships were rather well protected for their size and armament for close battles in the North Sea. The main battery, however, was too weak for a capital ship of World War II. The torpedo protection did not meet expectations, largely due to inconsistencies in structural alignments. The armor-deck system in the vicinity of the firerooms was very weak, and the lack of an upper citadel belt would allow heavy shells to penetrate to the firerooms as happened in the last action of the *Scharnhorst*. Speed was excellent, although it did not give these ships the same measure of protection that World War I German battlecruisers had had. The testimony to the design of these ships, however, was demonstrated during the

Scharnhorst's last battle against a numerically superior British force. She fought for over three hours until all main-battery ammunition was expended and she had absorbed considerable punishment. Some 11 torpedoes were required to sink her. Very crucial in her loss was the lack of an adequate destroyer escort; this could have prevented or greatly interfered with the torpedo attacks that slowed the ship and forced her into the fatal gunnery battle. Destroyers might also have been able to attack the British force and impair the effectiveness of the *Duke of York*'s fire. Also, because the forward radar of the *Scharnhorst* was damaged, destroyer radar sets might have warned of the approaching British force, enabling her to either be ready with correct bearings and ranges, or avoid battle altogether. The presence of the British units had been detected by the Luftwaffe; however, by the time that information had been received aboard the *Scharnhorst*, it was so garbled as to be misleading. The outcome of the battle was decided by the 356-mm shell hit in Number 1 boiler room, which slowed the German battlecruiser sufficiently for the British to resume fire at point-blank range and finally launch devastating torpedo attacks.

The importance of radar in the operation of these battlecruisers was summed up by the final commanding officer of the *Gneisenau*. Captain (later Admiral) Wolfgang Kähler wrote the authors:

> Perhaps I should stress the importance of our radar set in the *Gneisenau*. We could always rely on it; we spotted ships and other targets at great distances; and we received good and reliable data from it for fire control purposes, which in misty weather and at dawn served us better than information gleaned from optical rangefinders.

The loss of the *Gneisenau* resulted from a calculated risk. Contrary to regulations, cartridges and shells had been kept aboard when the ship was once more dry-docked before her scheduled departure for Norway. The only dock available was floating dock V of the Deutsche Werke Shipyard in Kiel, which lay near a pier where the *Scharnhorst* was undergoing repairs. The situation in which two capital ships would be in close proximity to each other underwent heated discussion. It was suggested that the *Scharnhorst* be moved, but this was rejected because it would interrupt her repair work. It was suggested that the ammunition aboard the *Gneisenau* be unloaded, but this too was rejected since it would involve four extra days for the ship to be in the shipyard, and thus it would mean that the ship would miss the new moon period for her passage to Norway. The shipyard appealed to the OKM, but Admiral Raeder directed that the *Gneisenau* be docked *with* her ammunition. The entire blame for the decision does not rest solely on Admiral Raeder because Hitler was also pressuring the German Navy to deploy all its large ships to Norway, where they could attack convoys to Russia.

The *Scharnhorst* and *Gneisenau*, the first of the new German capital ships, demonstrated throughout their careers the remarkable achievements of the German naval constructors in the development of displacement-limited warships with satisfactory protection and speed, although having only a barely adequate armament. When the German Naval Staff became aware of the basic design's limitations, a number of improvements were made at the cost of additional displacement (an increase of some 4,000 tons). This additional displacement increased the draft and trim, which, without the addition of either beam or depth, decreased their seakeeping abilities. Whatever successes these ships enjoyed, their careers were overshadowed by the exploits and dramas of the careers of the battleships *Bismarck* and *Tirpitz*.

Name:	*Scharnhorst*	*Gneisenau*
Builder	Kriegsmarine Werft Wilhelmshaven	Deutsche Werke Kiel
Laid down	16 May 1935	6 May 1935
Launched	3 October 1936	8 December 1936
Commissioned	7 January 1939	21 May 1938
Reconstruction	—	Conversion begun in March 1942
Disposition	Sunk in the Battle of the North Cape at 72°16′ North and 28°41′ East. 1,932 men were lost.	Conversion halted in 1943 and sunk as a block ship in Gdynia harbor on 28 March 1945. Raised by Polish salvage firm and scrapped 1947–51.

Displacement

Scharnhorst

37,224 tons (37,822 mt) — Full load — 1935
38,092 tons (38,703 mt) — Full load — 1943
31,847 tons (32,358 mt) — Standard — 1943
39,017 tons (39,643 mt) — War overload — 1943

Gneisenau

31,132 tons (31,632 mt) — Standard — 1935
37,303 tons (37,902 mt) — Full load — 1935

Dimensions

Scharnhorst

772′ 4″ (235.4 m) Length overall
753′ 11″ (229.8 m) Waterline length
98′ 5″ (30.0 m) Beam
32′ 6″ (9.93 m) Mean draft at 38,101 tons (38,713 mt)
46′ 1″ (14.05 m) Depth to main deck

Gneisenau

741′ 6″ (226.0 m) Waterline length
98′ 5″ (30.0 m) Beam
31′ 9″ (9.69 m) Mean draft at 37,303 tons (37,902 mt)

Armament

Nine 11.1″/54.5-caliber guns (Mark 1934) (283 mm)
Twelve 5.9″/55-caliber guns (Mark 1928) (150 mm)
Fourteen 4.13″/65-caliber guns (Model 1933) (105 mm)
Sixteen 1.46″/83-caliber guns (37 mm)
Sixteen 0.79″ (20 mm) *Gneisenau*
Thirty-eight 0.79.030 (20 mm) *Scharnhorst*

Torpedo tubes

Six 21.0″ (533 mm) deck-mounted torpedo tubes (triple mounts)

Aircraft

Three Arado-196 floatplanes with one catapult

Machinery Schematic

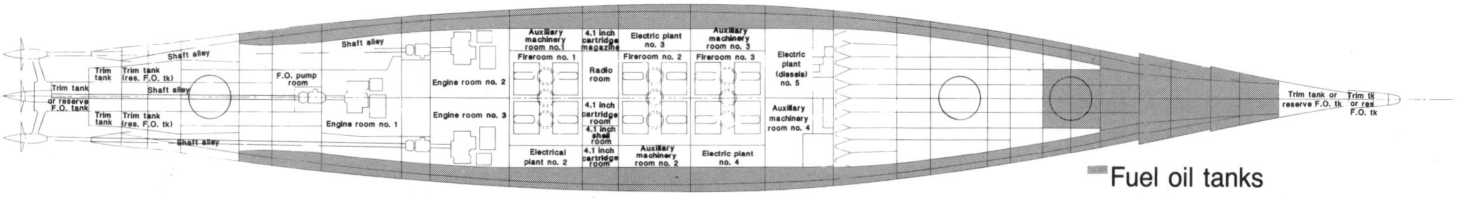

Frames

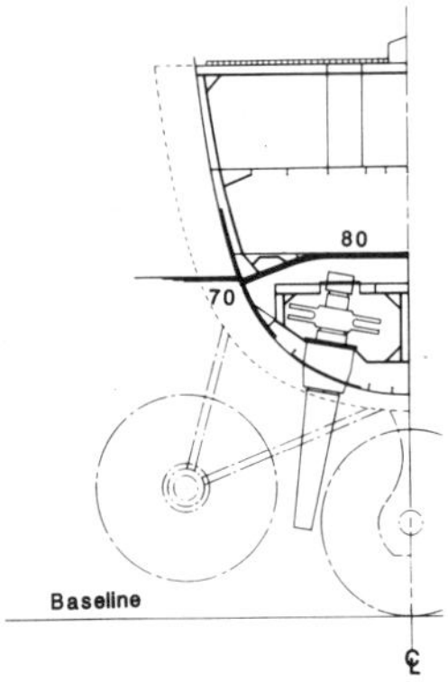

Section at frame 12.5 steering-gear room

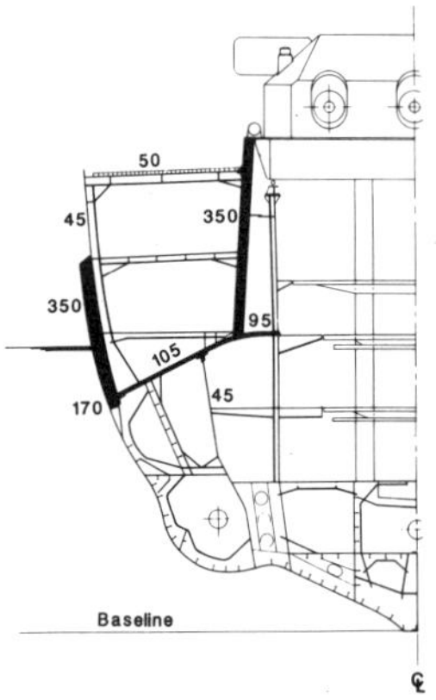

Section at frame 45.2 turret Caesar

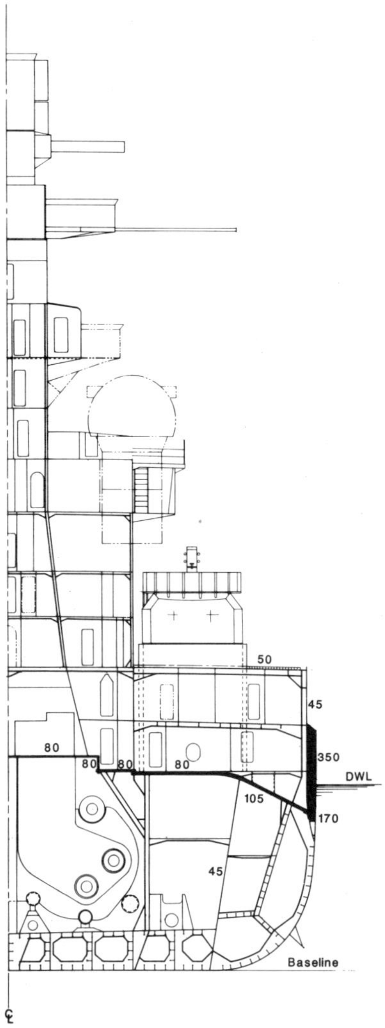

Section at frame 126.25 forward command tower

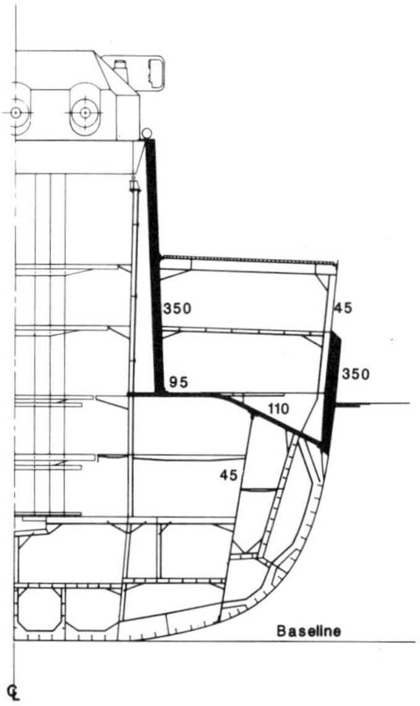

Section at frame 158.3 turret Bruno

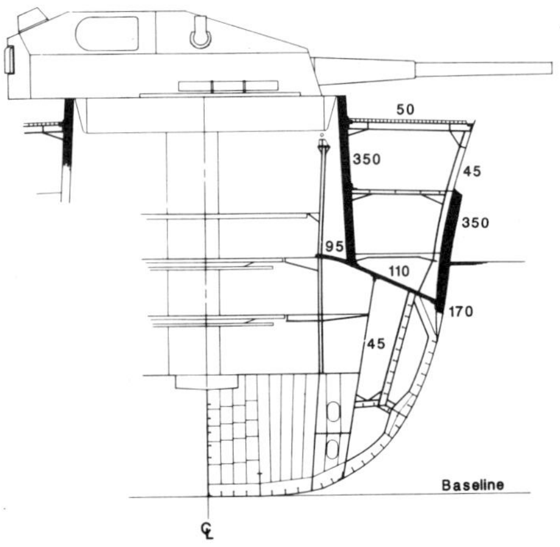

Section at frame 175.5 turret Anton

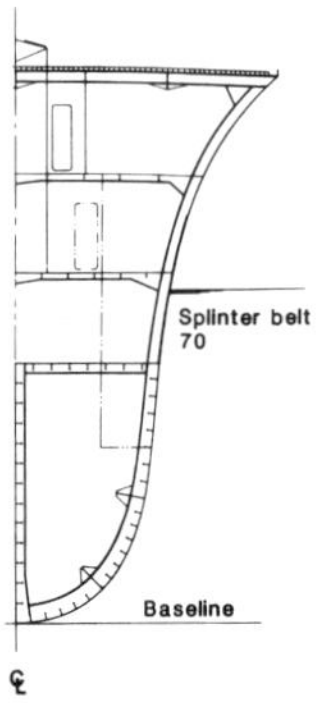

Section at frame 199.5

Midship Section

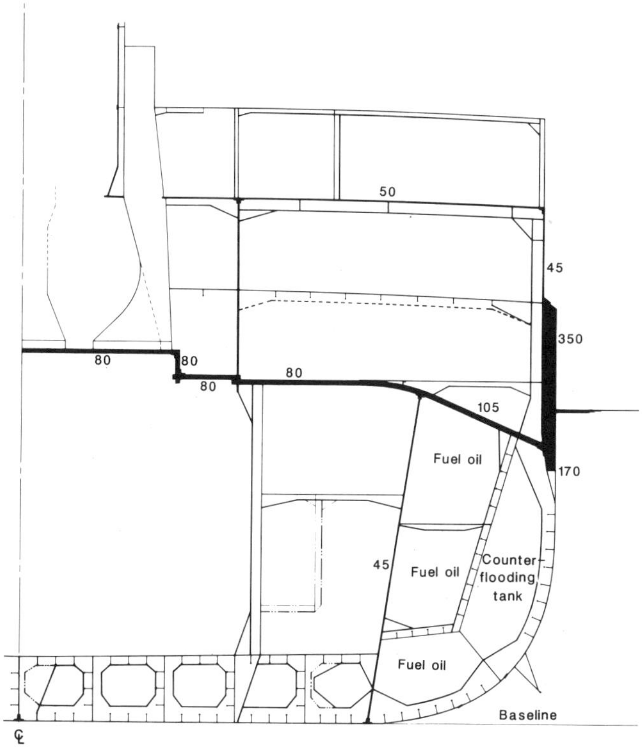

Armor protection (refer to plans for arrangement details)

Amidships		
Upper belt	1.77″ (45 mm)	
Main side belt	13.78″ (350 mm) tapered to 6.69″ (170 mm)	
Deck armor	(centerline)	(outboard)
upper deck	1.97″ (50 mm)	1.97″ (50 mm)
armor deck	3.15″ (80 mm)	4.14″ (105 mm)—slopes
total	5.12″ (130 mm)	6.11″ (155 mm)

Barbette armor	
Centerline forward	7.87″ (200 mm)
Sides	13.77″ (350 mm)
Centerline aft	7.87″ (200 mm)

Turret armor	
Face plates	14.17″ (360 mm)
Sides	7.09″ (180 mm)
Back plates	13.77″ (350 mm)
Roof plates	7.09″ (180 mm)

Secondary gun armor — twin turrets	
Face plates	5.52″ (140 mm)
Sides	1.97″ (50 mm)
Back plates	1.97″ (50 mm)
Roof plates	1.97″ (50 mm)

Secondary gun armor — single mounts	
Gun shield	0.98″ (25 mm)

Conning tower armor	
Centerline forward	13.77″ (350 mm)
Sides	13.77″ (350 mm)
Centerline aft	13.77″ (350 mm)
Roof plates	8.67″ (220 mm)
Communications tube	8.67″ (220 mm)

Body Plan

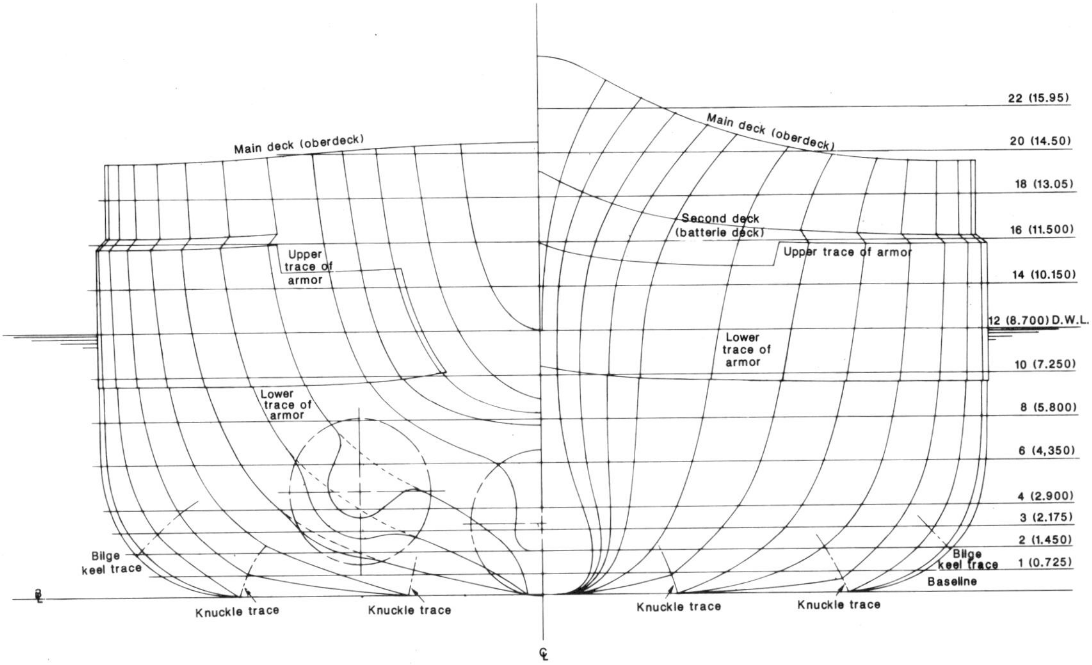

Underwater protection			
Designed resistance	550 pounds (250 kgs) of TNT		
Side-protection system depth	14.76′ (4.50 m) @ half-draft amidships		
S. P. S. design loading void	liquid		
Total bulkhead thickness	2.09″ (53 mm)		
Bottom protective system depth	5.58′ (1,700 mm) — one layer		
Tank capacities			
Scharnhorst		*Gneisenau*	
Fuel oil	6,012 tons (6,108 mt)	Fuel oil	5,610 tons (5,700 mt)
Diesel oil	144 tons (146 mt)		
Aviation gasoline (benzine)	26 tons (26 mt)		
Reserve feed water	463 tons (470 mt)		
Potable water	464 tons (471 mt)		
Lube oil	152 tons (154 mt)		
	7,261 tons (7,375 mt)		

Outboard Profile *(Scharnho neisenau*

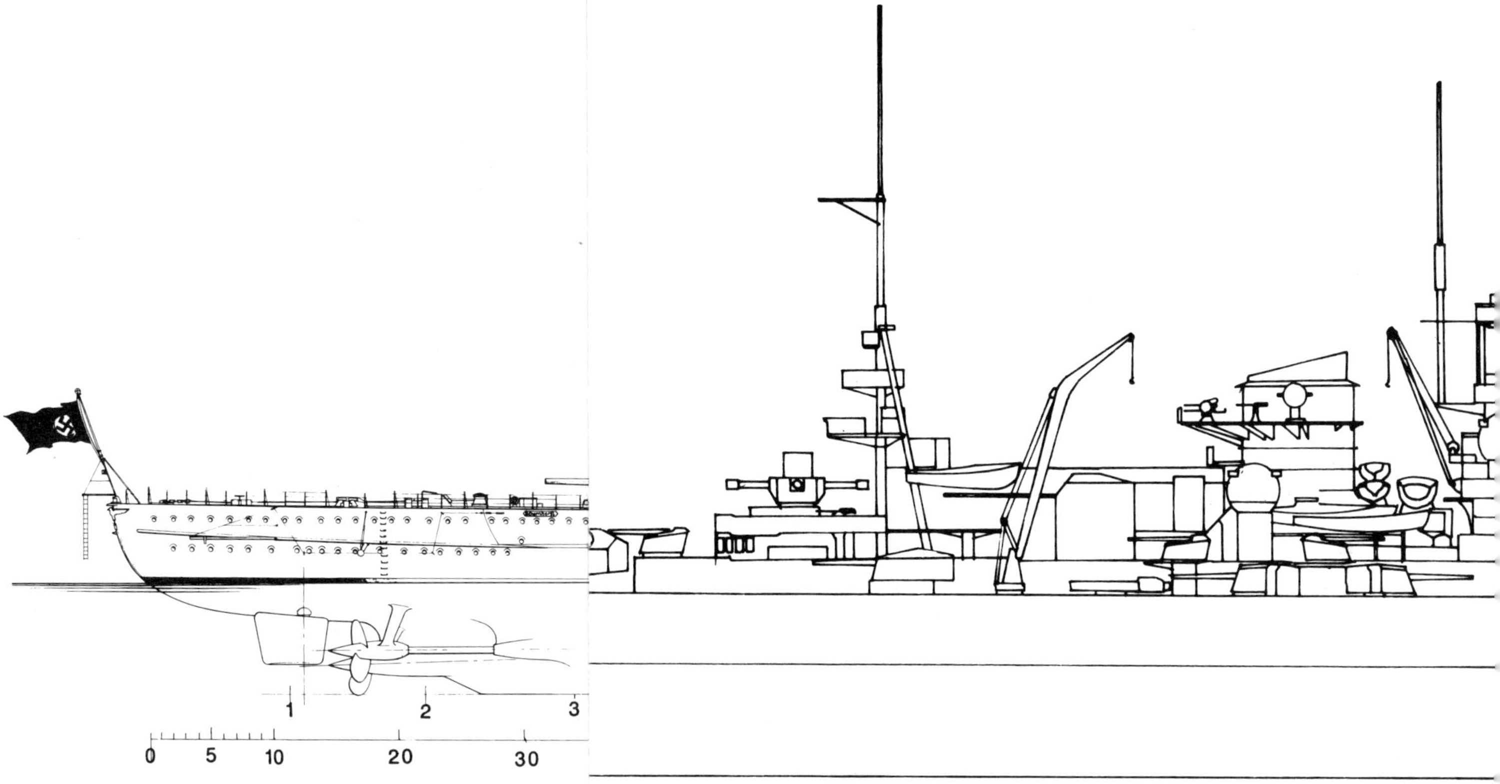

Topside View

2) Gneisenau

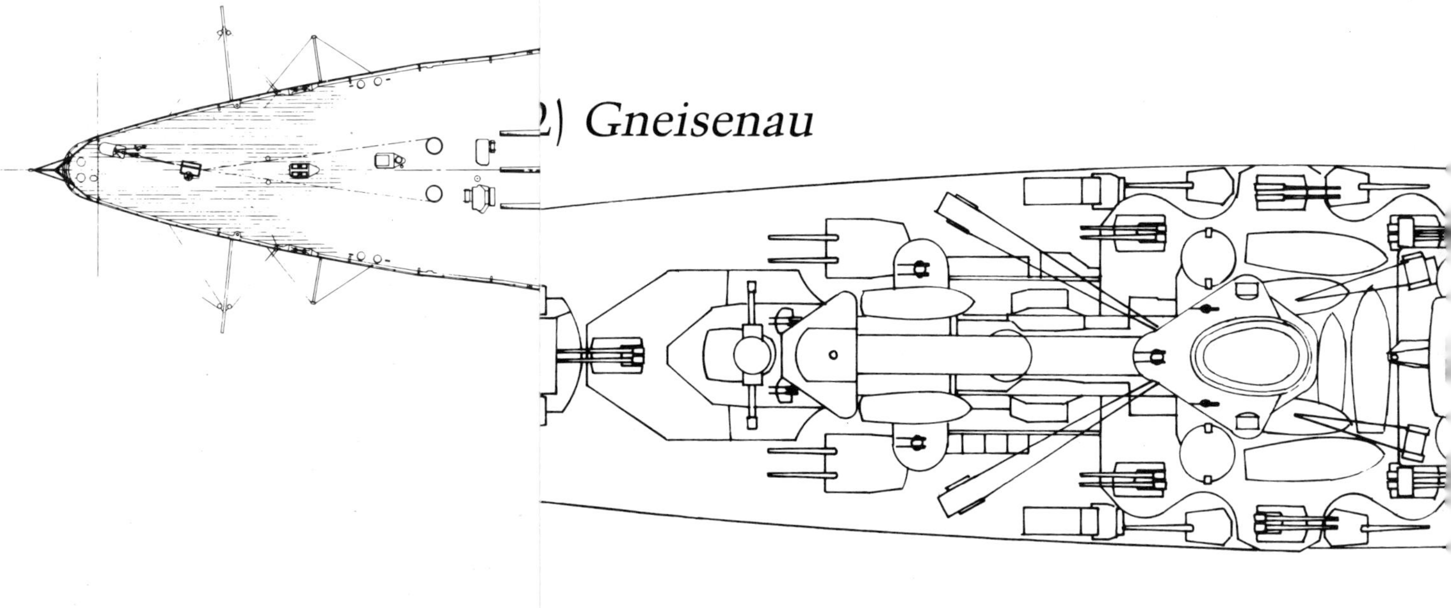

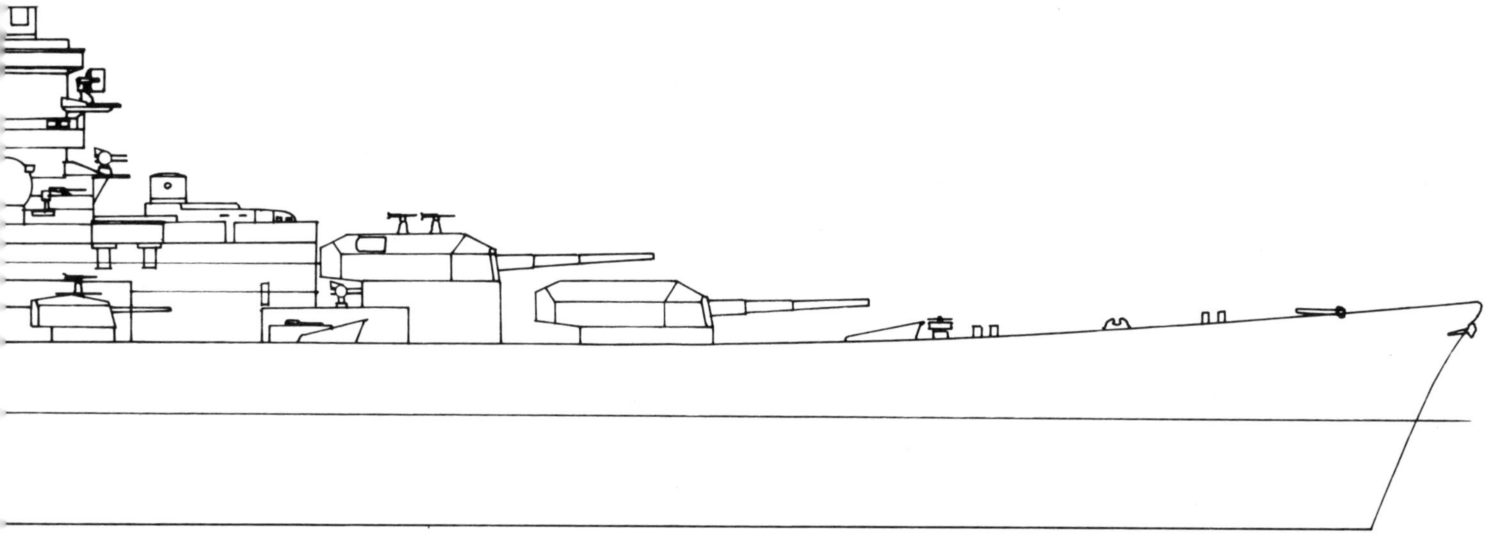

Section in way of 15″ Turret Anton (Gneisenau)

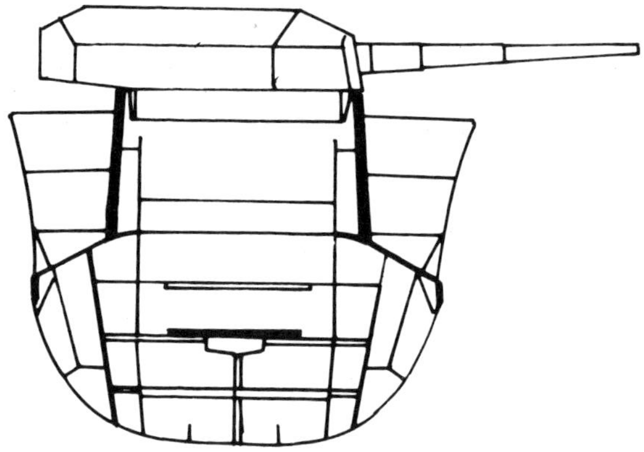

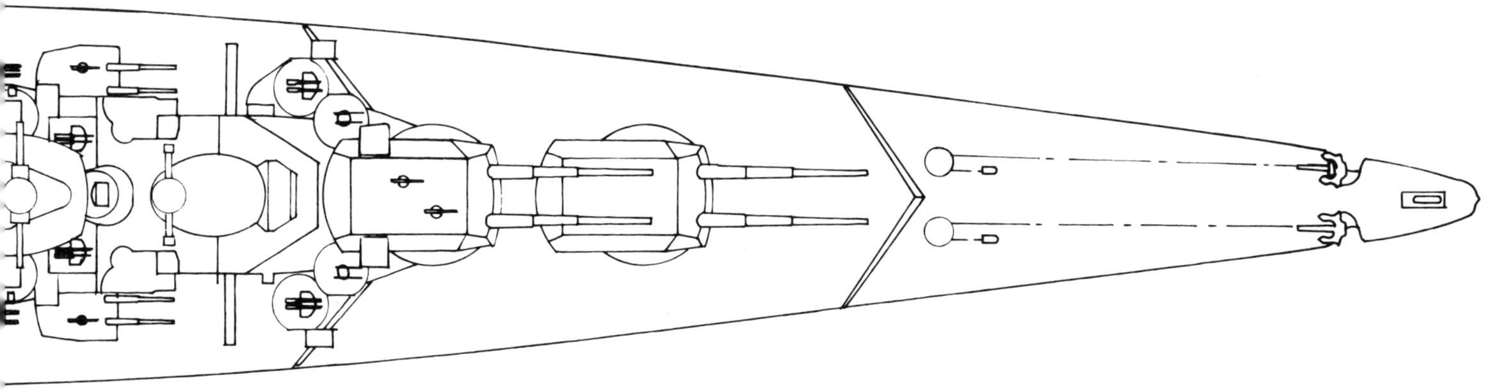

Machinery	
Boilers	twelve Wagner Höchstdruck small watertube type pressure 740 psi (52 kg/cm²) temperature 842°F. (450°C.)
Turbines	*Scharnhorst* — Brown Boveri and Co., Parsons-type geared turbines with single reduction *Gneisenau* — Germania geared turbines with single reduction
Shaft horsepower	design overload — 157,811 (160,000 mhp) design normal maximum — 123,290 (125,000 mhp) *Scharnhorst* service overload — 161,164 (163,400 mhp) service normal maximum — 125,756 (127,500 mhp) *Gneisenau* service overload — 151,893 (154,000 mhp) service normal — 107,509 (109,000 mhp)
Maximum speed	*Scharnhorst* at overload — 31.65 knots at normal maximum — 29.90 knots *Gneisenau* at overload — 30.70 knots at normal maximum — 28.00 knots
Nominal endurance	*Scharnhorst* 10,100 miles at 19 knots on 2 shafts *Gneisenau* 8,400 miles at 19 knots on 2 shafts
Generators	six turbo-generators at 460 kw two diesel generators at 150 kw two diesel generators at 300 kw two turbo-generators at 430 kw total ship's service capacity — 4,520 kw normal electric load — 3,620 kw at 220 volts, DC
Propellers	three 15.75′ (4.8 meter) diameter
Rudders	two

Miscellaneous	
Complement	1,968 (60 officers/1,908 enlisted) — *Scharnhorst* — 1943 1,669 (56 officers/1,613 enlisted) — *Gneisenau*

Weights

Scharnhorst	Weight Summary — 1943		
Hull	7,835 tons	(7,961 mt)	24.6%
Outfit and provisions	1,808 tons	(1,837 mt)	5.7%
Machinery	2,863 tons	(2,909 mt)	9.0%
Armor	14,025 tons	(14,250 mt)	44.0%
Armament and ammunition	5,316 tons	(5,401 mt)	16.7%
Light ship, provisions and ammunition	31,847 tons	(32,358 mt)	100.0%
Liquids from tank capacity tables	6,245 tons	(6,345 mt)	
Full load	38,092 tons	(38,703 mt)	

CHAPTER FOUR

The *Bismarck* Class

The *Bismarck* and *Tirpitz* were the largest battleships ever built for the German Navy and the heaviest capital ships ever completed by any European naval power. Due to their exploits in World War II, they have become legendary. The German Navy unofficially began conceptual studies for a 35,000-ton battleship in 1932, but it was not until 1934 that the Construction Office began preliminary and contract design work on Battleships "F" and "G." After Germany signed the 1935 Anglo-German Naval Treaty, which allowed her to build, but obligated her to observe certain international naval treaties that had been or would be concluded, Battleship "F" was officially ordered by the German government. It was, therefore, entirely permissible in 1935 for Germany to build a 35,000-ton standard-displacement battleship armed with 406-mm guns and still fulfill treaty commitments. The escalator clause in the 1936 London Naval Treaty permitted 406-mm guns and 45,000-ton standard displacement unless Italy and Japan had signed it by 1 April 1937. Since Japan steadfastly refused to sign the 1936 agreement unless granted parity in battleship displacement with the United States and the United Kingdom, the German Naval Staff was convinced that they could build ships with the larger displacement and gun size. Thus, it was entirely legal for the Germans to complete battleships with a standard displacement of 45,000 tons by 1940–1941. When the British government offered an agreement to Germany in 1938 for a limitation on battleship displacement of 40,000 tons, it was refused on the grounds that Germany did not want to be limited as long as Italy and Japan were not parties to the same agreement.

Some early conceptual designs were begun in 1932 to determine what armament, protection, and speed would be possible for a battleship of 35,000 tons. It was recognized that an armament of 305 mm would be entirely feasible for such a displacement, but the only advocate of such an armament was Great Britain, and the French were proceeding to construct a battlecruiser featuring an armament of eight 330-mm guns. The Germans concluded that an armament of eight 330-mm guns, a speed of 30 knots as well as good protection was possible in a 35,000-ton ship. A battleship with an armament of 350-mm guns produced a ship of 41,000 tons. When Italy and France decided to proceed with the construction of 35,000-ton battleships armed with 380-mm guns, the die was cast as far as Adolf Hitler was concerned. He personally intervened in the planning and decreed that the armament would have to be 380 mm. German naval constructors had already concluded that a modern German battleship built in the 1930s should be one armed with 380-mm or 406-mm guns. A better balanced design could be achieved with 380-mm guns, but no firm shipyard commitments could be made as long as the restrictive terms of the Versailles Treaty were in force. Anticipating this, the German Navy ordered the design, construction, and testing of 380-mm and 406-mm guns in 1934, since ordnance was the controlling factor in battleship construction.

The *Bismarck* in December of 1940 en route to Blohm & Voss in the Elbe River to complete her shipyard work. The ship has a slight list to port. (Blohm & Voss)

Several important factors influenced German warship design. Because the German Navy was inferior numerically, its ships had to be faster and more strongly armed than foreign counterparts. To achieve a speed superiority, the early studies evaluated different types of propulsion plants, namely:

- diesel geared drive
- steam drive
- turbo-electric drive

Although the turbo-electric drive produced a heavier propulsion plant, it was the first choice because of its success in the American carriers *Lexington* and *Saratoga* and the French passenger liner *Normandie*. Electric propulsion appeared to have greater reliability than other forms of propulsion. A new German superliner was ordered to be built at the Blohm and Voss shipyard in Hamburg, and it was to feature turbo-electric propulsion. Although diesel propulsion was preferred for its fuel economy, available engines did not have the power necessary to propel a 35,000-ton battleship at a speed of 30 knots.

When the preliminary design commenced in 1934, the very first task was the establishment of a standard displacement that would not excessively limit offensive and defensive qualities. Naval constructors had gained valuable experience and insights in the planning for the *Scharnhorst*-class battlecruisers and the *Deutschland*-class armored ships. They did not want to repeat the faults of these ships in their battleship design. More emphasis would be given to underwater protection, because the larger displacement would permit better protection against mine, torpedo, and near-miss bomb explosions than was possible in the displacement-limited *Scharnhorst*-class battlecruisers. Available battleship designs and studies based upon German experiences in the Battle of Jutland indicated a trend towards ships larger than these battlecruisers. Furthermore, the Construction Office had been continuously following and analyzing trends and developments among the major naval powers— trends in such things as gun size, speed, and standard displacement. All of these contributed to the design of the *Bismarck* class. For example, the French Navy decided to arm its 35,000-ton battleships with 380-mm guns, and during 1934 Italy had laid down two 35,000-ton battleships to be armed with nine 381-mm guns. With British naval opinion favoring the 305-mm gun in 1933 and later the 356-mm gun in 1934–1935, when other navies expressed no interest in smaller-caliber guns for battleships, the new French and Italian battleships had a great influence on the German decision to mount 380-mm guns in their new battleships. It was also decided that the main armament should feature a twin-turret arrangement that would provide equal firepower forward and aft and simplify the firecontrol requirements. The main-battery arrangement closely resembled that of the *Baden* and *Bayern* of World War I. This development has resulted in some speculation that the *Bismarck*-class battleships were mere copies of these older ships. This is false; the new ships had to be faster and have more protection, range, and gunpower. The percentages allocated to armor protection, propulsion, and armament were not the same. The triple-shaft arrangement and the distribution of the main armament and its caliber were the same, but these were the only similarities.

It was decided in 1935 to proceed with the contract plans and specifications for two ships that would replace the predreadnoughts *Hannover* and *Schleswig-Holstein*. During contract design, it was decided to increase the endurance, horsepower, and protection. Although the 380-mm guns would be smaller than those permitted by the Washington Treaty, their ballistic qualities were designed to ensure the penetration of armor of any battleship built under treaty limitations in a North Sea engagement, where the battle ranges would be short. The greatest emphasis, therefore, was on protection and stability because of the possibility of fighting a numerically superior enemy; these ships were to have an upper citadel belt unlike the armor arrangement of the *Scharnhorst*-class battlecruisers. This upper belt armor and more extensive splinter protection in the bow and stern areas resulted in an increased armor weight of over 1,000 tons. These ships had to be larger and more powerful than many of their contemporaries.

The requirements for the *Bismarck*-class battleships are tabulated in table 4-1.

TABLE 4-1

Basic Design Requirements

Standard displacement	35,000 tons (35,562 mt)
Waterline length	820.2 feet (250 m)
Waterline beam	124.67 feet (38 m)
Maximum draft	32.81 feet (10 m)
Armament	15-inch (380-mm) in twin turrets
	5.9-inch (150-mm) gun—secondary armament
	4.1-inch (105-mm) antiaircraft gun—long range
	1.46-inch (37-mm) machine gun—short range
Maximum speed	30 knots
Type propulsion	Steam, turbo-electric drive
Endurance	8,000 nautical miles at 19 knots plus 20% reserve for battle
Protection	to resist 15-inch (380-mm) gunfire at 20,000 to 30,000 meters and a 550-pound (250-kg) TNT charge in contact with the ship's side

The development of the design from these requirements presented a critical, complicated problem. The maximum draft had to be restricted to 10 meters, since the average depth of water in the Kiel Canal was 11 meters. The beam had to be restricted to 38 meters and the length to 250 meters to permit the construction and docking of the ship in the Wilhelmshaven Navy Yard. Even though the number and caliber of the main-battery guns, protection, speed, and endurance were set at minimum permissible levels, the standard displacement, with no fuel and reserve feed water, was estimated to be 43,900 tons on 3 May 1935. With fuel and reserve feed water, the displacement was 51,300 tons with a mean draft of 10.4 meters. Some reduction in draft was necessary to provide an adequate margin for maneuvering in the Kiel Canal and the rather shallow German harbors, particularly the Jade estuary.

An important design consideration was the speed of these new dreadnoughts. The French and Italian designs featured speeds of 30 knots, and it was decided that the German ships should have the same speed. This meant that the propulsion spaces would take up a significant portion of the hull volume and, in combination with the four-turret requirement, would dictate a ship length of at least 240 meters. Such a length had a disadvantage in that the armor required to protect the magazines and propulsion plant would be much more extensive than if triple or quadruple turrets were chosen instead, which is why many of the last-generation battleships were equipped with these. Even the 70,000-ton *Yamato* class was so equipped to save weight. However, in Germany there was no question of twin versus triple or quadruple turrets. There were to be eight 380-mm guns in four twin turrets.

The question of endurance was yet another consideration that could not be compromised. Distances from German ports to the Atlantic Ocean and the problems of the High Seas Fleet in World War I operations emphasized the need for German capital ships to have greater endurance than most foreign warships. Germany had no overseas

The *Bismarck*, Germany's most famous warship, was the most powerful battleship in the world at the time of her completion in August 1940. This photograph of *Bismarck* was taken in 1940 during her trials in the Baltic. (Blohm and Voss)

bases, and greater fuel capacity was sought from the earliest designs. Protection or offensive power, however, could not be reduced, since this would be self-defeating.

From the inception of design work, it was evident that the standard displacement would have to be larger than that permitted by the treaties. On 11 February 1937 the Construction Office advised Admiral Raeder:

> In view of the difficult conditions in our harbors, a displacement of 42,000 tons should be the limit of ships to be built, unless we widen the harbors and channels and also the curves of the Kiel Canal. . . . Politically, it must be decided if the displacement should be still further increased above the Treaty limits.
>
> The Construction Office considers that no further increase should take place as long as we are obliged to the 35,000-ton limit which we have already exceeded by 7,000 tons, so that a further increase could hardly be concealed. . . . The Construction Office would like to build an extra ship rather than exceed the 35,000-ton limit.
>
> Even if other navies renounce the limits of the London Naval Treaty or choose calibers above 380 mm, we should be reluctant to increase them owing to the state of our (harbor) channels and our Treaty obligations . . . a decision cannot be reached until the various aspects including the military factors and the clauses of the Treaty have been considered.

These two views of the *Tirpitz* were taken in the Baltic in 1941. The ship was destined to be part of a "fleet-in-being," which would tie up a number of Allied warships needed in other theaters in World War II. (Imperial War Museum and Albrecht Schnarke Collection)

Later Admiral Werner Fuchs, in charge of the General Command Office in the OKM, recommended to Admiral Raeder and Adolf Hitler changes that would reduce the standard displacement before shipyard contracts were signed. Both approved the changes, but only on the understanding that Italy and Japan would ratify the London Naval Treaty by 1 April 1937. Otherwise, the permissible standard displacement would become 45,000 tons, and no changes would be required. The standard displacement was finally established at 41,400 tons when the contracts to construct the ships were signed with the two shipyards. This figure was to increase slightly with changes in the superstructure and sheer line (Atlantic Bow). Design work for the *Bismarck* was completed on 16 November 1935 and that of the *Tirpitz* on 14 June 1936. Although they were sister ships, their detailed characteristics did vary. Those of the *Bismarck*, as commissioned, are tabulated in table 4-2.

TABLE 4-2
Characteristics of the Bismarck

Standard displacement	41,673 tons (42,321 mt)
Design displacement	45,202 tons (45,928 mt)
Full-load displacement	49,136 tons (49,924 mt)
Battle-load displacement	50,129 tons (50,933 mt)
Waterline length	792.48 feet (241.55 m)
Waterline beam	118.11 feet (36 m)
Draft (standard)	28.54 feet (8.70 m)
(design)	30.51 feet (9.30 m)
(full load)	32.78 feet (9.99 m)
(battle load)	33.37 feet (10.17 m)
Depth	49.21 feet (15 m)
Armament	Eight 15″/47 twin (380 mm)
	Twelve 5.9″/55 twin (150 mm)
	Sixteen 4.1″/65 twin (105 mm)
	Twelve 0.79″ single (20 mm)
	Four Arado-196 floatplanes
Speed (maximum)	30.12 knots
(normal)	29.0 knots
Shaft horsepower	
(overload)	136,200 (138,000 mhp)
(normal maximum)	147,900 (150,000 mhp)
Endurance	9,500 nautical miles at 19 knots
Fuel-oil capacity	8.167 tons (8,294 mt)
Protection	12.60″ (320 mm)—main side belt
	3.74″ (95 mm)—armor deck (magazines)
	3.15″ (80 mm)—armor deck (machinery)

The minor differences in the characteristics of the *Tirpitz* and the *Bismarck* are tabulated in table 4-3.

In order to achieve a successful design, it was decided to employ the most modern equipment and techniques, a practice that had begun with the *Deutschland*-class armored ships. Technical developments, experience, and the characteristics of likely opponents (the new French battleships *Richelieu* and *Jean Bart*) directly affected the choices of speed, ordnance, and protection. Decisions regarding fleet expansion as affecting the development of the *Bismarck*-class battleships were made with concurrent consideration of the military and political situation, the technical facilities available, and the fact that German industry was engaged in war production for the army and Luftwaffe.

TABLE 4-3

Specific Characteristics—Tirpitz

Standard displacement	42,343 tons (43,344 mt)
Full-load displacement	48,794 tons (49,947 mt)
Battle-load displacement	49,778 tons (50,954 mt)
Draft (full load)	32.78 feet (9.99 m)
(battle load)	33.37 feet (10.17 m)
Armament	Eight 21-inch (533 mm) torpedo tubes—quadrupled
Shaft horsepower	136,200 shp (138,000 mhp)—normal maximum
	162,800 shp (165,000 mhp)—overload
Speed	30.8 knots at overload power
Endurance	9,125 nautical miles at 19 knots
Fuel-oil capacity	8,170 tons (8,297 mt)
Protection	12.40″ (315 mm)—main side belt
	3.94″ (100 mm)—armor deck (magazines)
	3.15″ (80 mm)—armor deck (machinery)

The length, beam, draft, and stability limitations all influenced the design displacement. During the preliminary design, several schemes were considered to meet the basic requirements. As the Naval Staff advocated a 30-knot speed and would not compromise, a greater hull length than in the *Scharnhorst* became an absolute necessity. The beam was governed by the need for stability, underwater protection, and sufficient width combined with a relatively shallow draft to offer adequate buoyancy and to make it possible to navigate the Kiel Canal and most German shipyards and harbors, which were shallow. In order to attain the desired draft, it was decided to increase the length and beam. An increase in beam within a satisfactory length-to-beam ratio was considered a more desirable solution than increasing the length and maintaining the beam at a constant value. A longer ship would have been more costly, while a larger beam was better for underwater protection and stability after damage. The 36-meter beam was ultimately determined by the requirements of the propulsion plant, magazines, and the side-protection system.

The superstructure and the sheer profile of the final design were substantially altered after the yards commenced construction. It was decided to carry four Arado-196 floatplanes in hangars, rather than three out in the open on individual catapults. This rearrangement also permitted a better distribution of the secondary armament and the antiaircraft battery, while eliminating the possibility of blast damage to the aircraft from guns in the area. Stowage of aircraft in hangars was also preferred because salt spray could lead to serious corrosion problems in the fuselages of aircraft stowed in the open. The forward command tower was reconfigured so that there was greater distance between the bridge tower and turret Bruno. Test firings of a 380-mm gun at the naval proving grounds indicated that the arrangement first sought would have proved troublesome from the aspect of gun blast. Finally, the bow contour was changed, based on early operational experiences of the battlecruiser *Gneisenau*. This involved the adoption of the "Atlantic Bow" and changes in the anchoring arrangement.

Bismarck—Operational History. The keel for Battleship F was laid down on 1 July 1936 at the Blohm and Voss Shipyard in Hamburg, and the ship, christened by Frau Dorothea von Loewenfeld, granddaughter of Prince Otto von Bismarck, with Adolf Hitler in attendance, was launched on 14 February 1939. The remaining outfitting work proceeded on schedule, despite the war. The *Bismarck* was

This is the moment of launch of the battleship *Bismarck*. Having been christened by Adolf Hitler and her name unfurled over the bow, Germany's new battleship begins her slide down the launching ways of the Blohm and Voss Shipyard in Hamburg. A large number of government dignitaries, including the Navy's commander-in-chief, Admiral Erich Raeder, were in attendance. (Blohm and Voss)

commissioned on 24 August 1940 and on 15 September departed Hamburg for trials in Kiel Bay. The ship returned to Blohm and Voss for some post-trial adjustments and completion of some outfitting on 9 December 1940. Her return to Kiel for further training was delayed from 24 January 1941 until 6 March by a sunken merchant ship that blocked the Kiel Canal. Unusually severe winter weather made it difficult to remove the wreckage.

Preliminaries for Bismarck's first action. In view of the successful forays by German surface ships against Allied shipping in the Atlantic Ocean in 1940–1941, the German Naval Staff had anticipated further dividends in the early resumption of commerce raiding by using the *Bismarck, Tirpitz, Scharnhorst,* and *Gneisenau* in an operation that was to climax a series of surface-ship attacks on merchant shipping in the Atlantic Ocean. At the same time, U-boat activity would be intensified. The battleship-battlecruiser force was to break out into the North Atlantic; the battleships would leave from Baltic ports, while the battlecruisers would sortie from Brest. They were to rendezvous at a prearranged position and then attack convoys north of the equator. They were to coordinate their activities with the U-boats and stay at sea as long as the situation permitted. As the main aim of the entire operation would be to destroy merchant ships, action with enemy warships (other than light convoy escorts) was to be avoided.

The proposed operation was scheduled for the new-moon period around 25 April 1941 when conditions would be particularly favorable. It was anticipated that the battlecruisers would have completed their overhauls by that time and could participate in a combined attack on British merchant shipping, which, in combination with U-boat activity, would have seriously taxed the full resources of the Royal Navy. Moreover, it was thought that within certain reasonable risks, neither the Brest squadron nor the battleship force would have much difficulty with convoy escorts. It was a boldly imaginative plan, but was due to be limited by unforeseen events.

The first complication was that the *Tirpitz* had been commissioned on 25 February 1941, and the earliest she would be ready for combat duty would be the late fall. The Royal Navy had also moved to meet this challenge, since the moment when the *Bismarck* and *Tirpitz* were ready for battle had been anticipated. It was feared that if the *Scharnhorst* and *Gneisenau* broke out of Brest, coincidental with a move into the Atlantic by the *Bismarck* or *Tirpitz*, there would be a severe disruption of convoy traffic. Losses to U-boat warfare had reached a record level during March 1941 when the two battlecruisers were prowling the North Atlantic. Therefore, air attacks on Brest were intensified, with successful results. The *Gneisenau* was torpedoed in the stern, and while dry-docked for repairs, she sustained further bomb damage. The defects in the *Scharnhorst*'s boilers were more serious than the Germans had originally thought, so that neither ship could take part in the operation. Although the British were confident that a battlecruiser attack against their convoy traffic was not a likely eventuality for several months, Brest was bombed on a regular basis. On 22 April 1941 an operation that planned for the use of only the *Bismarck* and *Prinz Eugen* was then decided upon.

Bismarck sorties to Atlantic. On 23 April 1941 the *Prinz Eugen*'s main shaft couplings were damaged and a turbo-generator was displaced off its foundation by a magnetic mine explosion and shock response, repairs which delayed the operation. On 14 May one of the *Bismarck*'s cranes needed to be repaired, and this further delayed the operation by three days. Meanwhile, on 5 May 1941 Adolf Hitler had inspected the *Bismarck* and *Tirpitz* in Gotenhafen (Gdynia). During his tour of the two German battleships he was escorted by Admiral Günther Lütjens. Admiral Raeder was conspicuously absent. In a conference on board the *Bismarck* the morning of 18 May, Admiral Lütjens, in charge of the operation now coded Rheinübung (Rhine Exercise), outlined the operation to the captains of the *Bismarck* and *Prinz Eugen*. He intended to use the Norwegian Sea and the Denmark Straits, between Iceland and Greenland, to reach the Atlantic Ocean, and if the weather was favorable (i.e., fog, reduced visibility, etc.), not to stop in Norway (the operation called for refueling in Norway), but to continue into the Norwegian Sea and refuel from the tanker *Weissenburg* before their dash to the Denmark Straits.

After both ships left their moorings in the inner harbor, they proceeded to the outer roads, where they anchored and took on more fuel. The *Bismarck* was not refueled to her capacity because one of the fuel hoses broke, whereupon refueling ceased. The ship was filled to within 200 tons of her capacity of 8,294 tons. The *Prinz Eugen* left the outer roads first, conducted degaussing trials offshore, and was joined briefly by the *Bismarck* later that afternoon. At dusk the *Prinz Eugen* steamed into the Baltic and at 0200 on 19 May the *Bismarck* followed her course. The only indication that British

The *Bismarck* is shown leaving the Blohm and Voss Shipyard for the last time in early 1941 to begin intensive training in the Baltic for possible Atlantic Ocean operations. The *Bismarck* had returned to the shipyard for post-trial adjustments and a minor overhaul. (Imperial War Museum)

The Kiel Canal is a major waterway between the North Sea and the Baltic. The *Bismarck* is shown here in 1940 on her way for trials in the Baltic. Note the tugs deployed at the bow and stern to ensure that the battleship remains in the channel. (Albrecht Schnarke Collection)

This starboard view of the *Bismarck* was taken during her training in the Baltic. Note the camouflage scheme with the bow darkened and the false bow wave. Such a scheme was designed more to confuse the enemy's perception of the ship's heading than to prevent sighting the ship at all. The carmine turret tops stand out. (Albrecht Schnarke Collection)

A view of the *Bismarck* from *Prinz Eugen* during their voyage to Norway during May 1941. The hull camouflage scheme would be painted out while the ship was in Grimstadfjord near Bergen on 21 May 1941. (Imperial War Museum)

intelligence had of the Rhine Exercise by 18 May was some intercepted and decoded Luftwaffe radio traffic discussing delivery of nautical charts to the *Bismarck*.

The two ships joined up off Cape Arkona (Rügen Island) on Monday 19 May and two destroyers and a minesweeping flotilla also arrived to provide an escort. A third destroyer would arrive later. It was at this time that Captain Ernst Lindemann of the *Bismarck* informed his crew that they were headed for a three-month operation in the Atlantic Ocean to hunt and intercept British convoys. The German battlegroup continued northwestward into the Baltic without incident and during the night of 19 to 20 May began the passage of the Great Belt. Around noon on 20 May they reached the Kattegat in clear weather. The Luftwaffe provided some air cover.

By early afternoon, the two ships, three escorting destroyers, and the minesweepers had cleared the Kattegat and were off the west coast of Sweden. They encountered a number of Danish and Swedish fishing boats, and concern mounted in Lütjens' staff that the German fleet's presence in these waters could be reported to the Royal Navy. At 1300 they came upon the Swedish cruiser *Gotland* off Marstrand, Sweden, and she sent a routine signal to Stockholm, "Two large ships, three destroyers, five escort vessels, and 10–12 aircraft passed Marstrand, course 205°/20'." The signal was intercepted and decoded by an intelligence team on the *Bismarck*. This development prompted Admiral Lütjens to signal Group North that his position had been compromised. A Swedish informant would later notify the British naval attaché in Stockholm of the *Gotland*'s signal.* The German task force was slightly delayed by a convoy of eleven German and neutral merchant ships at the minefield that blocked the entrance of the Kattegat from the Skaggerak. They cleared this minefield around 1600† and were visible from the Danish and Swedish coasts. The minesweepers were detached. As the German task force hugged the Norwegian coast, it was spotted by a Norwegian resistance group off Kristiansand between 2100–2200, and this sighting was also passed on to British intelligence. At 0640 on 21 May radio intelligence personnel on board the *Prinz Eugen* decoded British orders to coastal aircraft to be on the lookout for two battleships and three destroyers. Just after sunrise planes were spotted low on the horizon, but they made no effort to attack. In bright sunshine and with clear skies, the German ships entered the Bergen fjords without incident and by midmorning were at anchor.‡

Admiral Lütjens requested the latest intelligence on the deployment of British warships, and a Luftwaffe officer was flown up from Stavanger to brief him on air-reconnaissance photos of the Denmark Straits and Scapa Flow. He was concerned whether British warships were at Scapa Flow or at sea. In the meantime, the *Prinz Eugen* and the three destroyers refueled from the tanker *Wollin*. The *Bismarck* did not refuel since this was not in her operational orders—a fact that the authors consider a fatal oversight. Additionally, there was not enough time to refuel all five ships in Norway.

British response was swift. Two Spitfire reconnaissance planes were dispatched from Scotland to determine the identity of the two German warships. Around 1315 on 21 May, a British Spitfire overflew and photographed the *Bismarck* in Grimstadfjord and the *Prinz Eugen* in Kalvanes Bay where she was refueling. After the German

---

*This was accomplished through an intermediary—Colonel Roscher Lund, who had become a trusted confidant of the British naval attaché, Captain Henry W. Denham (RN).

†This is German summer daylight savings time, which coincided with British double summer time. To make it easier, all times in this narrative will be in British time.

‡The authors believe that the clear skies finally convinced Admiral Lütjens that he should stop over in Norway before proceeding to the Norwegian Sea. It was important that the operation be concealed by bad weather.

presence in Norway had been confirmed, Admiral Tovey alerted the heavy cruisers *Norfolk* and *Suffolk*, already on patrol in the Denmark Straits; the light cruisers *Birmingham* and *Manchester* were also warned and were later to be joined by the *Arethusa*. The battleship *Prince of Wales*, battlecruiser *Hood*, and six destroyers were ordered to patrol the passages into the Atlantic, particularly those north of 62° latitude. This force, under Vice Admiral Lancelot Holland, sailed before midnight on 21 May.

At 1930 on 21 May, the *Bismarck* weighed anchor and headed for Kalvanes Bay to join the *Prinz Eugen* and the three destroyers. By 2300 the German force had cleared the fjords around Bergen and shortly before midnight were on a course towards the Arctic Ocean. During that same evening, the Coastal Command sent high-level bombers to attack the German ships in Bergen. The planes bombed a cloud-covered empty anchorage.

After the German force was at sea, Adolf Hitler was advised of the operation for the first time, by Admiral Erich Raeder. Even though he had been assured of the ships' intentions, Hitler still had very considerable reservations. In the end, he allowed Admiral Raeder to let the operation continue, but he did not want it to cause any problems with the United States, since final preparations were under way for the assault against the Soviet Union on 22 June 1941, and an operation against Crete had already commenced earlier that day.

Admiral Tovey received the news of the aircraft raid on Bergen, and under his orders on 22 May a British reconnaissance plane was dispatched to fly over the anchorage where the *Bismarck* had been spotted to determine if the *Bismarck* was still there. The plane crew returned to report that the anchorages were empty! At 2215 on 22 May the battleship *King George V*, aircraft carrier *Victorious*, light cruisers *Galatea, Aurora, Kenya, Neptune*, and *Hermione*, and six destroyers left Scapa Flow for the North Atlantic. The battlecruiser *Repulse* was ordered to leave the Clyde and join this British task force north of the Hebrides during the morning hours of 23 May.

At 0415 on 22 May the three German destroyers were detached off Trondheim. The weather remained misty, and the battleship/cruiser force maintained a speed of 24 knots during most of the day. Around 0930 Admiral Lütjens was advised by Group North that the latest intelligence information indicated that the British were not aware of his departure from Bergen. Around noon, the two German warships turned to port and began their dash to the Denmark Straits. Admiral Lütjens signaled the *Prinz Eugen* that he would not refuel from the tanker *Weissenburg* unless the weather cleared. After 2300 Lütjens received three important messages:

- One confirmed another received earlier that morning, which said, "Assumption that breakout has not yet been detected by the enemy confirmed."
- The second, based on Luftwaffe aerial surveillance of Scapa Flow on 22 May, indicated the presence of four battleships, possibly one aircraft carrier, apparently six light cruisers, and several destroyers.
- The third reported that no operational commitment of enemy forces had been noted and, in view of the Crete Operation, observed that an early Atlantic operation might inflict serious new damage on the British position at sea.

These radio messages probably influenced Admiral Lütjens in his earlier decision to run through the Denmark Straits, and at 2322 he ordered his ships to alter course accordingly. The report was in error, because the *Hood* and *Prince of Wales* had already left by the time the reconnaissance plane had flown over.*

---

*This was later acknowledged officially by German analysts when they reviewed the radio messages sent to the *Bismarck*.

At 0400 on 23 May the two ships increased speed to 27 knots, the *Bismarck* leading the *Prinz Eugen* by some 700 meters, with visibility down to 3,000 to 4,000 meters in light mist. Around 1000 some ice was encountered, and speed was reduced to 24 knots. About noon they were due north of Iceland. At 1821 they reached the limit of the ice barrier and headed south into the Denmark Straits, after having a false sighting at 1811. From then on, a zigzag course was necessary to avoid the ice floes.

At 1922 the German ships, using hydrophones and radar, made contact with the *Suffolk* at approximately 12,500 meters. The British cruiser, ahead of them to starboard and on a course to the southwest, went on the air to report "one battleship, one cruiser" and gave their course as 240 degrees. The Germans had a brief glimpse of the *Suffolk* as she turned into a fog bank. The German ships passed the British cruiser, which then began to follow astern. About one hour later, the *Bismarck*'s radar detected a second ship, and immediately this information was passed on to the *Prinz Eugen*. This was the *Norfolk* moving to port of the German force. Suddenly this British cruiser emerged

The *Bismarck* was sighted in Grimstadfjord near Bergen early during the afternoon of 21 May 1941. This photo shows the *Bismarck* at anchor, with two merchant ships deployed off her bows. The irregular shape at top center is a peninsula projecting into Grimstadfjord. (Imperial War Museum)

A view of the starboard side of the *Bismarck* in Norway on 21 May 1941. Note the discoloring in the side shell below turret Caesar. We believe this discoloration to be from the crew painting the sides a darker grey to match the British standard grey. Also note the turret tops are now painted yellow instead of carmine red. The structure immediately forward of this discolored area is an accommodation ladder for personnel transfer by boat. (Albrecht Schnarke Collection)

The battlecruiser *Hood* is shown in Scapa Flow in 1941 before her fateful duel with the *Bismarck*. The mighty *Hood* was in battle for eight minutes before an explosion in her after magazines tore her apart. (Imperial War Museum)

The *Prinz Eugen* was a handsome warship, the only surviving large ship in the German Navy in World War II. She was in action during the famous battle between the *Bismarck, Hood,* and *Prince of Wales.* (R. O. Dulin Collection) Shell splashes from the doomed *Hood* land near the *Prinz Eugen.* A few moments later the British battlecruiser would be destroyed by shells from the *Bismarck.*

from a fogbank and came into sight of the *Bismarck* at a range of 6,400 meters. With her guns trained to port, the *Bismarck* immediately engaged the British cruiser, firing five salvos from her 380-mm guns. Three of these straddled the *Norfolk*, one falling astern, and another missing completely. Some shell fragments, however, hit the ship, and the *Norfolk* turned away into a fog bank and increased speed to open range. The *Bismarck* had her forward radar set disabled by gun blast, and Admiral Lütjens ordered the *Prinz Eugen* to take the lead and make the radar sweeps forward.

The *Suffolk* positioned herself starboard astern of the *Bismarck*, near the limit of her radar range, which was 24,000 meters, and the *Norfolk* was stationed to port of the German squadron. The pack ice prevented any extensive maneuvers to starboard. Occasionally a snow or rain shower would obscure the German ships from their British observers, but the radar set on board the *Suffolk* kept track of the German ships. At one point around 2200, the German ships steamed into a heavy rain shower and increased speed to 30 knots. Then the *Bismarck* executed a 180-degree turn to starboard and headed for the *Suffolk*. This maneuver caused the British warship to alter course to the east and lose contact. The radar signals, however, were still being received aboard the German ships. In fact, as time passed by, Admiral Lütjens became very concerned with the efficiency of these radar waves, not knowing that his latest maneuver had placed him out of *effective* radar range of the British radar.* During the very early morning hours of Saturday, 24 May, the four ships, *Bismarck, Prinz Eugen, Suffolk*, and *Norfolk*, kept on a parallel course to the southwest.

To the southeast, the *Hood, Prince of Wales*, and four destroyers were on an intercepting course, having received the *Suffolk*'s position reports, and all crews were at battle stations. At 2400 this British battlegroup was 100 miles away. At their present course and speed, they projected that they would cross ahead of the German ships by some 60 miles. By altering their course, interception could be possible sooner, but this would mean a night action. By altering course to the northwest, Admiral Holland estimated that he might intercept the German ships in such a way that they would be silhouetted against the fading twilight (sunset at these latitudes occurred early in the morning). Since the *Hood* and *Prince of Wales* would be approaching from the east, it would allow some time to close the German force before the British ships could be seen.† The loss of radar contact by the *Suffolk* foiled this plan. Holland ordered a course change to the southwest, and the two battlegroups were on parallel courses some 20 miles apart at 0141. Shortly thereafter the German ships altered course to the west to follow the ice barrier. This resulted in the divergence of the two battlegroups—a situation unknown to Holland. It was not until 0300, when the *Suffolk* re-established contact, that the British admiral had a position for the *Bismarck*. By that time, the German ships had opened range to 35 miles. At this time Admiral Holland made his fateful decision to intercept. Instead of a swift closing head-on approach, the ships' courses would converge at a wide angle that would expose the *Hood* to the *Bismarck*'s penetrative fire for a longer period of time.

Action with the Hood and Prince of Wales. At 0515 on 24 May the *Prinz Eugen* made a sound contact identified as high-speed propellers. By 0545 smoke was sighted off the

*This continued encounter with the British ships was something German planning had wanted to avoid. An undetected passage was extremely important to success in intercepting convoy traffic.

†The German ships were equipped with both radar and sound-detection gear, and the authors consider it doubtful that tactical surprise would have been achieved.

port beam, and warship masts soon appeared on the horizon. The range decreased rapidly, and the German ships went to battle stations.

To officers on the *Bismarck* and *Prinz Eugen*, ship identification was uncertain. Some of them correctly identified the lead ship as the *Hood*, while others believed it to be the *Exeter*. By 0552 the opposing forces had closed to 26,000 meters, and the *Hood* opened fire. She was followed at 0553 by the *Prince of Wales*, which the Germans correctly identified as a *King George V*-class battleship. The chief gunnery officer on the *Prinz Eugen* still believed the lead ship to be a cruiser, and ordered his gun crews to load high-explosive 203-mm ammunition, perhaps the best ammunition for a cruiser to use in engaging a capital ship. *Prinz Eugen* opened fire at 0555.

Although the *Bismarck* was straddled by the fourth salvo from the *Prince of Wales*, Admiral Lütjens did not permit his ship to return fire. Several anxious minutes passed until Captain Lindemann gave the word, "Permission to fire."* Meanwhile, the *Hood*'s first salvo had fallen harmlessly around the *Prinz Eugen*;† her 381-mm shells dropped 140 to 460 meters abeam in the wake of the cruiser. Shortly after this, a large jagged shell splinter landed near the base of the cruiser's funnel. This was the nearest that the *Prinz Eugen* would come to receiving battle damage during the entire engagement. The splinter came from a 356-mm shell of the *Prince of Wales*, which had overshot the *Bismarck*.

It was difficult for the British ships to use their entire main battery because of their acute approach angle to the German battlegroup. A few minutes after opening fire, the British ships began a 20° port turn, which would bring their after turrets into action. Both German ships concentrated on the *Hood*, the leading ship, firing four-shell salvos in quick succession. Within a minute the *Prinz Eugen* made a hit with a 203-mm high-explosive shell that struck the base of the *Hood*'s mainmast near the hangar and the light-steel UP ammunition lockers. A large fire flared up on the *Hood*, but was quickly extinguished. Nonetheless, this helped the gunnery officers on the *Bismarck*, as their firing directors and stereoscopic rangefinders were quickly able to gain an accurate bearing and range. By the third salvo from the *Bismarck*, the *Hood* had been straddled by 380-mm shell splashes. After the third salvo from the *Prinz Eugen*, she was ordered by the *Bismarck*'s gunnery officers to shift fire to the *Prince of Wales* at 0558.‡

When the fire broke out in the *Hood*'s UP lockers, Commander Adalbert Schneider, the gunnery officer of the *Bismarck*, ordered a full eight-gun salvo, the fourth salvo to be fired at the *Hood*. A fifth salvo was fired, but it was one or more shells in the fourth salvo that hit the *Hood* near her after turrets, causing a large explosion at 0601. Flames leaped hundreds of meters into the air, followed by a large glowing ball of white fire from which orange-like flares showered down.§ One of the after turrets and the mainmast flew into the air. A giant smoke cloud towered thousands of meters high. Through the dense smoke, broken sections of the bow and stern could be observed. As they slowly

*There has been much speculation on what happened during these crucial moments on the *Bismarck*'s bridge. Baron von Müllenheim-Rechberg has stated in his book, ***Battleship** Bismarck, A Survivor's Story*, that Captain Lindemann was heard to mutter before giving the order to open fire, "I will not let my ship be shot from under my ass. Permission to fire."

†Admiral Holland realized his error and unsuccessfully tried to inform his gunnery officer. The gunnery officer of the *Prince of Wales* had realized his admiral's error and opened fire on the rear ship, the *Bismarck*.

‡The Germans were convinced that she was the *King George V*, never thinking it possible that the new *Prince of Wales* would be sent into action with so little time for training. The *Tirpitz* was not included in the operation for precisely this reason.

§Captain Collin McMullen, chief gunnery officer of the *Prince of Wales*, told the authors that his position became bathed in orange light after the *Hood* exploded.

These three photographs are views of the *Bismarck* engaging the *Prince of Wales*. Note the huge size of the gunpowder smoke clouds from the 380-mm guns.

Shells from the *Prince of Wales* fall short of the *Bismarck*. However, the German battleship has already been struck by a 356-mm shell forward, and she is down by the bow. This photo was taken around 0556–0559, when the *Prinz Eugen* was between the three battleships, just before the *Hood* was hit by a 380-mm shell that devastated her after magazines.

This is, perhaps, the most famous photograph of the *Bismarck-Hood-Prince of Wales* battle. The *Bismarck* was engaging the *Prince of Wales* at this moment. The photo is somewhat underexposed.

sank below the surface, the forward guns fired one last salvo, like a defiant gesture. The *Hood*, in her second sea battle, was hit, blown up, and sunk in an interval of only eight minutes!* British destroyers later picked up only three survivors from a crew of 1,419 officers and enlisted men.†

By the time the sixth salvo could be fired, the *Hood* was already sinking, so fire was shifted to the *Prince of Wales*, which had to maneuver radically towards the German ships to avoid colliding with the debris from the *Hood*. This made the fire-control problem easier for the gunnery officers in the *Bismarck*. Although the seventh salvo landed short, two hits were made during the eighth and ninth salvos around 0602. As the range decreased to 18,300 meters, the *Bismarck* opened fire with her 150-mm guns. Shell splashes were now ringing the *Prince of Wales*. By 0605 the range decreased to 14,000 meters, and the torpedo tubes of the *Prinz Eugen* were poised for action should the opposing forces come much closer. They were not used, because the British battleship began a turn to port and opened the range. The British cruiser *Suffolk* placed several 203-mm shells well astern of the *Bismarck* around 0600, but ceased fire a few minutes later because of the extreme range. A large number of shells from the German ships landed around the *Prince of Wales*, which ultimately had to withdraw from the action and lay down a protective smokescreen, a maneuver to consolidate position.‡

At 0609 the *Bismarck* and *Prinz Eugen* ceased fire, after scoring three 380-mm and four 203-mm hits on the *Prince of Wales*. The *Bismarck* had fired 93 armor-piercing shells, while the *Prinz Eugen* had fired a total of 179 armor-piercing and high-explosive shells. The *Prince of Wales* had been the target of many of these shells, and she sustained heavy topside damage, moderate flooding from shell waterline hits, plus several malfunctions in her main-battery turrets and the temporary loss of the after turret as she made a sharp port turn away from battle. About that time a Coastal Command Catalina attempted to determine the damage to the *Bismarck*, which was trailing an oil slick, but the aircraft was driven off by intense and accurate antiaircraft fire.

---

*The *Hood*'s first sea battle was against the French fleet in Oran. She damaged the battlecruiser *Dunkerque* and pursued the *Strasbourg* for over two hours. Commander J. Wellings, USN, was aboard her four months before the *Bismarck* action and stated to the authors that her fire-control system was no better than that of the old battleship USS *Florida*, which he had served aboard in 1927.

†Boards of inquiry were convened in June and September 1941 to determine the cause for the loss of the *Hood*. The official report attributed the cause to a 380-mm shell that penetrated the side armor and exploded in or near the after main-battery magazines. It was also stated that the shell landed in the 100-mm gun magazines just forward of turret "X" and caused an explosion of this ammunition, which then set off the 381-mm powder magazine. The authors' research indicates that it might also have been possible for a shell with an underwater trajectory to have struck the ship deep underwater and penetrate to the powder magazines, whereupon they exploded.

‡This cessation of action with the *Bismarck* was severely criticized by officers and enlisted men in other British naval ships and throughout the Royal Navy. The *Prince of Wales* was unjustly nicknamed, "the ship that ran away," and her commanding officer, Captain John Leach (RN), was threatened with a court-martial by Prime Minister Winston Churchill. This is an unfair criticism, and the authors believe Captain Leach should have been decorated for taking an unprepared ship into battle and achieving an outstanding and remarkable performance from her crew. A ship only two weeks out of a shipyard had taken seven hits, had her bridge hit by a major-caliber shell, had a number of malfunctions in two main-battery (quadruple) turrets, and scored two decisive shell hits on the *Bismarck*. It should be remembered here that the Germans held back the *Tirpitz* because they believed it required some 6-8 months before she would be ready for combat duty.

Shell damage. Early in the encounter the *Bismarck* had been hit by two 356-mm shells and took a glancing blow from a third; all were fired at ranges of between 20,000 to 22,000 meters with low angles of incidence. The most severe damage was caused by a projectile that penetrated the bow and an armor-piercing shell that detonated in the port underwater protective system, just abaft the main conning tower (see figure 4-1).

The crucial hit came from a 356-mm shell that passed through the 60-mm splinter belt in compartment XXI (two compartments forward of the forward armored bulkhead) without exploding, and opened a 1.5-meter exit hole in this belt, above the waterline but below the bow wave. The shell trajectory carried it through a main transverse bulkhead between compartments XX and XXI. Water began to flood two watertight compartments (including the anchor windlass compartment) and began to displace the fuel oil in bunkers there. These spaces eventually filled with approximately 1,000 to 2,000 tons of salt water. At first, there was only a small intake of water. Damage control recommended a reduction in speed and trimming down by the stern, which would bring the shell entry and exit holes above the bow wave so that they could be made watertight. Admiral Lütjens would not accept the reduction in speed. As a result of the high speed during the morning hours, the ship occasionally plunged into the sea; the area of damage increased, and more water entered the ship. The *Bismarck* was making 28 knots at the time and could not, for tactical reasons, decrease her speed.

Additionally, a main watertight bulkhead had been damaged; the main line from the forward fuel tanks was cut; and a bilge and flooding pump and a fuel-oil transfer pump were underwater and could not be reached. This meant that almost 1,000 tons of fuel was cut off, some beginning to be contaminated with salt water, and still more was leaking into the sea. The ship was down 2° by the bow and listing slightly to port.

After the battle with the *Prince of Wales* ended the damage control party tried to control the intake of water. The splinter belt damage could not be repaired without stopping the ship. With the forward pump room under water, it was found that the pumps in compartment XVII could not provide sufficient suction. Moreover, the manifolds in the fuel lines under water were no longer serviceable. To prevent the complete loss of fuel oil in compartments XXI and XXII, damage control parties attempted to bypass the pumps and valves by rigging a fueling hose to transfer oil to settling tanks in the boiler rooms. These efforts failed. To prevent further damage to the shell plating from the motion of the ship into the heavy seas, divers placed collision mats over the holes inside the ship, but these merely slowed the seepage. Speed was reduced to 22 knots while this operation was completed by the divers.

The second shell caused no significant structural damage. It demolished the captain's motor launch on the port side and went overboard to starboard without explod-

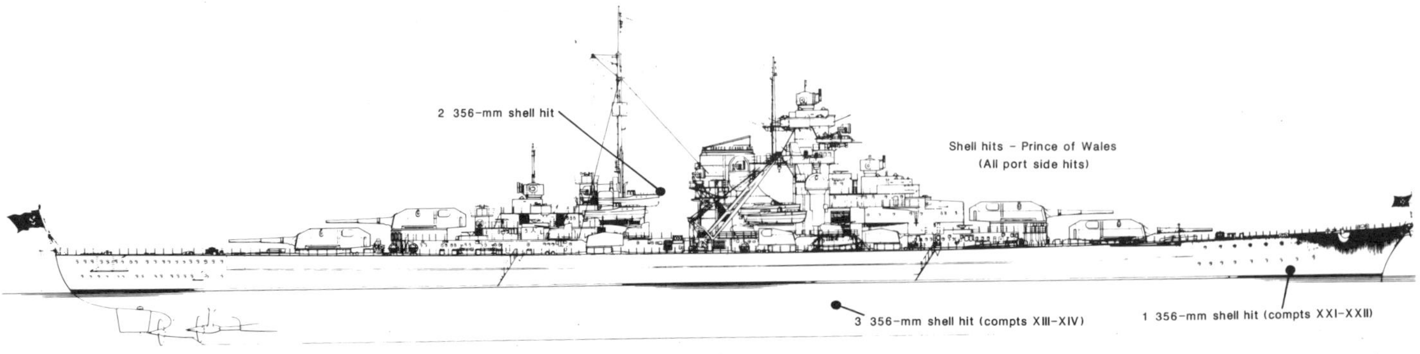

*Figure 4-1. Shell damage—*Bismarck, *24 May 1941. Despite her crew's incomplete training,* Prince of Wales, *using her type 284 radar, obtained three hits on the* Bismarck *(numbers one and three were crucial ones).*

These three sequential photographs document the destruction of the *Hood* as viewed from the *Prinz Eugen.* The upper photograph shows the beginning of the deflagration which split the *Hood* apart. The near vertical smoke cloud at the right is from the fire caused by a shell hit in the aft 102-mm powder magazine which then spread to the 381-mm powder magazines. The black cloud just to the left is from the fires of the propellant for the unrotated projectiles stowed on the boat deck. The *Prince of Wales* can be seen to the left of the *Hood* and has fired a salvo at the *Bismarck.* The middle photo shows the smoke cloud from *Hood*'s explosion with the *Prince of Wales* having moved just to the right of this cloud as a *Bismarck* salvo falls short. Note the close spread of the shells at this distance. The cruiser *Norfolk* is seen in the center with a salvo from *Prince of Wales* falling short. The lower photo shows the *Norfolk* in the center with the smoke from *Hood*'s demise now dissipating.

ing. Fragments of the motor launch damaged the catapult swing and severed or bent a compressed-air pipe, which prevented the launching of aircraft. No men were injured. The full nature of this damage was not discovered until 27 May when the attempted launching of an Arado-196 floatplane met with failure.

The third shell had a slight underwater trajectory before penetrating the side shell below the main side belt in compartment XIV. It exploded against the torpedo bulkhead, outboard of turbo-generator room 4 on the port side and tore holes in the 45-mm armored bulkhead. It struck near the main transverse bulkhead between compartments XIII and XIV, and large cracks appeared on the outboard strakes of this bulkhead, which was the forward boundary of the port number 2 boiler room. Splinters from the shell also severed a main steam line in the turbo-generator room, scalding the occupants and necessitating the shut-down of the generators. Splinters also passed into several adjacent wing and double-bottom tanks. The hit was in an area where the torpedo defense system began to diminish in width. The loss of this electrical generating station was not serious, since the power supply of German battleships was so conceived that either the forward or the after station could supply the maximum power individually (100 percent reserve). Repair parties stuffed the cracks in the transverse bulkhead with canvas hammocks to keep the flooding in check.

After these hits, which flooded the wing compartments, the generator room, and the two forward compartments, the trim increased to 3 degrees by the bow and there was a 9-degree list to port. The combination of trim and list brought the starboard propeller blades out of the water at times. The cutoff of fuel forward also aggravated the trim problem, since fuel now had to be taken from the midship and after bunkers. Starboard void tanks between the steering-gear room and after armored bulkhead (compartments II and III) were flooded to reduce the trim and list. This also corrected the problem with the starboard propeller. Speed was limited to 28 knots because of the large bow trim.

One hour after the battle had ended, Admiral Lütjens and Captain Lindemann appear to have discussed which course of action to take. Admiral Lütjens decided to head the *Bismarck* to either Brest or St. Nazaire for repairs, which he believed would not be major. The damage the *Bismarck* had received was severe enough to interrupt her mission. The undamaged *Prinz Eugen*, however, was to conduct actions against convoy shipping. The decision to return the *Bismarck* to port was based upon these factors:

- The maximum speed of the *Bismarck* was only 28 knots, but this could be further reduced to only 26 knots if the two boilers in boiler room 2 were permanently disabled through flooding. Fifty percent of the reserve electrical power had already been lost through the flooding of turbo-generator room 4.
- Damage to the bow structure and the tanks outboard of the boiler machinery room was causing the loss of fuel oil. If all the fuel in the affected area were lost, the operational endurance would be reduced by 1,100 miles. If the ship continued her mission, the damage in this section of the ship to the piping and structure was of a nature that it could not even temporarily be repaired at sea.
- The British ships in pursuit had unexpectedly efficient radar gear that enabled them to maintain contact under the most adverse circumstances. The ship could not be refueled if contact was maintained.

Table 4-4 is a graphic representation of the fuel-oil problem existing between 0700 and 0800 on 24 May. It has been reconstructed through a review of known ship speeds during the course of the operation and the use of endurance curves derived from trials in the Baltic Sea. It can be seen that the German command aboard the *Bismarck* had little choice but to head for France.

TABLE 4-4

Fuel Situation Aboard Bismarck (0700, 18 May, to 0800, 24 May 1941)

May 18	*Bismarck* loaded with 8,100 tons of fuel oil, of which 7,800 tons is burnable. Some 200 tons more were not loaded because of a broken fuel-oil hose. Maneuvers with *Prinz Eugen* during the afternoon and early evening Total fuel expended—*150* tons.
May 19	At 0200 departs for mission. In Baltic, steaming at 18 knots. Total fuel expended—*400* tons.
May 20	German force arrives at Great Belt and Skaggerak. Speed between 0000–2100 averages between 17–18 knots. After leaving Skaggerak and off southern Norway, speed is increased to an average of 22 knots. Total fuel expended—*432* tons.
May 21	Off southern Norwegian coast and arrives in Bergen at 1030. Speed between 0000–0900 was 24 knots, reduced when entering Bergen fjords. At anchor from 1030–1930. Between 1930 and 2300 at slow speed leaving fjords, but between 2300 and 2400 builds up speed to 18 knots. Total fuel expended—*425* tons.
May 22	Ship is steaming towards Denmark Straits at 24 knots. Total fuel expended—*640* tons.
May 23	Steaming towards Denmark Straits and arrives there around 1821. Average speed is 24 knots. At 1922 ship is sighted by the *Suffolk*, and speed is varied between 24–30 knots. Total fuel expended—*710* tons.
May 24	The *Suffolk* and *Norfolk* are in contact and at 0553 are engaged with the *Hood* and *Prince of Wales*. Speed is almost a constant 30 knots. Fuel expended to 0700—*435* tons. Fuel contaminated and lost—*1,000* tons.

Fuel Summary

Burnable fuel (18 May 1941)	7,800 tons
Fuel consumed (+ 10%)	3,500 tons
Contaminated fuel	1,000 tons
Available fuel (0700, 24 May)	3,300 tons

Projected fuel rate at 30 knots—1,000–1,100 tons per day on 10 boilers. Hence, the available fuel on the *Bismarck* was sufficient for a range of slightly over 2,000 miles at 30 knots, presuming all burnable fuel was consumed—a very unlikely eventuality.

During the day a few British reconnaissance planes were driven off by intense antiaircraft fire. Despite all the *Bismarck*'s maneuvers and high speed, the *Suffolk, Norfolk*, and *Prince of Wales* maintained their pursuit, some 40,000 meters astern. Admiral Lütjens decided to detach the *Prinz Eugen* during the mid-afternoon. At 1420 speed was reduced to 24 knots in expectation of the separation maneuver. The first attempt to break contact was thwarted because the *Suffolk* was still visible from the *Bismarck*. At 1814 the two German ships successfully separated; the *Bismarck* turned 180° to starboard, closed on the *Suffolk*, and opened fire with her main battery, closing to 18,300 meters before that cruiser withdrew at high speed under a smokescreen. The *Suffolk* fired nine salvos before retiring. The *Prince of Wales* commenced fire at 1846 from 9,000 meters to the port side of the British cruiser. Meanwhile, several 380-mm salvos fell some 900 meters to the port side of the *Suffolk*. A total of twelve salvos were fired from the 356-mm guns of the *Prince of Wales* at ranges around 27,500 meters. The British observed one straddle. The *Bismarck* fired nine salvos in this skirmish at the *Prince of Wales*. No hits were scored. As a result of this encounter, all three British ships took up positions to the port side of the *Bismarck*.

This starboard view of the *Bismarck* was taken during the afternoon of 24 May 1941. Her main guns are trained in, and she is down by the bow from flooding caused by two 356-mm shells—one in the bow and the other to port abreast of the forward command tower and below the waterline. Note that the black camouflage is showing through the sea grey paint and that the two forward 150-mm turrets are trained to starboard. The red carmine turret tops have been painted out with yellow.

This is the last German photograph of the *Bismarck,* taken shortly before the detachment of the *Prinz Eugen* at 1914 on 24 May 1941. Note the heavy wake left by the cruiser's 28 knots and the seas washing over the bow of the *Bismarck*. The latter was caused by flooding and the resulting bow trim.

By 1830 on 24 May the British Admiralty had ordered 19 warships into the chase, but it would take some time before an effective concentration could be arranged. The British were not aware of the damage to the *Bismarck* or of her critical fuel situation, which had prompted Admiral Lütjens to signal that he intended to break off the mission. High-speed steaming during the hours since the *Hood*'s sinking had further reduced the fuel reserves. This prompted Admiral Lütjens to signal at 2056, "Shaking off contacts impossible due to enemy radar. Due to fuel, steering to St. Nazaire."

Torpedo damage (Victorious aircraft). The carrier *Victorious* was ordered to launch an air strike against the *Bismarck* to try and slow her down. At 2200 six Fulmar fighters and nine Swordfish torpedo bombers, manned by pilots who had no experience in taking off from a carrier at sea, were launched and flew towards the British task force that was in contact with the *Bismarck*. In their approach, they almost attacked the *Norfolk*, some 36,500 meters astern and to port of the *Bismarck*, and the U.S. Coast Guard cutter *Modoc*, which was between the two opposing forces. These misdirected air attacks alerted the Germans to the presence of aircraft, and their antiaircraft guns were ready for action when the first flight made radar contact with the *Bismarck* at 2327. The attack was made under severe weather conditions and intense gunfire; nonetheless, none of the British aircraft were shot down. Two Fulmars, however, later did not make it back to the *Victorious*. The torpedo planes attacked mainly from the port side, with one attacking from the starboard side. All the torpedo planes made their attacks while avoiding fire from the *Bismarck*, which by skillful maneuvering evaded all but one of the nine torpedoes. Strafing runs and bombing by the Fulmars were unsuccessful in diminishing the intensity of the antiaircraft fire.

The torpedo that hit had a depth setting of 9.45 meters, but it turned out to be a surface runner, striking the main side belt near the foremast (see figure 4-2). A large column of water shot up, followed by a burst of black smoke from the *Bismarck*'s stack, indicating a large shock response from the torpedo explosion.* The torpedo struck the

*Based upon torpedo hits of this nature on other warships in World War II, it seems likely that the side belt and its supporting structure were slightly displaced inboard. The USS *West Virginia* was struck on her armor belt by one or more torpedoes during the Japanese attack on Pearl Harbor. The 343-mm armor belt was displaced inboard, and seven armor plates and their keys had to be replaced. Since the vertical belt of the *Bismarck* was 320 mm and the torpedo charges in the Japanese and British aerial torpedoes were almost the same in intensity, the damage sustained could have been more, but probably not less, since the mass of the belt was less than that of the *West Virginia*. One of the survivors of the *Bismarck* stated that the torpedo hit close to the surface, struck the armor belt, and merely chipped the paint. The latter cannot be true, but such damage to an armor belt and internal structure could not be observed from the outside.

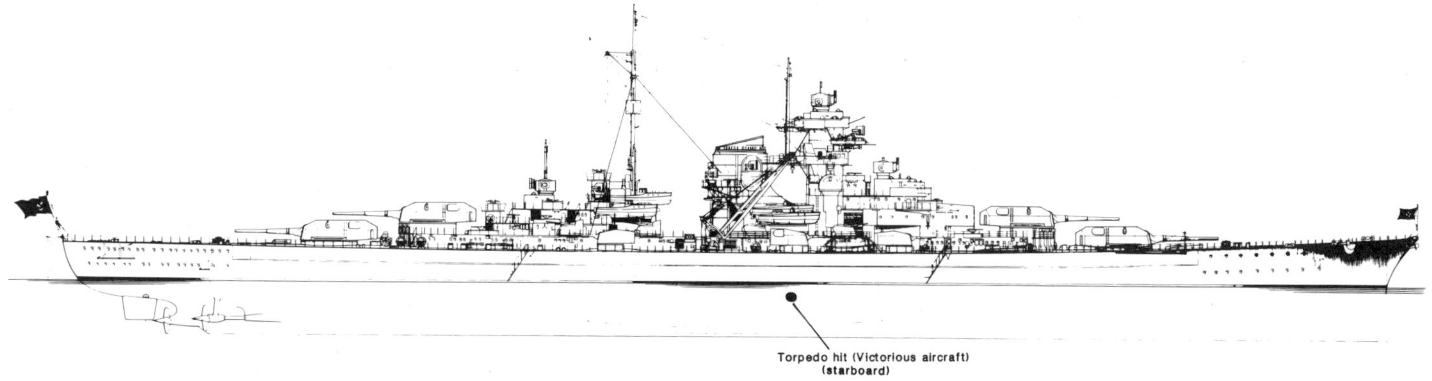

*Figure 4-2. Torpedo damage—*Bismarck*, 24 May 1941. Photography of Dr. Robert Ballard confirms that the armor belt was displaced some 25 to 50 centimeters, as noted in the footnote above.*

The *Bismarck* is about to be attacked by the *Victorious* aircraft on 24 May 1941 in failing light. The ship is in a heavy sea, and due to a bow trim is leaving a large wake. Her antiaircraft batteries have already commenced fire. (Imperial War Museum)

starboard side almost opposite the point where a shell from the *Prince of Wales* had plunged into the ship and exploded against the torpedo bulkhead. The shock response in the area of the detonation was very severe. A main control switchboard in the turbo-generator room on the starboard side was put out of action. One man topside near the point of impact was thrown against the ship structure and killed; five others from the turbo-generator room were injured. There was probably some minor flooding in some of the tanks in the area.

The radical maneuvers by the *Bismarck* and the shock response to the torpedo explosion loosened the collision mats forward, and the bow trim increased. There was no visual damage to the ship's side where the torpedo struck, but a combination of the slight displacement of the armor belt inboard or the shock response and the firing of the guns enlarged the cracks stuffed with canvas hammocks in the damaged bulkhead at the forward boundary of the port number 2 boiler room. All attempts to control the flooding there failed, and eventually the boiler room filled with water. With the loss of the two boilers on the port shaft, the low fuel reserves, the increased bow trim, and the heavy weather, speed was temporarily reduced to 16 knots. Divers were sent into the forward flooded compartments to readjust the collision mats. Speed was then increased to 20 knots after the work had been accomplished. Careful calculations had shown that under the prevailing wind and sea conditions, and presuming no further actions with enemy forces, this would be the most economical and safe speed for the ship to reach France with the remaining fuel oil on board.*

Meanwhile, the three shadowing British warships lost visual contact with the *Bismarck* about the time the British aircraft attacked. The *Bismarck*'s speed reduction and course alterations, however, allowed the range to close. Contact was reestablished about 0130 on 25 May, and a minute later the *Prince of Wales* fired two salvos at a range of 15,000 meters. No hits were scored. The *Bismarck* fired two salvos back, but overshot her target by a wide margin. The cease-fire was ordered due to failing light. Radar contact was being maintained by the *Suffolk*, now to port of the *Bismarck*.

Around 0310 the *Bismarck* began a long turn to starboard as the zigzagging British ships began a turn to port. By the time they began their turn back to starboard, the *Bismarck* had swung around 270° to a southeast course and was to the north, astern of her pursuers. The Germans apparently were still picking up radar emissions at a distance of 41,000 meters, and they still believed the British to be in contact. They did not know that the British radar set in use could not maintain contact at such a range.† Around 0700 and again at 0900 Admiral Lütjens sent radio messages to Naval Group West in Paris, the second of which was rather lengthy. British and American radio-direction-finder stations intercepted the transmission and were able to determine bearings, but these were erroneously plotted aboard Admiral Tovey's flagship, the *King George V*, and British warships were kept on wrong courses for some seven hours. By the time the error was noted and rectified, the *Bismarck* had escaped.

The long radio message sent by Admiral Lütjens brought a response from Group West to cease transmissions. The length of the radio message permitted the British,

---

*There are several factors to be considered here that to our knowledge no other author has addressed. The conditions calculated for the fuel rates and endurances of the *Bismarck* and *Tirpitz* are ideal; that is, they do not correct for in-service fuel factors such as fouling, roughness allowance of paints, increased resistance due to damage, and rudder corrections. In the *Bismarck*, in her trials in the Baltic, there was some directional instability due to the three-propeller hull form, and to correct for such a condition, the rudder needed to be used more than was normally the case in other capital ships of her size. It may very well have been that once Admiral Lütjens and Captain Lindemann received reports on the fuel rate of the *Bismarck* during the evening of 24 May, they became very concerned. It is a known fact in ship-motion theory that a ship in a following sea (as the *Bismarck* was to find herself in the dash to France) will have significant yaw response, which combined with the directional instability created by the triple-shaft arrangement of propellers, meant that more use of the rudder would be necessary, further increasing the rate of fuel burned. It is the authors' conclusion that if the *Bismarck* had continued on her course and speed in sea state 6 into the Bay of Biscay and had not been intercepted by British forces, her fuel situation may have been very critical, below 1/3rd fuel condition, which in the U.S. Navy is usually considered the minimum acceptable operating condition. Any further encounter with British battlegroups would have further aggravated the condition and certainly might have led to the need to refuel at sea to reach port. Furthermore, there was not enough fuel reserve to make a detour to the south; to do so would have necessitated using up the fuel-oil reserved for possible encounters with British ships or attacks by aircraft. Certainly the German intelligence group on the *Bismarck* were aware of the ships in pursuit and that there could be a possible encounter with Force H out of Gibraltar, even though those ships were reported to have left port with a convoy. The statement of Admiral Lütjens to the crew on his birthday on 25 May is interesting in view of the above, "On the way there, the enemy will gather and give us battle." It appears that the 20-knot speed, therefore, was the best possible speed under the circumstances of the sea conditions. If the sea had been calmer, it is our conclusion that the speed would have been greater.

†The British radar transmissions had sufficient energy to be detected by the *Bismarck*, but the range was such that the reflections of the signal could not be sensed by the British receivers. This is a common, easily understood phenomenon, but radar was quite a novelty in 1941, and the Germans aboard the *Bismarck* apparently were unaware of this factor.

Salvo! The *Bismarck* engages the *Hood* during the early morning hours of 24 May 1941. This photograph was taken aboard the *Prinz Eugen*. Note the intensity of the wake and the change in course.

with American assistance, to pinpoint the position of the German battleship during that period, but the code could not be broken, even with intensive work by expert British cryptologists. The British had broken some of the German secret codes, and they were able to decipher messages to Admiral Lütjens that preparations were being made to enter Brest and to Luftwaffe squadrons to redeploy to airfields around Brest.* Confirmation of some sort was furnished during the night of 25/26 May 1941 by the French resistance in Brest. While Ultra was trying busily to decode the *Bismarck*'s last signal, the British Admiralty began to redeploy its forces in the North Atlantic. By 0930 on 25 May, 46 British warships were committed to the hunt. The *Rodney* and *Ramillies* were closing on ships already in pursuit, and the *Renown, Ark Royal, Sheffield, London, Dorsetshire, Edinburgh,* and nine destroyers (five from Captain Philip Vian's flotilla) were hurrying to intercepting positions.

Although the British were unable to decode the portion of the long radio message concerned with battle damage, radar, and the fuel situation, their knowledge of the German ship's destination made it possible to locate and concentrate their forces in areas the German battleship would have to cross in her dash to safety. Another

*Note that the *Bismarck*'s primary port was St. Nazaire.

important available search force was a squadron of American PBY Catalina flying boats based at Lough Erne in Northern Ireland. These planes had been provided by the U.S. Navy to the British Coastal Command on direct orders from President Roosevelt for aerial reconnaissance of surface raiders. Great Britain had nothing to equal their radio equipment, their endurance, or their crews. All available Catalinas were dispatched to fly search patterns in areas where the German battleship might be located in her dash to the French coast.

On board the *Bismarck* there was some concern over the danger of the boilers salting up and the possible damage to the turbines caused by the flooding of port number 2 boiler room. It was feared that some salt water had entered the feed system, necessitating a complete purge of the feed water in the ten operational boilers. All the reserve feed water aboard the *Bismarck* also was primed by using an auxiliary boiler and four distillers. By the evening of 25 May that danger had been eliminated.

The effectiveness of British radar caused deep concern on the *Bismarck* and was an important factor in the decision to head for France, as can be observed from the following excerpt from a radio report made to Berlin around 0900 by Admiral Lütjens:

> Presence of radar on enemy vessels with radar range at least 35,000 meters impairs operations in Atlantic greatly. Ships maintained contact in fog in the Denmark Straits and never lost it. To escape was not possible, even though weather was favorable. Refueling impossible, unless high speed enables me to disengage. . . .

Early on the afternoon of 25 May, the *Bismarck* slowed to a speed of 12 knots to facilitate repair work being done in compartments XX and XXI. Divers were able to reach the manifolds in these flooded compartments and open valves, which permitted several hundred tons of fuel oil to be recovered. This was a significant development, for it provided the extra fuel reserve that Admiral Lütjens felt was needed in the *Bismarck*'s dash to St. Nazaire.

One of the naval constructors on board from the Naval Construction Office suggested after the fuel recovery that the anchor chains and anchors be jettisoned, reasoning that the loss of weight and moment would favorably alter the vessel's trim. This idea was rejected by the ship's command.

Bismarck located. The period of time between the *Suffolk*'s loss of radar contact and the re-detection of the *Bismarck*—more than 31 hours—was filled by an intensive British search and anxious German moments. At 1030 on 26 May a Catalina aircraft sighted her about 690 miles northwest of Brest, making a speed that would ensure the protection of U-boats and the Luftwaffe's heavy bombers in less than 24 hours.* No British ships strong enough to stop her were near enough to do so. The only chance of destroying the ship was to slow her down by an air attack so that superior forces could overtake and engage her. The *Renown*, *Sheffield*, and *Ark Royal*, steaming northward at flank speed, were the only British warships close to her position.

The *Ark Royal* flew off some search aircraft within minutes after the sighting report had been received. At 1154 a Swordfish bomber from the *Ark Royal* sighted the *Bismarck* and maintained contact for some time. Several more planes were launched in the early afternoon to confirm the identification; there had to be absolute certainty that the contact was the *Bismarck* and not the *Prinz Eugen*. The reports from these aircraft confirmed that the *Bismarck* was some 100 miles to the southeast of the *Ark Royal*'s

---

*This aircraft was piloted by an American, Ensign Leonard Smith, USN, and his part in the *Bismarck* affair was not publicized until well after the war. Many U.S. actions during this period were decidedly unneutral.

position. The aircraft reported the *Bismarck* making 26 knots at 1210, 20 knots at 1605, and 22 knots at 1800.* At 1350 the *Sheffield* was detached from Force H to establish radar contact with the German battleship and serve as a guide for air strikes. Between 1450 and 1500 the *Ark Royal* launched 15 torpedo planes with orders to attack the *Bismarck* in a position 40 nautical miles due south. They were not informed of the presence of the *Sheffield* in the area, and as soon as they had made a radar contact, they immediately launched 11 torpedoes at that cruiser before an urgent message from the *Ark Royal* warned them off. Many of the torpedoes exploded after short runs because of defective Duplex pistols (detonators), and the *Sheffield* was not hit.

A second attack was launched at 1910 in very rough seas. The pilots were directed to use the *Sheffield* as a guide, as she had been in visual and radar contact with the *Bismarck* since 1740. The planes carried torpedoes armed with contact pistols and set to run at a depth of only 7 meters. At 1724 the *Sheffield* had been sighted by the *Bismarck*, and battle stations were sounded on the German battleship.

The flight of Swordfish torpedo planes was sighted by the *Bismarck* around 2030, and at 2047 the first attack was made. The action comprised a series of three-plane assaults and took some 30 minutes, during which the German battleship had to make some violent course alterations to elude 13 torpedoes. The speed of the ship rapidly decreased, and the rough seas made sharp course alterations difficult. All the antiaircraft guns, even the 150 mm and 380 mm, were fired to break up the attack. About midway during the action, the *Bismarck* was struck slightly aft of midship by one or possibly two torpedoes. Near the end of the air attacks, two Swordfish torpedo planes approached the port side, so low that the 105-mm and 37-mm guns could not effectively maintain fire at them, and released torpedoes at a range of 450–500 meters. The *Bismarck* began a turn to port, but the torpedoes were too fast and too close to avoid at the speed the German ship now had left. One of these torpedoes struck in the stern area.

Torpedo damage (Ark Royal aircraft). The first torpedo (or torpedoes†) struck just outboard of the port engine room in the vicinity of compartments VIII and IX, slightly below the lower edge of the armor belt. The explosion was largely contained by the underwater protection system and armor belt, but structural damage inboard caused some flooding of the port shaft alley. Clouds of gas and smoke filled the port engine room until the ventilation cleared it away; there was some very minor flooding in that space. These leaks were quickly sealed off and the engine room was later pumped out. It seems reasonable to assume, however, that the port shaft alley was never emptied of its flooding water due to the extent of structural damage sustained during this attack (figure 4-3). Some floor plates in the middle engine room buckled upwards about 0.5 meters. A valve in the starboard engine room closed, and the turbines ceased turning until the personnel reopened the valve.

The torpedo hit aft was the most devastating the *Bismarck* had thus far received. Apparently, the torpedo hit the shell plating below the level of the steering-gear room

*Authors' note: These notations on speed reported only estimates. The 26-knot speed reported at 1210 was in error–it probably should have been 20 knots.

†Lt. Müllenheim-Rechberg has reported definitely that there were two torpedo hits before the critical one that disabled the rudder. Whether there were one or two torpedo hits will never be confirmed, since the more heavily damaged port side could not be observed by survivors when the ship sank. It is the opinion of the authors, based upon analysis of U.S. battleships damaged at Pearl Harbor, that if two torpedoes struck the ship, they would have hit the armor belt, causing severe inboard damage. Such damage would not have been easily visible from the outside. The heavy seas and the slight port list make this event likely. These conclusions are confirmed by those eyewitnesses who survived the action and were in spaces near these torpedo hits (or hit).

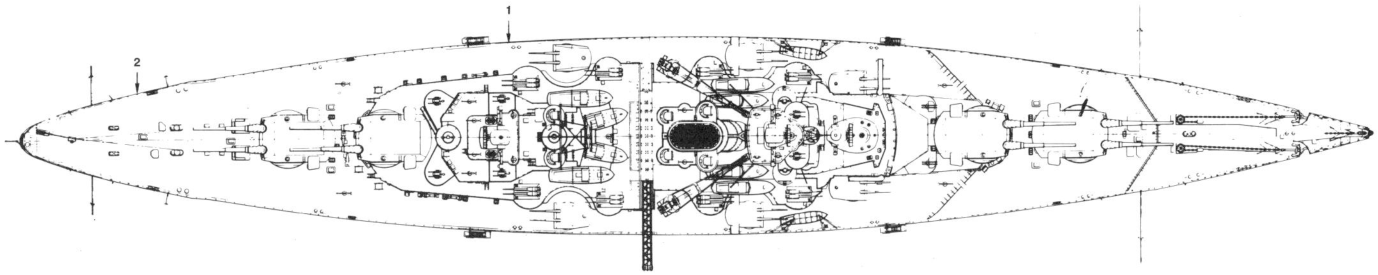

Figure 4-3. Torpedo hits—Bismarck, *26 May 1941.*

and near the port rudder shaft. The explosion opened a large hole in the shell plating just forward of the rudder shafts and severely damaged the port coupling of the rudder shaft to the steering engine, so that it was impossible to disengage it. The void tanks forward of the hit on the starboard side had been used to correct trim and list from earlier damage, and these had not been drained. The ballast water in these tanks increased the extent of the flooding and damage. Survivors attested to a severe shock response to this particular torpedo hit, considerably increasing damage to ship systems and structure. Any equipment not properly mounted or secured was subjected to a very severe shock response and possible damage.

The torpedo hit aft jammed both rudders at 12° to port and ripped such a large hole in the bottom shell that both steering compartments were flooded. Men there were quickly evacuated, and the armored hatch above the steering gear was secured. Water in these compartments rose and fell with the motion of the ship, causing the failure of a sounding tube. This led to the flooding of compartments on the main deck as well as the manual steering position. A flooding pump was started to dewater these spaces, but it soon was stopped by an electrical failure. A substitute circuit was provided, but water had seeped into the pump's automatic starter, and the pump motor was unusable.

While the *Bismarck* was trying to avoid the torpedoes, her course alterations caused her to close range with the *Sheffield*. After being torpedoed in the stern, the *Bismarck*'s erratic course carried her even closer to this British cruiser. Suddenly the German battleship loomed out of the mist to the south of the British warship. At once four 380-mm guns opened fire. The first salvo landed approximately 1,900 meters from the *Sheffield*, and as the range closed a second salvo from the 380-mm guns produced a straddle. A third salvo landed abeam, and fragments killed three men, wounded nine others, and rendered the radar gear useless. The *Sheffield* made smoke and headed north, while three more salvos fell around her. The *Bismarck* ceased fire. Before the *Sheffield* lost contact at 2155, she noted that the German battleship was steering northwest.

Aboard the *Bismarck* every effort was made to regain steering control. Eventually the starboard rudder was uncoupled from the steering gear, and the hand rudder was engaged to the rudder yoke, but the port coupling was so badly jammed that it could not be freed. Several divers attempted to enter the steering-gear room to detach the rudders and steering gear. This effort was abandoned when the divers collapsed after being pulled out of the port rudder room, which was awash with surging and swirling water. A young officer who was stationed aft suggested that a diver with a Momsen lung try to reach the damaged shaft and sever it with explosives. This idea was dismissed. Soon after this Captain Lindemann and Chief Engineer Lehman had a discussion over measures to restore steering control. It was agreed that it was impossible to lower divers over the side to work, as the seas were too rough. The suggestion to sever the rudder

The photo at left shows the *Bismarck* leading the *Prinz Eugen* during training exercises in the Baltic while preparing for Operation Rhienübung. Note the camouflage of the *Bismarck* showing the false bow wave. The photo at the right shows the *Tirpitz* in Norway as seen from the *Admiral Hipper* during exercises prior to Operation Rosselsprung. The cruiser is being towed by the *Tirpitz* and note disruptive camouflage scheme then being used in German ships. (Albrecht Schnarke Collection)

shafts by the use of explosives was also ruled out by Lütjens who said, "We cannot endanger the ship with measures of that kind." With the loss of steering control, the *Bismarck* was unable to maintain a course to safety and her fate was sealed.

Meanwhile, slow flooding occurred in the port outboard spaces of watertight compartment III through torn plates and a damaged cable stuffing tube in the main transverse bulkhead, or through tears in the side shell. There was flooding in the upper and lower passages on the port side in compartment III. More asymmetrical flooding occurred in compartment VII. The ship began listing to port.

Lieutenant Burkard von Müllenheim-Rechberg, the senior surviving officer, who was stationed in the aft main-battery director, knew the rudders were damaged by the action of the rudder-position indicator. Within minutes after the hit he learned of the flooding there, and knew that all repair work would have to be done under water. He wrote the authors:

From the time of the hit until the sinking of the ship the next morning, it was never possible again to steer the ship (except with propellers, which, understandably, were of little avail, particularly in the weather situation the *Bismarck* was encountering).

Efforts to steer the ship that night were described by Lieutenant (Engineering) Gerhard Junack:

> After trying for hours, we finally were able to engage the clutch of the hand rudder. To have divers make a descent from the outside part of the ship to survey the damage was impossible with the rough sea. A suggestion to dynamite the rudder shafts was dismissed because of the possible damage to the propellers, which were close to this installation. The use of auxiliary makeshift rudders was also dismissed as impractical.
>
> It was not possible to keep the ship on a southeast course with the propellers alone. Therefore, it was necessary to slow the ship, and in the prevailing seas the ship took an unfavorable course to the northwest, towards the enemy.

Everything possible had been done to regain steering control. Captain Lindemann finally resorted to various combinations of rpms to the propellers, and according to Lieutenant von Müllenheim-Rechberg, "The *Bismarck*'s hull shook violently . . . in attempts to bring us back on course. His orders from the bridge came in rapid succession: 'Port and center engines half ahead, starboard engines stop! Port engines full ahead, starboard engines stop!' The engine room crews responded to every order, and some safety measures were ignored. These measures brought the ship away from a northwest course, but the combination of wind and waves turned the ship around once again."

At 2115 Admiral Lütjens reported that the ship was unmaneuverable, and at 2140, ". . . we shall fight to the last shell . . . ." Shortly after midnight all attempts to rectify the rudder situation were halted. A damage-control party shored up the bulkhead forward of the steering-gear rooms and sealed the broken sounding tube. Two members of this party, equipped with underwater escape apparatus, tried to make an entry through an emergency escape trunk that led to the armored hatch over the steering-gear compartment. Once they had opened the hatch, seawater shot up through the opening as the ship plunged into a wave trough. The water level fell when the ship rode up on a wave crest. After some consultation, this effort was halted.

The 15 Swordfish planes had returned to the *Ark Royal* around 2100, four of them seriously damaged by antiaircraft fire. The fact that they had obtained at least one hit was reported to Vice Admiral Somerville in the *Renown*, and he passed this information on to Admiral Tovey. Shortly after midnight, Swordfish reconnaissance planes returned to report that the *Bismarck* seemed out of control and almost dead in the water. This was verified by a Coastal Command Catalina—the ship observed to be turning in circles; however, the plane had to leave the area quickly because of intense antiaircraft fire.

British destroyers attack Bismarck. At 2238 Admiral Vian's flotilla of destroyers *Zulu, Piorun, Cossack, Sikh,* and *Maori,* made visual contact with the *Bismarck.* Four minutes later the *Bismarck* fired three main-battery salvos at *Piorun,* which had closed range to 12,500 meters, and obtained a straddle. The main-battery rangefinders barely could detect these warships in the dark. The Polish destroyer relentlessly continued to close range to almost 12,000 meters and kept up a spirited exchange with the German battleship for 30 minutes before a near miss forced her to open range. In the meantime, the *Bismarck,* headed into the wind at speeds varying from 6 to 10 knots, attempted to use her propellers for steering. During the early morning hours of 27 May, the destroyers, individually or in pairs, fired 16 torpedoes at ranges of 3,000 to 8,000 meters, under intense and accurate fire. At 2342 the *Cossack* had her radio antennae sheared away by splinters from a near-miss salvo of 150-mm shells. Some eight minutes later, the *Zulu* had a similar experience when straddled by a salvo of 380-mm shells. A nose cap of one shell was later found on her forecastle, while splinters of another sliced into her gunnery control tower. None of these destroyers' torpedoes were seen to hit by German survivors. The *Maori* and *Cossack* reported one hit each; these would have occurred around 0130 on the *Bismarck*'s port side. However, the hits reported by the British seamen may have been either 150-mm gun flashes from the *Bismarck* or star shells that periodically came down near the German battleship.

The destroyers were ordered to keep in contact with the *Bismarck* during the night and to fire star shells periodically to illuminate her so that the *Rodney* and *King George V* would not inadvertently come in contact with her during the night. Some of these star shells fell near the *Bismarck* after the destroyers had fired their torpedoes. Lieutenant von Müllenheim-Rechberg indicated to the authors that one of these star shells fell on the bow shortly after 0100 and started a fire. It was quickly extinguished. It is not surprising that the *Bismarck*'s erratic course and fluctuating speed resulted in no torpedo hits.

An attempt between 0500 and 0600 to fly off an Arado-196 floatplane with the ship's log, films of the action with the *Hood*, and other important documents met with failure. The 356-mm shell hit that had splintered the captain's motor launch and damaged the catapult had also created fragments that severed the compressed-air line to activate the catapult. Since there was no time to repair it, the crew shoved the fuel-laden aircraft overboard.

By 0640 on 27 May the *Bismarck* was under way again with a slight port list, making only 7 knots due to the use of her propellers for steering. With the port rudder still jammed (the starboard rudder coupling was disengaged), the ship kept a heading to the northwest, varying some eighty degrees of compass. Although she had almost no maneuverability, the guns and propulsion plant were ready for action. Around this time the last torpedo attack was made by the British destroyers; the *Maori* fired at a range of 8,000 meters, but missed. She was straddled several times by salvos from the *Bismarck*'s 150-mm guns. This scenario had occurred repeatedly during the night when the British destroyers tried to torpedo the German battleship. At 0708 the *Bismarck* increased speed to 12 knots, but the ship's course was still so erratic toward the northwest that Captain Lindemann ordered the engines to stop. They were later ordered to start again.

Bismarck's last battle. The scene aboard the *Bismarck* was one of quiet desperation and despair. As the minutes passed, tension and apprehension grew, as everyone knew that the British battleships were coming, and a final battle had to be fought. At 0753 the *Norfolk* had the *Bismarck* in sight at a range of 15,000 meters, and after determining it was the *Bismarck* and not the *Rodney*, radioed Admiral Tovey her position and heading. Shortly after 0815 a rain squall covered the German battleship, and when it cleared the *Rodney* and *King George V* were in sight off the port bow. Shortly after 0830, battle stations were sounded aboard the *Bismarck*. Adalbert Schneider announced that the target was to be the *Rodney*. It took some 15 minutes before the engagement started, and it began with the *Rodney* at 0847 when she fired the initial rounds, followed a minute later by the *King George V*. The range at this instant was 18,000 to 20,000 meters. The *Bismarck* opened fire at 0849 and her first salvo fell some 900 meters short of the *Rodney*, the second was 900 meters over, and the third was a straddle. Succeeding salvos were over, until 0858 when a straddle was again attained. The British ships relentlessly continued their approach. The *Bismarck* was to fight bravely, but hopeless-

TABLE 4-5

Ammunition Expenditure of British Ships on 27 May 1941

Ship	*406mm*	*356mm*	*203mm*	*152mm*	*134 mm*
Rodney	380	—	—	716	—
King George V	—	339	—	—	660
Norfolk	—	—	527	—	—
Dorsetshire	—	—	254	—	—

Note that British gunfire commenced at 0847 and ceased at 1022. During the most intense part of the engagement (0900–0930), shells were falling at a rate of almost one per second.

ly, against a hail of gunfire (see table 4-5) that was finally delivered at point-blank range. That gunfire, while not instrumental in sinking her, did render the *Bismarck* defenseless and led to a German decision to scuttle. The thick armor belt was not penetrated by the main-battery fire of the two British battleships, even though the *Rodney* closed to ranges of less than 2,400 meters. The upper citadel belt, however, was pierced many times by 203-mm, 356-mm, and 406-mm shells. The reason why the main belt probably was not penetrated was due to the close ranges of the engagement. The *Rodney* fired most of her 406-mm ammunition between 2,400 and 9,000 meters, necessitating main-battery fire at maximum depression. Several of her guns jumped their cradles due to prolonged fire at low angles. This caused some projectiles to hit the water surface and ricochet, thus losing some velocity and energy upon impact. Such short-range fire was virtually incapable of striking the *Bismarck* below the waterline. The *Bismarck*'s main battery was silenced at 0931, and her superstructure was riddled with shell hits. Exposed equipment, such as searchlights and gun mounts, were either carried away or reduced to smoking undistinguishable wrecks. At 1015, the British were convinced that the *Bismarck* could not be sunk by gunfire and by 1022 they had ceased fire. Their fuel reserves were also low, so the cruiser *Dorsetshire* was ordered to sink the *Bismarck* with torpedoes. The entire gunnery engagement lasted one hour and 35 minutes.

In the opening moments of the battle, after the *Bismarck*'s fifth salvo straddled the *Rodney*, the German battleship's forward fire-control director was knocked out by a 203-mm shell from the *Norfolk*, which had closed on the *Bismarck*'s starboard bow. The British battleship *Rodney* closed to 14,000 meters on the port bow and altered course to allow the *King George V* to fire full salvos. At 0902, a 406-mm shell from the *Rodney* heavily damaged turrets Anton and Bruno. Although one of these two turrets was to fire one last salvo at 0927, both turrets were effectively out of the battle. At this time, the *Dorsetshire* appeared some 18,000 meters off the *Bismarck*'s starboard quarter and at 0904 commenced fire.* At 0910, the *Bismarck* shifted fire control for turrets Caesar and Dora to the aft fire-control station, which was under the command of Lieutenant Burkard von Müllenheim-Rechberg, the senior surviving officer. He chose the *King George V*, which was 11,000 meters to port, to be his target, since he had a clear view of her. The *Rodney* probably was forward and in the blind space for this position. After firing four salvos, at 0913 a 356-mm shell hit the rotating dome of the after fire-control station, destroyed the aft rangefinder, and carried away the protruding parts

*The *Dorsetshire* played a minor role in the *Bismarck* gunfire engagement, even though she fired 254 203-mm shells. Her part in the action records gunfire action at 0904–0913, 0920–0924, 0935–0938, and 0954–1018. In Captain Martin's report he did not claim any hits before 1002, and in view of the long range at which his ship fired during the action, the spotters on the *Dorsetshire* had difficulty in spotting falling shells. It seems plausible that all of her gunfire produced misses.

of the firing directors. The after two turrets were still in action, having good placements of salvos until the aft fire-control station had been hit. They now had to fire under local control. At this time (0916) the *Rodney* had completed her firing run down the port side of the *Bismarck* and then launched six torpedoes at a range of 10,000 meters. The turret commanders then switched to her as she had now closed to 7,500 meters.

The *Bismarck*'s initial fire was very accurate. By 0858 the *Rodney* had been straddled, but thereafter the salvos became more erratic and less frequent. The only damage to the British ships, other than the blast effects of their own shellfire, was a jammed sluice valve door on one of the *Rodney*'s underwater torpedo tubes, caused by a near miss and a shell splinter that penetrated the starboard side of the antiaircraft director.* The *Norfolk*'s hit on the *Bismarck*'s forward director greatly reduced the German battleship's gunnery accuracy. By 0930 only turret Dora and some 150-mm guns were left in action; turret Caesar had dropped out before. Turret Dora was finally silenced at 0931 from a hit on her port barbette.

British gunfire systematically destroyed the *Bismarck*. Her radar and rangefinders were destroyed first. The fire-control system was disrupted by damage to the superstructure and destruction of equipment. Many shells from the *Norfolk, Rodney*, and *King George V* penetrated the upper citadel belt because they were fired at such low angles of elevation that they were probably deflected into this region after striking the water surface. As the range decreased, turrets, armored portions of the superstructure, and the upper works were wrecked by everything from 134-mm to 406-mm guns, with the most devastating damage occurring between 0915 and 0945 when the *Rodney* and *King George V* closed to 2,500 and 10,000 meters respectively. Table 4-6 is an approximate summary of those shell hits that can be readily identified. The list is incomplete, and there is some disagreement among the survivors concerning certain hits. It would appear that the side armor may have been penetrated at least once or twice. This possibility is enhanced by the fact that the *Rodney* was at very close range and firing full nine-gun salvos at the maximum angle of depression. The upper side belt was vulner-

TABLE 4-6

Shell Hits on the Bismarck—27 May 1941†

Hit No.	*Approximate Time*	*Shell Caliber*	*Description of Damage*
A	0859	16″ (406 mm)	Shell fired at range of 20,000 yards (18,300 meters) and burst on upper deck near turrets Anton and Bruno, putting both of them out of action, apart from a salvo at 0927 by turret Anton.
B	0859	8″ (203 mm)	Shell hit and demolished the forward fire control director.
C	0900–0905	16″ (406 mm)	Shell hit below catapult platform and exploded 4.1-inch (105-mm) ready ammunition. This caused heavy casualties among those crews taking shelter nearby.
D	0900–0915	8″ (203 mm)	Shell struck the forward starboard 5.9-inch (150-mm) turret, jamming its access hatch. All of its crew were imprisoned.
E	0902	16″ (406 mm)	Shell hit damaged turret Anton, and blew off pieces of turret Bruno, which killed almost all men in exposed portions of the bridge.

* *Rodney* sustained damage to her turret mechanisms and deck structure near the main turrets from the prolonged fire at maximum depression. Repairs were made at the Boston Navy Yard.
† See figure 4-4.

F	0912	14″ (356 mm) or 16″ (406 mm)	Shell destroyed the bridge and damaged the forward conning tower.
G	0913	14″ (356 mm)	Shell hit the aft rotating arms of the aft rangefinders and carried away the protruding parts of its optics and instruments.
H	—	14″ (356 mm) or 16″ (406 mm)	Shell (or shells) hit electrical power plant port side aft, starting a transformer fire.
I	—	14″ (356 mm) or 16″ (406 mm)	Shell hits on main deck and 01 level on port side exploded ready-service ammunition.
J	—	14″ (356 mm) 16″ (406 mm) and 6″ (152 mm)	Shells struck the port forward 5.9″ (150 mm) turret and its magazine, causing internal explosions that tore off the after roof plate and riddled the adjacent superstructure with shell fragments and debris.
K	0927	14″ (356 mm)	A salvo of shells (observed by the *Norfolk*) hit near turret Anton. Range was between 8,000–11,000 yards (7,300–10,000 meters). Turret lost hydraulic power and guns ran down to maximum depression.
L	0931	—	The port barrel of turret Dora burst, but the starboard barrel fired two more rounds. There was smoke and fire in the turret and the gun crew had to abandon it.
M	0930–0935	16″ (406 mm)	Shell of undetermined caliber (probably 406 mm) fired at a range of 4,000 yards (3,700 meters) hit the armor deck and penetrated to port engine room.
N	0940	16″ (406 mm)	Shell or shells, fired at 7,500 to 8,000 yards (6,900 to 7,400 meters), caused a large explosion just abaft turret Bruno, blowing off most of its rear armor plate over the bridge. A large notch was made in the port beam segment of the barbette from shell and a small fire was reported in the turret.
O	0930–0935	16″ (406 mm)	Probable shell hit on upper main-battery director, which subsequently toppled over to port.
P	0921	14″ (356 mm)	A 14″ shell struck the face plate of turret Caesar. No damage within turret, but left gun would not elevate. Shell fragments penetrated into decks below starting small fires which were easily extinguished.
Q	—	—	Several shells struck the bow region. One believed to be 16″ (406 mm) tore an oblong hole in main deck and started a fire in berthing spaces below. Other shells carried away the bow casting and forward anchor as well as the two side anchors.
R	—	—	Three hits near after control tower tore huge holes in upper deck.
S	0950–1000	—	Foremast knocked down.
T	—	—	Radio Room in compartment XV was wrecked.
U	—	—	Shell hit in hangar, aircraft burned (compartment X or XI).
V	—	—	Two or three shells killed all men in radio room (compartment XV).
W	1000–1005	16″ (406 mm)	Shell exploded in forward canteen, killing some 200 men.
X	—	16″ (406 mm)	Shell penetrated armor belt below and forward of bridge. This hit was reported by a junior officer,

			later denied by other survivors. *Rodney* moved into close range (2,500 meters) during her last few salvos.
Y	1011	16″ (406 mm)	Shell hits in compartments I and II weaken already severely damaged stern structure, leading to collapse and loss of stern some 15 minutes later.
Z	—	—	Shell hit in compartment IV silenced the ventilation system for the main deck aft.
AA	0950–1000	16″ (406 mm)	Two shell hits in compartment IV, damaging access trunks. One shell penetrated barbette of turret Dora on its aft port segment. Red-hot splinters started a fire in lower turret. Quick flooding of D turret magazines prevented any further conflagration.

Alone and defenseless, the *Bismarck* can be barely seen under fire from the *Dorsetshire* near the conclusion of the gunnery battle on the morning of 27 May 1941.

The battleship *Rodney* is seen from the *Dorsetshire* as she fires a salvo at the *Bismarck* around 0950–1000. At this point *Rodney* was 2,500–3,000 meters to the starboard quarter of *Bismarck*. This photo was taken when most of the devastating shell hits occurred. (Imperial War Museum)

This is a view from the *Dorsetshire* as one of the *Rodney*'s salvos falls astern of the *Bismarck,* which is trailing very black smoke from her damaged funnel. (Imperial War Museum)

able to 356- or 406-mm projectiles at the ranges *King George V* and *Rodney* were fought. This belt was vulnerable to the 203-mm shell within 14,000 meters. The 50-mm upper deck could have been penetrated by the 356- and 406-mm shells outside of 11,000 to 13,700 meters. These close ranges, however, also enhanced the tendency of these projectiles to ricochet off the water before striking the ship. The spaces above the armor deck were continuously exposed to damage from projectiles entering in one of these ways. Testimony to the severity of damage can be gained from the fact that large portions of the main and second decks were without electrical power due to cables being severed by direct shell hits or splinters. Access ladders and light fixtures were also destroyed. For penetration of both the 50-mm upper deck and the heavier armor deck by 356-mm and 406-mm shells, the ranges would have had to have been:

- deck over machinery 20,000 to 23,800 meters
- deck over magazines 25,600 to 27,400 meters

This shows why no penetrations of the deck over the magazines occurred during this action, but does indicate that the machinery was very vulnerable in the opening stages of the battle on 27 May 1941. The hits designated (M) and (X) occurred at close ranges. At such short distances the main side belt, plus the sloping armor deck over the machinery, are believed to have been penetrable by the 406-mm shells of the *Rodney* at an approximate target angle of 90°. Target angles were considerably less than 90° throughout most of the engagement, and the target angle changed rapidly throughout the short-range portion of the action.

It should be noted that the damage to unprotected or lightly armored upperworks, such as director towers, hangars, etc., was very severe. The forward command tower had sustained a number of direct hits, and the funnel was riddled with shell holes. Of the 2,876 shells fired at the *Bismarck*, it would appear that a devastating number, perhaps as many as 300-400, struck the ship, since the range was so short. Between 0910

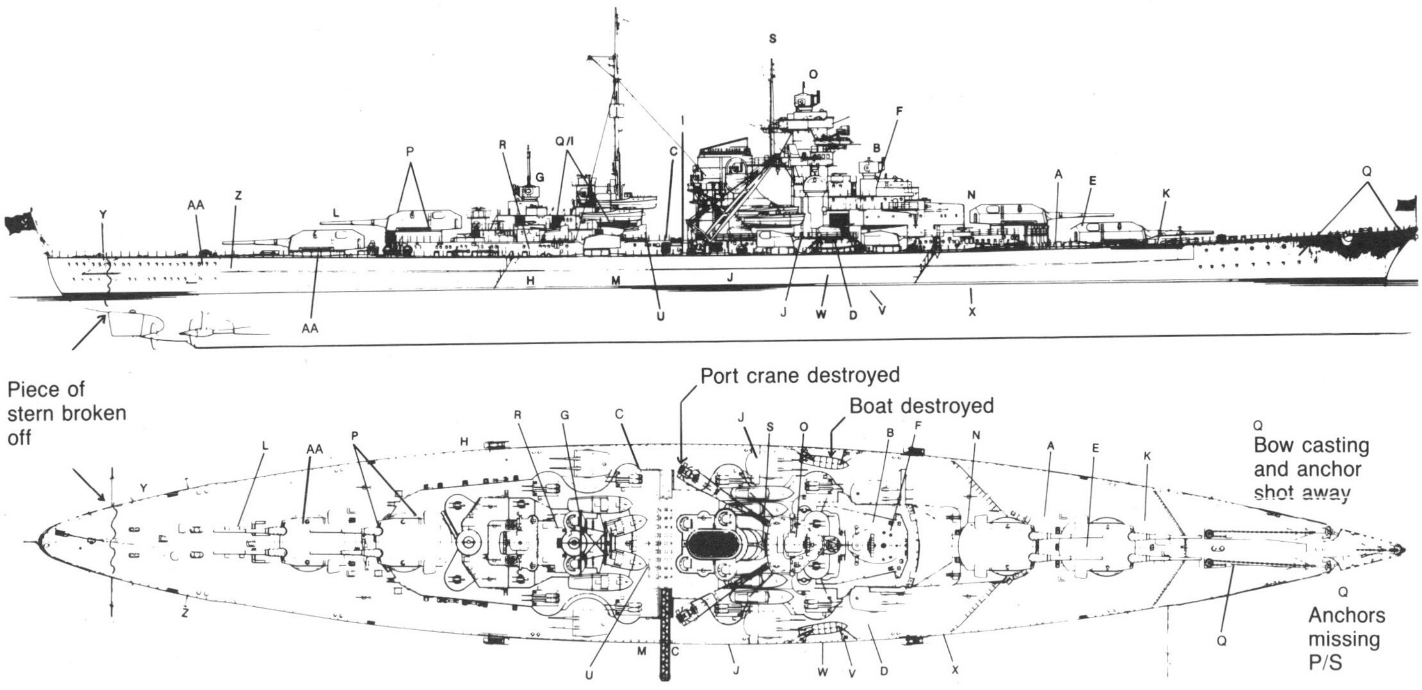

Figure 4-4. Approximate location of shell hits on Bismarck, *27 May 1944.*

and 0920, 356- and 406-mm shells were falling all around the *Bismarck,* making it difficult for the two British battleships to identify hits. By this time, however, most of the ordnance of the *Bismarck* had sustained severe damage and was silent.* After this most guns fired sporadically. The hits in table 4-6 hardly give a full picture of the damage sustained by the *Bismarck,* and a true accounting will never be possible.

Lieutenant Müllenheim-Rechberg described the scene as he saw it as follows:

> The last one to leave the station, I went forward, towards the searchlight-control station or, rather, towards where it had been. The scene that lay before me was too much to take in at a glance and is very difficult to describe. It was chaos and desolation. The antiaircraft guns and searchlights that once surrounded the after station had disappeared

---

*The status of the ordnance of the *Bismarck* at 1022 when the British ceased fire was as follows:

- Turret Anton had a small fire on its gun platform, and its barrels were at maximum depression. (This may have been due to damage to its elevating system.)
- Turret Bruno, with its guns positioned to port and at maximum elevation, had large portions of its armor backwall blown away.
- Turret Caesar, with its guns at maximum elevation and trained to port, had no apparent external damage, but had sustained a direct hit from a heavy-caliber shell on its face plate. This disabled the turret.
- Turret Dora, trained to port and blackened by smoke, had several holes in its sides and its right gun barrel torn to shreds. There were several shell penetrations in its barbette on the port and starboard structure.
- All of the 150-mm turrets had been struck by shells and were pointed in assorted directions. The forward port turret had its rear roof plate torn away completely.
- All of the 105-mm mounts were damaged with some missing their gun barrels.
- All of the 37-mm gun mounts were damaged, and most 20-mm machine guns were demolished or carried overboard.
- There was a large hole in the 01 level outside the hangar (port) and just aft of the catapult from a combination of shell hits (406 mm) and the explosion of 105 mm ordnance.

> without a trace. Where there had been guns, shields, and instruments, there was empty space. The superstructure decks were littered with scrap metal. There were holes in the stack, but it was still standing. . . .*

Conditions within the hull of the *Bismarck* were equally as bad between 0915 and 0930, and the horror scene was later described by Warrant Officer Wilhelm Schmidt, who was in charge of an aft repair party (Damage Control Team No. 1):

> About five heavy shells penetrated the upper deck in my area of responsibility and exploded on the battery and main decks. As each shell exploded, you could feel the sharp tremors in the ship. Two of these shells exploded in compartments I and II. The one in compartment II generated a large amount of nitrous gas and started fires in the decks above. The gases from the explosion seeped through the closed armor door of the aft armored bulkhead in the vicinity of my action station in compartment III. A third shell landed on the battery deck in compartment III on the starboard side. The fragments which flew around knocked out the lights and silenced the ventilation system for the main deck aft.
>
> The fourth and fifth shells exploded in compartment IV. The access ladders were torn from their foundations, a portion of the lighting was knocked out, and a large amount of nitrous gas was generated. A messenger arrived to ask me to send some men to put out a fire on the superstructure deck aft. Some men were sent, but did not return. Several of my party were already lost to shell splinters. A chief gunner rushed up with news that there was a fire in turret Dora. At once, I switched flooding pump number 2 to flood the ammunition spaces in the area of the stricken turret.

By 0915 the shell hits in the region of the forward turrets and command tower had killed a large number of the complement there. Also, due to the severe shell damage, communications were disrupted between the forward command tower and the rest of the ship. Commander Hans Öels, who was the executive officer, assumed command and around 1000 ordered scuttling charges to be set, as the main and secondary batteries had been silenced by enemy shells, and the ship above the waterline was a wreck. Commander Öels led a large party from his command post, situated at the base of the communications tube on the lower armor deck. They could not go forward because fires were raging there, so they proceeded aft to the forward canteen, which was located near the forward uptakes. Soon after they reached this space, a 406-mm shell (hit W in table 4-6) exploded within the space, killing the executive officer and some 200–300 men who had gathered in this space awaiting the order to abandon ship. At least two men (Able Seaman Staat and Machinist Blum) survived this carnage and were later rescued by the British.

The *Rodney* and *King George V* continued gunfire in hopes of sinking the crippled *Bismarck*. Although torpedo aircraft from *Ark Royal* were in the area standing by, they were ordered to stay away by Admiral Tovey. At 1011 *Rodney*, whose guns were firing continuously throughout the action, loosed a salvo at the *Bismarck* and this struck in the stern. The shells blew off chunks of steel and set off a fire with greyish-white smoke. Earlier a *Rodney* shell had hit the barbette armor of turret Dora and the shell exploded after penetrating the 340-mm armor and its backing structure. A 380-mm cartridge on the gun platform ignited and there was a heavy loss of life in this turret's crew. At 1013 another 406-mm salvo struck amidships near the catapult platform (hit W). A final *Rodney* salvo at 1014 struck turret Anton causing a bright fire to flare up. A salvo from the *King George V* hit the forward command tower at 1014, severely weakening the upper bridge structure. With the fuel situation in both British battleships now very

---

*Baron Burkard von Müllenheim-Rechberg, *Battleship Bismarck: A Survivor's Story* (Annapolis, Md.: Naval Institute Press, 1980).

serious and the *Bismarck* offering no resistance, Admiral Tovey ordered both battleships to cease fire. The *Rodney* fired her last salvo at 1014 and the *King George V* at 1022. The German battleship was a wreck, with guns silenced, superstructure demolished, funnel filled with shell holes, fires burning everywhere, and a large number of crew casualties. At 1005 the *King George V* was within 3,000 meters of the German ship, while the *Rodney* was at 2,500 meters. From these vantage points fires could be seen burning in all parts of the *Bismarck*. There was no resistance being offered.

Earlier, the *Rodney* had fired 12 torpedoes, with one hit being observed by a lookout at 0958, although German survivors have denied this. The *Norfolk* fired 8 more (one possible hit). Torpedo hits were difficult to ascertain because of the numerous shell splashes around the ship. Around 1020 Admiral Tovey ordered the *Dorsetshire* to torpedo the *Bismarck* and she moved in to fire two torpedoes at the starboard side, one of which was seen to hit. Then, moving around to the port side, she fired a third torpedo, which also hit. The *Bismarck* by then had a noticeable port list.

In the middle engine room soon after 1000, Lieutenant (Engineer) Junack received an order to set scuttling charges (Measure V). These were placed in the condenser intakes and sea chests in the engine room. All watertight doors in these spaces were opened, including the one to the shaft alley. Charges were also set in the starboard engine room and boiler rooms, but could not be placed in the bilges of the port engine room because of water that had been ingested down the ventilation trunks from near-miss shells. The explosives detonated shortly after 1020, and the ship took a heavy port list around 1035. She capsized slowly and sank by the stern, with her bow disappearing around 1040. Lieutenants Müllenheim-Rechberg and Junack have maintained that they observed no torpedo damage to the starboard hull. It was impossible to observe the port side, because that was the side that the *Bismarck* capsized to.

Only 115 of the *Bismarck* crew were rescued—85 by the *Dorsetshire* (one later died of injuries on 28 May), 25 by the *Maori*, three by a German U-boat, and two by a German trawler. The severity of the battering the *Bismarck* sustained is attested by the fact that there were no survivors from the forward superstructure. Pilots in Swordfish torpedo planes overflying the *Bismarck* before she sank reported that of the fires on the ship, the largest was concentrated where the forward command tower had been! The British ships had to leave some 500 to 600 men behind in the water when a submarine periscope or exhaust was sighted. The next day the Spanish heavy cruiser *Canarias* searched in vain for survivors. The German Navy lost 2,211 excellently trained officers and enlisted men. Moreover, any technical lessons that could have been gained from the performance of equipment on board the *Bismarck* was also lost. The loss of the ship was a heavy blow to Admiral Raeder and his naval strategy, because it led to Hitler's insistence on so cautious a deployment of major German warships, for fear of their loss, that they would never again be the formidable threat to Great Britain that they had been prior to this operation. Hitler's attitude was a great detriment to the fighting abilities of all the surviving capital ships, particularly the *Tirpitz*. Admiral Raeder later wrote in *My Life* that the sinking of the *Bismarck*

> . . . was to have lasting effects upon the whole naval warfare for the rest of the war. Hitler's attitude towards any suggestions or proposals advocated by me was now completely different. Where before this he had given me a free hand as long as government policies or the other armed services were not involved, he now became extremely critical and very apt to insist on agreement with his own personal views. He had previously preferred not to be worried with too much advanced briefing about the sorties of the large ships, as he always felt anxious about them, but now he issued directives to me that radically restricted the movements of these major units. He forbade their sorties into the Atlantic. The successes we had even with our inferior forces through bold initiative and the taking of calculated risks was to be a thing of the past. For us the naval war was given an entirely new face.

Tirpitz—Operational History. The keel for the *Tirpitz* (Battleship "G") was laid on 30 October 1936 at the Wilhelmshaven Navy Yard. The ship was christened and launched on 1 April 1939, and was commissioned on 25 February 1941. During the time the ship was completing in Wilhelmshaven, the shipyard was attacked at different times by a total of 1,042 bombers that dropped approximately 670 tons of bombs. None of these hit the ship, but the repeated attacks did delay her completion. During the severe RAF air attack on German naval bases in Wilhelmshaven and Kiel, the *Prinz Eugen* was hit twice and the *Lützow* once.

On 9 March 1941 the *Tirpitz* left Wilhelmshaven for Kiel, and on 16 March she commenced sea trials in Danzig Bay and the Baltic Sea, and at the same time underwent an intensive training schedule that continued through 20 July. After the ship was transferred to Kiel to be safe from air attacks, trials and further ship tests were directed by personnel at the naval dockyard there.

While the *Tirpitz* was at Kiel for additional installations and modifications, she was temporarily made the flagship of the German Baltic Fleet during Operation Barbarossa (the German invasion of the Soviet Union). In company with the armored ship *Admiral Scheer*, light cruisers *Köln, Nürnberg, Emden*, and *Leipzig*, and escorting destroyers, plus two minesweeping flotillas, she steamed to the Aaland Islands to establish a blockade in the event any Russian warships attempted to leave Leningrad for internment in neutral Sweden. The Luftwaffe bombed the Russian Fleet in Leningrad on 22 September, sinking the battleship *Petropavlovsk* (ex-*Marat*) and damaging the battleship *Oktyabrskaya Revolutsia* and the heavy cruiser *Maxim Gorki*. The *Tirpitz* and the German task force remained off the Aaland Islands from 23 to 26 September before

The *Tirpitz* was launched on 1 April 1939 at the Wilhelmshaven Navy Yard. With her name now prominently displayed on her bow, she begins to enter the water. (Albrecht Schnarke Collection)

dispersing. The *Tirpitz* continued training in the Baltic — the war at sea had been going badly for Germany. After the loss of her sister ship *Bismarck*, Hitler forbade any sortie of the *Tirpitz* into the Atlantic.

Tirpitz sails to Norway. On 13 November 1941 Admiral Raeder proposed that the *Tirpitz* be sent to Norway where she could attack Russia-bound convoys. Hitler agreed, and she entered the dockyard at Gdynia to prepare for such duty, which included augmenting the antiaircraft battery, principally the 20-mm weapons, which were increased to 61, and the mounting of two quadruple 533-mm torpedo tubes for use against merchant ships. The 105-mm twin mounts on the superstructure near the catapult were moved outboard to increase their arcs of train forward and aft, and the aircraft cranes were moved inboard. The *Tirpitz* sailed on 11 January 1942 to Wilhelmshaven, a feint to cover her real destination. At 2300 on 14 January she sailed for Trondheim.

The decision to send the *Tirpitz* to Norway, which Hitler termed the "zone of destiny," centered upon his personal concern over a possible Allied invasion there. That is why he eagerly agreed with Admiral Raeder on her deployment to Norwegian waters. There she could discourage raids by the Royal Navy. It also illustrated the "fleet-in-being" concept that was adopted after the loss of the *Bismarck*; the *Tirpitz* became of great importance to the Royal Navy, and merely by her presence was able to play a major role in the Battle of the North Atlantic. In 1941–1942, the Royal Navy was prepared for another series of attacks by German surface raiders on its Atlantic convoys similar to that planned in the Rhine Exercise.

The movement of the *Tirpitz*, destined to be known as the Lonely Queen of the North, brought a swift response from the Royal Navy. On 16 January she was discovered by air reconnaissance at anchor in Trondheim. A later flight tracked her to

The *Tirpitz* is shown firing her forward 380-mm guns in May 1941. The heavy clouds to the stern are from an earlier salvo. Note the shock wave on the water from the blast of these heavy guns. (Albrecht Schnarke Collection)

The *Tirpitz* spent much of her service in the stark beauty of the lonely fjords of northern Norway. "The Lonely Queen of the North" was far from home and friends. The top view shows the ship in Faettenfjord near Trondheim in 1943. The bottom view vividly shows the steep slopes typical of the terrain in these fjords. (Albrecht Schnarke Collection)

Fottenfjord, a small fjord to the north of Trondheim. In her new anchorage she was moored near some cliffs, which protected her from air attacks from the southwest. These cliffs were overgrown by trees that also helped to camouflage the ship. The crew also cut down some of these and placed them aboard the *Tirpitz* to further conceal her. During her stay here from 16 January to 5 March, she was overflown three times by Spitfire reconnaissance planes.

Constraining factors in the early operations of the *Tirpitz* were the great fuel shortage in the German Navy and the transfer of her escorting destroyers to France for the Channel dash of the *Gneisenau, Scharnhorst,* and *Prinz Eugen.* A planned assault against Convoy PQ-8 at the end of January 1942 was therefore postponed. The fuel situation was an important factor in the decision to deploy the *Tirpitz* to Norway. Fuel had to be rationed in the German and Italian navies to maintain minimum operations. In addition, many of the replenishment ships the German Navy had maintained in the Atlantic Ocean in early 1941 had been sunk by the Royal Navy after the loss of the *Bismarck.* On 28 March 1942 British commandos blew up the gate of the dry dock at St. Nazaire that was to be used if the *Tirpitz* needed shelter or repairs—or so thought the Royal Navy. Hitler had told Admiral Raeder that the *Tirpitz* could not be sent into the Atlantic Ocean and that her operations would have to be confined to the Arctic.

The *Tirpitz* conducted two unsuccessful actions against Russia-bound convoys—one in March and the other in June 1942. On 4 March 1942, Group North requested that the *Tirpitz, Admiral Scheer,* and three destroyers attack convoys QP-8 and PQ-12.* This request was later changed by the German high command to include only the *Tirpitz* and three destroyers, on the grounds that the armored ship was too slow for operations with the German battleship. German reconnaissance aircraft sighted convoy PQ-12 in the vicinity of Jan Mayen Island on 5 March, but the Germans were unaware that this convoy had a close escort of the *Duke of York, Renown,* and six destroyers. There was a distant converging force of the *King George V, Victorious, Berwick,* and six destroyers some 100 miles behind convoy PQ-12. As the German force steamed west of Trondheim, the British submarine *Seawolf* spotted the four ships and radioed a position report on 6 March. Since Ultra had been able to read the operational orders for the *Tirpitz,* the two convoys were moved out of reach and were narrowly missed by the *Tirpitz* and her consorts. The destroyer Z-25, however, came upon the Russian merchant ship *Ijora,* a straggler from convoy QP-8 en route to Iceland, and sank it at 1632 on 7 March. The *Ijora,* however, was able to send an SOS, which provided a possible position of the *Tirpitz* to the searching British forces. To aid in locating the *Tirpitz,* Admiral Tovey dispatched six of his escorting destroyers to search in the area of the *Ijora*'s last signal. Around midnight Admiral Tovey was convinced that the *Tirpitz* had ended her operations and was heading for port. Since he had no escorting destroyers for the ships directly under his command, he reluctantly headed the *King George V, Victorious,* and *Berwick* to a position off Bear Island. At 1730 on 8 March the British admiral decided to change his mind, as Ultra intercepts of radio messages from the *Tirpitz* during the day indicated that the German battleship would be at sea and within range of torpedo bombers on the *Victorious.* Admiral Tovey issued orders for an air strike by the *Victorious* on 9 March, and decided to pursue the *Tirpitz* without destroyers. The British ships were on an eastward course at 0240 on 9 March, steaming at a speed of 26 knots.

---

*Convoys to Russia were designated PQ while those returning were termed QP. On this occasion convoy PQ-12 had sailed from Reykjavik, Iceland, on 1 March while convoy QP-8 sailed from Murmansk. It was planned to have them pass each other southwest of Bear Island on 7 March 1942.

In the meantime, on board the *Tirpitz,* Admiral Ciliax had decided to return to base at 2025 on 8 March, and by 0240 on 9 March the German battleship was on a course southward to Trondheim. At 0630 the destroyer *Friedrich Ihn* rejoined the *Tirpitz* after having been refueled at Tromsö;* the other two destroyers that were sent to refuel at 0400 on 8 March did not return. At 0640, the *Victorious* launched reconnaissance aircraft, followed some 50 minutes later by 12 torpedo-bearing Albacores. At 0802 the reconnaissance aircraft sighted the *Tirpitz* and one escorting destroyer some 70 to 80 miles west of the Lofoten Islands.

At 0810 lookouts on the *Tirpitz* spotted the reconnaissance aircraft astern and low on the horizon, and at 0815 Admiral Ciliax ordered a turn to port towards Vestfjord and safety. A strong 35-knot wind was blowing out of the northwest. With the ship headed into the wind and prevailing seas, her best speed was approximately 28 knots. By 0830 two Arado-196 floatplanes were launched, one to provide air cover and the other to provide submarine patrol. When the British planes began to slowly close the *Tirpitz,* one Arado-196 attempted to drive the reconnaissance plane away. The German aircraft headed for shore and landed at Bodø. The Albacores, however, were able to close the *Tirpitz,* despite the ship's speed and the adverse head wind. The attacking force split up into three groups. The first two (in groups of three) attacked from the ship's port side, while the third group of six attacked from starboard. When the first wave made its low-level approach, they were taken under fire by the antiaircraft guns of the *Tirpitz* and the destroyer *Friedrich Ihn.* At 0920 the aircraft released their torpedoes, and the German battleship began a turn to port to evade them. While these torpedoes were running parallel to the ship, the second group of three aircraft began their approach and released their torpedoes at 0921. All the torpedoes were armed with a 211-kg warhead with a speed of 40 knots and a depth setting of 7.6 meters. Because of the ship's turn and change in speed these torpedoes passed fore and aft of the ship, the closest being some 140 meters.

At 0923 the German battleship began her turn to starboard when the third group of six planes began their attack from starboard. At this time, all the guns of the *Tirpitz,* except the 380-mm, fired on these aircraft.

The last air strike commenced around 0925, with the aircraft attacking in groups of two at altitudes of 80 meters or so. The 150-mm starboard guns began barrage fire, while the 105-mm, 37-mm, and 20-mm guns opened direct fire on individual aircraft. Two aircraft were shot down before they could reach their torpedo release points. Only four torpedoes were dropped, but these were in a fan-shaped pattern. One of them narrowly missed the *Tirpitz* by 10–20 meters. As the planes began to turn away, they approached the *Tirpitz* close enough to strafe her with their machine guns. The command tower and the exposed antiaircraft crews were their targets. Miraculously, only three German sailors were wounded. Due to the inexperience of the British airmen in working with their flight leader, the attacks had failed. In addition, the airmen had not had enough torpedo-launch drills against high-speed ships. As a result, the German battleship escaped damage. The battleship was maneuvered very skillfully in avoiding the torpedoes that were launched. Some of the turns were so violent that the chief engineer notified the bridge that the steering gear would not hold up to such excessive strain.

The 105-mm and machine-gun batteries had been in constant action during the 11-minute attack. The antiaircraft gun positions were ankle deep in empty cartridge cases, but these gun crews had helped to save the ship from serious damage. The action

*At this point the German ships were 115 nautical miles east southeast of the British battlegroup and 100 miles west of the Lofoten Islands.

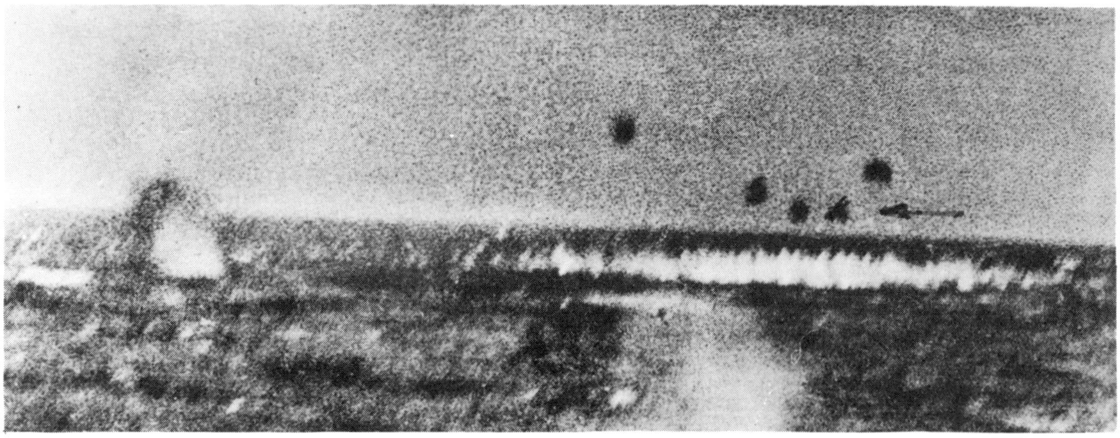

These three photographs were taken on 9 March 1942 during an attack by biplane torpedo bombers from the British aircraft carrier *Victorious*. Note the heavy wake and the violent turns the battleship has executed to evade torpedoes. The middle view shows two torpedo bombers approaching the *Tirpitz*. Bottom view shows a direct hit on one of these aircraft. (Albrecht Schnarke Collection)

The *Tirpitz* is in the Lofoten Islands shortly after the unsuccessful air attack on 9 March 1942. This view shows two of the 105-mm twin AA mounts and the twin 150-mm secondary guns. (Albrecht Schnarke Collection)

over, she headed for the Lofoten Islands, anchoring in Vestfjord during the afternoon of 9 March. She then went on to Trondheim, where she arrived on 13 March at 2100.*

The Royal Air Force made several unsuccessful attacks on the *Tirpitz* after this operation. On 30–31 March and 27–28 and 28–29 April, the British lost a total of 12 Halifaxes and one Lancaster in raids in which they dropped specially equipped mines. The air-dropped mines were caught in the trees and hills adjacent to the *Tirpitz*'s anchorage. The dropping of mines in the slopes above the ship caused the Germans to conclude that the mines were supposed to fall down to the seabed and explode there. From this time on, the ship was equipped with seabed safety nets to prevent such an occurrence.

The narrow escape of the German battleship during the March operation so alarmed Adolf Hitler that he would not permit her to be deployed against any future convoy unless its supporting carrier was sunk or immobilized. Furthermore, the *Tirpitz* and her destroyers had expended some 8,230 metric tons of fuel, and it was not until early June that the German Navy would have accumulated enough fuel for another major operation.

Convoy PQ-17, which left Iceland on 27 June en route to Russia, was the target. The Allied covering force consisted of the battleships HMS *Duke of York* and the USS *Washington* and the British carrier *Victorious*. Two German forces were to be involved—the *Lützow, Admiral Scheer*, and six destroyers from Narvik, and the *Tirpitz,*

*An Ultra intercept indicated the departure time and destination of the *Tirpitz*, so the Admiralty dispatched a flotilla of destroyers from Scapa Flow to intercept. As they would have to operate too close to shore, the attack was canceled.

An overhead view of the *Tirpitz*, showing the complex of anti-torpedo nets that complicated the British efforts to attack her with midget submarines. The nets and the character of the terrain surrounding the fjords precluded torpedo attack. (Keystone)

This photo was taken by Norwegian resistance in August 1942 and shows a low-level view of the *Tirpitz* in the same fjord as on the facing page. The *Admiral Hipper* can be seen in the left background.

Admiral Hipper, and six destroyers from Trondheim. The entire German plan was known to the British from sources in Sweden. The *Lützow* and three destroyers of the Trondheim group hit some uncharted rocks and had to be left behind. Within 12 hours after the Trondheim group had sailed, a nervous British Admiralty ordered the convoy to disperse. At 2132 on 5 July the German Admiralty canceled the mission by the surface ships when it became apparent that they had been sighted. The element of surprise had been lost, and there was a possibility of a carrier air strike. With the surface ships now recalled, the Luftwaffe and U-boats combined efforts to sink 21 of the 34 virtually defenseless ships in the dispersed, unprotected convoy. None of the German surface ships had sailed a great distance from the Norwegian coast. The *Tirpitz* went to the Lofoten Islands and then to Altenfjord.

The recall of the German ships was, to a limited degree, due to the nervousness of Hitler and the admirals in charge of the operation. The torpedo attack on the *Tirpitz* of 9 March was all too vivid a memory. In the meantime, shortly after leaving the Norwegian coast, the *Tirpitz* was sighted by the Russian submarine K-21, which fired two torpedoes, but missed.

During the summer the *Tirpitz* was moved to Bogenfjord near Narvik, but it was becoming increasingly evident that the ship was in need of a major overhaul. Hitler forbade her return to Germany, so the navy planned an extensive overhaul schedule in Norway. This involved moving the ship from Narvik to Trondheim during October and equipping the anchorage with an extensive network of smokepots and antiaircraft guns. A double torpedo net was also arrayed around the ship. On 23 October the *Tirpitz* left Bogenfjord, Narvik, and returned to Foettenfjord, Trondheim. Repairs were conducted in planned stages so that the ship would never be immobilized for an extended period. The extensive overhaul included the replacement of boiler tubing and the construction of a caisson to permit the removal of the rudders. These were remarkable

The *Tirpitz* and the armored ship *Lützow* operating in Norwegian waters. (Albrecht Schnarke)

This is a view of the *Tirpitz* under repair in Altafjord during the fall of 1943. Note the number of craft surrounding the battleship. The repairs to this ship were one of the most difficult naval engineering feats of World War II. (U.S. Naval Institute)

Shore facilities in Spitzbergen are ablaze as seen from *Tirpitz*. The *Scharnhorst* and *Tirpitz* conducted a shore bombardment in September 1943. Shortly after this, the *Tirpitz* would be immobilized by a midget submarine attack. (Albrecht Schnarke Collection)

feats without dry-dock or shipyard facilities. On 28 December, with the limited overhaul completed, the *Tirpitz* began her sea trials. On 4 January 1943, she began gunnery trials in Trondheimfjord.

While she was undergoing her overhaul, the British made another attempt to disable the *Tirpitz.* In October 1942 a Norwegian fishing cutter, adapted to carry two concealed Mark I human torpedoes (Chariots), left the Shetland Islands for Trondheim.* On 30 October this craft passed inspection by a German patrol, but rough seas caused the loss of the torpedoes. The ship was scuttled, but one of the crew escaped to Sweden.

After the Battle of the Barents Sea on 31 December 1942, in which the *Admiral Hipper* and *Lützow* failed to intercept a convoy and were foiled by a small force of British cruisers and destroyers, Hitler was enraged by their poor performance and ordered all the surface ships to be laid up. This included the *Tirpitz.* After Admiral Dönitz took command, however, these orders were rescinded. In March 1943 the *Tirpitz* joined with the *Scharnhorst* at Altenfjord, but there were no Atlantic convoys to Russia during the spring and summer of 1943. Admiral Dönitz decided to give the battle group an opportunity to work together by bombarding port installations and coal mines in Spitzbergen, which was occupied by Norwegians. The island also served as a weather station and refueling base for the British. On 6 September 1943 the *Tirpitz, Scharnhorst,* and ten destroyers headed for Spitzbergen. On 8 September, for the first time, the *Tirpitz* fired her main battery at an enemy target. The two capital ships destroyed all their assigned targets and at 1100 began their return to Altenfjord.

After the unsuccessful fishing-cutter episode, the Royal Navy decided to use a newly developed type of midget submarine—X Craft—which could be towed to Norway. There they would slip underneath the torpedo nets and attach time-fuzed 2-ton mines to a ship's bottom plating. The crippling damage from such a large underwater explosion was recognized by the experience of the heavy cruiser *Belfast,* which had been mined in some 24.4 to 27.4 meters of water, causing heavy damage to her bottom plating and propulsion plant. It was believed that such damage could be increased by greater explosive charges attached at special points on the bottom plating. A planned attack on German anchorages in Norway using midget submarines commenced in August 1942. Six X-craft were to be involved in the operation on the *Tirpitz,* and the date of 20–25 September 1943 was chosen for the attack. Ten X-craft were towed by ten oceangoing submarines, but only eight reached Norway, and two of these successfully deployed their mines.

The attempt was made on the morning of 22 September, when three X-craft successfully approached the torpedo net of the *Tirpitz.* At 0500 the X-6 passed through the net in the wake of a coaster. Inside, it hit a submerged rock, was pushed to the surface and spotted by lookouts, who mistook it for a porpoise. Submerging once again, the craft headed for the *Tirpitz,* when it was recognized to be a submarine. None of the guns aboard the battleship could be trained on the midget submarine, as she was too close to the battleship. The little ship, however, was damaged by grenades and small-arms fire. Undaunted, X-6 reached the *Tirpitz* near turret Bruno, whereupon she attempted to attach her two mines to the bottom plating. They subsequently dropped to the seabed. An alarm was sounded on the *Tirpitz,* and all watertight doors were closed. The X-6 went out of control and broke surface, where her crew decided to scuttle. A German picket boat attempted to tow the sinking craft ashore, but the tow

*These were torpedo-like craft manned by two men with underwater breathing equipment. The object was to approach the target and attach a charge to its underwater hull.

had to be cut loose, as the little craft started to go down astern. The X-6 sank close to the charges she had laid.

Around 0400 the second craft, X-7, became stuck in the nets reserved for the *Lützow*, which had returned to Germany, but she finally managed to free herself through careful maneuvering. At 0710 she headed for the *Tirpitz*. Reaching the port side, approximately abeam of turret Bruno, X-7 slid under the keel, released one mine there and another some 45–60 meters farther aft. There were now three 2-ton mines on the bottom some 30 meters below the keel and beneath turret Bruno, with a fourth positioned somewhat forward of turret Caesar and some 25 meters below the keel.

In the confusion aboard the *Tirpitz*, no submarine alarm was sounded. In order to conserve fuel, the *Tirpitz* was being fed shoreside power, and the boilers were shut down. It would take at least twenty minutes to raise steam so that the ship could be taken to sea. Captain Hans Meyer, however, was reluctant to take the ship outside the torpedo nets after a second X-craft was sighted around 0740. This was X-7 trying to escape. The *Tirpitz* was moored by six heavy lines, and the captain ordered her swung to starboard using these lines, a procedure that moved the ship slightly forward of the last position of X-6 and the three charges laid below turret Bruno. The fourth charge remained unexploded somewhere between turret Caesar and the port engine room. The operation was not successful; it took too long and could not move the ship a safe distance from the charges. In fairness to the Germans, however, they were not aware of the danger they were facing. Divers were preparing to check the hull for limpet mines, and the crew was in the process of drawing a wire along the ship's bottom when, at 0812, the log of the *Tirpitz* recorded:

> Two heavy simultaneous detonations to port at a tenth of a second interval. Ship vibrates strongly in the vertical direction and sways slightly between the anchors.

A third midget submarine (X-5) arrived on the scene at 0843 after the explosions. When she surfaced, some 200 meters from the nets, the light antiaircraft crews opened fire. This was halted on orders of Captain Meyer. The craft submerged, and a German destroyer dropped depth charges over her. The X-craft was later to sink from the damage caused by this attack.

Mine damage. The first explosion was abreast of turret Caesar, about 5–7 meters from the side; the second was about 45–55 meters off the port bow. The ship was lifted by the force of the detonations, shuddered violently, and swayed from side to side before coming to rest with a slight port list. Small cracks and tears appeared in the shell plating, and some supports in the inner bottom buckled.

When the decks of the *Tirpitz* suddenly rose under the feet of men handling lines on the quarterdeck, they were tossed into the air like loose paper. Lights went out. The bow and stern vibrated violently.* The mainmast plating was buckled and torn in places. Anchor shackles were broken, and the chains slithered about the forecastle. Shattered glass rained down from the bridge and topside structure. Watertight doors jammed, and some carbon dioxide fire extinguishers were ripped from the bulkheads and their contents discharged. An oil slick alongside amidships indicated a fuel oil tank rupture.

Damage to the ship was considerable and extended over her entire length. One explosion was centered below watertight compartments VII, IX, and X, somewhat to

---

*The technical term for this phenomenon is "transient whipping."

A unique view of the bow waves generated by the *Tirpitz* at high speed in a calm sea. (Albrecht Schnarke)

port of the centerline. Another explosion was centered a few meters to port of the centerline below compartments XX and XXI.

The hull absorbed the forces without any major structural damage. The shell plating was torn in places for lengths of approximately 1.5 meters over one explosion center. There was a large indentation, 35 meters long and 12 meters wide, in the underbottom in the areas of the tears. Some welds in the plating of the sea chests of the port engine room failed. Floors in the inner bottom tore or buckled, causing some flooding in the fuel tanks and voids of the underwater protection system. Due to deformation of the underbottom, valves for piping on the torpedo bulkheads could not be closed. The deformations in the bottom plating nowhere exceeded 12–20 cm. Several floors in the inner bottom were so weakened that it was feared, later, that even if the turbines were placed back in operation, the ship could not steam at full power. Weldments for the condenser scoops were broken and the plates were bulged out.

Total flooding amounted to 1,430 metric tons, mostly in the fuel-oil tanks and double bottom where piping was severed. There was water seepage through small tears in the bottom plating. Of the operational spaces, the port rudder compartment, the after machinery operating station, and turbo-generator room 2 were under water. The middle engine room, electric switchboard room number 2, and the port engine room were taking on water. By 1500 the flooding was contained. The *Tirpitz* listed one or two degrees to port, but this was checked by flooding some starboard voids, which took some time to accomplish.

One of the most serious problems occurred in turbo-generator room 2 where all the generators were knocked out by flooding resulting when the automatic release for the bilge and ballast pumps was damaged by a shock response. (The space was later sealed off and finally pumped out by 1415.) The electrical load situation became critical by 0940 when all electrical power had to be supplied by only one generator in turbo-generator room 1; all other generators were disabled by broken steam lines or could not supply power due to severed cables. Within two hours full power was restored to essential services when two ships came alongside the *Tirpitz* and rigged power leads to essential power distribution switchboards. Damage control parties continued to rig temporary power leads around cables snapped from the response to the explosions.

Much equipment in electrical, fire control, radio, and radar systems was severely damaged, and only partial service was possible. All auxiliary machinery within the engineering spaces was damaged due to failures in housings or components. In addition, most of the bolts in the machinery foundations were severed. Gunnery circuits were severely disrupted, and a complete overhaul would be required. The fire control computer for rangefinding aft was heavily damaged and required total replacement. The protective domes of the 10-mm rangefinders were displaced off their bearings.

The ship's armament was seriously damaged by the shock response to the explosions. Turret Dora, which weighed around 2,000 metric tons, had been lifted from its roller tracks and had then crashed back down onto them, severely damaging the roller bearings. The turret was jammed and could not be trained. Turrets Bruno and Caesar were temporarily out of action; damage could be inspected only after they had been ventilated. The bearings in these two turrets, however, were unaffected. Number 3 150-mm turret on the port side was completely jammed. One group of 105-mm guns lost one axis of stabilization and the electric drive for training and elevating. The automatic fuze-setting devices for all mounts in this battery were affected by the explosion. However, all 105-mm guns were ready for action in the manual mode. Damage to turret Dora could not be repaired in Norway, as there was no heavy-lift crane of sufficient capacity to lift the turret and reposition it.

The *Tirpitz* at high speed. (Albrecht Schnarke)

Loose equipment that was not securely fastened down was thrown about. Excellent examples of this were the two Arado-196 aircraft. They sustained heavy damage from being thrown against the ship's structure and were a total loss.

A great amount of hull damage occurred in compartments VII to X, where the propulsion plant was situated. Bolts in the line bearings of all three shafts were sheared, and the shafting was cracked in a number of places. Damage to the three propulsion turbines and their foundations made it impossible for the *Tirpitz* to move under her own power. Some of the steam connections from the boilers and also some of the fuel oil service piping to them were severed.

More shell plating damage occurred in the bow in compartments XX and XXI on the port side. The rupture in the shell plating was considered of secondary importance to that which took place in compartments VIII and IX. Naval constructors who examined the damage attributed the tears in the shell to defective plating that had not been renewed during previous overhauls. Some of the foreframes were also weakened because of local buckling of their flanges and webs from the whipping response of the ship. A secondary response had occurred aft in the port rudder trunk, where there was a local rupture of the shell plating. This damage was caused by the rudder clearances necessitating a fine afterbody shape. The steering gear was damaged, the rudder shaft bent, and the port rudder structure was extensively damaged. Seepage through tears in the shell plating finally caused the flooding of the port steering-gear room.

Shortly after the attack, the German destroyer *Erich Steinbrinck* (Z-15) signaled, "Heavy explosion, 60 meters to port of *Tirpitz* at 1012 . . . 500 cubic meters (of water in ship) . . . ." This radio message was intercepted by Ultra, and the British knew that the *Tirpitz* had been seriously damaged. Confirmation of the intercept was obtained the following day when a Spitfire reconnaissance plane overflew the fjord and reported the *Tirpitz* immobile behind her nets and an oil slick some two miles long floating down thc fjord.

Admiral Dönitz conferred with Adolf Hitler on the condition of the *Tirpitz* on 24 September, although he knew that a full report on whether the ship could be repaired would take about a week. They agreed that the ship repairs would have to be done in Norway, since it was impossible to return the ship to Germany. She could not steam under her own power, and there was excessive risk involved in taking such a totally disabled ship to sea under tow in wartime. The repairs were made in Kaafjord by personnel from the repair ship *Neumark* using the limited facilities on the *Tirpitz*. The accommodation ship, the SS *New York* (formerly of the Hamburg-American Line), provided living quarters for the extra skilled workmen and technicians.

The means at hand to repair the *Tirpitz* were far from adequate for making major repairs. Some of the material had to be shipped from Germany to Alta. These repairs to the *Tirpitz* during 1943–1944, without sufficient shipyard facilities and under severe weather conditions, were one of the most notable feats of naval engineering during the Second World War.

The most critical task was the reconditioning of the propulsion plant, as all three shafts had to be returned to service. There were cracks and dislocations in major machinery foundations, in addition to distortion of the turbine rotors, cracks in the coupling flanges of the shafting, and broken valves.

A serious problem developed in trying to realign the tail shaft. It was necessary to align the shafting forward of the tail shaft and accept the shifts resulting from such an operation, since it was not possible to work on the tail shaft out of dry dock. Some of the shafting had to be replaced and structure in its area renewed or altered. After the alignment was completed, the foundation of the turbine, gears, and thrust block had to be adjusted accordingly. The greatest adjustment in the three shafts was 13 cm. The

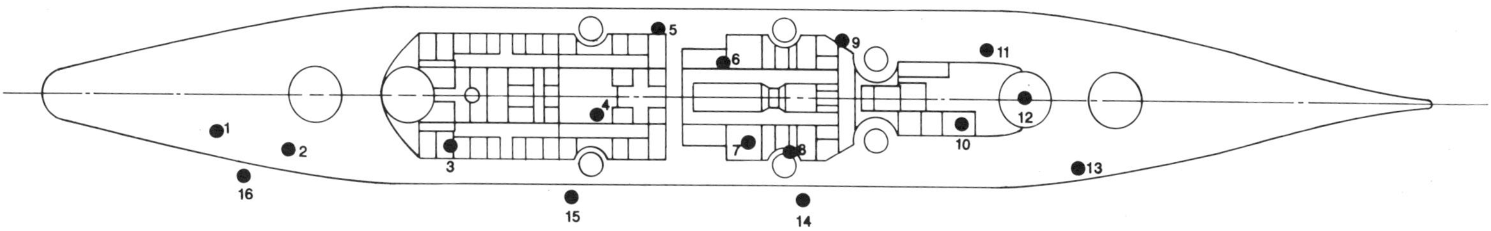

*Figure 4-5. Location of bomb hits—*Tirpitz, *3 April 1941.*

the antiaircraft crew, and the engine-room detail were at their stations when an air alert was sounded. At 0529 the British fighters attacked, strafing the superstructure to diminish the antiaircraft fire for the dive bombers that were to follow. The planes achieved relative surprise, as it took from 12 to 14 minutes for *Tirpitz* to man all antiaircraft batteries, and therefore the first attack met with sporadic fire. Nonetheless, one Barracuda was shot down.*

A second attack at 0635 met stiffer resistance, although only one more Barracuda was shot down. The attacks were made from an altitude of only 430 to 900 meters, and the armor-piercing bombs failed to achieve velocities required to penetrate the lower armor deck.† The two attack groups carried the following bombs:

	First Attack	*Second Attack*
227 kg (semi-armor-piercing—SAP)	24	39
227 kg (general-purpose)	12	9
272 kg (antisubmarine bombs)	4	2
726 kg (armor-piercing—AP)	7	2

The antisubmarine bombs were dropped for their mining effect. Bomb damage is described in table 4-7, and bomb hit locations are shown in figure 4-5. The most severe damage occurred in the superstructure and the spaces between the armor decks amidships. The ship was never in danger of sinking, although compartments were riddled with bomb splinters, the decks ripped open, cables and piping severed, and equipment destroyed. Holes in the upper deck averaged about 3 meters in diameter, and the battery deck was bulged from detonations on the lower armor deck. Because of the incomplete security of the ship (the crew rushing to their action stations), 122 men were killed, and 316 more were injured. Bomb blasts and fire went almost unimpeded through open watertight doors and hatches.

Some concrete patches to the damage from the underwater explosions of September 1943 were loosened or destroyed. About 2,000 metric tons of water from fire extinguishing or flooding increased the draft almost 0.3 meters. Two near misses that showered the side shell with splinters were responsible for most of the flooding.

Bomb hits in the superstructure destroyed the starboard catapult and crane. The number 2 starboard 150-mm turret was demolished, and number 3 port turret was

*The best time achieved for a full air alert aboard the *Tirpitz* was during the summer of 1944—8 minutes.

†The plan was to drop these bombs from an altitude of 1,000 meters, but the pilots' eagerness to obtain hits resulted in a shallower attack.

TABLE 4-7
Bomb Hits on Tirpitz—3 April 1944

1	A 227-kilogram bomb exploded on contact with upper deck, aft and to starboard of turret Dora. Upper deck slightly deformed. No ruptures in the plating; damage minimal. (first attack)
2	A 227-kilogram bomb pierced upper deck, exploded in chief petty officer and crew living spaces; damage confined to the bulkheads and equipment therein. The bomb casing may have fractured after penetration, and the explosive charge merely burned out. (first attack)
3	A 227-kilogram bomb struck the superstructure on centerline near the after main director, exploded on the upper deck in officers' after living quarters. Deck dished down in area. (first attack).
4	A 227-kilogram bomb passed through starboard airplane crane and detonated in officers' messroom. Structure in the area was badly damaged. (first attack)
5	A 726-kilogram (AP) bomb penetrated the port side of the upper deck just aft of the port catapult, but outboard of the superstructure on the upper deck. The bomb's explosion on the lower armor deck caused severe damage to decks and bulkheads in the vicinity. An outboard longitudinal bulkhead on the port side was demolished; an inboard splinter bulkhead was pierced by numerous splinters; and up to 8 meters of the upper deck was rolled back from the point of penetration. The port outboard portion of the main watertight bulkhead above the lower armor deck between compartments X and XI completely disintegrated. All ventilation ducts and power cabling in the area of the bomb burst were damaged or destroyed. The exhaust trunks from the boilers were holed by splinters in compartment X. A fire broke out in compartments X and XI, but was extinguished by a large amount of water, which collected on the lower armor deck. (first attack)
6	A 272-kilogram bomb hit and mutilated the funnel, destroyed the searchlights, and heavily damaged the 20-mm guns on the same platform. (first attack)
7 and 8	These two hits, in close proximity, completely destroyed the officers' quarters in the area where they exploded. A 227-kilogram bomb detonated in the starboard superstructure inboard of number 2 150-mm turret, and a 726-kilogram bomb hit the port superstructure forward of the funnel, exploded on the upper deck, and caused heavy damage. (no. 7 second, and no. 8 first attack)
9	A 726-kilogram bomb exploded on the upper deck, port side, forward of number 2 150-mm turret. Shell fragments pierced the armor deck; small enclosed spaces below and around the impact area were completely destroyed. (first attack)
10	A 227-kilogram bomb exploded on the starboard side of the superstructure near the base of the forward command tower and damaged the forward secondary and main-battery gun directors. However, this was not a serious loss because the aft directors for the 150-mm and 380-mm guns could be used instead. One director or a combination could control all gun positions. (second attack)
11	A 227-kilogram bomb exploded on upper deck, port side, near the main conning tower and forward of number 1 port 150-mm turret, and over the longitudinal splinter bulkhead. Upper armor deck was warped. Bomb fragments heavily damaged the surrounding area. (second attack)
12	A 726-kilogram bomb hit the roof of turret Bruno, demolished a 20-mm gun mount, and caused some dishing of armor plate. (second attack)
13	A 227-kilogram bomb detonated on contact at edge of upper armor deck, starboard, and slightly aft of turret Anton. Deck warped and was pierced by fragments. The surrounding structure was damaged by splinters; damage to compartments below decks was negligible. (second attack)
14	A near miss, probably a 272-kilogram bomb, starboard side near compartments XII–XIV. "Mine effect" noted for considerable extent; outer skin and framing distorted as much as 0.90 meters, plating torn in two places with maximum extent of 0.50 meters. (first attack)

15	A 726-kilogram bomb struck the water opposite watertight compartment IX, and its trajectory carried it through the shell plating under the armor belt. The bomb exploded in a fuel-oil bunker. The inner structure of the side protection system was totally destroyed in the area of compartments IX and X. The outer skin was penetrated in two or three places by small splinters around the entrance hole, which was about 0.50 by 1 meter in diameter. (first attack)
16	A near miss, astern. (first attack)

Note: This table is largely based upon a German naval constructor's assessment of the damage and his official report.

damaged and out of action. Fragments from bomb hits in and around the funnel had heavily damaged all the intake ducts to and uptakes from the boilers, which themselves were not damaged. The two Arado floatplanes were total losses to fires started by bomb hit 5 as shown in table 4-7.

The starboard turbine was disabled by shock damage to its new foundation. Although the decks were ripped open, some main transverse bulkheads above the armor deck damaged, and the ship was full of smoke, her vitals were intact. Some salt water that had been used to put out the fire in sections X and XI reached the boilers through damaged uptakes, but the ship was later able to steam to a repair area.

Further inspection revealed flooding in the side protection system in compartments IX and X on the starboard side from tears in the shell plating (hit number 15). A near miss (hit 14) also did considerable damage to the shell plating and bilge keel. The blast from the bomb's explosion traveled up the sea-water inlet pipe to the starboard boilers in compartment XI. Temporary patches in the plating outside the port rudder room failed, and tears around the starboard shaft bracket permitted some flooding aft.

Admiral Dönitz ordered the ship to be repaired, no matter how much work and manpower was involved. The ship was to remain in Norway, where her presence tied up Allied forces; he informed Hitler of this on 13 April 1944. It was understood that the *Tirpitz* would not take part in further sea actions, as that would have required air support that the Luftwaffe could not provide, but great importance was attached to restoring her to battle readiness.

Repairs began early in May with three shifts of work. Personnel and important equipment were brought by destroyer from Kiel to Alta in 2–3 days. The work was aided by the long hours of daylight, and by 2 June the ship was able to operate under her own power. At the end of June she was ready for gunnery practice. The number of 20-mm machine guns had steadily increased and now numbered 78. The 150-mm guns were adjusted for use in air defense, while specially fuzed shells were provided the 380-mm guns for barrage antiaircraft fire. Repair work concluded in mid-July; the aircraft complement was now reduced to one, and the starboard shaft was limited to ahead operation only.

On 17 July the *Formidable, Indefatigable,* and *Furious* launched 45 Barracudas, which were covered by 50 fighters, to attack the *Tirpitz,* but effective smokescreens and a heavy antiaircraft barrage foiled the assault. The *Tirpitz* had received ample warning of the strike, but in spite of the artificial cover and air barrage, one bomb fell close to her. Two British planes were lost. On 31 July and 1 August the *Tirpitz* sortied into the Norwegian Sea with the Fourth Destroyer Flotilla for what was to be her last fleet exercise.

Bomb damage—August 1944. In August and September the British made five more air attacks on the ship. In two separate attacks on 22 August, three planes were lost, but the *Tirpitz* received no damage. Two days later 29 fighters and 40 bombers with 726-, 554-, and 227-kilogram bombs attacked, making only two hits. Six planes were lost.

One 227-kilogram bomb hit the roof of turret Bruno and demolished the 20-mm gun mount on top. The bomb exploded on contact and the armor prevented any penetration into the gun chamber below. A 726-kilogram bomb hit on the port side of compartment XV, penetrated the upper and lower armor decks, and came to rest in number 4 electrical switchboard room. Its trajectory carried it through radio room B, where it killed one man, and into the forward secondary gunnery switchboard room. The fuze was damaged, and the bomb did not explode. The Germans considered this an immeasurable stroke of good fortune, as an explosion there would have disabled the entire ship for an extended period.

On 29 August another 67 planes attacked the ship, but smokescreens and intense antiaircraft fire foiled them. Two planes were shot down.

Bomb damage—September 1944. On 15 September RAF Lancasters attacked the *Tirpitz* with the revolutionary "tallboy" bombs. The "tallboys" were bombs especially designed for deep penetration into the ground and maximum earth shock. They had an overall weight of 5,454 kilograms, contained 1,724 kilograms of TNT, and could penetrate 3.66 meters of reinforced concrete when dropped from 4,572 meters. The planes dropped their bombs into a haze created by the smoke pots around the fjord in which the *Tirpitz* was moored.

A "tallboy" struck the forecastle, passed through the bow overhang and exploded in the seabed. A vast amount of water and mud shot up in a tall column, which rolled back the overhanging portion of the upper deck on its way. A 30-meter section of the bow was devastated by the explosion. The bottom plating was ruptured or buckled for approximately 15 meters abaft the heavily damaged forward end. Subsequent inspection of the ship revealed that the longitudinal bulkheads in the forward part of the ship had sustained tears in the plating, and the connection between the bulkheads and the shell had been severed. The bow was flooded from the stem to a point 36 meters aft, below the upper platform, and flooding occurred for almost 24 meters on that deck (some 800 to 1,000 tons). Damage in the bow area further weakened some of the side frames that had been badly strained by the mine attack in September 1943. The anchor windlass was put out of action, and there was light shock-response damage to the main propulsion turbine foundations, but heavy damage to sensitive fire-control instruments. Fuel oil and lubricating oil tanks were also damaged.

The *Tirpitz* had a heavy bow trim, and her maximum permissible speed was reduced to 8-10 knots, making her unseaworthy. Captain Wolf Lunge proposed to Admiral Dönitz that the *Tirpitz* be taken out of service, but at a meeting in Berlin on 23 September, it was decided that the ship would not be repaired for use at sea, but for service as a floating gun emplacement. It had been estimated that it would take some nine months of repairs to make the *Tirpitz* able to go to sea, but patching the damage to her bow would take only a few weeks. It was also decided to move her to Tromsø Fjord, which was farther south. On 15 October 1944 the *Tirpitz* made her last sea voyage of approximately 200 miles under her own power. The new location made it possible for British bombers of the 617th Squadron to reach her with their large bombs.

The first British air attack at Tromsø occurred at 0850 on 29 October by 27 Lancaster bombers from the famous 617th "Dam Buster" Squadron. Each plane carried one "earthquake" bomb filled with 2,541 kilograms of torpex instead of the normal 1,724 kilograms of TNT. The torpex had nearly twice the explosive power of TNT. There were no direct hits, but a near miss damaged the port shaft and rudder. There was flooding in way of the rudder and damaged shafting. Even though the main battery had been equipped with time-fuzed shells that could explode some 20,100 meters from the ship, they proved to be ineffective in breaking up this attack. One plane was damaged by shoreside antiaircraft fire.

This is a view of Tromsø, Norway, in November 1944 prior to the raid by the RAF Dam Buster Squadron, which destroyed the *Tirpitz*. The German battleship is moored to the right, off the island of Aaroy. Note the number of her crew on liberty. (Albrecht Schnarke Collection)

That same month Admiral Dönitz ordered the *Tirpitz* to be used as a floating gun battery for the defense of northern Norway. It was no longer possible to maintain the battleship in a battle-ready condition, as the damaged bow structure prevented this. Only necessary gun crews and a limited number of machinery personnel remained aboard. Of the 2,300 crew in 1941, only some 1,600 were aboard at this time. To prevent her capsizing from underwater damage, she was berthed in as shallow water as possible, with torpedo nets rigged to prevent attack by submarines or small surface craft. Air cover was increased. Dredgers placed a sand bank directly under and around the *Tirpitz* to reduce the possibility of her capsizing. Fuel was reduced to that necessary to maintain generators for domestic and gunnery electric current. Thus, only 3,000 tons of fuel oil and fresh water—about one third of the full-load capacity—remained on aboard. The ship had no trim, but there was a one-degree list to port. The list was not removed because it involved the counterflooding of starboard tanks, which would reduce the reserve buoyancy. Stern and bow anchors were rigged to keep her from moving into an unfavorable defensive position.

Destruction of the Tirpitz. The end came on the morning of 12 November 1944. The day was clear and there was a light breeze. At 0738 18 British Lancasters were sighted well south of the *Tirpitz* and by 0800 12 more planes had been detected by radar. Both flights were at heights of 3,500 to 4,900 meters. Around 0840 an air-raid alert was sounded on the *Tirpitz*, and the local airfield at Bardufoss was assigned to provide her air cover. At 0855 a fighter alarm was sounded in Tromsø, as an Army coastal artillery unit some 50 km to the northeast had reported sighting seven aircraft. This later proved

The *Tirpitz* can be seen here on 18 October 1944 inside the net enclosure of her final anchorage at Tromsø Fjord. (Imperial War Museum)

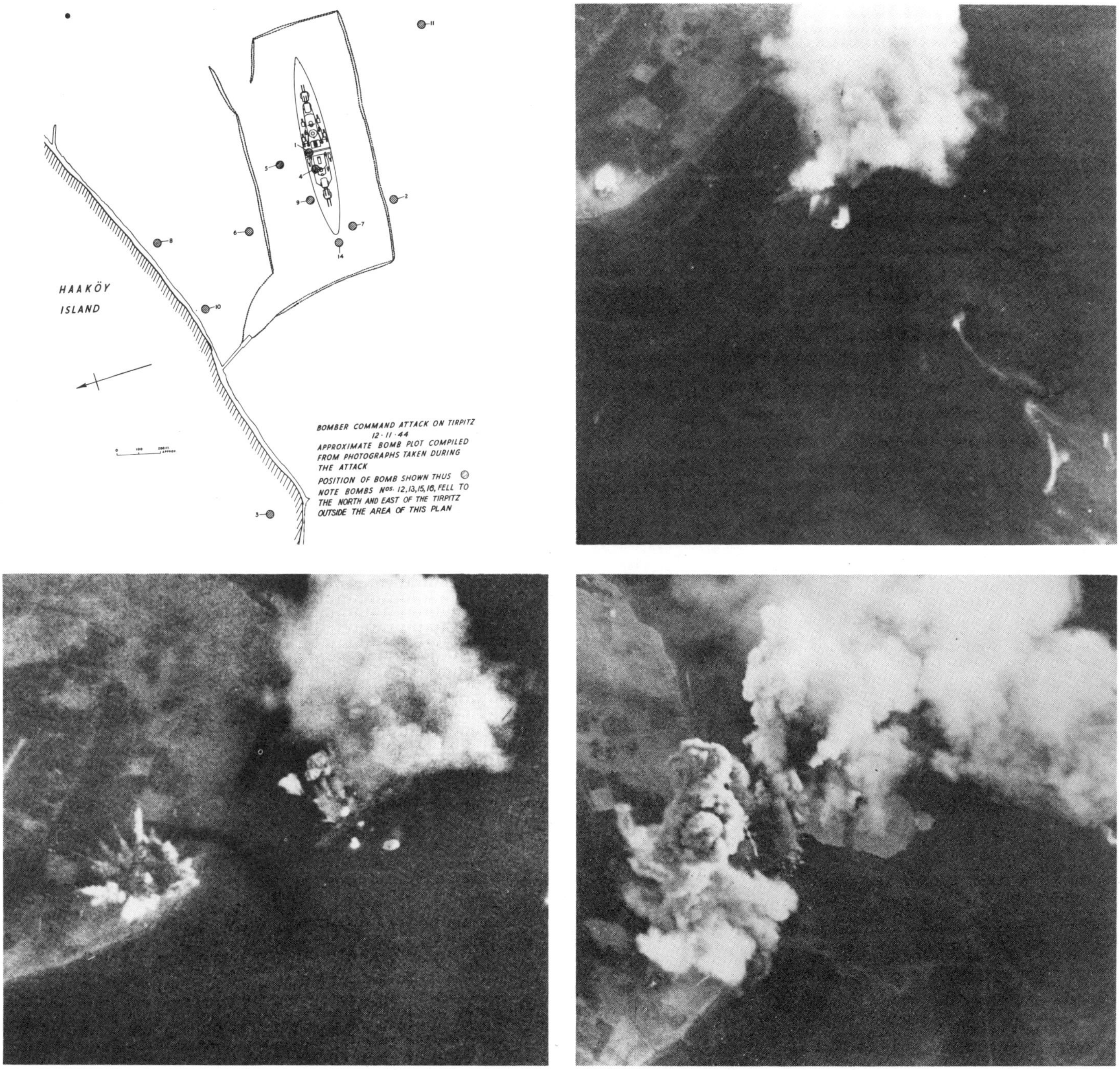

This sequence of views shows the bombing attack on the *Tirpitz* during the raid of 12 November 1944. The *Tirpitz* was moored 240–300 meters off Haaköy Island in Tromsø Fjord and enclosed with a double torpedo net. The diagram at top left shows the position of the bomb hits, compiled from photographs taken during the attack. Confirmed bomb hits are shown by circles 1 and 4. Four bombs fell to the north and east of *Tirpitz*, outside the area of this diagram. The near-miss bombs (5,7,9, 14) could have done extensive underwater hull damage to the German battleship, since the bombs contained 2,541 kgs. of torpex, approximately twice the strength of TNT. (Top right) A blockbuster hits the portside of *Tirpitz* (hit number 1 in previous diagram). (Bottom left) Smoke and flames rise from hit 1 as other bombs fall around the *Tirpitz*. Note the large explosion of one blockbuster on Haaköy Island. (Bottom right) Smoke obscures the German battleship, but the force of the explosions can be observed in the behavior of the water around her. To starboard, the large light-colored area is from the sand and mud brought to the surface by the powerful explosions. Fuel oil flowing from the ruptured plating shows on the port side. (Imperial War Museum)

false, but effectively kept the German fighters grounded*. Battle stations was sounded at 0902. By 0905 the flight of Lancasters was reported 120 km south of the ship. The captain urgently requested fighter protection, but he was informed at 0915 that such coverage could not be provided, as the Lancasters were already over the airfield at Bardufoss.

At 0935 the *Tirpitz* opened fire on the bombers with her 380-mm guns at a range of 11,200 meters. The explosion of the time-fuzed projectiles momentarily dispersed the formation. Next, the secondary battery, and finally the 105-mm and 37-mm guns, joined the action. By this time the main battery had ceased fire because the planes were too close. None of the 30 planes were shot down, and in an eight-minute attack, commencing at 0942, the planes dropped 29 "tallboy" bombs, each filled with 2,541 kilograms of torpex, scoring two hits, one probable hit, and seven near misses. Two or three of the latter were very close, inside the torpedo nets, all on the port side. Those that did not strike the shore blew large craters in the seabed, making it possible for the *Tirpitz* to capsize. The ship was drenched by the immense water columns thrown up by the near misses. The two direct hits and the one probable hit were all to the port side of the ship's centerline (see figure 4-6)—one between turrets Anton and Bruno that pierced both armor decks but failed to explode, a second between the catapult and funnel, and a probable third to the port side of turret Caesar that may have started a fire near its powder or shell magazine. After the first bomb hits there was a marked reduction in the antiaircraft fire due to the material damages, the large amount of water that came aboard the ship, and crew casualties.

The damage caused by the bomb hit near the catapult on the port side was severe; the bomb went through all decks and armor and detonated in a filled wing tank near the torpedo bulkhead of the port engine room. Shell plating and waterline armor in the immediate area completely disappeared for approximately one-third of the ship's breadth, and the bilge keel and ship's bottom plating were missing up to the cooling water discharge sea chest. The longitudinal extent of this damage was over four main watertight compartments (10, 11, 12 and 13). The torpedo bulkhead was bent inboard and torn open; the remaining portion of the demolished shell plating was spread outboard. The engine room, boiler auxiliary engine room, two boiler rooms, and electrical generator room number 4 on the port side, including all spaces beneath the lower armor deck and to the port side of the centerline, flooded immediately. The explosion also destroyed the watertight integrity of the adjoining spaces on the upper

*No one will ever know what transpired at Bardufoss airfield that morning. The local commandant was later to appear before a military tribunal.

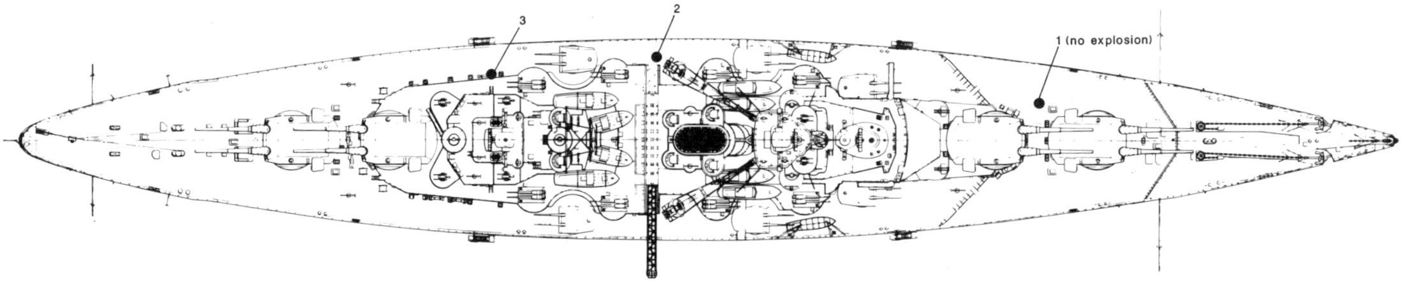

Figure 4-6. Earthquake bomb hits on Tirpitz—Bismarck, *profile).*

deck. Seconds after the bomb hit, antiaircraft fire became sporadic, mostly from 20-mm and 37-mm machine guns that continued to fire until the ship listed heavily to port.

The near misses threw a great quantity of shrapnel and water against the port side aft from the stern to the funnel. Small leaks through the shrapnel holes contributed to the subsequent port list. Although seven bombs fell within the torpedo nets, ten craters were clearly visible from the air at various distances on the port side of the ship. There was no report of any bombs exploding anywhere along the starboard side.

The ship rapidly listed 15–20 degrees to port, and counterflooding was ordered, but the controls had been abandoned. Even if the controls had been manned, the sudden inrush of water on the port side was beyond the counterflooding capability of the ship's bilge and ballast system. The initial overturning moment to port was very large because of the lack of the requisite fuel oil and fresh water normally carried aboard. The main drainage pumps were not used. The list reached 30–40 degrees by 0945, and all hands were ordered out of the lower decks. At this point the ship's watertight integrity was disrupted, as the crew hurried to abandon ship, leaving hatches and doors open behind them. Serious progressive flooding then began. Shortly after his order to abandon ship, the captain ordered all guns unloaded through the muzzles.

Around 0950 the ship listed to 60 degrees, and the port side of the superstructure was partially under water. The ship seemed to balance at that angle, perhaps because her bottom was resting against what was left of the sand bank. At 0958 another explosion shook the *Tirpitz.* This apparently originated in the powder magazine of turret Caesar, and it blew the roof and part of the rotating structure from the barbette 25 meters into the air and overboard into a crowd of men attempting to swim ashore, killing all of them. The list began to increase rapidly, and the ship suddenly turned over on her side, capsized, and remained lying at an angle of 160 degrees with most of her superstructure and the entire forward command tower buried in the mud and sand. The sudden alteration in the ship's displacement and her center of gravity with the loss of turret Caesar brought about the final phases of the capsizing. The latter took place so quickly that the men stationed in the forward command tower could not open the armored doors. All personnel in the main conning station except the second navigation officer perished. About 950 officers and crew lost their lives in the *Tirpitz*; approximately 680 survived. Some 87 men were rescued from the capsized ship within 12 hours after the attack by cutting torches that opened holes in the bottom structure so they could escape.

The turret explosion was probably caused by a shell or bomb splinter or a fire caused by a short circuit that ignited the cartridges. The *Tirpitz* had an unusually large quantity of ammunition in her magazines, since she was to be a floating battery. The fire could have possibly started from a glowing splinter hurled through bulkheads and decks from the bomb hit amidships, by a possible bomb hit near turret Caesar that was not noticed by the crew, or an electrical fire in the magazines caused by a short circuit resulting from the shock damage and flooding from the large number of near misses that occurred to the port side of the turret. The normal burning of cartridges, even with the funneling effect from the burning of the explosive inside the casing, would not have sufficed to explain why the turret was blown out. When the *Gneisenau* experienced a similar ignition of cartridges in Kiel after a bomb hit, her turret Anton was lifted 0.5 meters, but came back to the same position as before the explosion. On the *Tirpitz,* when the turret flew out of the ship, a large flame shot out of the barbette, followed by a black cloud of smoke, as if a bomb or shell had exploded on or above the cartridge platform. Finally, this was followed by the brown-colored smoke characteristic of German cartridges burning under high pressure and heat. The force of this explosion was severe; after the attack, roller-path ball bearings from turret Caesar were found on shore some 300 meters away. The cause of the explosion in the magazine of this turret will always remain a mystery.

The ship capsized in three stages. Immediate flooding caused a 20-degree list. Initial stability was impaired by the lack of sufficient liquids in the double bottom and wing fuel-oil tanks. It was not customary in the German Navy to ballast fuel tanks with salt water, but no compensation was made for the fuel being removed. Furthermore, the underwater protection system was designed so that the space adjacent to the shell was used for counterflooding. When these empty tanks were ruptured on the port side, the listing moment was much greater than if they had been inboard of the wing fuel-oil tanks. Moreover, some of the wing tanks were empty or partially filled prior to rupture; their flooding could have added substantially to the list.

Counterflooding would have sunk the ship, but salvage would have been possible with her in that position. For that reason, repair parties had orders to open sea valves on all counterflooding tanks on the undamaged side when the list reached 5 degrees, but this was never done. It is doubtful whether this planned counterflooding would have had any effect in controlling the list. The valves were too small to handle the required amount of water, and when the list reached 35 degrees, the sea connections were above the water surface and therefore useless.

When the lower decks were evacuated, the resulting loss in watertight integrity was also a contributory cause of the capsizing, as seawater could enter the unflooded spaces through open hatches and scuttles. As much as 17,000 metric tons of water flooded in on the port side—more than enough to capsize and sink the ship in deeper water. No design calculations ever included the effect of the "earthquake bombs," which were simply too powerful for the armor system or the underwater protection system to withstand.

The captain told the crew before the attack that another major air raid was imminent and that he counted on the crew to do their best to repulse the attack. The crew responded bravely in spite of the tremendous effect of the bombing and the heavy damage. The water columns thrown up by the near misses were so huge that they literally drenched the ship fore and aft. The crew's conduct in such terrible circumstances was completely disciplined, and there was no panic during or after the capsizing.

Armament.

The principal characteristics of the guns are summarized in table 4-8.

Main battery. The main armament of both ships featured the 380-mm/47-caliber gun, model 1934. The new guns had the same bore as those in the *Baden* and *Bayern* of World War I, but longer barrels and improved design permitted greater ranges.

The guns and turrets were designed and constructed by Krupp. After a conference with navy representatives on the particulars and overall design, Krupp determined the turret armor plate thickness and composition. The guns had Krupp sliding-wedge breech plugs; powder was in brass cartridge cases. The cartridges were 420 mm in diameter, larger than the shells. A main and auxiliary charge was rammed with a single stroke of the rammer. The rate of fire was one round every 18 seconds, which included shell-cartridge placement into the barrel, firing of the gun, and ejection of the cartridge.

The weight of a turret, including its foundation and gun carriage, but minus the substructure and ammunition hoists, was 1,082 to 1,097 metric tons. About 1,000 rounds of ammunition were carried for the main-battery guns, permitting 125 shots per

TABLE 4-8
Gun Characteristics

Gun	14.96 (380 mm)/47	5.9" (150-mm)/55
Shell weight	1,764 lbs (800 kgs)	99.8 lbs (45.3 kgs)
Muzzle velocity	2,690 fps (820 mps)	2,871 fps (875 mps)
Maximum range	39,590 yds (36,520 m)	25,153 yds (23,000 m)
Elevation	30 degrees	40 degrees
Gun	4.1" (105 mm)/65	1.46" (37-mm)/83
Shell weight	33.3 lbs (15.1 kgs)	1.64 lbs (0.745 kgs)
Muzzle velocity	2,952 fps (900 mps)	3,281 fps (1,000 mps)
Maximum range	19,357 yds (17,700 m)	7,382 yds (6,750 m)
Elevation	85 degrees	80 degrees
Gun	0.79" (20 mm)/65	
Shell weight	0.291 lbs (0.132 kgs)	
Muzzle velocity	2,952 fps (900 mps)	
Maximum range	5,249 yds (4,800 m)	
Elevation	90 degrees	

barrel. The armor-piercing shells weighed 800 kilograms, lighter than those of most World War II 380–381-mm guns. However, improved ballistic qualities gave the German gun increased muzzle velocity and additional range. The increased muzzle velocity caused a rather flat trajectory, suitable for combat in the North Sea. Therefore, a modest 30-degree elevation was selected because the Germans expected that combat conditions there would not require elevations greater than 20 degrees. The extra 10° was thought to be a good allowance for rolling in the Atlantic Ocean. Actions, however, with the *Bismarck* indicated that this limited elevation actually hindered her shooting. The first salvo in her long-range gunnery engagements was extremely accurate, but thereafter, accuracy fell off due to rolling. A larger gunport might have alleviated this

The *Tirpitz* was attacked and sunk by 29 Lancaster bombers of the famous "Dam Buster Squadron" on 12 November 1944. The super-heavy 5,444-kilogram bombs carried by these aircraft overwhelmed the ship's protective systems. The *Tirpitz* is shown capsized after the attack. Note the rescue operations underway on the starboard bilge. (Albrecht Schnarke Collection)

These photographs, above and on facing page, taken in the Blohm and Voss offices in Hamburg, show a very detailed model of the *Bismarck*. The upper view shows details of the forward part of the ship. The lower view depicts the superstructure in great detail, including the boat arrangement, an Arado-196 poised for launch, the arrangement of the fore bridge tower, and the radar antennae. Note the location of the 150-mm guns on the main deck and the 105-mm guns on the 01 level. The facing photograph shows the stern, including the propellers and rudders. The centerline shaft is enclosed in a bossing, while the two outboard shafts have semi-bossings and struts. It is believed that the crippling torpedo hit occurred just forward of the rudder post and below the line of the boot-topping at a sharp angle to the ship's waterplane. (Blohm & Voss)

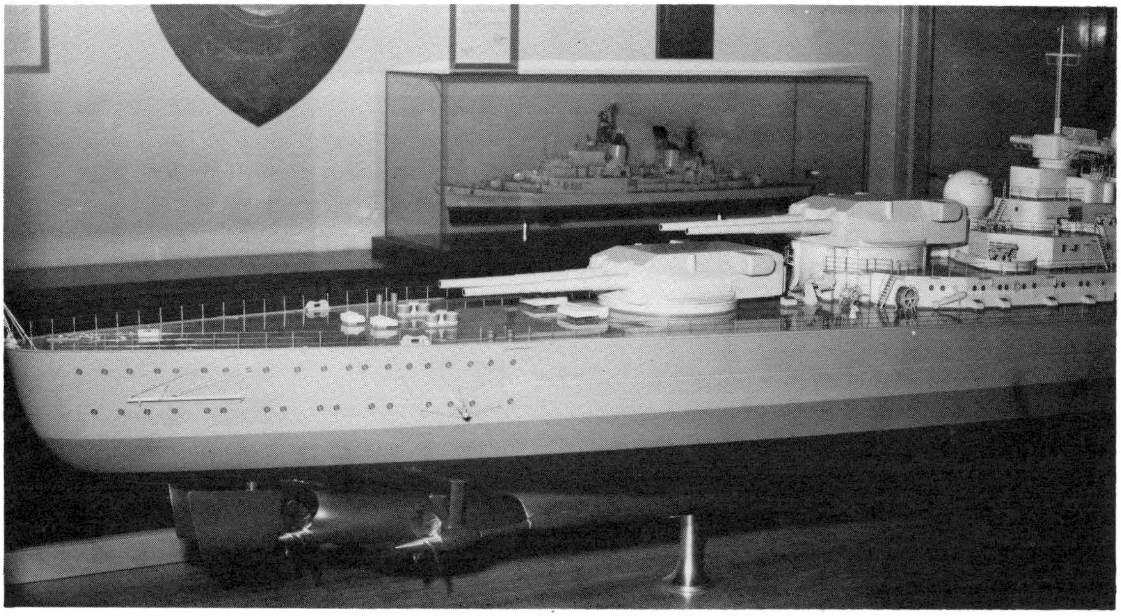

problem, but such an arrangement would have complicated the turret machinery, been more expensive, and would have weighed more. Nevertheless, the *Bismarck* and *Tirpitz* could fire at ranges that compared favorably with their contemporaries, and the guns, with stereoscopic rangefinders, were very accurate. At the time the turrets were designed, German ordnance engineers could not foresee that the guns would be later used against aircraft and that radar would make it possible to engage an opponent at longer range with greater accuracy.

In designing and selecting the main armament, primary consideration was given to the fact that the French and Italian navies were known to have authorized 35,000-ton battleships with 38-cm guns, even though the United Kingdom favored a smaller gun. In their evaluation of gun caliber, German constructors considered the armor protection of possible adversaries. The gun distribution was determined by the desire for good gun coverage.

Although 406-mm guns could have been used, more time would have been needed to develop turrets and do the necessary gunnery trials. Firing main-battery guns on a fast-moving ship at a fast-moving target at long ranges produced special problems, and a gun with high performance and muzzle velocity was necessary. However, as gun size increases, so does displacement, draft, construction cost, and magazine size. Definite limits had been established for the displacement, and any change to a larger gun caliber would have increased the weight and diminished the underwater protection. Accordingly, it was concluded that the 380-mm gun best met the requirements for optimum ship size and protection.

In the preliminary design, various combinations of twin- , triple- , and quadruple-mounted main-battery guns were considered. The main battery was finally grouped into four twin turrets, two forward and two aft, for the most effective salvo fire. Such an arrangement had these advantages:

- Ammunition supply in the gun chamber was considerably simpler than in triple turrets.
- Two twin turrets aft would have a larger number of rounds available than one triple turret. A triple turret forward made it difficult to stow powder and projectiles in the slender hull.

- Four twin turrets permitted better organization of firepower; either the two forward or two after turrets, or one of each, could form a battery.
- There was less chance of firepower loss as a result of turret damage.
- It was technically possible to fire one barrel from each turret in a salvo, and fire the other barrels in another salvo, but this was not provided for, as alternate firing of the barrels in each turret would disrupt firing procedures and reduce the salvo rate.

The four twin turrets required more weight than any three-turret arrangement, and a longer and heavier citadel. Triple turrets would have allowed nine barrels instead of eight and reduced the length of the armor belt; for these reasons that arrangement had been used for the *Scharnhorst*-class ships, and because a workable, design-proven 28-cm gun turret was already in use in the *Deutschland*-class armored ships. Weight could have been saved by using quadruple turrets, but German designers were opposed to these because they would have required a larger cut in the strength deck to accommodate the barbettes. Such a concentrated battery, aside from its devastating blast effects on adjacent structures, would have increased the risk of losing more guns due to damage to any one turret. Accordingly, it was decided to limit gun size to 380 mm and accept the twin-turret arrangement.

Longitudinal separation of the guns was determined by the lower-turret overhang, the upper-turret barbette diameter, and the ammunition stowage arrangement. Heights for the lower turrets were determined by the maximum depression of the upper-turret gun barrels. There was also a requirement that these guns have sufficient clearance for downward height adjustments when firing ahead with approximately zero elevation. Once this had been set, the arrangement and contour of the lower turrets was established. In addition, there had to be enough clearance for recoil at those firing elevations. The turrets had electric azimuth control and hydraulic elevation drives. The latter could be remotely controlled. Guns had to be loaded at an elevation of 2.5 degrees. Under favorable conditions, the main battery could fire an average of approximately three salvos per minute per gun.

The *Bismarck* in the Elbe River en route to the Baltic, via the Kiel Canal, for Builder's and Acceptance Trials. The main mast is not yet complete, and the forward rangefinder over the conning tower has not been fitted. (Blohm & Voss)

These two August 1940 views of the *Bismarck* at her outfitting pier show the ship shortly before her commissioning. Note the single gun at elevation in turret Dora. These 380-mm guns could be elevated to a maximum of 30 degrees. (Blohm & Voss)

Secondary battery. The secondary battery consisted of a split battery of twelve 150-mm guns, Model 1928, in six twin turrets, Model 1934, and sixteen 105-mm/65-caliber guns in eight stabilized twin mounts.

The 150-mm turrets were developed from those used in the *Gneisenau* and *Scharnhorst*. The four single mounts in the *Scharnhorst*-class ships lacked the operational freedom and protection required in a battleship, and the ammunition supply through a centrally positioned hoist was considered a poor arrangement for guns of this caliber. The guns, with a maximum elevation of 40 degrees and a depression of 10 degrees, were primarily designed for use against surface ships; their rate of fire was approximately six rounds per barrel per minute. The *Tirpitz* was eventually given a fuze-setting provision for her 150-mm and 380-mm shells because of the increased threat of bombing attacks.

Turret location was thoroughly studied, evaluating lessons derived from the combat experiences of battleships in World War I. The secondary battery was positioned on the upper deck; four turrets fired directly forward and two directly aft. The two forward turrets on each side could train through 150 degrees; the after turrets could train through 135 degrees. Thus, all guns could cover a sector of the forward or after quadrants.

Antiaircraft battery. The long-range 105-mm guns were identical to those used on the *Scharnhorst*-class ships. These were located on the first superstructure deck and had as much operational freedom as the superstructure would permit. Four guns could fire directly forward and four aft, and all four mounts on a side could provide barrage fire. After the loss of the *Bismarck*, the two 105-mm mounts directly forward of the *Tirpitz*'s catapult were moved farther outboard to avoid obstructions.

The 105-mm guns were provided with three axes of stabilization. This enabled the trunnion axis to remain horizontal, which in the case of a fast-moving ship meant a real reduction in the training rate. All three axes were remotely controlled by a director that was similarly stabilized. There were four of these on each ship, with two abaft the forward command tower, one abaft the mainmast, and a fourth forward of turret Caesar. Protective domes were fitted in the *Tirpitz* only. The 105-mm gun was first tried out in the Atlantic during the world cruise of the light cruiser *Leipzig* in 1933. The ammunition for these guns was supplied by a centralized hoist and then hand-carried to the gun. This and the fact that the gun had no overhead protection made the gun crews vulnerable to strafing attacks and shrapnel from bombs exploding in the superstructure. Fuze-setting was done by a device attached to the breech. Incorporation of automatic fuze-setting was not accomplished for the 105-mm gun mounts of the *Bismarck* and *Tirpitz.* One serious drawback was the possibility of the mount's electrical failure due to insufficient attention to watertightness in its design.

Turret Caesar firing during the main armament trials. The muzzle blast effects of the 380-mm guns are seen on the water surface. The 380-mm gun fired a 800-kg (armor-piercing, semi-armor-piercing, or high-explosive) shell to a range of 36,520 meters at a 30-degree elevation and could pierce 741-mm of face-hardened armor with the armor-piercing projectile at the muzzle.

A view of the after part of the *Tirpitz* showing turrets Caesar and Dora during her builder's trials in 1941. The after protective dome and equipment for the antiaircraft director have not yet been mounted, and the aft rangefinder is only a temporary one. Four 20-mm pedestals are visible to port and starboard of turret Dora. Also, the barrels for the four after 37-mm gun mounts have not been installed.

Sixteen twin-mounted 37-mm/83-caliber machine guns (Model 1930) were mounted in the superstructure of each ship. They had great flexibility in train—almost 360 degrees for the after mount—and provided effective close-range antiaircraft protection. Their stabilization was similar to that of the 105-mm guns. In addition to the hand-operated elevating and training axes, a third axis was provided for stabilizing the carriage when the ship rolled or pitched. An automatic stabilizing mechanism enabled the gun to track an airplane without interference from the motion of the ship.

The *Bismarck* had twenty 20-mm machine guns mounted in twelve single and two quadruple mounts. When the weak antiaircraft armament of the *Bismarck* proved ineffective against British aircraft during her operation in the Atlantic Ocean, the Germans realized that the antiaircraft battery of the *Tirpitz* should be strengthened. The *Tirpitz* originally had sixteen single mounts, but this battery was progressively increased, ultimately to a total of 78 guns in single and quadruple mounts. The quadruple-mounted 20-mm gun was capable of 800 rounds per minute per gun, and the mount was fully stabilized. It would appear that this multiple mount gave much better performance than the single mount. These weapons were intended to augment the antiaircraft defense against close-in attacks. The number of 20-mm guns progressively increased, while the remaining 105-mm and 37-mm batteries were unchanged.

Torpedo tubes. A torpedo armament was not included in the final design requirements for these ships. After the loss of the *Bismarck,* Hitler wanted all heavy surface ships to be based in Norway to attack Russia-bound convoys. To equip the *Tirpitz* for such duty, two 533-mm quadruple torpedo tubes were added during the fall of 1941. The installation resulted from a report made by Admiral Lütjens in April 1941 on the cruise of the *Gneisenau* and *Scharnhorst.* Additional torpedoes were stowed in deck lockers forward of the tubes, and eight warheads were stowed in a magazine below deck.

Aircraft. To aid in spotting long-range salvos, four Arado-196 aircraft were to be carried in four hangars that were arrayed around the catapult. Two more planes could be carried on the catapult. The athwartship catapult, aft of the funnel, was 32 meters long, but had extensions that allowed an additional 16 meters on each side if two planes had to be launched simultaneously. Two 12-ton cranes, port and starboard, could handle aircraft or boats. Each ship carried 18 boats of assorted sizes. Bomb stowage for the aircraft complement was in a below-deck magazine.

Protection. In the design of the *Bismarck* and *Tirpitz,* there was great emphasis on close-in protection for engagements in the North Sea as well as provision for adequate torpedo protection and subdivision. With a larger standard displacement and beam than was possible in the *Scharnhorst* class, designers could develop a more extensively armored and protected ship. The underwater protection system, however, was nearly identical to that of the battlecruisers, which had a smaller beam. Furthermore, the smaller *Scharnhorst*-class battlecruisers had armor thicknesses that were in some instances greater than or equal to that of the larger ships.

Around 1930 the German Navy requested that the Krupp concern investigate the quality of existing armor plate and determine whether it could be improved. The studies and tests made by Krupp resulted in armor plate that was some 25 percent superior to that used in World War I German capital ships. By adding molybdenum to the older chrome-nickel alloys, the cemented and noncemented armor plate had greater resistance against shells striking at all angles. Only this new composition plate was used in the *Bismarck* and *Tirpitz*.

Official German data concerning the immunity zones of the *Bismarck* and *Tirpitz* have been lost, but some indication of the effectiveness of their armor protection can be gained from the U.S. Navy's calculations on the performance of the *Bismarck*'s armor in resisting the American 406-mm/45-caliber gun firing a 1,018-kg armor-piercing shell. According to these calculations, the propulsion plant had an immunity zone from 11,000 to 21,000 meters with the resistance of the belt and slopes combined, and magazines had even better protection, from virtually point-blank range out to 25,000 meters. When consideration is given to the slopes and thicker horizontal protection, the immunity zone for the magazines was much superior to that for the propulsion plant. At ranges above 11,000 meters, the upper armor deck was penetrable, and upper citadel belt was vulnerable at all ranges. In general, the U.S. Navy felt that the magazines and propulsion plant were well protected against this 406-mm gun, although there was little of these ships above the waterline that was not vulnerable to this gun at all ranges.

The system of armor protection was the conventional "turtle-deck" type used in some World War I battleships and in the *Scharnhorst*-class battlecruisers. The main side belt was vertical, with the bulge incorporated into the lower underwater hull structure and a thinner upper citadel belt superimposed between the two upper decks. Horizontal protection was provided by a two-deck armor system, with an intervening nonballistic deck.

The vertical main side belt was divided into an upper citadel belt and a lower mainside belt. Both extended the length of the armored box with undiminished thickness. The armor belt, 68 percent of the waterline length, represented the most extensive use of side armor up to that date. It also indicated what was required as protection for a fast battleship with four turrets in a close-in engagement. The heavier lower belt was 320-mm case-hardened nickel steel in the *Bismarck* and 315-mm steel in the *Tirpitz*. The armor had a vertical height of 3.6 meters and extended 1.6 meters below the design waterline. It tapered to a thickness of 170 mm over a vertical distance of 1.6 meters. The main side belt had a wood backing; the armor plates were bolted to the side structure by armor bolts 50 to 70 mm in diameter.

Questions have been raised as to the wisdom of providing a vertical side belt of less than the traditional standard—thickness equal to the bore of the main battery. Gunnery tests of the new composition KC n/A 320-mm armor plate showed that its thickness and resistance capabilities equaled an optimum thickness of 360 mm, but as thickness increased, little was gained (the point of diminishing return). Krupp considered the thickness of 320 mm to be optimum and 400 mm neared the limit of effective production. Admiral Karl Witzell, chief of the German Navy's Bureau of Ordnance (Marine Waffenamtes), wrote the authors:

> The old rule of thumb that the armor belt should correspond in thickness to the caliber of the main battery originated from a philosophy developed around the turn of the century and had long since lost its meaning. Such limited-range battles were forecasted that at greater distances the heavy shell hits would strike the armor plate almost vertically. Meanwhile, the development of armor-piercing capped shells had so far-outdistanced armor plate that it was not feasible to provide a battleship with extensive waterline armor that could withstand short-range vertical hits. On the other hand, the distance of battle had increased so much that the angle of incidence in firing departed further and further from the horizontal. Similarly, the likelihood of hits on the armor belt grew less with this greater striking angle.

This can be demonstrated graphically even better than with a few test shots—for a given number of hits at a given angle the main armor can demonstrably reduce the calculated risk from a large number of salvos. It, therefore, was incumbent for us to determine what combination of armor (belt, armor deck, and bulkhead) would best protect the vital parts of a ship at the most critical distances, and find out where the particular weaknesses were that had to be reinforced. If there was a weakness in the armor belt, the adversary must be kept abeam but somewhat forward or aft the beam, but not so far that the heavy shell hits striking above the side armor had only the deck armor to traverse, a danger that was not to be feared at *short* range because of the small angle of incidence. Indications of the particular strengths and weaknesses of the ship were appropriate to the battle-handling instructions of the ship.

The upper citadel belt, which extended from the battery deck to the upper deck, was 145-mm nickel steel, provided to keep medium-caliber shells from entering spaces above the battery deck where the uptakes and some ammunition-handling spaces were located. The German naval constructors were convinced that this belt was necessary in close encounters with destroyers and light cruisers in the North Sea.

Beyond the armored citadel, armor plate was provided around the waterline, but instead of erecting armor on the side shell, these splinter belts were made a part of the side plating. The butts had an inside strap and were riveted; connections to the shell plating above and below these belts were rabbeted and riveted. The 80-mm lower side belt aft extended past the steering-gear room. This face-hardened armor belt extended from 0.6 meters above the lower armor deck to 1.5 meters below the design waterline. Forward of the citadel, a 60-mm splinter belt of face-hardened armor extended to the stem. This thickness was maintained for a vertical distance of 3.9 meters. Its use stemmed from the Battle of Jutland, where the bow of the battlecruiser *Lützow* was hit by fragments of a 381-mm shell that tore open a large area of the unprotected side shell and resulted in extensive flooding of bow compartments and a heavy bow trim.

Deck armor. The improved range of heavy guns and the reduction of the side armor from 350 mm in the *Scharnhorst* and *Gneisenau* to 320 mm in the *Bismarck* class necessitated a change in the deck armor system used in the *Scharnhorst* class. The armor slopes were increased to compensate for the reduction in thickness of the main side belt so that the same protection was guaranteed for the vitals. A comparison of the deck armor thicknesses of the *Bismarck* and *Tirpitz* is shown in table 4-9. It should be noted that the deck armor thicknesses of the *Bismarck* and *Tirpitz* were slightly different. The deck armor thicknesses in the *Tirpitz* followed the latest thoughts on deck protection during her design period.

TABLE 4-9
Deck Armor Thicknesses

	Centerline		*Slopes*	
Deck	*Machinery*	*Magazines*	*Machinery*	*Magazines*
Bismarck				
Upper	1.97″ (50 mm)	1.97″ (50 mm)		
Armored	3.15″ (80 mm)	3.74″ (95 mm)	4.33″ (110 mm)	4.33″ (110 mm)
	5.12″ (130 mm)	5.71″ (145 mm)		
Tirpitz				
Upper	1.97″ (50 mm)	1.97″ (50 mm)		
Armored	3.15″ (80 mm)	3.94″ (100 mm)	4.33″ (110 mm)	4.72″ (120 mm)
	5.12″ (130 mm)	5.91″ (150 mm)		

The armor around the secondary battery barbettes was increased to as much as 80 mm on the upper deck. The barbettes were 4.95 meters in diameter and were close to the stringer plates. It was necessary to increase the plating sizes for strength considerations. This was an unusual use of armor, but mandatory, since the barbettes extended above the upper deck and had to be welded to it. The outboard portion of the main armor deck sloped about 23 degrees from the horizontal. The sloping portion and the flat part of the lower armor deck were reinforced by double straps of 300-mm-wide armor plate where it intersected the torpedo bulkhead. Special care was taken with the structural arrangements because of the importance of stress flow and the elimination of stress concentrations. The upper part of the armor deck was bolted; the slopes were flush-riveted. Where the straps pierced the torpedo bulkhead armor, they were welded all around, an unusual construction feature based on earlier underwater explosion tests.

A small armor or splinter deck of moderate thickness extended beyond the fore-and-aft armored bulkheads, as continuations of the lower armor deck. It was 60-mm thick forward and 80-mm thick aft because of the additional protection required for the rudders, shafts, and steering gear.

The armor slopes considerably reinforced the moderately thick side armor belt. Such an arrangement was intended to increase the protection against high-angle hits on the side armor belt, where a shell would be exploded, ricocheted, or rendered inert. It also provided increased protection against bombs for the wing fuel-oil tanks and side-protection system. As a result, the *Bismarck*-class ships were exceptionally well protected against close-range shell fire.

Special consideration was given to the design of the armor gratings in way of the funnel and air-intake shafts. World War I gratings were considered ineffective because their long openings let shell fragments through to damage steam lines and exhausts from the boilers. In the *Bismarck* and *Tirpitz*, homogeneous armor plate of increased thickness was perforated with circular holes, with no weight increase. Firing tests showed that the thicker gratings gave reasonable protection even against direct hits.

Based upon proving-ground tests, German naval constructors believed that the 50-mm upper armor deck would be thick enough to initiate fuze action and detonate bombs above or on the lower armor deck. Between the upper and lower armor decks,

Another view of the *Tirpitz* in 1943. She still carries the old radar equipment. It was not until 1944 that the Wurzberg equipment was added. (Drüppel)

The *Tirpitz* at anchor with destroyer *Z-28* in Norway, probably in June 1942 before the planned attack on convoy PQ-17. (Drüppel)

splinter protection was provided for vital spaces and equipment such as uptakes, ammunition hoists, and shell-handling rooms. During an air attack on the *Scharnhorst* at La Pallice in 1941, however, two 454-kilogram armor-piercing bombs penetrated the entire armor deck system and exited the ship through her bottom plating.

Conning tower. There was heavy armor protection for the fire-control position and the conning tower. The sides and roof were constructed from face-hardened armor steel; the plates were rabbeted and bolted together. From the forward control tower, a face-hardened armor communication tube, one meter in diameter, extended to the lower armor deck. The after conning tower had smaller thicknesses, with a communication tube of 0.8-meter inner diameter.

Armored bulkheads. Three armored bulkheads were provided, one at each end of the armored citadel and one at the after end of the steering gear room. The armored bulkheads enclosing the citadel extended from one meter below the lower armor deck to the level of the upper deck. Armor thicknesses were equivalent to those in most modern battleships, and for armor more than 100 mm, only face-hardened plate was used.

Turret armor. The main-battery barbettes and turrets were the most heavily protected structures on these ships. Turret protection was not as extensive as in some modern battleships, but the face plate, sides, and rear armor were sloped to reduce thicknesses, save weight, and afford the same protection. The thicknesses of the rear walls were not determined solely for the reasons of protection, but to compensate for a short turret length, so that the center of gravity was as close as possible to the center of rotation. The foremost section of the inclined roof, over the gun chamber, was thicker than that of the roof (180 mm vs 130 mm). The lower barbette ring, below the upper deck, was 220-mm non-cemented armor; the upper ring, which was exposed to direct shell attack, was 340-mm face-hardened armor. The barbette armor was uniform in thickness throughout its circumference on the *Bismarck*, but on the *Tirpitz* was reduced to 220 mm for a small segment on the centerline. Barbette armor was reduced in thickness below the upper deck because of the protection afforded by that deck.

Secondary armament protection in modern battleships was often weak. The *Bismarck*-class battleships, however, had better turret armor protection than other capital ships of the period with the exception of the Japanese *Yamato* and the Italian *Vittorio Veneto*-class battleships. The barbette thickness of 80 mm was limited to a vertical height of 1.64 meters above the upper deck; below, this became a cylinder of 20-mm high-tensile steel plate that originated at the level of the lower armor deck.

Side protection. One of the most important features of the *Bismarck*-class battleships was the torpedo defense system, an essential factor in the determination of the beam. It was designed to resist 250 kg of TNT, which was slightly less than the charge used in British ship-launched torpedoes, but greater than that used in their aerial torpedoes. The underwater protection system was much weaker near the ends of the armor box, where the hull taper decreased the ship's breadth. It appears from German experience that a well-protected fast modern battleship with four turrets was not possible in a 35,000-ton design. Only a more concentrated battery would provide the additional breadths at the forward turret locations. However, torpedo protection in way of the after turrets was deeper and of better structural arrangement than the comparable area of a *Scharnhorst*-class battlecruiser.

The underwater protection scheme originated from earlier tests on the old predreadnought *Pruessen*. In World War I battleship designs, small fore-and-aft passages adjacent to the torpedo bulkheads were used for cable runs. Since the number of cables in the *Bismarck*-class ships would increase significantly, it was decided to make such wireways part of the protective system. Also, cables would have to be installed on the torpedo bulkhead where, in case of damage, many would be severed by the movement of structure. Attachments were made on the inner bulkhead of the wireway and on the upper and lower portions of the torpedo bulkhead, where the deflections would be at a minimum.

This arrangement of side protection was chosen because the additional width that could be gained in a torpedo-defense system would provide more margin. If a damaged torpedo bulkhead leaked, then the inboard space would prevent water from reaching vital spaces.

In experiments on the *Pruessen*, the torpedo bulkhead withstood a contact explosion of 250 kilograms of TNT. The additional longitudinal bulkhead also stood up, and there was no flooding in the wireway. In areas of high stress flow, the cable mountings came loose. Better structural design would have enabled them to withstand stress. Therefore, the *Bismarck* and *Tirpitz* were given this better structural design, which contributed to the selection of a beam of 36 meters. The *Pruessen* trials also provided information on better design of supporting structures for cables and relays from explosion-induced accelerations. These were the first systematic experiments on "shock response" in the German Navy.

During later stages of the design of these ships, more volume was necessary in the boiler and engine rooms, and the vertical extent of the wireways also had to be reduced. This reduced the effectiveness of the torpedo-protection system and had grave consequences in the *Bismarck*'s action with the *Hood* and *Prince of Wales*, when a 356-mm shell hit outside a boiler room.

The torpedo protection was divided primarily into two zones. The outer void space, which gradually increased in depth in lower sections of the ship, was the expansion zone where the gases from the torpedo explosion would undergo a rapid drop in pressure and be spread over a wider area. This was the most important principle in this design. In addition, as little structure as was possible, other than the protective bulkheads, was permitted in these areas, and the shell plating was made as thin as the structural theory would allow. In order that this zone would remain empty at all times, the counterflooding tanks were located in the compartments below the wing fuel tanks. A bunkering space was arranged inboard of the void space, and provision was made for oil and water to fill 70 to 75 percent of the maximum volume. German naval constructors had anticipated that much of the explosive energy would be absorbed in the destruction of the tank boundary, the hydraulic resistance of the oil and water, and the expulsion of some gas to the air spaces above such fluid surfaces. The remaining energy was to be expended in the elastic and plastic deformation of the torpedo bulkhead. In order to accomplish this, the wing fuel-oil tanks were located adjacent to the torpedo bulkhead, which was 5.5 meters inboard of the side shell. As the hydraulic effect of the liquid absorbed some explosive energy, the liquid medium would cause the remaining energy to be evenly distributed over the bulkhead surface and also provide a dampening effect for the deflection of the torpedo bulkhead. This bulkhead was stiffened to resist static water pressure after it had been plastically deformed.

This 45-mm-thick bulkhead, the boundary between the vitals and outer protective compartmentation, ran the full length of the citadel and extended from the bottom shell, where it connected to a longitudinal girder, through the lower armor deck to the upper armor deck. The thickness was determined by underwater explosion experiments; too thin a bulkhead would have been easily damaged; too thick a bulkhead would not have been flexible enough for elastic and plastic deformation and might have been displaced from its connections to the armor deck or inner bottom. Several feet above the lower armor deck the thickness decreased as the bulkhead became a splinter defense boundary in areas between the two armor decks. The essential features of the side-protection system are tabulated in table 4-10.

TABLE 4-10

Characteristics of the Side Protection System

Location	*Depth at ½ draft*	*Total bulkhead thickness*	*Torpedo bulkhead thickness*
Turret Anton	10.00′ (3.05 m)	2.09″ (53 mm)	1.77″ (45 mm)
Turret Bruno	11.50′ (3.51 m)	2.09″ (53 mm)	1.77″ (45 mm)
Amidships	18.04′ (5.50 m)	2.09″ (53 mm)	1.77″ (45 mm)
Turret Caesar	11.00′ (3.35 m)	2.09″ (53 mm)	1.77″ (45 mm)
Turret Dora	10.00′ (3.05 m)	2.09″ (53 mm)	1.77″ (45 mm)

Near the turrets where the side-protection system narrowed, the effective depth was increased locally by the arrangement of a wireway and shaft alley aft and a wireway and compartment forward. The distance between the magazine bulkhead and torpedo bulkhead was approximately 2.4 meters. This was localized and not a continuous structure. There was considerable difficulty in design with the conflicting requirements of the twin turrets, the three shafts, and the need for torpedo protection.

In 1936 the German Navy began a series of tests (the *Falk* experiments) to evaluate ship response to mine explosions deep beneath a ship. (Data on shallow explosions had already been obtained.) The experiments on an old cargo ship with charges of approximately 400 kilograms of TNT at distances of 6 to 14 meters below the keel indicated a need to reinforce certain portions of the shell plate, underbottom, and deck longitudinals. However, the Germans overlooked equipment response to shock effects in their analyses. Such experimental data would have been valuable in the design of structurally sound shock-proof foundations for electrical, electronic, and mechanical installations. The tests proved that extensive compartmentation was the best means of confining damage and flooding resulting from an explosion beneath a ship.

The *Bismarck*-class battleships had a standard double bottom 1.7 meters deep. No other arrangement was seriously considered; the vital portions within the armor box were extensively subdivided, and a triple bottom structure would have increased the ship's structural weight unacceptably and raised the vertical center of gravity of the propulsion plant. Subdivision was to provide a measure of protection. Since these ships were larger than the *Scharnhorst* class, only two boilers were installed in each fireroom instead of four. This was considered a better arrangement and entailed less risk of flooding disruption in the propulsion plant.

Subdivision and stability. A thorough damage stability analysis for the *Bismarck*-class ships closely paralleled the principles used in such an analysis for the *Scharnhorst*-class ships. As a result, they were divided into 22 main compartments by 21 transverse bulkheads, including the armored bulkheads. Compartments were further subdivided by longitudinal torpedo bulkheads, 12.5 meters off the centerline, forming 16 wing compartments on either side. The design allowed for a loss of buoyancy in any two main compartments at the ends of the ship without the submergence of the main deck. A large amount of water could enter the wing compartments from a torpedo hit, and such flooding would have been at a faster rate than caused by shell hits. The larger beam and increased volume in the side-protection system made it necessary to control flooding in the wing tanks and required that effective countermeasures be taken quickly to reduce lists. This involved trim tanks with large-capacity pumps.

The larger beam and extensive subdivision resulted in a corresponding increase in the structural weight of the hull. With the large beam, the moment of transverse inertia

This is the *Tirpitz* on trials after her overhaul in December 1942 and January 1943. It is a very rough sea and thus very windy as the German battleship glides along the fjord. Note that the crow's nest on the mainmast has been removed and that the cranes for the aircraft and boats were attached to the funnel in the *Tirpitz*, whereas in the *Bismarck* deck-mounted cranes were used. The latter were very troublesome, and this might thus explain the change. (Drüppel)

TABLE 4-11
Stability Characteristics

Displacement		*GM*		*Angle of maximum stability*	*Range of stability*
39,565 tons	(40,200 mt)	11.81′	(3.60 m)	35 degrees	53 degrees
43,010 tons	(43,700 mt)	11.65′	(3.55 m)	34 degrees	55 degrees
46,455 tons	(47,200 mt)	12.19′	(4.00 m)	33 degrees	59 degrees
52,360 tons	(53,200 mt)	13.41′	(4.40 m)	31 degrees	65 degrees

was increased, creating a large metacentric height and thus greater stability. The calculated stability characteristics of the *Bismarck* are listed in table 4-11.

The ships combined enough static stability with a significant range of stability to satisfy damage-control requirements. The *Tirpitz* had similar metacentric heights to those of the *Bismarck*; at the displacement of 50,956 metric tons it was 4.23 meters; at 45,951 metric tons it was 3.87 meters. There were more additions topside in the *Tirpitz*, which raised the vertical center of gravity.

Radar.

At the beginning of World War II, the German Navy had a fairly good radar fire-control system, although there were problems with its directional accuracy and height-determination capability. Intensive design work during 1940 eliminated the problem of directional accuracy, and the system installed aboard the *Bismarck* performed very well. The decimeter radar, however, installed in the *Bismarck*-class battleships was primarily for search, but only limited fire control. The equipment was adequate for immediate target determination, but could not provide target course and speed for an accurate fire-control plot. This was due to the lack of display units and a plotter, and a broader radiated beam.

The radar operated on an 82-cm wavelength with a power output of 14 kw, later increased to 100 kw in the *Tirpitz*. There were radar antennas fitted to all three of the directors that controlled the main and secondary guns. They were 6 by 3 meters, with six rows of dipoles, sixteen to a row. The radar was the Funkmessortung 25 and 26 Seetakt Gema set, standard equipment for German surface ships of the period. The display was accomplished by a pointer. Therefore, the *Bismarck* could not use her radar for rangefinding, and it was the use of the stereoscopic rangefinders, with their ability to measure accurately great distances in conditions of adequate visibility, that resulted in the quick destruction of the *Hood*. Some improvements were later made in the *Tirpitz* that permitted some radar-assisted gunnery, but in 1944 the *Tirpitz* still had her three Gema sets for surface search and fire control. To improve the height determination, a parabolic Wurzberg radar set was fitted. This was a stop-gap measure and did not completely solve the problem of height determination.

Propulsion Plant.

The *Bismarck*-class battleships were designed in an era when battleship and battlecruiser functions merged. Admiral Raeder wanted to develop a fast battleship of great size and power, which was not possible using a diesel geared drive. It was decided from the outset of the preliminary design to use steam propulsion, although several alternatives would be studied.

It was originally proposed that these ships should be the first German naval vessels with a turbo-electric plant operating under high-pressure high-temperature steam conditions. German marine engineers had long been fascinated by the success of such an arrangement and, after evaluating technical literature on the U.S. carriers *Saratoga* and *Lexington*, were convinced of its merits. This interest was further intensified when the new French passenger ship, *Normandie*, set a speed record on her first transatlantic voyage in 1935. A new transatlantic liner was designed by Blohm and Voss during this period to rival the *Normandie* and *Queen Mary*, and it was planned to use turbo-electric propulsion for her. To incorporate such a power plant in a German battleship, it had to be arranged in three separate groups athwartships, each with boilers, turbines, generators, and electric motors within longitudinal bulkheads bounding the machinery plant. Propellers would have to be driven by electric motors, since the turbines were not coupled to reduction gears and could only drive generators. The construction of turbines could be simplified, as they would turn in only one direction. Electric drive would also permit a more precise control of propeller speeds.

Shifting from ahead to astern operation could be done faster and more efficiently with electric drive than by any other type of propulsion. The propeller shafts could be shorter than in a comparable geared-turbine drive, and the struts and bossings could be reduced in size accordingly. Shorter shafts offered the possibility of reduced vibration at high speeds. The electric motor could be located as far aft as the ship's arrangement would allow, while the location of the turbines was influenced by a range of other factors.

Electric propulsion also had some serious disadvantages. Power cables from the generators to the motors were a weak link in a warship, where minor battle damage could disable a shaft. In addition, the four main machinery components (i.e., boilers, turbines, motor, and generator) were more ponderous in weight, and the type of electric motor desired was more difficult to design, as compared to steam geared drives.

After the turbo-electric propulsion plant had been decided upon, the German Navy proceeded with its design. Siemens-Schuckert Werke in Berlin would not accept the responsibility of constructing it in the time period specified in the contracts because engineering and deck officers requested that the interval for shifting from full speed ahead to full speed astern be less than 20 seconds. This would have been better than the interval claimed for the American turbo-electric drives. Siemens withdrew, as the firm was afraid of an inadmissible warming of the motor windings in making such a shift.

German naval constructors also realized that the maintenance of a turbo-electric drive would be more complicated than that of a comparable geared-turbine plant. Besides, from a design viewpoint, the weight saved in the propulsion plant could be used in protection or armament. If the turbo-electric machinery did not perform to expectations or was damaged, it would be difficult to repair, rebuild, or replace major machinery components in a ship with heavy side and deck armor.

Since the German Navy stressed reliability, a high-temperature, high-pressure geared turbine drive, patterned after that in the *Scharnhorst*-class battlecruisers, was chosen instead. The propulsion plant of the *Bismarck*-class battleships was the most advanced geared-steam turbine drive used in a European capital ship to that date. This type of propulsion was adopted to obtain high speed, more power, better fuel economy, and reliability.

Boilers. There were 12 Wagner watertube boilers in six watertight compartments forward of the three engine rooms. The firerooms were divided into groups of three athwartships by a boiler auxiliary machinery space. This arrangement, for damage-control purposes, represented a more extensive subdivision of the propulsion plant than had been possible in the *Gneisenau* and *Scharnhorst*.

The boilers were oil-fired, with natural circulation. They were of a quick-start type—designed to be ready for service within 20 minutes after light off—and in comparison with other marine boilers, were lighter in construction. Steam at the superheater outlet was at a temperature of 465 degrees centigrade and a pressure of 58 kilograms per square centimeter, to improve the heat drop in the propulsion turbines.

The number of boilers required at various steaming levels was similar to that of the *Admiral Hipper*-class heavy cruisers. Primary consideration was given to maintaining high speed after battle damage to the propulsion plant. As a result, the plant offered many possible boiler combinations. The normal procedure was to have two boilers per shaft at cruising speeds of 19 to 21 knots; as power demands increased, more boilers were placed on line, until at speeds above 27 knots all were in operation.

Turbines. Three separate lightweight turbine sets of the Curtiss type were in three watertight athwartship compartments. The turbines (Brown Boveri in the *Tirpitz* and Blohm and Voss in the *Bismarck*) had the high-pressure and intermediate-pressure units located abaft the single reduction gear driving the side pinions, and the double-flow low-pressure turbine ahead of the gear driving the center or top pinion. High-pressure and intermediate-pressure turbines had two-row Curtiss first stages, but the remaining stages were of the reaction type. The astern turbine consisted of a two-row Curtiss high-pressure element in a separate casing located aft of the reduction gear driving the center pinion, and exhausting to a four-row reaction element in the after end of the low-pressure ahead casing. The lightweight geared turbines had a reduction gear of cast steel instead of welded steel plates. In 30 years of experience with lightweight steel castings, the Germans had become skilled in their metallurgy and design. When the *Tirpitz* was damaged by the X-craft in 1943, the use of cast steel in the turbine housings simplified repairs, as they were suited to normal, simple welding.

This high-pressure high-temperature steam cycle was not accompanied by high-speed turbines coupled to double reduction gears. Such turbines would have been smaller, have required less steam and fewer boilers, and would have offered better fuel economy at high speeds. The low-speed turbines, single reduction gears, and the resulting poor use of energy in the heat cycle was responsible for a comparatively high fuel-consumption rate at cruising speed—0.32 kilograms per metric horsepower - hour.

The *Bismarck* had a maximum metric horsepower of 150,000—the *Tirpitz*, 165,000. The normal designed power of 138,000 produced a speed of 29 knots. The *Bismarck*, on trials at the full-load displacement of 51,700 metric tons, reached 30.12 knots at maximum 150,000 metric horsepower. The *Tirpitz*, at a higher normal power of 163,000 mhp, made 30.81 knots and a 31.0-knot maximum at a displacement of 53,200 metric tons. Great emphasis had been placed on the underwater hull form and reduced resistance to produce such an excellent performance from a broad beam and shallow draft. These ships had a very low prismatic coefficient of 0.56, indicative of a fine hull form.

Considerable thought was given during the design period to the cruising speed and good endurance qualities. The *Bismarck* could carry 8,046 metric tons of fuel, sufficient for a range of 4,500 miles at 28 knots. In the *Tirpitz*, capacity was increased to 8,818 tons. These ships had a slightly better endurance than the *Scharnhorst*-class battlecruisers, mainly because the fuel capacity was increased.

Electric plant. The electric plant had a total output of 7,910 kw, with eight 500-kw diesel generators, five 690-kw turbo-generators, and one 460-kw turbo-generator.

These generators were distributed into four spaces, the forward ones containing the steam generators while the aft spaces had the diesel generators. One diesel generator of 550 kva and the 460-kw turbo-generator were equipped to generate alternating current. The normal electric load was 3,950 kw at 220 volts (direct current), but under battle conditions this increased to 5,920 kw. The basic ship power and lighting was direct current, with a few alternating-current circuits for special equipment. The electric plant was much larger than any previous one installed in a German warship, but resulted from the power needs of electric fans and blowers, commissary outfit, steering, mooring, and gunnery. There was no dedicated emergency power system, but there was a reserve electric power of almost 2,000 kw (four diesel generators). In case of damage to normal power cabling, power could be supplied by portable electric cables attached to switchboards.

Rudders. The design of the rudders of the *Bismarck*-class ships has drawn much attention because the *Bismarck* was lost on her first mission by a chance hit near or on the port rudder. The shafts of the twin spade rudders were inclined at 8 degrees from the vertical and were connected to the steering motors by a transverse shaft and coupling. A cross connection was provided so that either steering motor could maneuver both rudders if one was damaged. The steering gear consisted of a left- and right-

This is a stern view of the *Bismarck* in the Baltic in 1940. The stern view of battleships is not always glamorous, but this photo does show the immense beam of the *Bismarck* and *Tirpitz* as well as the after superstructure detail and the stern anchor. (R.O. Dulin Collection)

handle spindle driven through a center worm shaft controlled electrically by the Ward-Leonard system. The lower edges of the rudders were situated in the horizontal plane of the propeller shafts and centered midway between the centerline and outboard shafts. The torpedo that struck the aft port side of the *Bismarck* is believed to have hit the port rudder, just below the steering gear room, where it tore a large hole in the hull and severely damaged the rudder shaft and couplings. The greater part of the explosion was vented into the ship, damaging the steering motors and their foundations. There was a severe shock response in the ship. Damage to the steering system and structural damage to the hull prevented the rudders from resuming a normal position and made it impossible to disengage the rudders from the steering motors so the hand-steering device could be engaged.

Hull Characteristics. With the progressive increase in the use of electrical arc welding in German warship construction, the *Bismarck* and *Tirpitz* were 90-95 percent welded. Welding was used in some connections in the non-heat-treated armor plate, since a special electrode for this type of armor steel had been perfected. In both ships, the entire hull and all the non-face-hardened armor plate, with the exception of the torpedo bulkhead and the lower armor deck, were welded. The structure of the principal decks was also welded. Seams and butts were welded, but specifications allowed for handling shrinkage by permitting riveted seams on each side of the ship.

It was calculated that welding, instead of riveting, contributed much saving in hull weight. Another factor was improved and higher-strength steel. A comparison of the *Baden, Hindenburg,* and *Tirpitz* in table 4-12 shows this.

TABLE 4-12

Comparison of Baden, Hindenburg, and Tirpitz

	Baden *	*Hindenburg* †	*Tirpitz*
Length overall	548′ (180 m)	647′ (212 m)	832′ (253.6 m)
Beam	91′ (30 m)	88′ (29 m)	118′ (36 m)
Depth	40.94′ (12.48 m)	40.35′ (12.3 m)	46.0′ (14 m)
Displacement (Design)	28,080 tons (28,350 mt)	26,961 tons (26,945 mt)	45,220 tons (45,950 mt)
Hull weight	8,460 tons (8,600 mt)	8,403 tons (8,538 mt)	11,506 tons (11,691 mt)
Percent hull wgt of displacement	30.1%	31.5%	25.5%
Armor weight	11,160 tons (11,340 mt)	9,194 tons (9,342 mt)	16,820 tons (17,089 mt)
Percent armor wgt of displacement	40.0%	33.6%	37.3%
Hull and armor wgt	19,620 tons (19,940 mt)	17,597 tons (17,880 mt)	28,326 tons (28,780 mt)
Percent of hull and armor of displ	70.1%	65.1%	62.8%
Side-armor thickness	13.77″ (350 mm)	11.81″ (300 mm)	12.60″ (324 mm)
Deck-armor thickness	1.18″ (30 mm)	1.18″ (30 mm)	3.94″ (100 mm)

*Completed in 1916, the *Baden* was the last battleship to join the High Seas Fleet in World War I.
†Completed in 1917, the *Hindenburg* was the last battlecruiser to join the High Seas Fleet in World War I.

Atlantic bow. Both ships had "Atlantic bows" fitted during their building period. These bows proved suitable for use in the North Sea and Atlantic Ocean. With adequate freeboard and no overweight problems, both ships performed well in heavy seas. At no time during the action with the *Hood* and the *Prince of Wales* did the *Bismarck* have any problems caused by sea conditions (green water over the bow). As the bow area was particularly susceptible to damage from shell hits and mine explosions, a moderate amount of flare and high freeboard was incorporated to provide sufficient reserve buoyancy in case of damage. This would prevent dangerous trims that might immerse the bow, slow the ship, and make her difficult to steer. Even with heavy flooding of the bow region, the *Bismarck* was able to steam at 26 knots.

The clipper stem provided a novel anchor arrangement, with one located at the stem and the other two hauled up on deck instead of into external hawse pipes. There they were laid flat and suitably secured. To anchor, the anchors were pushed overboard mechanically. This left the sides of the bow free from any major projections or recesses that could lead to spray formation. A stern anchor was also provided.

All structural materials were carefully selected. Steel 20-mm thick or greater was of a high-tensile type, unless otherwise specified, to save weight. For lesser thicknesses, mild steel was used. Longitudinal framing and high-strength steels were used in upper portions of the ship to save weight and maintain sufficient longitudinal strength. High-strength steel was ultimately used in all primary structural members and contributed to savings in weight.

Joiner bulkheads in living spaces were of a special lightweight material. Water closets and wash rooms had steel partitions. Furnishings in the staterooms and mess areas were aluminum, excepting chairs where the use of aluminum would not save enough weight to warrant the extra cost. There was no structural aluminum in partition bulkheads.

Turret supports for the 150-mm twin turrets had an interesting design. The turret foundations on the armored deck were octagonal in section, but gradually changed to a circular form at the point where they supported the roller paths for the turrets. The main barbette armor was supported below the lower armor deck by two transverse and two longitudinal bulkheads. The four longer sides of the octagonal foundation were supported by these bulkheads. The rectangular structure formed by the intersection of the longitudinal and transverse bulkheads served as the boundary for the shell and cartridge rooms. Shells were not stowed in the rotating part of the turret structure. The foundations of the main-battery barbettes were similar.

Accommodations. The complement was 1,927 men without a staff for the flag officer, or a total of 2,106 men with a flag embarked. Provision was made for an additional 2,500 men for a single-day accommodation, but sleeping quarters could be provided for only 1,600 additional men. The berthing for enlisted men and petty officers were in the same rooms. Bunks were of the type supported in tiers of three by wire stays (turnbuckle cables) fastened to the overhead and deck structure.

Airconditioning was provided for the sick bay, photographic laboratory, and the wardroom. A total of 230 electrically driven exhaust and supply blowers were installed, of which 33 served the machinery spaces. Eighteen supply and three exhaust blowers were equipped with air coolers. Watertight closure devices were fitted in the ventilation ducts, so that watertight integrity was maintained up to the bulkhead deck (battery deck). Below the armored deck the ventilation ducts were arranged vertically to each watertight division, so that no penetrations were made in the main transverse watertight bulkheads. In the machinery spaces, gas-tight closures in the ventilation ducts were provided for use in the event of oil fires.

Summary. The loss of the *Tirpitz* marked the end of a long line of distinguished battleships and battlecruisers in the German Navy that started in 1897 with the appointment of Admiral Alfred von Tirpitz as State Secretary of the German Navy. The *Bismarck* and *Tirpitz* exemplified the battleship at the end of its epoch. The destruction of the *Bismarck* was a classic example of the contest of gunfire and torpedoes versus armor and torpedo protection. The airplane, however, had played a decisive role in her sinking. Some three years later, the destruction of the *Tirpitz* by heavy bombers was but another demonstration of the obsolescence of the battleship in the era of air power. The loss of the *Bismarck* did revive interest in carriers, but this did not last. It took the near-torpedoing of the *Tirpitz* in March 1942 to finally convince the Germans of the importance of aircraft carriers and naval aviation in war at sea. The Luftwaffe approved the use of an aircraft complement for the carrier *Graf Zeppelin* and began to train carrier pilots. The German Navy also decided to convert the heavy cruiser *Seydlitz*, which was nearing completion, to an aircraft carrier and made studies of the conversion of three passenger liners—the *Europa, Bremen*, and *Potsdam*. All of these plans were later set aside after Admiral Raeder resigned as commander-in-chief in January 1943.

The German Navy had no reason to change its philosophy in the design of large warships after its experiences in the Battle of Jutland, where it was proven that excellent subdivision, reserve buoyancy protected by extensive armor, a good combination of belt and inclined armor, and a large distance between the torpedo bulkhead and side shell were features necessary in a successful battleship design. This philosophy was also followed in the design of the *Bismarck*-class battleships. From all indications, the following criteria were confirmed by the experiences of the *Bismarck*:

- The minimum distance between the torpedo bulkhead and side shell should be 5 meters amidships and 3.5 meters at the fore-and-aft armored bulkheads.
- The lowest part of the side armor under the design waterline should only emerge with a list of 5 degrees.
- The horizontal armor deck should have a height of 3.8 to 5.3 meters above the design waterline.
- No openings should be made in the watertight longitudinal and transverse bulkheads inside the armored citadel and below the lower armor deck. Additional protection of openings and vertical access in the armor deck should be attained with the use of very high watertight trunks.
- The external layer in the side-protection system should be kept void, if possible.
- There should be extensive subdivision of machinery spaces, distribution of boilers, and associated equipment in as many compartments as possible, and very extensive arrangements for damage control and switching of vital functions.
- Cable trunks should be provided outside the vital spaces to augment the underwater protection system and should be symmetrically distributed on both sides of the ship.

Several weaknesses were revealed in the *Bismarck* and *Tirpitz* during their wartime careers. Among these were the rudder arrangement, the secondary armament, and the thicknesses and arrangement of the horizontal armor protection.

Rudders are the "Achilles heels" of all warships, and no effective means for their protection have yet been found. Rudder vulnerability and its consequences were demonstrated in the loss of steering in the *Bismarck* and the damage sustained by the *Tirpitz* during the X-craft attack. After the loss of the *Bismarck*, German naval constructors concluded that it was not possible to protect the rudders against weapons then available. Studies were more concerned with the separation of a jammed rudder from the ship's hull by special explosive charges. Since the direction of the explosive force could be controlled, the proposal gave some assurance that the propellers would not be damaged. A rudderless ship could be maneuvered by engines and propellers. Since there was no material available for a full-scale test, further investigations ceased.

The antiaircraft armament exhibited some weaknesses, as the historical narratives have shown. The use of a dual-purpose armament would have possibly increased the number of antiaircraft guns, but might have weakened the defense against destroyer attack, which German naval experts deemed more important. Although the 105-mm gun gave good performance against Fairey Swordfish torpedo bombers and Albacores, it was no match for the Martlets, Hellcats, Barracudas, and heavy British bombers that were used against the *Tirpitz* in Norway. It finally became necessary to use the 150-mm and 380-mm guns for the air defense of the *Tirpitz* in 1944. If the 150-mm gun had been converted for dual-purpose use earlier in the war, the air defense of these ships would have been greatly augmented.

The lack of foresight demonstrated in the air defense of all German capital ships, including the uncompleted carrier *Graf Zeppelin*, is perhaps understandable when one considers the German Navy's predilection for the gun as the primary offensive weapon of naval warfare and the control of all German operations (land, sea, and air) by the Luftwaffe. The Luftwaffe commander, Hermann Göring, never appreciated the airplane as an effective weapon of war at sea. It is possible that this indifference was due to the German lack of experience in the problem of air defense in the prewar period, when the

Passive defense, such as the use of smokescreens to mask the moorings of the *Tirpitz*, became particularly important after the ship was virtually immobilized by severe bomb damage on 3 April 1944. (Albrecht Schnarke Collection)

These three photographs, above and facing, of the *Bismarck* were taken at Blohm and Voss Shipyard in Hamburg between 10–15 December 1939. Note that the upper 145-mm citadel belt is not in place, turret Caesar is under construction, and the forward command tower is very incomplete.

importance of a dual-purpose weapon could have been demonstrated. If Germany had been able to operate an aircraft carrier during the design period of the *Bismarck*-class ships, fleet exercises would have probably shown the need for a dual-purpose armament to deal effectively with the airplane and destroyer on equal terms.

The main-battery guns held the same superiority over their British counterparts as had been the situation during World War I. What one navy can do, another can, but for some reason British gun development lagged behind that of Germany during the interwar period, except for the development of a workable and reliable dual-purpose medium-caliber gun and a successful multi-barrel light machine gun mount. The most noticeable event during the 24 May 1941 battle between the *Hood, Prince of Wales, Prinz Eugen*, and *Bismarck*, however, was the poor performance of many of the German shells. Although the shell that struck the *Hood* gave a flawless performance, those that struck the *Prince of Wales* did not perform as well.

German armor-piercing shells did not change all that much from the basic design of World War I, except that the armor cap was enlarged, the nose was blunted somewhat, and a ballistic windscreen was added. Even the acceptance specifications did not change—normal impact against caliber-thickness cemented armor and 30-degree obliquity from the normal against half-caliber thickness cemented armor—being well below the specifications of American and British World War II armor-piercing projectiles. The base plug and fuze design and the size and shape of the cavity for the explosive hardly changed. There was a slight drop in the filler weight as a percentage of the overall shell weight. The same filler was used, however—TNT blocks pre-formed to the cavity shape. These blocks were wrapped in cardboard, paper, and felt and were inserted into the projectile with a large wooden cushion in the upper end of the cavity to reduce the impact shock on the TNT. This resulted in the cavity being, perhaps, up to 20 percent larger than an equivalent American shell cavity would have been for the same filler.

American projectiles used a far less sensitive ammonium picrate filler that needed no cushion. The larger cavity in the German shell weakened the projectile body, which explains the poor effectiveness after impact of the two 203-mm shells that hit the *Prince of Wales* in the stern. It appears that Krupp decided that it had a good projectile design aside from the nose shape and armor-piercing cap and that no real design or specification change was required. German projectile design was superior until the British realized that the performance of their shells was less than satisfactory during the Battle of Jutland, and enormous improvements were made in British projectiles (outmatching the developments in gun design). While the penetration ability at low-to-medium obliquity against cemented armor was equal to or better than that of other navies, proper emphasis was not given to the structural adequacy of the shell so that it remained in effective bursting condition after penetration.

The upper deck armor thickness of 50 mm was defeated by non-armor-piercing bombs dropped during a low-level attack on the *Tirpitz* in Altenfjord on 3 April 1944. The giant earthquake bombs greatly magnified the weaknesses of the horizontal protection, but to have strengthened the ship against the vertical penetration of such a bomb would have required an armor deck of impossible thickness. The attack on the *Tirpitz* on 3 April 1944, however, demonstrated the weakness of German armor protection against bomb attack, while it also served as an example of the superiority that aircraft and the aircraft carrier had achieved at this time in the war. The British pilots were over-anxious to obtain hits, and their bombs were not dropped from the prescribed altitude of not less than 1,000 meters, so the *Tirpitz* was able to escape more serious damage.

The side-armor protection of the *Bismarck* and *Tirpitz* compared very favorably with other battleships constructed during the same period, with 68 percent of the waterline length protected by heavy side armor. A splinter belt extended to both ends of the ship to prevent splinter damage to the side shell. The only weak point in the side armor was the upper citadel belt, which, in the *Bismarck*'s last battle, was easily penetrated by British 152-mm and 203-mm shells.

In view of the apparently low priority that German authorities assigned to the construction of the *Bismarck* and *Tirpitz*, it is not too surprising to find that their turbine and gear design did not match the developments in boiler technology. This was also aided by a last-minute edict by Admiral Erich Raeder to use high-pressure and high-temperature steam. This choice of boiler led to higher fuel-consumption rates, higher turbine and steam rates, lower reduction-gear efficiencies, and higher weights per unit of power than was possible in double-reduction gear machinery. These ships used single-reduction gears with three pinions meshing with the main gear. The low turbine speeds resulting from this limitation on gear ratio contributed significantly to the higher weights encountered, as did also the large number of turbine cylinders used in order to allow more pinions to bear on the main gear.

Steam conditions were chosen to be 540 degrees centigrade and 58 kilograms per square centimeter, which, if properly utilized, should have produced high propulsive efficiencies. Because of the relatively poor turbine performance, higher than necessary gear losses, and a poor use of heat in the heat cycle, however, the specific fuel rate was 0.32 kilograms per metric horsepower per hour at the full-load displacement. When it became necessary to limit the initial temperature to 500 degrees centigrade because of the lack of proper materials for the boiler superheaters, the fuel rate became even worse. All these problems, although serious, were not major factors in the loss of the *Bismarck*. The critical fuel situation during her short, but eventful, deployment was brought about by tactical errors. The loss of fuel resulting from shell damage was more serious than the Germans first thought. The decision not to stop and repair the damage further aggravated a very bad situation.* From the time the *Bismarck* broke visual contact with the British ships around 0300 on 25 May until the time she was rediscovered by the Catalina at 1030 on 26 May, her speed was limited to only 20 knots. This was the most economical speed, and Admiral Lütjens was hoping to maintain his fuel reserve when he entered the Bay of Biscay, where he might have to fight off British ships to enter port. So ardent was this belief that one of his dispatches called for a tanker to refuel the *Bismarck* some 200 to 400 miles from the French coast. Certainly the fuel crisis eased slightly on 25 May with the recovery of some 200 tons of fuel from the damaged forward compartments. The wind and wave conditions of 25 and 26 May and added displacement from flooding were such that a higher than normal fuel rate occurred, but the poor performance cited in the discussion of the main propulsion units was also a minor contributing factor.

In reviewing the damage sustained by both ships, the *Tirpitz* sustained some extraordinary damage during the course of her service, while the *Bismarck* absorbed considerable shell and torpedo damage before she sank. The most spectacular damage sustained by the *Tirpitz* was from non-contact explosions. The 2-ton mines laid by the X-craft created explosive forces far in excess of any design loads that had been derived from the *Falk* experiments. It seems that the second pressure pulse was the most severe, and this might be attributed to the explosives' amplification on the seabed. Various sections of the ship were set in motion at different periods of time, since there were two major sources of explosive loadings. The ship, as a whole, was deformed momentarily by the forces of the explosions, causing local oscillations of plates and frames, while larger items such as main bulkheads tended to vibrate. In turn, the whole ship vibrated as a beam with two or more nodes. This phenomenon is termed *whipping*, and its magnitude will vary considerably with the size of the explosive charge and position as well as the distance of the explosion from the ship. Although whipping involves lower accelerations of the ship and its components than ship motions, it is the phenomenon most identifiable by observers. Whipping threw men about on the decks and caused damage to loose objects and equipment on foundations. The tiedown bolts for equipment on board the *Tirpitz* were not designed to withstand the large oscillations generated by this size of explosion. The *Tirpitz* was the largest warship put out of action in World War II through a shock response to a non-contact explosion.

The Royal Navy intended to destroy the *Tirpitz*, not merely immobilize her, and time fuzes were to be set to go off simultaneously against the shell plating. There is little doubt that the resulting damage would have been greater if all the mines had been placed under the bow and stern as specified in the attack plan. The immediate water pressure of such a detonation would have exerted the largest force on the ship's bottom, and the closer the explosion was to the hull, the greater the damage. These conditions would ensure that the maximum shock response in the form of whipping would take place. There might have been severe damage to the hull girder, and the great pressures developed would have extended over a larger portion of the underbottom. Flooding would have been more widespread the closer the explosions took place to the hull.

The ineffectiveness of British aerial torpedoes to cause extensive internal damage to the *Bismarck* was the result of the small warhead charges of 204–211 kilograms of TNT. The underwater protection system in these battleships was designed for a TNT charge of 250 kilograms and thus contained the damage within the system. There is no

---

*Under the circumstances, this was not possiblc, but there were other suggestions that the ship and fleet command ignored, such as dropping the anchors and chain—a measure that would have decreased displacement, but more importantly would have reduced bow trim.

A view of the *Bismarck* from the *Prinz Eugen* during her work-up period in early 1941. The heavy cruiser is towing and refueling the battleship. The astern fueling technique was vastly inferior to the beam-to-beam method now used in most navies.

question to the damage caused by the torpedo that hit the port rudder, just below the steering gear room. The rudders were jammed at a 12-degree position so that steering was impossible. After many hours of work, efforts to disengage the starboard coupling were abandoned. The rudders were immobilized, and not even using the propellers to counter the rudder position was successful. British ship-launched torpedoes, however, carried a greater explosive charge of 340 kilograms, and those fired by the *Dorsetshire* would have had greater effect if the *Bismarck* had not been in a sinking condition at the time the torpedoes hit. Because of the port list and deep draft, they probably struck the side armor belt at its maximum thickness. A great portion of their destructive energy was expended in displacing thick armor plate. Thus, the area of damage would have been substantially reduced. Those torpedoes striking to starboard also did not cause large holes in the hull, as two officers who survived—Lieutenants Junack and Müllenheim-Rechberg—testify.

How and why the *Bismarck* sank cannot be attributed to a single factor. Examination and evaluation of the ship's loss, the design of the underwater protection system, the damage stability calculations for the *Bismarck*, and a critical assessment of the survivors' testimony has established beyond doubt that the torpedo hits by the *Dorsetshire* hastened the *Bismarck*'s end. Several torpedo hits and a number of 356-mm and 406-mm projectiles had caused flooding, but the decisive factor that hastened the sinking resulted from scuttling charges detonated in the propulsion plant. There might have been more torpedo hits, although it was difficult for observers on the British warships in the battle to distinguish between shell splashes and the plume from torpedo hits against the ship's side, but the authors do not believe this to be the case.

One of the actual stability conditions considered during the damage analysis has been reproduced in appendix A because it closely resembles the condition of the ship prior to the torpedoing by the *Dorsetshire*. It can be seen from figure A-3 that even with a large amount of flooding the draft was not greatly increased. Assuming the ship to have a GM of 4.4 meters at a displacement of 53,000 tons, the GM increases to 5.25 meters because of the upward shift in the center of buoyancy due to flooding of low spaces. This also means that the ship could resist off-center flooding much better. The *Bismarck*, however, did have a fuel problem arising from the bow shell damage. Thus, several of the fuel tanks in the side-protection system were probably empty, improving the stability somewhat before damage occurred. Flooding of high tanks would have worsened the condition summarized in appendix A. After the *Dorsetshire*'s torpedoes struck, the *Bismarck* began to list to port, and with subsequent progressive flooding on that side and with the large flooding surfaces in the engine and boiler rooms, she gradually went over to port and finally capsized before disappearing below the water surface.

Two torpedo hits could not have produced sufficient flooding to have caused the ship to settle so quickly. Previous torpedo hits, while diminishing the ship's buoyancy, were evenly spaced, port and starboard, forward and aft, making the problems of counterflooding easier for the German damage-control teams. The two torpedo hits from the *Dorsetshire* certainly would have damaged the armor belt more extensively than the damage caused by the *Victorious* aircraft carly in the chase. There would have been a greater inboard shift of the side armor belt and flooding of the outboard void tanks and possibly the wing fuel-oil tanks in the side-protection system. If the torpedo hit close enough to a transverse bulkhead, there might even have been an inboard displacement of the torpedo bulkhead, loosening the watertight backing structure and supports. The subsequent flooding from this type of torpedo hit, however, is less than that produced by a torpedo striking an unarmored steel side shell or the lower one-third of submerged side shell. Thus, the authors' thorough study of the *Bismarck*'s sinking confirms Lieutenant Gerhard Junack's claim that he was ordered by the executive officer to detonate scuttling charges and that this act proved to be the decisive factor in the loss of the ship's buoyancy. British shells and torpedoes, however, had defeated and doomed the ship to destruction; the scuttling merely hastened the end.

Without question the *Bismarck* and *Tirpitz* were formidable opponents for the Allied navies. Their endurance qualities were quite good for their period of design, but could have been improved by the use of higher turbine speeds and double reduction gearing. The boilers had the highest temperature and pressure used in a World War II capital ship; credit for this is attributed to Admiral Raeder, who ordered these conditions over the objections of many senior officers.

The *Bismarck* and *Tirpitz*, while not the most powerful battleships ever built, were good fighting ships, with many design innovations and well-trained crews. Despite their deficiencies, the exploits and dramatic careers of the *Bismarck* and *Tirpitz* have made these ships near legends in modern naval history.

Name	*Bismarck*	*Tirpitz*
Builder	Blohm and Voss, Hamburg (Yard No. BV 509)	Kriegsmarine Werft, Wilhelmshaven (Yard No. S–128)
Laid down	1 July 1936	30 October 1936
Launched	14 February 1939	1 April 1939
Commissioned	24 August 1940	25 February 1941
Disposition	Sunk in action with British cruisers and battleships 27 May 1941 at 48°10′ North and 16°12′ West 2106 men lost, 115 survivors (110 captured)	Sunk in Tromsø Fjord by British aircraft on 12 November 1944 at 69°36′ North and 18°59′ East 950 men lost 1948–1957 partially salvaged by British, Norwegian, and German firms.

Displacement

Bismarck—August 1940
38,892 tons (39,517 mt) Light ship
44,734 tons (45,451 mt) Design
48,626 tons (49,406 mt) Full load
49,609 tons (50,405 mt) Battle load

Tirpitz—February 1941
48,648 tons (49,429 mt) Full load
44,755 tons (45,474 mt) Design
38,915 tons (39,539 mt) Light ship
49,628 tons (50,425 mt) Battle load

Dimensions

Bismarck
820′ 4″ (250.5 m) Length overall
792′ 6″ (241.55 m) Waterline length
118′ 1″ (36.00 m) Beam (maximum)
49′ 3″ (15.00 m) Depth (midships)
33′ 6″ (10.20 m) Draft at full load 48,626 tons (49,406 mt)

Tirpitz
832′ 0″ (253.60 m) Length overall
793′ 1″ (241.72 m) Waterline length
118′ 1″ (36.00 m) Beam (maximum)
49′ 3″ (15.00 m) Depth (midships)
34′ 9″ (10.61 m) Draft at 52,054 tons (52,890 mt)

Hull characteristics at D.W.L.

Bismarck	
Displacement	44,733 tons (45,451 mt)
Hull depth amidships	49′3″ (15.00 m)
Prismatic coefficient	0.56
Actual metacentric height	11.81′ (3.60 m) @ 39,565 tons (40,200 mt) 11.65′ (3.55 m) @ 43,010 tons (43,700 mt) 12.19′ (4.00 m) @ 46,455 tons (47,200 mt) 13.41′ (4.40 m) @ 52,360 tons (53,200 mt)
Tirpitz	
Displacement	44,785 tons (45,474 mt)
Hull depth amidships	44′3″ (15.00 m)
Prismatic coefficient	0.56
Actual metacentric height	12.70′ (3.87 m) @ 45,225 tons (45,951 mt) 13.88′ (4.23 m) @ 50,151 tons (50,956 mt)

Frames

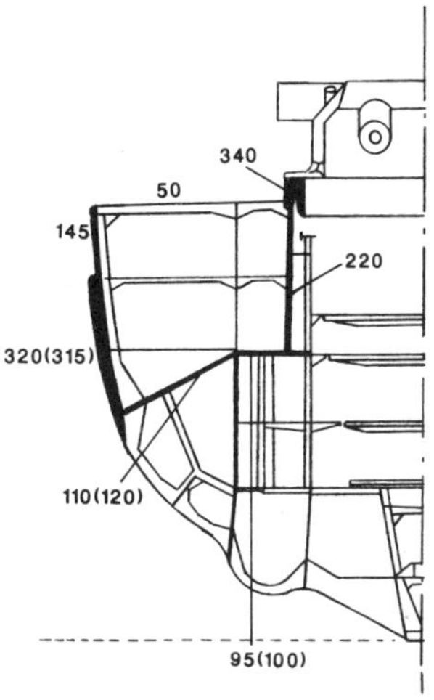

Frame 46.15

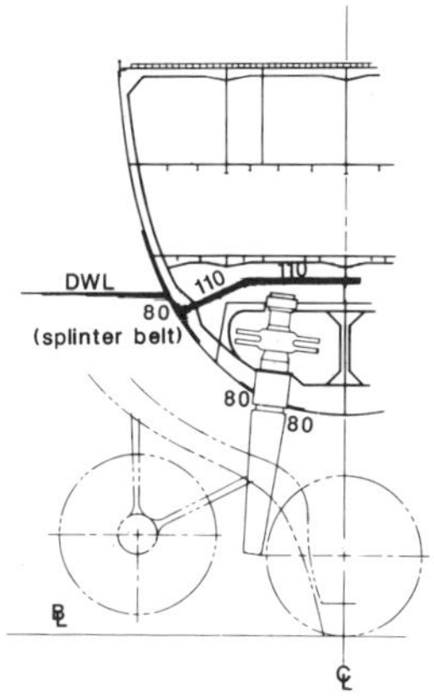

Frame 12.5

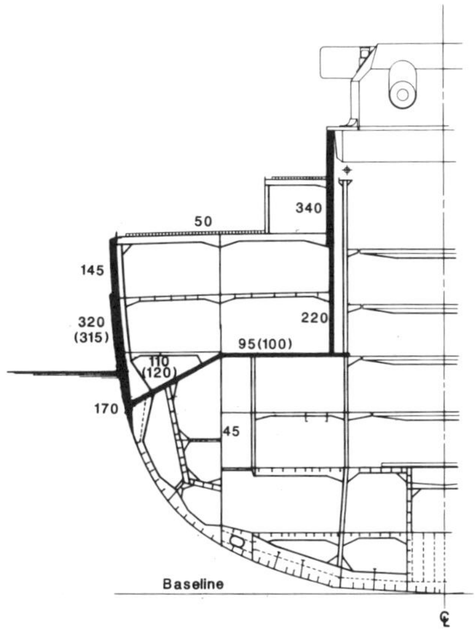

Frame 64.35

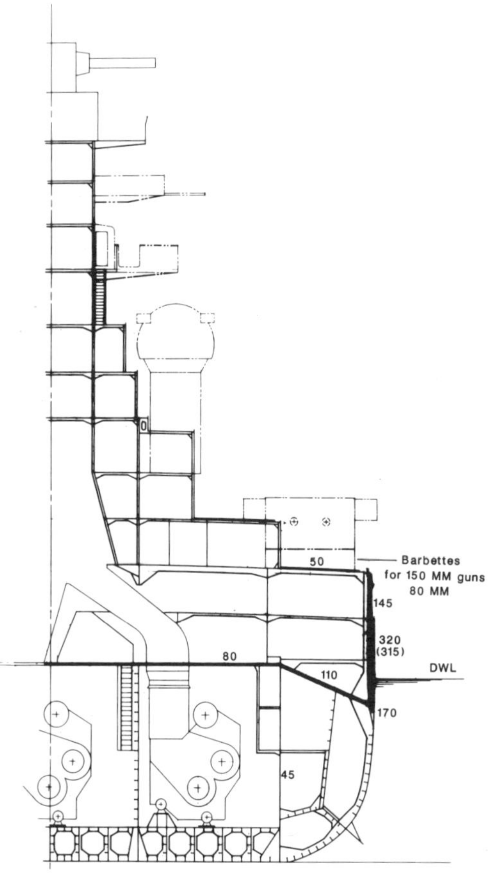

Frame 135.6

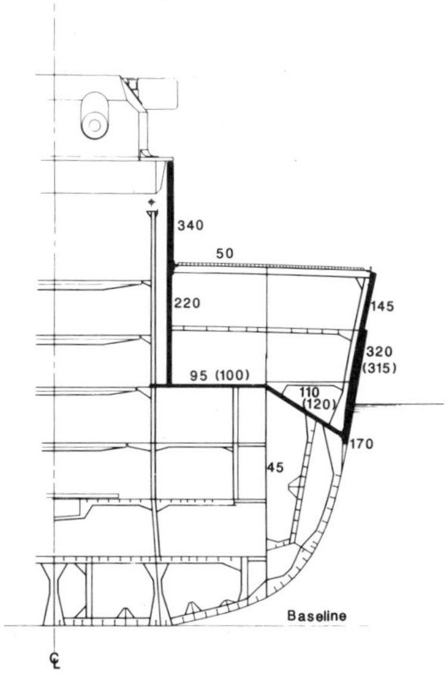

Frame 174.35

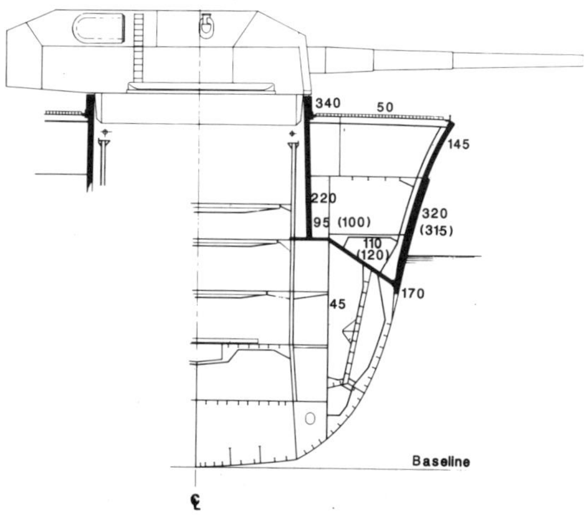

Frame 192.55

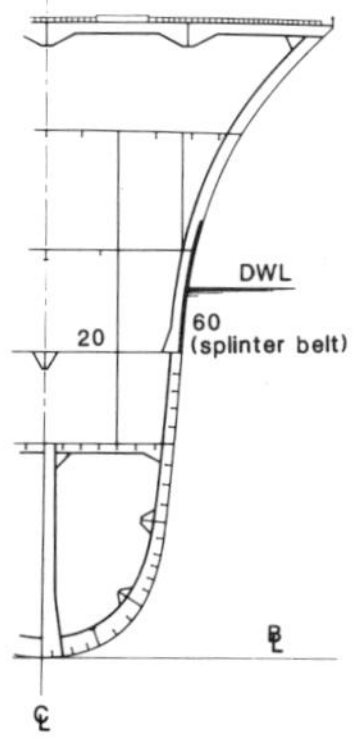

Frame 216.5

306

Midship Section

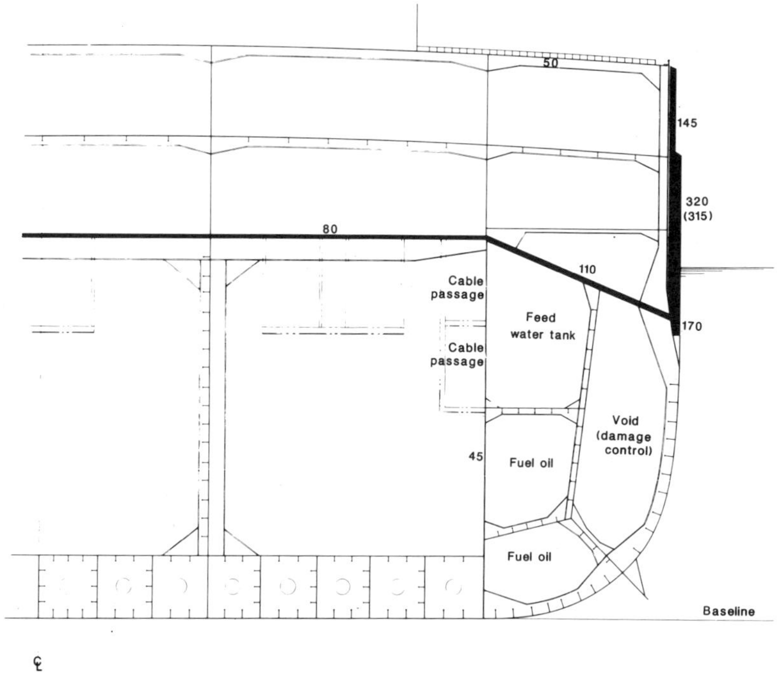

Armament
- Eight 15″/47-caliber (Mark 1934) (380 mm)
- Twelve 5.9″/55-caliber (Mark 1928) (150 mm)
- Sixteen 4.1″/65-caliber (Mark 1933) (105 mm)
- Sixteen 1.46″/83-caliber (Mark 1930) (37 mm)
- Sixteen 0.79″ (20 mm)—*Bismarck* (May 1941)
- Seventy-eight 0.79″ (20 mm)—*Tirpitz* (July 1944)

Torpedoes
- Eight 21.0″ (533-mm) deck-mounted torpedo tubes—*Tirpitz* only—1941

Aircraft
- Four Arado-196 floatplanes with one double catapult

Armor protection
- Immunity zone
  - Machinery—12,000 to 23,000 yards (10,973–21,031 m)
  - Magazines—(Point blank) 25,500 yards (23,319 m) against16″/45-caliber gun firing a 2,240-pound (1,016 kg) shell

Machinery Schematic

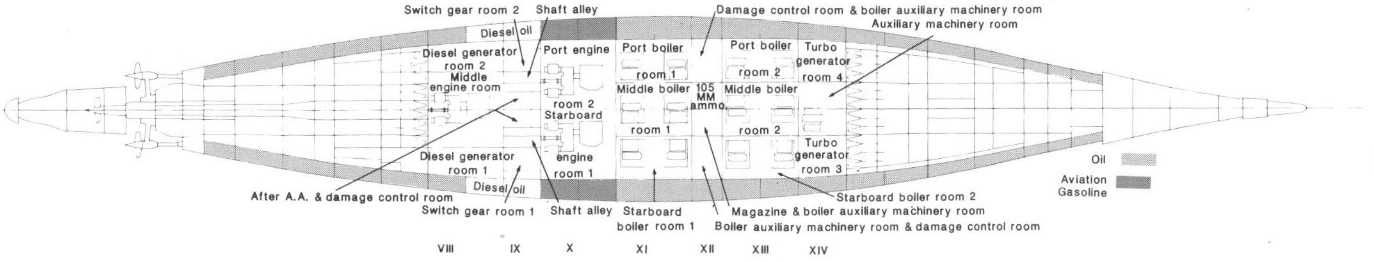

Amidships and ends

Belt Armor

Upper belt	5.72″ (145 mm)
Main side belt	12.60″ (320 mm)—*Bismarck*
	12.40″ (315 mm)—*Tirpitz* tapered to 6.69″ (170 mm)

Deck armor

Bismarck

	Centerline		*Slopes*	
	Machinery	*Magazines*	*Machinery*	*Magazines*
upper deck	1.97″ (50 mm)	1.97″ (50 mm)		
armor deck	3.15″ (80 mm)	3.74″ (95 mm)	4.33″ (110 mm)	4.33″ (110 mm)
	5.12″ (130 mm)	5.71″ (145 mm)		

Tirpitz

	Centerline		*Slopes*	
upper deck	1.97″ (50 mm)	1.97″ (50 mm)		
armor deck	3.15″ (80 mm)	3.94″ (100 mm)	4.33″ (110 mm)	4.72″ (120 mm)
	5.12″ (130 mm)	5.91″ (150 mm)		

Turret armor

face plates	14.17″ (360 mm)
sides	8.66″ (220 mm)
back plates	12.76″ (320 mm)
roof plates (fwd)	7.09″ (180 mm)
(aft)	5.12″ (130 mm)

Secondary gun protection

face plates	3.94″ (100 mm)
sides	1.57″ (40 mm)
back plates	1.57″ (40 mm)
roof plates	1.57″ (40 mm)

Barbette armor

centerline forward	13.39″ (340 mm)
sides	13.39″ (340 mm)
centerline aft	13.39″ (340 mm)—*Bismarck*
	8.66″ (220 mm)—*Tirpitz*

Conning-tower armor	forward tower	after tower
sides	13.78″ (350 mm)	5.91″ (150 mm)
roof	7.88″ (200 mm)	
communications tube	8.66″ (220 mm)	1.97″ (50 mm)
deck	2.76″ (70 mm)	

Body Plan

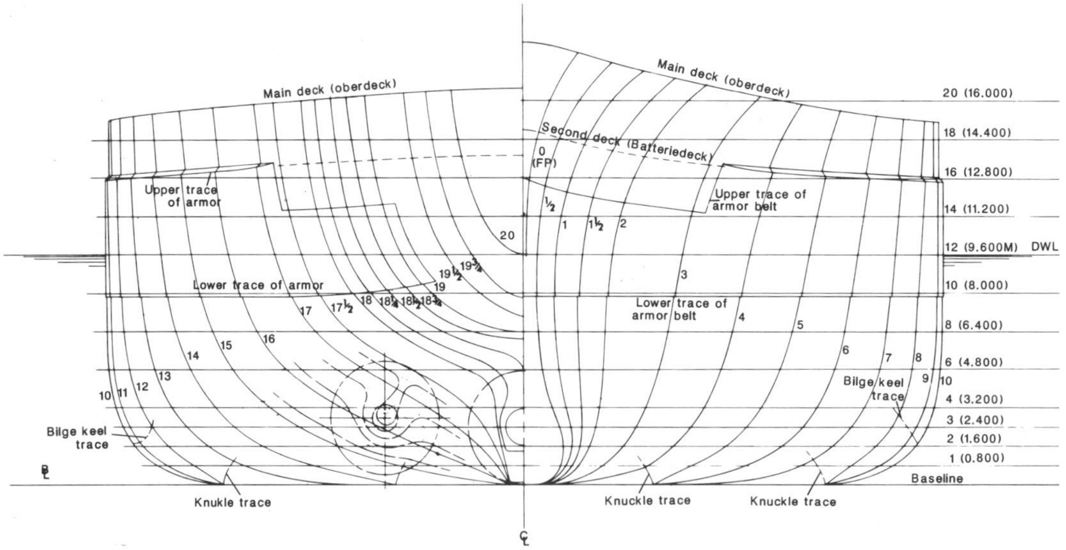

Underwater protection

Designed resistance	550 pounds (250 kgs) of TNT
Depth of side protection	18.04′ (5.50 m)
S.P.S. loading	void-liquid
Total bulkhead thickness	2.09″ (53 mm)
Bottom protective system depth	5.58′ (1,700 mm)—one layer

Tank capacities	*Bismarck* (1941)	*Tirpitz* (1941)
Fuel oil (plus reserve)	8,167 tons (8,294 mt)	8,170 tons (8,297 mt)
Diesel oil	190 tons (193 mt)	
Aviation gasoline	33 tons (34 mt)	
Reserve feedwater	369 tons (375 mt)	
Fresh water (potable)	301 tons (306 mt)	
Fresh water (reserve)	383 tons (389 mt)	
Lube oil	157 tons (160 mt)	
	9,600 tons (9,751 mt)	

Outboard Profile (May 194position *(Bismarck)*

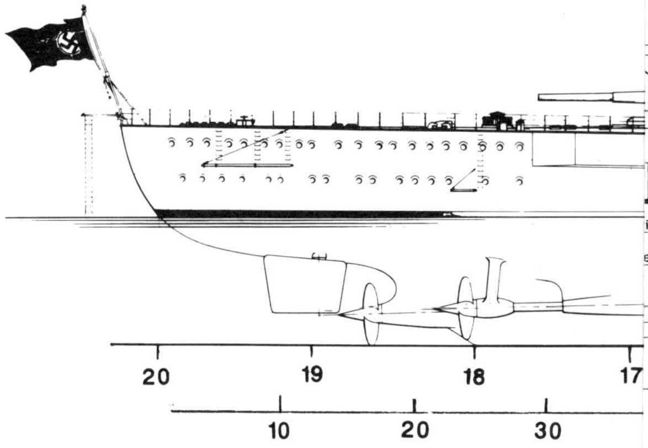

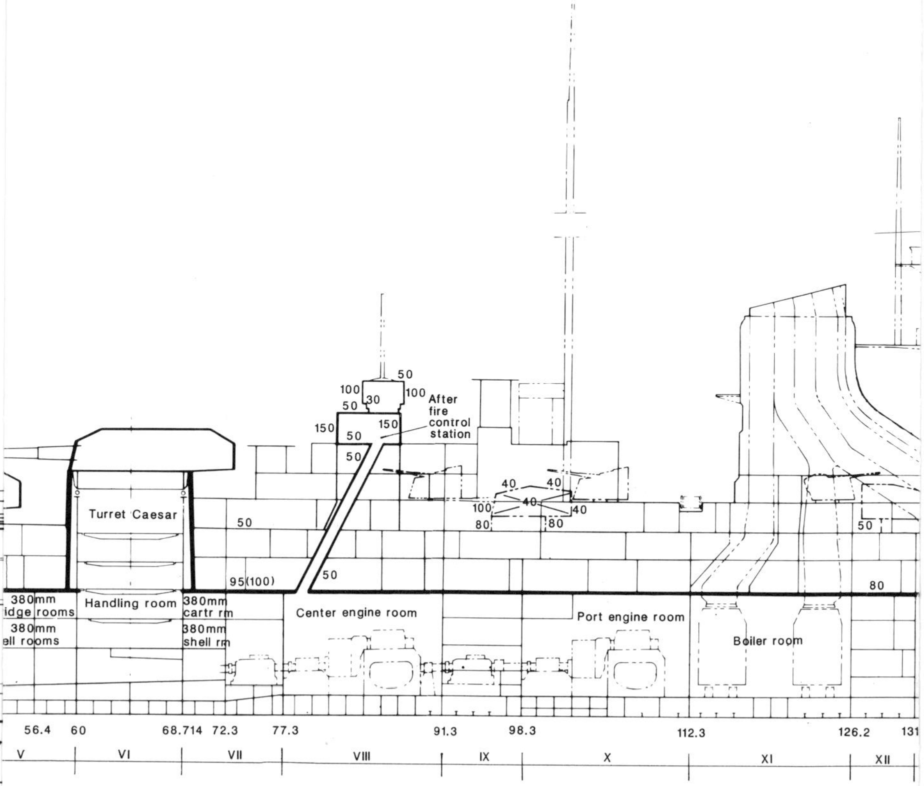

Topside View

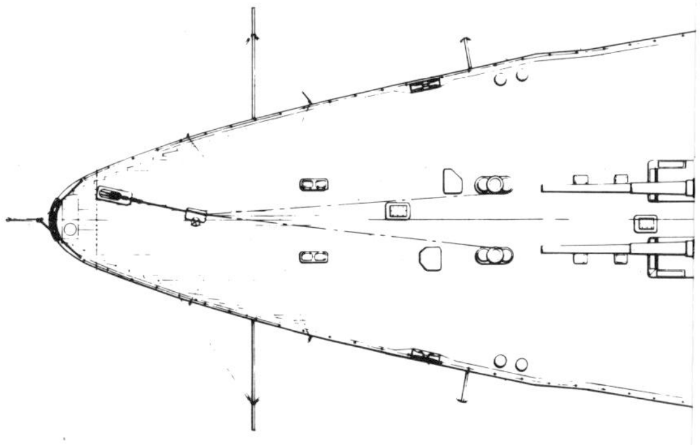

© 1985
Allison

CHAPTER FIVE

Germany: The "H"-Class Battleships

In 1937 the German Supreme Naval Command (Oberkommando der Kriegsmarine—OKM) ordered that studies begin for a new battleship class that would feature a heavier caliber main battery than the two battleships then currently under construction (*Bismarck* and *Tirpitz*). German naval constructors considered that the design of those ships, while good, was by no means perfect, as the final contract design involved many compromises. It was also recognized that the battleship was still a prestige naval weapon, and the OKM and Adolf Hitler wanted to match any ship that any possible opponents might construct. It appeared that Japan would not ratify the Washington Naval Treaty, and by a clause in the 1936 London Naval Treaty, the permitted caliber of the main battery would become 406 mm. In fact, when Japan did not sign the 1936 London treaty by April 1937, the United States proclaimed its intention to arm the new battleships of the *North Carolina* class with 406-mm guns.

Admiral Werner Fuchs, heading the OKM section responsible for defining the military requirements of the new battleship class, conferred with Hitler on the operational requirements of that class. Although the German leader had a keen interest in technical problems relating to ship design, he lacked an understanding of their interdependencies, and when he proposed that each new ship be armed with larger caliber guns than those of any possible enemy, Admiral Fuchs carefully explained that such heavy armament would result in a displacement of 80,000 to 120,000 tons. Ships with such large displacements would have had extreme difficulty in entering German North Sea ports at that time, unless channels were widened and deepened. After much discussion Hitler was convinced that in 1937 the best gun caliber for the new battleship class would be 406 mm. When the design work on the "H"-class battleships began, the Construction Office had a clear understanding that the maximum gun size would be limited to 406 mm. No other gun bore was considered.

During 1937 Hitler had a persistent idea that he could reach an accommodation with the United Kingdom with regard to spheres of influence in Europe. In discussions with Admiral Raeder in May 1938, the German leader suddenly shifted his views and intimated that the German Navy should consider the possibility of France and the United Kingdom becoming opponents if the current political situation should further deteriorate. Based upon these discussions, the Operations Section of the OKM made three proposals for future naval construction, but they were not thought to be technically feasible. Several alternatives were considered during the latter half of 1938. One, which did not meet all the requirements of the Operations Section but could be accomplished in an eight-year period, was altered slightly and officially termed the "Z"

Despite their remarkable superficial resemblance to the *Bismarck* and *Tirpitz*, the "H"-class battleships were of a much different design, which featured a main-propulsion system with twelve diesel engines and a main-battery armament of eight 406-mm guns. (Painting by Richard Allison)

plan. It proposed that the six large "H"-class battleships, on which design work had already commenced, would be the nucleus of the new German fleet.

On 23 August 1938 Hitler again discussed future naval construction with Admiral Raeder. Since the United Kingdom and France had vigorously opposed his political aims in Europe, he now considered them as possible foes. He insisted that battleships "F" (*Bismarck*) and "G" (*Tirpitz*) be completed by the fall of 1940, and that six building ways for large ships be made ready so that fleet expansion would not be impeded by the lack of adequate shipbuilding facilities. This included an understanding that the large passenger ship (*Vaterland*) under construction at Blohm and Voss in Hamburg should be launched as soon as possible to clear the ways for a battleship of the "H" class. In January 1939 Admiral Raeder presented a naval construction program (the "Z" plan) to Adolf Hitler for his approval, adding that the completion of the entire program would take until 1946. Hitler, however, was very agitated and demanded that the construction of the six "H"-class battleships be accomplished by 1944, adding, "If I could built up the Third Reich in six years, the Navy could likewise build these six ships in six years." Again, Admiral Fuchs was dispatched to confer with Hitler on the difficulties of building such large ships in such a short period of time with the industrial means at hand.* On 18 January 1939, in an act unprecedented in the Third Reich, Hitler gave the German Navy unlimited power to complete the construction program. New procedures were adopted. Only four shipyards had slipways long enough to accommodate the construction of such vessels; therefore, two battleships ("H" and "M") were to be built by Blohm and Voss, Hamburg; "J" and "N" by Deschimag in Bremen; "L" by the Kriegsmarine (Navy Yard) at Wilhelmshaven; and "K" by the Deutsche Werke at Kiel. Even with the long slipways at these yards, the hulls would have to be supported by four poppets. The first keel laying was for Battleship "H" at Blohm and Voss on 15 July 1939, that for "J" on 1 September 1939, and the keel for "K" was scheduled for 15 September 1939, but the war intervened and all construction was canceled. Given the priority assigned these ships, the six-year deadline probably could have been met.

In other conversations with Admiral Raeder in 1939 concerning German warship design, Hitler stressed one point: each new German warship had to be stronger, defensively and offensively, than her counterpart in the Royal Navy. When Raeder advised Hitler that the German Navy would not be ready to fight the Royal Navy until at least 1945, Hitler assured him that the German Navy would not be needed to back Germany's political objectives until 1946. Without consulting with Admiral Raeder, he renounced the 1935 Anglo-German Naval Treaty on 28 April 1939. In May Hitler declared that the shipbuilding program was to continue as planned and friendly relations were to be maintained with the United Kingdom.

Preliminary Designs.

In any conflict with the Royal Navy, the decisive factors would have been the endurance, speed, and armament of the ships. During 1937, when the design work on these ships began, the requirements, tabulated in table 5-1, were established by the German Naval Staff.

*The steel production in Germany had become critical at this time, and there were delays in the construction of the *Bismarck* and *Tirpitz*. This was partially answered on 25 October 1937 when Hitler promised Admiral Raeder an increase in the Navy's steel quota from 40,000 to 75,000 tons. Krupp's steel mills were to be enlarged for this purpose. With the construction of the ships in the "Z" plan in 1939, this quota would have had to be increased even more.

TABLE 5-1

Design Requirements—"H" Class

Standard displacement	Greater than 50,000 tons (50,816 mt)
Armament	
Main	Eight 16" (406 mm)
Secondary	Sixteen 5.9" (150-mm) guns in twin turrets
Antiaircraft	4.1" (105-mm) in the new "flakturmen" (a special enclosed mount) 1.46" (37-mm) antiaircraft machine guns, including two "flakturmen"-type mounts
Aircraft	Four
Torpedo tubes	Six 21" (533 mm)
Speed	30 knots
Endurance	16,000 nautical miles at 19 knots
Protection	The armored citadel to withstand 16" (406-mm) gunfire at normal battle ranges, and torpedo defense system to resist a torpedo warhead with a 551-pound (250-kg) charge of TNT

The "H"-class design was to progress in five distinct phases because of the additional requirements of Adolf Hitler, the size of the ships, and wartime developments.

H-39 design. The initial design effort was complicated by questions regarding the ship size and the gun caliber. Hitler wanted these ships to be armed with the largest gun possible, but development work had begun on a 406-mm gun in 1934, and an entire series of tests, range table firings, and test salvos had been completed with that gun. By 1939 only seven barrels had been produced, since Krupp had been given industrial priority to produce 380-mm guns for the *Bismarck* and *Tirpitz* and 283-mm guns for the *Scharnhorst* and *Gneisenau.* With the signing of the nonaggression pact with the Soviet Union in 1939, sixteen 380-mm guns, plus spares, were to be delivered to the Soviets. As a result, despite pressure from Hitler for larger guns, it was impossible to design, build, and test a larger-bore naval gun, and it was decided that the H-39 design would be armed with the 406-mm gun. This gun bore would be the same for the 1940 redesign.

Another important design decision had to do with endurance and the type of propulsion. Since the Royal Navy was now considered a potential foe, the endurance of any future capital ship would be a critical factor in its operation and design. Because of Germany's lack of overseas bases and her difficulty in getting ships to the Atlantic Ocean in a potential war with Great Britain, a large cruising radius was necessary. To attain a long endurance, a diesel propulsion plant was preferred because it had a more economical fuel rate and thus required less fuel. Aside from the economical fuel consumption, the diesel propulsion plant was capable of a high-performance operation in a matter of minutes, whereas a ship with a steam-propulsion plant required at least 20 minutes to get under way. It was established early in design that the diesel-fuel capacity would have to be based on a 1,000-hour supply at a cruising speed of 19 knots.

Although battleship "H" had a much larger standard displacement than the *Bismarck* and *Tirpitz,* the German naval constructors adhered to the classical arrangement of four twin turrets. The Supreme Naval Command felt that the advantages of such a gun arrangement seemed to be self-evident from experiences in World War I. Triple and quadruple turret arrangements were considered undesirable.

Early in the preliminary designs, any plans for a larger secondary battery than that of either the *Scharnhorst* or *Bismarck* classes had to be abandoned, because it was difficult to find sufficient space in the superstructure and to provide effective training arcs for additional turrets. The all-diesel propulsion plant also required more hull volume than a steam plant of comparable horsepower, and there was difficulty as well in finding space for the magazines to service additional gun positions. Moreover, the requirement for four aircraft meant that considerable superstructure volume and deck area were needed for the hangars and launching/recovery facilities. Because it was impossible to provide a single stack to house the intakes and uptakes for the twelve propulsion diesels and the twelve diesel generators, a twin-stack arrangement had to be used. The space requirements for the intakes and uptakes, plus their mufflers, and boat stowage made it impossible to have an amidships hangar and catapult, as in the *Bismarck* class. Therefore, the aircraft facilities had to be located at the after end of the superstructure, where they took up considerable space between the after funnel and turret Caesar. Better aircraft facilities and additional 150-mm guns might have been possible if the German naval constructors had been willing to adopt a triple-turret main battery.

Structural and armament arrangements were patterned closely after the *Bismarck*-class ships; the "H"-class ships were essentially more heavily armed, better protected, and enlarged *Bismarck*s. Outstanding differences between the two designs were that the "H" class had 406-mm guns and diesel propulsion, while the *Bismarck* and *Tirpitz* had 380-mm guns and steam propulsion.

In the spring of 1939 the initial "H"-class ships had the characteristics tabulated in table 5-2. Beam and draft restrictions limited the ships to a full-load displacement of 63,596 metric tons.

Design work progressed very slowly because of the numerous changes and differences of opinion within the OKM, and Hitler became impatient with the delays. To avoid further delays and arguments, Admiral Raeder felt it necessary to give one individual the responsibility for directing the project and the necessary authority to

TABLE 5-2

Design Characteristics—Spring 1939

Standard displacement	52,643 tons (53,489 mt)
Design displacement	57,617 tons (58,543 mt)
Full-load displacement	65,592 tons (63,596 mt)
Waterline length	872.7′ (266.0 m)
Waterline beam	121.39′ (37.0 m)
Draft (design)	32.87′ (10.02 m)
(full-load)	36.75′ (11.2 m)
Armament	eight 16″ (406-mm)/50-caliber guns in four twin turrets
	twelve 5.9″ (150-mm) 55-caliber guns in six twin turrets
	sixteen 4.1″ (105-mm)/65-caliber guns in eight twin mounts (special flakturmen mounts—an all-enclosed type)
	sixteen 1.46″ (37-mm)/83-caliber guns in two twin armored mounts and six twin mounts with shields
	thirty-two 0.79″ (20-mm) machine guns
	six 21.0″ (533-mm) underwater torpedo tubes
	Four Arado-196 floatplanes
Speed	30.4 knots
Shaft horsepower	147,950 (150,000 mhp)
	162,750 (165,000 mhp)—overload
Fuel	9,839 tons (10,000 mt)
Endurance	16,000 miles at 19 knots
Protection	
	5.91″ (150 mm)—upper side belt
	11.81″ (300 mm)—lower main side belt
	3.94″ (100 mm)—lower armor deck
	1.97″ (50 mm)—upper deck

implement policy. On 27 January 1939 Admiral Raeder created the 15-man "Special Group New Construction" under Admiral Werner Fuchs, and the construction of the six ships was to be coordinated by the Blohm and Voss Shipyard in Hamburg. Admiral Fuchs at once moved his staff to this yard, where they proceeded with the plans at such a pace that the keel for battleship "H" was laid on 15 July 1939. Design work had been concluded by July; much of the hull armor and steel had been delivered, and towing tests at the Hamburg Model Basin were finished.

H-40 design. Although the outbreak of war had resulted in the postponement of the construction of the six "H"-class battleships, design work continued. In July 1940 Hitler instructed the navy to investigate new designs, since France now had been defeated and Germany was facing only the United Kingdom. The OKM requested that studies be made as to how wartime experiences would influence the design of the "H"-class battleships. At the conclusion of these studies on 15 July 1940, certain design modifications were recommended—including increased double-bottom depth, greater separation between the torpedo bulkhead and side shell, improved relationship of the lower armor deck to the full-load waterline (recommended weight increases would have raised the waterline above the armor deck level), more freeboard, and increased horizontal protection (particularly against bombs). There were indications that the design draft would be exceeded and that this would adversely affect the freeboard (already at a minimum) and the speed. Two design alternatives were proposed during 1940 (see figures 5-1 and 5-2), with the characteristics in table 5-3.

Scheme A kept the planned speed and displacement of H-39, but reduced the main battery to six 406-mm guns in twin turrets. This proposal was offered as a means of retaining the original displacement and at the same time increasing the armor protection, which was only 39.2 percent of the total displacement in H-39. Most important, however, the reduced number of guns made it possible for Krupp to produce the required number of 406-mm barrels in time. The scheme bears a remarkable likeness to that for battlecruiser "O," which had only three main-battery twin turrets. It was also

TABLE 5-3
Design Characteristics—1940

	Scheme A	*Scheme B*
Displacement		
Standard	54,830 tons (55,700 mt)	62,816 tons (63,000 mt)
Designed	57,586 tons (58,500 mt)	64,969 tons (66,000 mt)
Full load	64,575 tons (65,600 mt)	68,906 tons (70,000 mt)
Dimensions		
Length	885.83′ (270.00 m)	941.60′ (287.00 m)
Beam	123.36′ (37.60 m)	128.61′ (39.20 m)
Draft	32.87′ (10.02 m)	32.87′ (10.02 m)
Armament		
Main	six 16″ (406-mm) guns in 3 twin turrets	Eight 16″ (406-mm) guns in 4 twin turrets
	(all other armament same as in the 1939 design)	
Propulsion	two shafts with MZ65/95 engines and two shafts with steam turbines	two shafts with VZ42/58 engines and two shafts with steam turbines
Shaft horsepower	226,850 (230,000 mhp)	236,700 (240,000 mhp)
Speed (approx. max.)	30.4 knots	30.4 knots
Armor (percent of std. displacement)	43.7	43.8

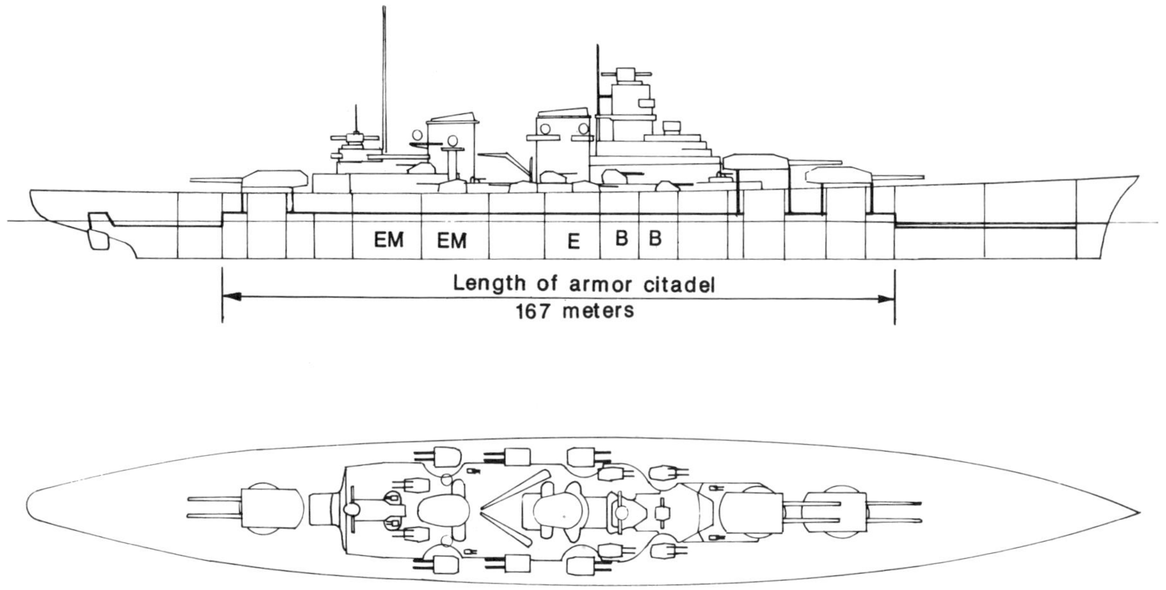

Figure 5-1. Battleship "H"—Scheme A.

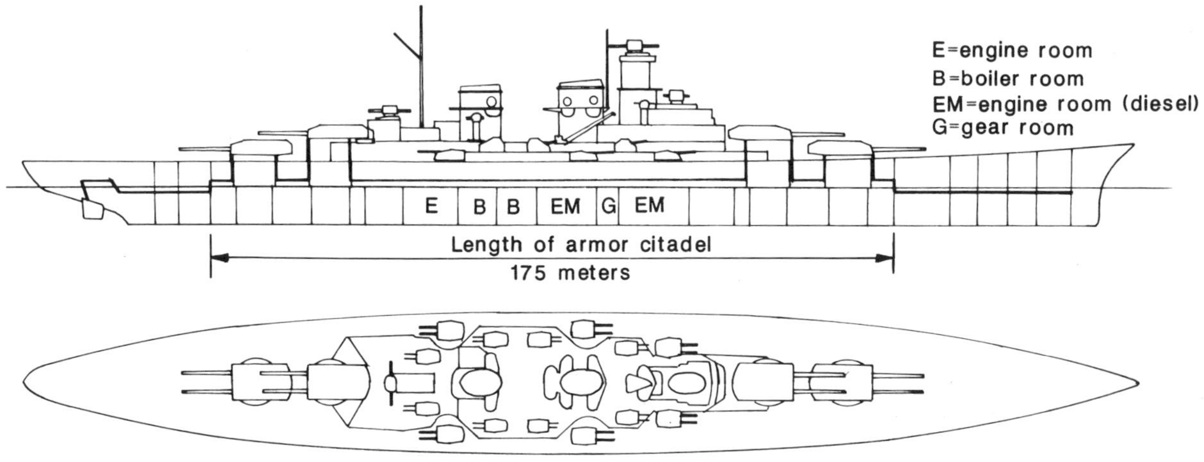

Figure 5-2. Battleship "H"—Scheme B.

necessary to increase the total shaft horsepower by 64,000 mhp to maintain the speed of the original "H" design. However, this increased power requirement meant that it became necessary to use four shafts instead of three, since a maximum of 58,000 mhp could be delivered on a single shaft. The introduction of a fourth shaft also complicated the choice of propulsion units, since a four-shaft, all-diesel propulsion plant would have been impossible to install within a beam of 38 meters and still provide the necessary torpedo protection. The larger machinery space required by sixteen MAN

(Maschinenfabrik Augsburg-Nürnberg) Type MZ64/95 diesels would have increased the length of the armored citadel. To prevent this, it was decided to use a propulsion plant such as that contemplated for battlecruiser "O"—a diesel and steam-turbine combination.

Scheme B was a four-turret ship, with the increase in armor protection and power reflected in a much larger displacement. Again it was decided to accept four shafts and a mixed propulsion plant. Both schemes represented what would be necessary in ship size to incorporate wartime experiences and industrial realities into the design of an "H"-class battleship. To reduce the length of the armored citadel in Scheme B, it was decided to use a VZ42/58 MAN 24-cylinder diesel engine, which was slightly more powerful than the larger MZ65/95 engine and occupied less space. With these diesels and steam propulsion on the inboard shafts, it was possible to increase slightly the installed power but limit the size of the plant to develop 240,000 metric horsepower.

One of the most important objectives in these two designs—the provision for more depth in the torpedo-defense system—can be seen in figures 5-3 and 5-4. Figure 5-3 shows the same type of torpedo-defense system used in previous designs, but with an additional longitudinal bulkhead in the side-protective system. Figure 5-4 shows a sloping armor-belt arrangement, whereby some weight could be saved. This arrangement was somewhat similar to that used in the armored ship *Graf Spee* (figure 5-5). The 250-mm and 170-mm upper belts were to be sloped at 13 degrees, and the sloping armor would have been equivalent in thickness to the 200-mm and 300-mm thickness shown in figure 5-3. In both arrangements the system would have been 6 meters deep; each featured a void space one meter wide separating the torpedo-defense system and the vitals over the length of the citadel.

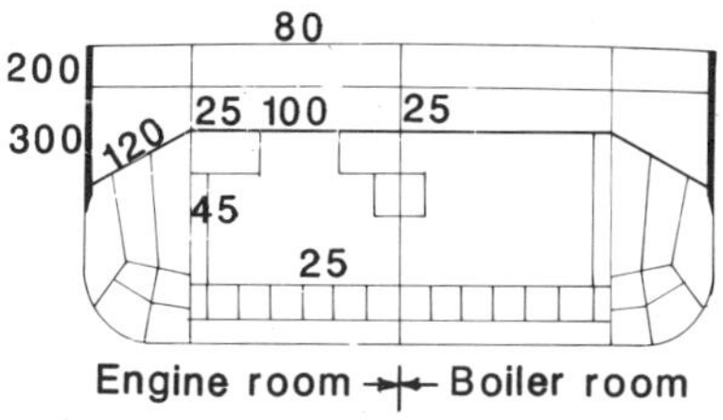

Figure 5-3. Torpedo defense system—Alternate I.

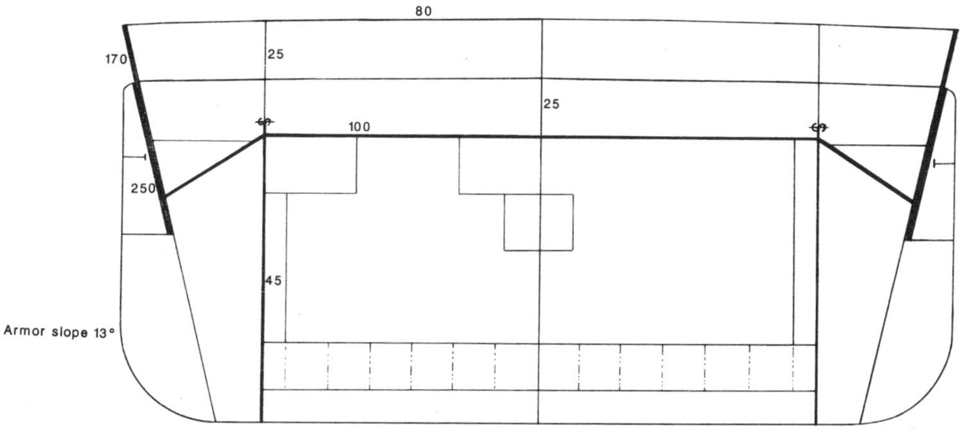

Figure 5-4. Torpedo defense system—Alternate II.

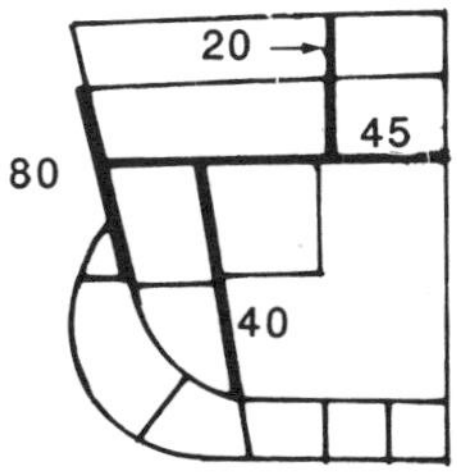

Figure 5-5. Midship section
—Admiral Graf Spee.

These H-40 design studies were set aside in 1941, when Hitler ordered that Battleships "H" and "J" would be built immediately after the end of the war. As a result of this directive, the Germans decided to concentrate their efforts on enhancing the protection of battleship "H," accepting the degraded performance that would result from the added weight.

H-41 design. Analysis of the July 1941 bomb damage sustained by the *Scharnhorst* indicated that the horizontal armor protection had to be substantially increased. It was also found impossible to increase the armor and maintain the full-load draft of 11.5 meters necessary for shallow waters in the North Sea, unless the fuel-oil capacity was reduced. This would have cut the endurance at 19 knots from 20,250 miles to 15,250 miles. The OKM was against such an alternative, and the full-load draft was permitted to increase, with the proviso that deep Atlantic anchorages would be available in the near future. The following increases in armor protection were, therefore, proposed:

- 50-60 mm for the entire length of the upper deck
- 120 mm for the lower armor deck
- 130 mm for the slopes
- 150 mm for the upper citadel belt, which was also to be lengthened
- Freeboard to be increased by 0.5 meters, beam increased by 0.7 meters, and barbette heights increased

With the keels of battleships "H" and "J" still on the ways, there were plans to continue other construction. Studies made by the Special Group New Construction and the Blohm and Voss Shipyard were completed for a redesign of the H-39. The armor, armament, and underwater protection were major items of interest. The displacement was to be increased only 5,000 tons, and a decrease in the speed to 29 knots was to be accepted. Of the additional 5,000 tons of displacement, 2,000 tons were for additional deck-armor protection, and the remainder was for the increase in the main-battery gun bore. This improved "H" was still designated as S525 by Blohm and Voss, as were later battleship studies through H-41.

A complete redesign of the H-39, redesignated H-41, proceeded without delays and was fairly well completed. Preliminary studies were made into longitudinal strength, and launching calculations were developed. Battleship H-41 had the characteristics summarized in table 5-4.

It was decided that the main-battery bore should be changed to 420 mm. When the main-battery gun size was discussed with the Krupp concern at the Navy Weapons Office, it was determined that the gun barrels and turrets could be so constructed that the original 406-mm gun design could be retained. The 406-mm gun barrels were oversized, so that it was possible to have the 406-mm barrel relined to the larger 420-mm caliber without any major modification. Studies of the ammunition hoists and loading equipment indicated that the necessary changes could be easily made to accommodate the larger-diameter shell. No 420-mm gun was produced from a 406-mm barrel, but ammunition tests were made with a special shell to be fired from these guns. It was anticipated that a greater range would be possible with special powder and a sub-caliber shell. The upgraded gun size was in keeping with Hitler's 1939 directive that each German warship must be stronger than any enemy counterpart. A 420-mm gun bore had also been proposed for the 1917 battleship designs. It was expected that 420-mm shells would cause serious damage to any ship, since most new battleships were armed with 380- or 406-mm guns. As German naval guns performed as well as or better than British guns in World War I, ordnance designers believed that the correct way to make use of this superior technology would be to develop a gun larger than any

TABLE 5-4
Design Characteristics of H-41

Standard displacement	62,989 tons (64,000 mt)
Design displacement	67,712 tons (68,800 mt)
75% load displacement	71,846 tons (73,000 mt)
Full-load displacement	74,799 tons (76,000 mt)
Waterline length	901.2′ (275.0 m)
Waterline beam	127.95′ (39.0 m)
Draft (design)	36.45′ (11.1 m)
(full-load)	39.86′ (12.15 m)
Armament	eight 16.54″ (420-mm)/48-caliber guns in four twin turrets (all other armament same as H-39)
Speed	28.8 knots
Shaft horsepower	147,950 (150,000 mph) 162,750 (165,000 mph) overload
Fuel	11,810 tons (12,000 mt)
Endurance	20,000 miles at 19 knots
Protection	
	7.87″ (200 mm)—upper belt armor
	11.81″ (300 mm)—lower main side belt
	3.15″ (80 mm)—upper deck
	4.72″ (120 mm)—lower armor deck

opposing weapon. Such a gun would be unmatched for some time. The Germans knew in 1938 that the Royal Navy was committed to the 406-mm gun for its proposed new battleships of the *Lion* class, and even if the British found out about the actual gun caliber of Germany's newest ships, it would have taken them some time to counter with new battleships with an equal or larger size gun.

The increased size of H-41 allowed heavier horizontal protection—120 mm for the lower armor deck throughout the armored citadel and 80 mm for the upper armor deck in the same section. Armor on slopes was increased from 120 mm to 130 mm to provide better protection of the wing fuel-oil tanks, but more importantly, to further reinforce the main side belt.

The torpedo-defense system was to be strengthened by an increase in the distance of the torpedo bulkhead from the side shell in the stern portion of the ship, since torpedo damage in the stern of the *Scharnhorst* and *Gneisenau* had indicated that these areas in German capital ships were weak. This was accomplished by increasing the beam, particularly at the ends of the armored citadel where the torpedo-defense system decreased in width due to the fining of the ships lines. The German naval constructors also had data and plans for the *Richelieu*-class battleships to guide them in the analysis of their underwater-protection system. The review of the underwater protection based on torpedo damage to the *Scharnhorst* and *Gneisenau* indicated, too, that improvements had to be made in the structural arrangement of the stern portion of the system to adequately contain a torpedo explosion. Amidships, the torpedo-defense system reached its maximum breadth of 5.5 meters, the same distance as in the H-39.

As the H-41 would have been a longer ship, it became mandatory to make all the longitudinal bulkheads serve as part of the hull girder. There was a poor relationship between the length and depth of the ship due to draft restrictions and dockyard limitations, so that these bulkheads would have proven indispensable in reducing the stresses in a seaway or during dry-docking. The German Navy disliked concentrating all power for one shaft in one large centerline compartment as had been done in French battleship design, because its arrangement would not permit the necessary longitudinal bulkheads to strengthen the hull girder. Large compartments also resulted in undesirable flooding effects (i.e., sinkage), resultant loss in buoyancy, and damaging free-surface effects on the residual metacentric height if such compartments were not completely filled with flooding water. Therefore, the H-41 was to be subdivided into 22 watertight compartments, one more than the subdivision employed in the 1939 design. With the power plant the same as that of H-39, the triple-shaft propulsion plant offered an ideal opportunity to subdivide the armored citadel transversely and longitudinally. As was typical in modern German battleship construction, subdivision was an important factor in the protective system of this design. The innerbottom protection was the same as that used in the 1939 design.

Since the H-41 was to have been much larger than the 1939 design, speed and endurance would decrease if the original shaft horsepower and fuel-oil capacity remained unchanged. The OKM no longer considered the 30-knot speed requirement as important as protection and better endurance. Thus, fuel-oil capacity was increased to 12,000 metric tons, and the increased draft and reduced speed were accepted.

Final speed and endurance characteristics are tabulated in table 5-5.

TABLE 5-5

Speed and Endurance Characteristics

Displacement	*Fuel*	*Max. Speed*	*Max. Continuous Speed*	*Endurance at 19 knots*
66,630 tons (67,700 mt)	5,610 tons (5,700 mt)	30.2 knots	29.4 knots	8,000 miles
70,862 tons (72,000 mt)	8,858 tons (9,000 mt)	29.8 knots	29.0 knots	15,000 miles
73,815 tons (75,000 mt)	11,810 tons (12,000 mt)	29.3 knots	28.5 knots	20,000 miles

The loss of the *Bismarck* in May 1941 had a profound effect on the design of shafts, rudders, and steering gear. The triple-rudder arrangement was retained, but twin skegs were introduced to provide partial protection to the shafts and also to better support the stern in dry-docking. The Germans concluded that the *Bismarck* would have undoubtedly reached Brest if her rudders had not been hit and jammed. To prevent this occurrence in an "H"-class battleship, it was also decided to design the rudder with an explosive charge in the hollow rudder post, which would allow blowing the rudder away without damaging the propellers.

German naval constructors were not fully satisfied with the side-protective system and the armor protection, and on 15 November 1941 Admiral Raeder approved modifications of the basic characteristics, as tabulated in table 5-6.

TABLE 5-6

Modifications to Basic H-41 Design

Full-load displacement	77,752 tons (79,000 mt)
Waterline length	925.20′ (282.00 m)
Waterline beam	132.87′ (40.50 m)
Draft (full load)	39.37′ (12.00 m)

The three-shaft diesel propulsion plant was retained, but due to the increased size, speed was reduced to 28.0 knots at the full-load displacement. The following improvements were attained:

- The lower armor deck was increased from 120 mm to 150 mm.
- The upper armor deck was increased from 50 mm to 80 mm.
- Double-bottom depth increased from 2 to 3 meters.
- Freeboard at full-load increased to 6 meters.
- Lower armor deck was raised, relative to full-load draft, from 0.25 meters below to 0.5 meters above the waterline.
- Torpedo bulkhead was moved from 5.5 to 6.65 meters from the side of the ship; one additional longitudinal bulkhead was provided.
- Heights between the middle and battery decks and the battery and upper decks were increased by 26 mm.

Much work was accomplished on the design of the torpedo-defense system. By 10 September 1941, Blohm and Voss had prepared drawings of a side-protection system that would have been subdivided into three zones by longitudinal bulkheads (figure 5-6). This system was based on the principles derived from the torpedo-defense system of the *Richelieu* class, but modified to suit German needs. The triple bottom would have been a continuous structure, extending through the torpedo bulkhead to its termination at the armored slopes. Two of the spaces in the side-protection system were to be voids or "working zones" for the expansion of explosive gases, while the central space would carry fuel oil and provide hydraulic resistance if the gases ruptured the outboard restraining bulkhead of the fuel-oil bunkers. The elastic and plastic deformation of the two outboard longitudinal bulkheads was to absorb a major portion of the explosive energy. The torpedo bulkhead was to be constructed of 45-mm non-cemented armor plate, while the two outer bulkheads would have a total thickness of 30-mm special-strength steel. The amalgamation of the underbottom with the side-protection system was very interesting; it provided a good flange for the lower half of the hull girder, but also provided good protection against underbottom explosions, about which the Germans were becoming more concerned.

After Admiral Raeder approved the design changes, a final side-protective scheme was drawn up by the OKM in November 1941. This system would have been similar to that used in the *Scharnhorst*- and *Bismarck*-class battleships, with three zones outside the 45-mm torpedo bulkhead, which would have been 6.6 meters from the side of the ship. The structure within this system was designed to counter charges of more than 500 kilograms of TNT. The torpedo bulkhead would still have been constructed of 45-mm noncemented armor plate. The innerbottom was no longer a part of the protective scheme, but it still provided excellent protection against mines. There was a working zone in the compartment below the innerbottom plating. The underbottom was to have been very strong, with two plated portions of special-treated steel with a total thickness of 45 mm. German designers were convinced that a noncontact under-

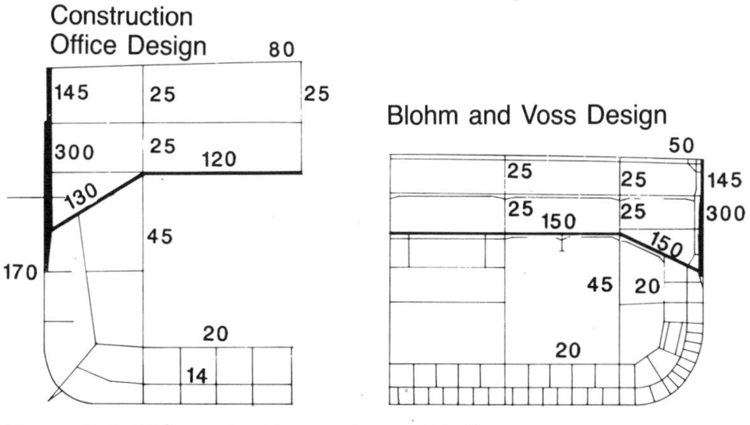

Figure 5-6. Side protective systems, H-41.

water explosion could be effectively resisted by increasing the thickness and strength of the resisting structure and making the distance between the torpedo bulkhead and the side shell as great as possible.

Design work continued at a slow pace after November 1941; most of the effort centered upon the application of the remaining 980 tons of margin. On 19 August 1942 Admiral Raeder decided that this weight should be used in augmenting the horizontal protection; the lower armor deck thickness was increased to 175 mm and the deck over the rudder installation was increased to 135 mm. Blohm and Voss was informed of all changes made, and modified plans and calculations were made as rapidly as possible. Thus, it would have been possible, six or seven months after demobilization, to have laid keels for a revised battleship "H" at Blohm and Voss, Hamburg, and Deutsche Werke, Kiel. A third ship could have commenced construction in 20 months, after Blohm and Voss completed its new dry dock in Hamburg. Shallow water at Bremen prevented the building of such a large ship there. Actually, due to the increased size of the ships, construction time would have increased from four to five years.

Large battleship design studies. The influence of the Construction Office, the technical section of the OKM, decreased in importance after Albert Speer became Reichsminister for Armament and Munitions on 8 February 1942 and assigned some of the battleship design technicians and engineers to concentrate efforts on the design of new U-boat types and other critical tasks, including all logistic necessities. An intermediary between Speer and the German Navy was created, the New Ship Design Commission under Admiral Karl Topp. Admiral Topp's group had the task of coordinating the requirements of the navy with the material contingencies and production capabilities controlled by Speer. It was this group that would make the later design studies, H-42, H-43, and H-44.

The Construction Office did not participate in those design studies; their work concluded with the H-41 design in August 1942. Soon thereafter, Hitler requested that the OKM study the feasibility of constructing very large battleships. With no limitations on gun caliber, and displacement no longer a governing design consideration, it was possible to develop design studies to fulfill the following requirements:

- Speed approximately 30 knots
- Horizontal protection sufficient against shell and bomb attacks
- Underbottom protection sufficient against mines, particularly in the machinery spaces
- Main-armament caliber in proper balance with other design features*

Several battleship studies prepared by the New Ship Design Commission between 1942 and 1944 reflected design trends resulting from combat experiences. The design characteristics were never discussed with Admiral Raeder or his successor, Admiral Karl Dönitz, the Naval Command, or the officers who were responsible for determining the military characteristics of warship construction. These men had other urgent tasks to perform and could not spare the time to address a problematical battleship-construction program that would take at least five years to accomplish. Surface-warship construction had been canceled or postponed in favor of U-boat construction.

Table 5-7 summarizes the principal characteristics of the H-42, H-43, and H-44 design studies. Figures 5-7, 5-8, and 5-9 illustrate the appearance of these design studies.

---

*Displacement increased from 41,400 metric tons in the *Bismarck* to 53,393 metric tons (standard) in H-39, but even the latter was considered too small to attain all the above conditions.

TABLE 5-7

The Last German Battleship Studies

Design	*H-42*	*H-43*	*H-44*
Displacement			
Designed	88,578 tons (90,000 mt)	109,246 tons (111,000 mt)	128,930 tons (131,000 mt)
Full load	96,451 tons (98,000 mt)	118,104 tons (120,000 mt)	139,264 tons (141,500 mt)
Dimensions			
Waterline length	1,000′ (305 m)	1,082′ (330 m)	1,133′ (345 m)
Waterline beam	140.4′ (42.8 m)	157.5′ (48 m)	169.0′ (51.5 m)
Depth	63′ (19.15 m)	66′ (20 m)	69′ (21 m)
Draft (design)	38.7′ (11.8 m)	39.4′ (12 m)	41.5′ (12.65 m)
(full load)	41.7′ (12.7 m)	42.0′ (12.8 m)	44.29′ (13.5 m)
Armament			
Main	eight 16.54″ (420 mm)	eight 20.06″ (508 mm)	eight 20.06″ (508 mm)
arrangement	2-2-A-2-2	2-2-A-2-2	2-2-A-2-2
Secondary	twelve 5.9″ (150 mm)	twelve 5.9″ (150 mm)	twelve 5.9″ (150 mm)
Antiaircraft	sixteen 4.1″ (105 mm)	sixteen 4.1″ (105 mm)	sixteen 4.1″ (105 mm)
	sixteen 1.46″ (37 mm)	sixteen 1.46″ (37 mm)	sixteen 1.46″ (37 mm)
	twenty-four 20 mm	forty 20 mm	forty 20 mm
Torpedo tubes	six 21″ (533 mm)	six 21″ (533 mm)	six 21″ (533 mm)
Aircraft	six	six	six
Armor protection			
Decks (total)	12.99″ (330 mm)	12.99″ (330 mm)	12.99″ (330 mm)
Main side belt	14.96″ (380 mm)	14.96″ (380 mm)	14.96″ (380 mm)
Torpedo bulkhead	1.77″ (45 mm)	1.77″ (45 mm)	1.77″ (45 mm)
Bottom plating	1.18″ (30 mm)	1.18″ (30 mm)	1.18″ (30 mm)
Torpedo bulkhead			
Distance from shell (amidships)	25.10′ (7.65 m)	31.17′ (9.5 m)	36.09′ (11.0 m)
Propulsion			
Normal maximum shp	266,300 (270,000 mhp)	266,300 (270,000 mhp)	266,300 (270,000 mhp)
Speed (knots)	31.9	30.9	29.8
Overload shp	296,000 (300,000 mhp)	296,000 (300,000 mhp)	296,000 (300,000 mhp)
Speed (knots)	32.2	31.0	30.1
Endurance at 19 knots	20,000 miles	20,000 miles	20,000
Type plant	Four shafts with four diesel engines (MZ 65/95) on each outer shaft and two steam-turbine-driven inboard shafts with the overload feature		
Speed with diesels (only)	24 knots	23 knots	22.5 knots

Note: To increase the endurance to 25,000 miles, the extra fuel-oil capacity would have caused an increase in the full-load draft of approximately 0.5 meters and would decrease the speed by 0.15 knots. The design displacement is taken at 40 percent of the fuel-oil capacity.

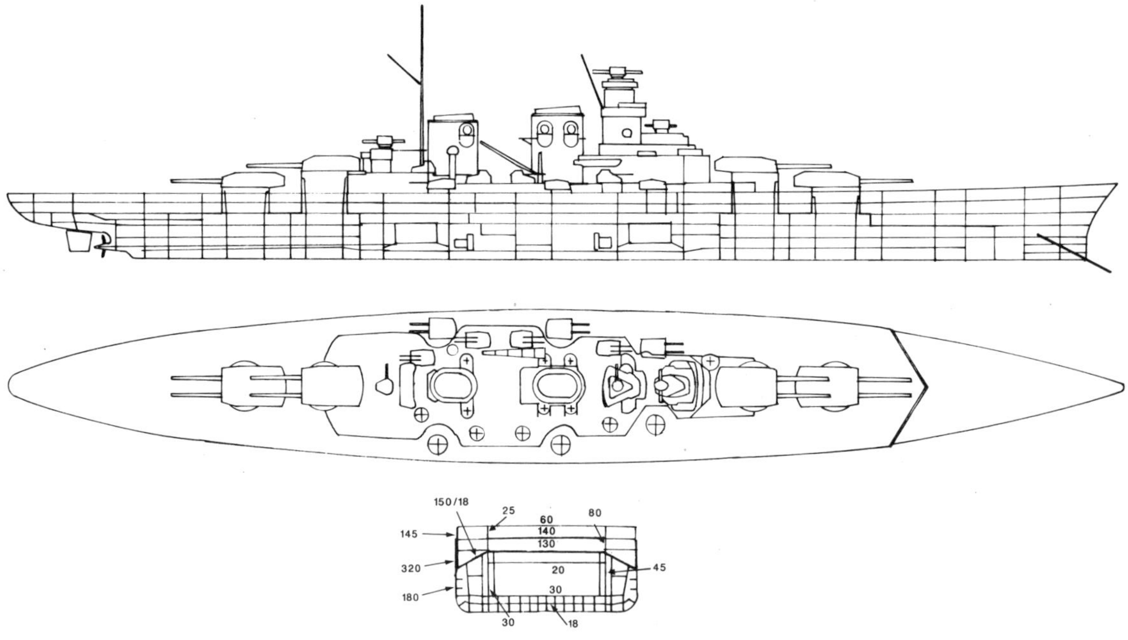

Figure 5-7. Sketch, 1942 design—profile, overhead view, and midship section.

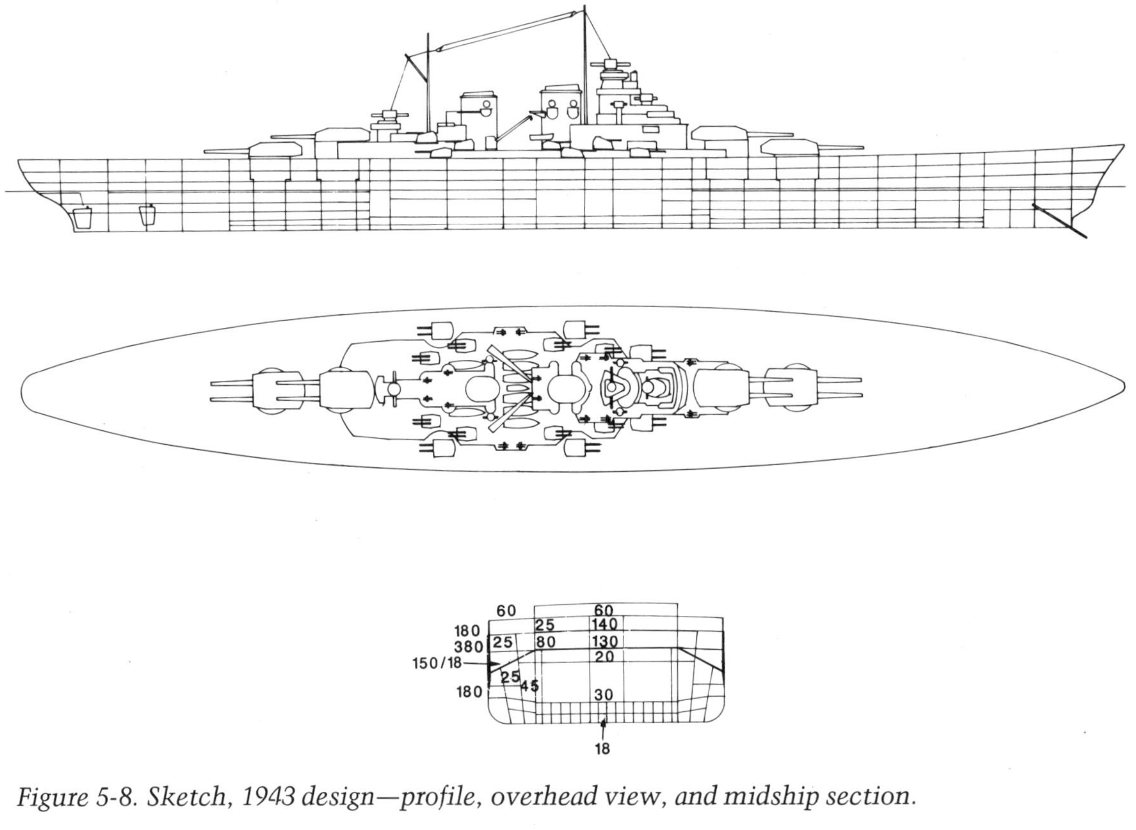

Figure 5-8. Sketch, 1943 design—profile, overhead view, and midship section.

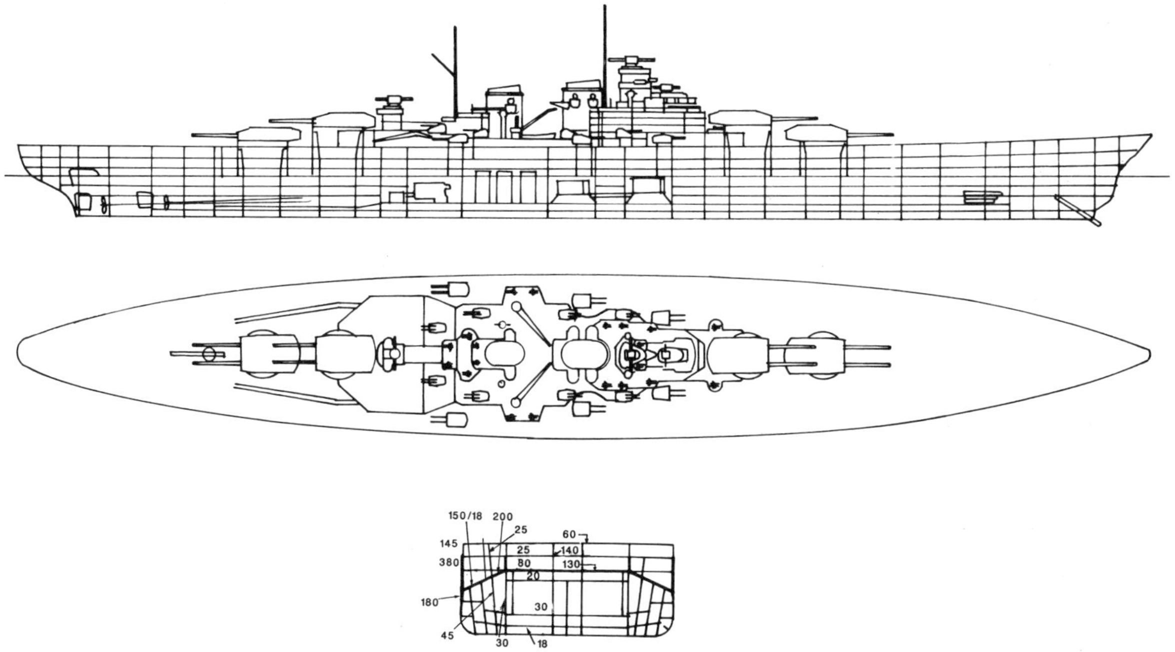

Figure 5-9. Sketch, H-44 design—profile, overhead view, and midship section.

The outstanding trends in these ships were the accelerating growth in size and displacement resulting from the heavier-caliber main armament, better armor protection, and startling increases in the transverse depth of the torpedo-defense system. Increased main-battery gun bores and the growing menace of air attack necessitated stronger armor protection.* It was believed, and later proven in tests, that 380-mm armor plate was the optimum thickness to employ for face-hardened armor, since greater thicknesses were not advisable due to the low probability of hits on a side belt. More important, however, was the fact that this armor was to be carried farther below the design waterline than in previous German battleships. This was probably done to counter diving shells, which had done serious damage to the *Bismarck* during her engagement with the *Hood* and *Prince of Wales*. From the dispatches received from the ship after the *Hood* engagement had concluded, German naval constructors were aware of this damage. Deck-armor thickness increased from a total of 200 mm in the two-deck arrangement of H-41 to 330 mm in the three-deck arrangements of H-42, H-43, and H-44. Most of the horizontal armor was distributed over two decks and concentrated inboard of the torpedo bulkhead. Outboard, the 60-mm upper armor deck and the 150-mm armored slopes provided horizontal protection for spaces outboard of the torpedo bulkhead. By late 1944, German designers recognized that no practical armor system could defend a ship against the very large "earthquake" bombs dropped on the *Tirpitz*.

The torpedo-defense system was the most notable feature of these ships. It was correctly recognized that the torpedo was the greatest single threat to the large capital ship. German naval constructors had recognized this when they reviewed the plans and data captured from the French concerning the *Richelieu* design. Extensive protective

*Conventional capital-ship design practice called for armor sufficient to withstand shellfire from guns equivalent to a ship's own main battery—the principle of corresponding protection.

systems demanded larger and wider hulls, with a resulting increase in displacement. There was a progressive increase in the percentage of beam allocated to the torpedo-defense system in German capital-ship design, beginning with the *Scharnhorst* and concluding with H-44, as shown in table 5-8.

TABLE 5-8

Comparison of Torpedo Defense Systems of the Last German Capital Ships

Ship	*Beam*	*Depth of Torpedo Defense System**	*Percent of Beam for Torpedo Defense*	*Full-load Displacement*
Scharnhorst	98.4′ (30 m)	29.52′ (9.00 m)	30.3%	38,101 tons (38,713 mt)
Bismarck	118.2′ (36 m)	36.08′ (11.00 m)	30.7%	49,135 tons (49,924 mt)
Battleship H-39	121.4′ (37 m)	36.08′ (11.00 m)	29.3%	62,592 tons (63,596 mt)
Battleship H-41	128.0′ (39 m)	43.64′ (13.30 m)	32.8%	74,799 tons (76,000 mt)
Battleship H-42	140.4′ (42.8 m)	50.16′ (15.73 m)	35.6%	96,451 tons (98,000 mt)
Battleship H-43	157.5′ (48 m)	62.34′ (19.00 m)	39.5%	118,104 tons (120,000 mt)
Battleship H-44	169.0′ (51.5 m)	72.18′ (22.00 m)	42.3%	139,264 tons (141,500 mt)

*Total, both sides of ship.

The H-42 design featured a multi-bulkhead torpedo-defense system with the double bottom and armored slopes forming a firm foundation for the elastic bending of the torpedo bulkhead. The oil was segregated into three separate tanks arranged vertically. In H-43 the breadth of the system was increased, and the structural arrangement was similar to H-42, except for the addition of another bulkhead 30-mm thick. It was further recognized that the protection might have to be increased, while the armament remained unchanged. In H-44 there was a substantial change. The torpedo bulkhead was located 11 meters from the side shell, and two noncemented armor longitudinal bulkheads, extending from the keel to the main deck, were the primary defense against torpedo explosions. The outer bulkhead was 45 mm below the armored slopes and 25 mm above. Inboard, the torpedo bulkhead was 30 mm below the lower armor deck, increased to 80 mm between the middle and lower armor decks, and then maintained at 25 mm for the portion between the middle and upper armor decks. The additional bulkheads provided better subdivision and additional longitudinal material for the hull girder. In addition, splinter protection above the upper armor deck was also improved and expanded. This resulted from the analysis of the battle damage sustained by the *Tirpitz* on 3 April 1944. When it is considered that the *Bismarck* and *Tirpitz* had good stability and subdivision in a damaged state, these later German designs would have shown a remarkable ability to withstand damage, as they had much greater beam and better underwater protection against torpedoes. The German naval constructors assumed that with increasingly larger charges in torpedo warheads, the extent of damage would also be greater. With increased watertight compartmentation, the stability and subdivision would have improved to counter this. Allied torpedo charges would have further increased in destructive power during the construction period of these ships, however, slightly offsetting some of the advancements made in the design of the torpedo-defense system. It is doubtful, though, that such torpedoes would have been very effective against a ship of the size of either H-43 or H-44. Only a massive torpedo attack of the type used against the *Yamato* and *Musashi* would have achieved the desired results.

The increasing importance of protection against mine detonations and noncontact underwater explosions of bombs demanded a deeper innerbottom. German designers believed that flooding through the innerbottom in either the magazines or machinery spaces could no longer be risked with air attack now being a serious threat. The noncontact explosions of bombs dropped on the *Tirpitz* produced serious structural damage to the underwater hull in several sections of the ship. A triple bottom was proposed for these last German battleship studies; the major features are summarized in table 5-9.

TABLE 5-9
Characteristics of Underbottom Protection—H-42, H-43, and H-44

Design Study	*Depth*	*Thickness of Tank Top Plating*
H-42	12.1′ (3.7 m)	1.18″ (30 mm)—Noncemented armor
H-43	13.5′ (4.1 m)	1.18″ (30 mm)—Noncemented armor
H-44	15.4′ (4.7 m)	1.18″ (30 mm)—Noncemented armor

Interesting developments also occurred in the design of the propulsion plants of H-42, H-43, and H-44. Like their immediate predecessor, H-41, they all had a mixed propulsion plant consisting of diesels and steam turbines and boilers. Although the diesels were more powerful than those used on H-39, the steam plants could provide more power per shaft. The limitations on the transmission of horsepower per shaft and the power limitation of the diesel engines kept the total power of the three large battleship studies constant. Therefore, as ship size grew, the maximum speed diminished. All the designs featured four shafts and four rudders, each located behind the propeller race for the greatest efficiency. The multiplicity of the rudders was yet another lesson learned from the exploits of the *Bismarck*.

Construction—Battleship "H."

When construction for battleship "H" was halted in September 1939, 14,055 metric tons of material had been ordered, 5,800 tons had been delivered, and 3,425 tons had been machined, but only 766 tons had actually been installed.

On 16 November 1939, the OKM issued sweeping decrees for the full-scale implementation of all contracts and directed that orders be given the priority needed to effect rapid delivery of all items near completion. On 15 April 1940, following the completion of the initial portion of all contract and purchase order items for battleship "H," the field construction offices of the OKM in Hamburg and Kiel were closed. On 25 November 1941, after work had been suspended for almost two years, the OKM ordered the hulls of battleships "H" and "J" scrapped and the steel diverted to other purposes. Contracts were annulled on 29 and 31 August 1942.

Construction—Battleship "J."

The keel for battleship "J" was laid down on 1 September 1939 at the Deschimag Shipyard in Bremen. Only 3,531 tons of material had been delivered for battleship J and only 40 tons had been installed on the slipway.

Construction of Battleship "H" had proceeded this far by 4 August 1939 on the ways at Blohm & Voss Shipyard in Hamburg. The outbreak of war the following month brought the construction of this ship to a halt, and her construction was later canceled.

Construction—Battleships "K," "L," "M," and "N."

Battleship "K" was to be constructed at the Deutsche Werke Shipyard in Kiel, and it was planned to lay her keel around 15 September 1939. When the war began, her construction was halted, with 12,921 tons of steel on order, 2,980 tons of steel delivered, and 1,503 tons machined. The steel in the yard was later released to submarine construction.

The last three ships, "L", "M", and "N," never advanced beyond the projected phase. Battleship "L" was to have been constructed in the naval dockyard at Wilhelmshaven, while the last two were to be follow-on ships at Blohm and Voss and Deschimag respectively.

Armament.

The main characteristics of the guns are tabulated in table 5-10.

TABLE 5-10

Gun Characteristics

Gun	16" (406 mm)/50	5.9" (150 mm)/55
Shell weight	2,271 lbs (1,030 kg)	99.8 lbs (45.3 kgs)
Muzzle velocity	2,657 fps (810 mps)	2,871 fps (875 fps)
Maximum range	40,250 yds (37,800 m)	25,153 yds (23,000 m)
Max. elevation	33 degrees	40 degrees
Gun	4.1" (105 mm)/65	1.46" (37 mm)/83
Shell weight	33.3 lbs (15.1 kgs)	1.64 lbs (0.745 kg)
Muzzle velocity	2,952 fps (900 mps)	3,281 fps (1,000 mps)
Maximum range	19,357 yds (17,700 m)	7,382 yds (6,750 m)
Max. elevation	85 degrees	80 degrees
Gun	0.79" (20 mm)/65	
Shell weight	0.291 lbs (0.132 kg)	
Muzzle velocity	2,952 fps (900 mps)	
Maximum range	5,249 yds (4,800 m)	
Max. elevation	90 degrees	

Main battery. The basic gun for the H-39 and the H-40 was the 406-mm/50-caliber (SKC 34) gun, which was to be upgraded to 420 mm for the H-41 and H-42 designs. As has been discussed in the design history, the heavy barrel of the 406-mm gun made it possible to upgrade it to a 420-mm bore. With the greater bore, the barrel length would have been about 48 calibers. In any event, all designs featured the main battery arranged in twin turrets, two forward and two aft. The Germans preferred this because of the equal distribution of firepower, more effective grouping of salvos, and because it reduced the number of guns lost if a turret sustained some form of battle damage or malfunction during an engagement. It was recognized that triple or quadruple turrets were more efficient in the use of weight and would reduce the length of the armor citadel, permitting thicker armor, but complications in the mechanical design of the triple turret would have required extra time for design, testing, and construction. Quadruple turrets would eliminate the need for superfiring turrets, reduce the armor citadel substantially, and eliminate two heavy barbettes, but ordnance engineers were concerned with the blast effects on ship structure and the probability of the loss of a large number of guns if the turret were immobilized. Structural engineers disliked the large openings necessary in the strength deck for a quadruple-turret barbette.

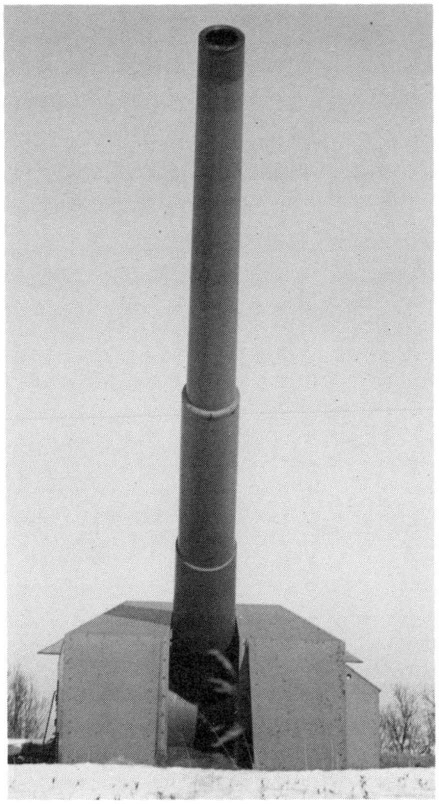

Although the "H"-class battleships were never completed, seven 406-mm guns out of the 48 that were to be mounted on these ships were constructed. Three of these were mounted in Norway, three were placed in railway gun mountings, and one was lost in shipment to Norway. This is one of the 406-mm/50-caliber guns that was mounted at Trondenes, Harstad, Norway, to guard the strategic port of Narvik. These guns survived World War II and were taken over by the Norwegian military. In 1968, they were sold for scrap. (Courtesy of the Royal Norwegian Navy)

The 420-mm upgrading of these guns was never accomplished. Although ammunition tests were made, it was planned to have the gun barrels rebored only when the guns were mounted on the ships. The 406-mm guns used in coastal defense and railway mountings had a barrel length of 20.3 meters and a weight of 157 tons. A special breech was fitted to these guns. The range was 43,000 meters, and the muzzle velocity was 940 meters per second. Two shells were used, one weighing 600 kg, the other 960 kg.

The 406-mm shell to be used aboard the "H"-class battleships would have weighed 1,130 kilograms. The twin turret was estimated to weigh 1,454 tons; rate of fire was estimated to be two shells per minute per gun. Krupp produced seven 406-mm barrels for battleship "H," but when it became apparent that the ship would not be completed during the war, they were assigned to coastal defense.

Four of the seven 406-mm guns were shipped to Norway in 1942, one of them being lost in shipment, but the remainder were mounted in Trondenes near Harstad to guard the strategically important port of Narvik. After the Germans surrendered, the Norwegians took over these guns, which were still usable in 1954. In 1968 they were placed on sale for scrapping.

The three remaining 406-mm guns were completed as railway guns. Two of these were initially assigned to the German occupation army in Poland and later taken to Hela to protect Danzig. All three 406-mm railway guns were subsequently taken to France during late 1941 to assist in the blockade of the English Channel to British shipping. These guns were located near Blanc Nez.

Secondary battery. The secondary armament, both in gun model and arrangement, was similar to that of the *Bismarck* class; twelve 150-mm/55-caliber guns (Model 1928) were mounted in six twin turrets. Elevation and rate of fire were the same as on the *Bismarck*—40 degrees maximum and six rounds per minute respectively—indicating that the battery's primary purpose was surface defense. These guns would have been ineffective in air defense, as their shells lacked a fuze-setting provision, and training and elevating rates were too slow for effective antiaircraft fire. Eight guns could fire directly forward, four could fire directly aft, and six guns on either beam could cover an arc extending some 45 degrees over the bow to 50 degrees abaft the beam.

Long-range defense against aircraft was to be provided by sixteen 105-mm/65-caliber guns (Model 1933) in eight new (Model 1938) mounts. The guns were to be stabilized, as in the *Scharnhorst*- and *Bismarck*-class ships, but the mounts would have increased protection for the gun crews, and faster training and elevating rates were achieved. One of the serious shortcomings in the mounts provided to the earlier ships was the fact that the gun crews could be hit with flying debris, shrapnel, or by the gunfire of strafing planes. Because of the greater beam, these mounts were farther from the ship's centerline so that there would be less superstructure interference with their operation. In fact, all eight mounts could fire ahead or astern, whereas in the *Bismarck* these guns could not fire in this manner. The 105-mm guns, however, lacked operational capability, as was proven in air attacks by the relatively obsolete and slow Fairey Swordfish torpedo bomber. They all were to have been mounted on the first superstructure deck, with ammunition supplied by hoist from magazines below for the forward and aft mounts and from centrally located magazines, port and starboard, for the other four mounts. Enclosed mounts would also end electrical failures experienced in open mounts.

Antiaircraft battery. The light antiaircraft defense was to be provided by sixteen 37-mm/83-caliber guns (Mode 1 1933) in eight twin mounts, two of them being the special "flakturmen" (a specially armored closed mount, Model 1937). The new 37-mm enclosed mounts were positioned between the two funnels and as far outboard on the first superstructure deck as space would allow, so that they could have uninterrupted fire over an arc of 180 degrees. These mounts featured a centrally located ammunition hoist. The antiaircraft battery also featured 24 20-mm guns in six quadruple mounts. Comparison with other battleships of the period indicates that this antiaircraft battery was very weak against high-performance aircraft, especially in destroying planes in close-range engagements. Undoubtedly it would have been augmented before the ships were completed.

Torpedo armament. The "H"-class ships were to be fitted with six 533-mm bow-mounted submerged torpedo tubes, diverging from the ship's centerline some 10° port and starboard. Torpedoes that required a large space in the forward part of the ship had been abandoned in all battleships built after the *Nelson* and *Rodney*. The tubes were mounted forward of the armored citadel, with some protection provided by a partial armored bulkhead forward of the torpedo-handling room and a side armor belt extending a distance of 1.825 meters below the design waterline of 10 meters. This protective arrangement was chosen because mine damage to the *Bayern* during the First World War had indicated the vulnerability of such spaces without some armor protection. The torpedo armament indicated that the Germans intended to use these ships as surface raiders on British merchant shipping, although that was not their only mission. In operations at long distances from home bases, where ammunition had to be conserved, a ship not quickly sunk by gunfire could be finished off with a torpedo, thus reducing ammunition expenditure.

Aircraft. These ships, like the *Bismarck* and *Tirpitz*, were to have carried four Arado-196 floatplanes. The hangar was located between the after funnel and turret Caesar. A rotating catapult was located immediately below the guns of turret Dora, and aircraft were to be carried aft on tracks on either side of the two after main-battery turrets. When the aircraft were to be launched, turret Dora had to be trained abeam or the main guns elevated to their maximum angle. The locations for the hangar and catapult were less than satisfactory, as the aircraft facilities were in a region of high gun blast, and the use of the catapult meant rotating a turret to clear the launch area. In a redesign during the fall of 1939, two fixed catapults were used instead; however, the entire aircraft arrangement was not adequate. German naval constructors did not want hangars in the stern because of the effects of gun blast on exposed equipment there and the possible obstruction to the firing of turret Dora.

Protection. The German concept of relatively close-range surface engagements in the North Sea led to the citadel-armor system first employed in the *Nassau*-class battleships of 1909 and continued through the *Bismarck* and *Tirpitz*. The vertically arranged main side belt was preferred over the inboard-armor concept favored by French and American naval constructors. Thus, the armor system of the "H"-class battleships was once again of the turtle-deck type, which was believed capable of effective resistance to the forms of attack envisioned by German naval experts.

There is no official data on the immunity zone of these ships, so that the protected zone has been estimated based on the U.S. 406-mm/45-caliber gun firing a 1,016-kilogram shell at a 90-degree target angle. The vitals had an immunity zone of approximately 11,000 meters to 21,000 meters, the inner limit being based on the penetration of the vertical side armor and the sloping deck armor. The spaces above the lower armor deck, which was near the waterline in the full-load condition, had practically no protection against the 406-mm projectile, since the upper deck (50 mm) is penetrable outside of 11,000 meters, and the upper side belt of 145 mm is penetrable at all ranges. Thus, while the magazines and propulsion plant were well protected against direct hits, except for spaces in the shadow of the main side belt, there was little of the ship above the waterline that was not vulnerable to this 406-mm projectile at all ranges.

In the "H"-class battleships, minimization and localization of damage replaced the principle of absolute protection against penetration. Draft and displacement considerations also determined the size and extent of the heavy side belt. The ships were to have range tables that would indicate which target angles to present to certain opposing battleships to prevent their shells from defeating the main side belt. That belt would have been vertical and divided into two thicknesses. The lower armor belt was to have been 300-mm thick and extend 1.825 meters below the design waterline, the draft of which was 10 meters. The belt tapered from 300 to 170 mm over a distance of 1.35 meters, to join the upper citadel belt, which was 145-mm thick. The lower belt maintained its thickness throughout the armor box; the upper belt decreased to 120 mm outside turret Anton.

Deck armor. The horizontal armor thicknesses were similar to the *Bismarck* and *Scharnhorst* arrangements in the region of the propulsion plant. In the region of the magazines, there were some notable departures. It was recognized that the 406-mm gun was more powerful than the 380-mm, and means had to be found to strengthen the armor, especially since the main side belt was thinner than in either of these two other battleship classes. Table 5-11 summarizes the main features of the deck-armor system.

TABLE 5-11

Deck-Armor Thicknesses

	Magazines	*Machinery*
Upper deck	3.15" (80 mm)	1.97" (50 mm)
Armor deck		
Inboard	4.72" (120 mm)	3.94" (100 mm)
Outboard	5.91" (150 mm)	4.72" (120 mm)
Total armor		
Inboard	7.87" (200 mm)	5.91" (150 mm)
Outboard	9.06" (230 mm)	6.69" (170 mm)

The outboard portion of the lower armor deck was sloped to provide additional protection against penetration and thereby protect the vitals below. It was recognized that sloping armor might result in some drainage problems, but this was a small price to pay for increased protection.

In the *Bismarck*-class ships, displacement was limited, and the only armor forward of the citadel was a small splinter belt. The "H"-class ships had greater displacements and waterline lengths, and German naval constructors made good use of splinter protection beyond the citadel. They employed 60-mm to 150-mm armor over much of the bow structure, forming what looked like an armored prow. Part of the need for a more extensive splinter belt was that a torpedo-handling room was located in a compartment just forward of the forward armored bulkhead, and some protection was required for that space. The extensive use of armor outside the armor box, greater than in any previous battleship design, was based on an analysis of the action of the battlecruiser *Lützow* during the Battle of Jutland. The bow was protected by 60-mm Wh armor plate for a depth of some 7.25 meters and a length of 12 meters. Between the forward armored bulkhead and the point where the bow splinter protection became 7.25 meters deep, the splinter belt was increased to 150 mm over the torpedo-handling spaces and reduced to 60 mm beyond. The splinter belt in this region was 4.25 meters deep. The stern was also provided with a splinter belt that extended to the armored bulkhead at the after end of the steering-gear room. This belt was 90-mm thick over a vertical distance of 4.25 meters. German naval constructors were concerned that portions of the hull not protected by armor could be seriously damaged by shell splinters, while protected sections would not share the same fate. Damage-control parties could easily cope with confined damage by a shell hit. These ships were to fight far from home bases, and extensive damage had to be prevented. A similar solution was desired for the *Bismarck* and *Tirpitz*, but this had been sacrificed to keep the standard displacement down to 42,000 tons.

Conning tower. German designers still believed that the main and secondary conning positions should be provided with substantial armor protection. The design of the main conning position was not a repeat of previous ones used in either the *Scharnhorst* or the *Bismarck*. The main command tower itself was some 5 meters wide by 8.08 meters long. The sides were of cemented armor 350 mm on the upper level and 200 mm on the lower level. The roof armor was 200-mm material, the lower deck of the tower being only 80-mm roof. The noncemented material was welded, while the cemented armor was of tongue-and-groove construction with armor bolts holding the sections together. The access and wireways were protected by 150-mm cemented armor shaped in the form of a horseshoe. This structure flared out from its beginnings on the lower armor deck to its termination at the lowest level of the command tower. In profile it resembled a giant trapezoid. The secondary conn was protected by 100-mm armor plate on all its sides and overhead. Again, cemented armor was used on the sides and noncemented armor on the horizontal portions.

Armor bulkheads. There were four armored bulkheads in battleship "H," unlike in the *Bismarck*- and *Scharnhorst*-class ships, but the arrangement of armor thicknesses was similar. The use of an additional armor bulkhead resulted from the need to protect the torpedo magazine. The forward armor bulkhead had 80-mm armor outside the torpedo bulkhead, but increased to 120 mm for the portion between. Below the lower armor deck, the thicknesses inside and outside the torpedo bulkhead were 220 mm, the armor on the bulkhead decreased to 150 mm inboard for the lowest strakes, and to 100 mm for the portion of the bulkhead outboard of the torpedo bulkhead. This was intended to protect the forward magazines from shell hits from the forward quadrants. The after armored bulkhead thicknesses were generally similar, except that the 220-mm armor protection was approximately half the height of that forward, and there was no armor below the 220-mm strakes. The bulkhead abaft the steering-gear room was protected by 200-mm armor plate. A partial armored bulkhead was fitted to watertight compartments forward of the forward armored bulkhead, to protect the torpedo-handling spaces. This bulkhead spanned the width of the splinter belt and was 100-mm thick. Forward of this transverse bulkhead, the splinter belt became 60 mm and aft it was 150 mm.

There were several important longitudinal armored bulkheads within the armored citadel. The torpedo bulkhead, armored over the entire length of the armored citadel, was 45-mm thick outside the propulsion plant and 60-mm thick in way of the magazines. Its upper extension (30-mm armor), between the upper and lower armor decks, was to provide splinter protection, and a 30-mm armor plate was used to protect the areas inboard. Parallel to the extension of the torpedo bulkhead, but nearer to the centerline, there was another splinter bulkhead. There were five transverse splinter bulkheads in the midships area of the ship to limit the longitudinal spread of bomb or shell fragments.

Turret armor. The heaviest armor protection was concentrated on the main turrets and their barbettes. The face plates of the main turrets were 400-mm thick; the sides were 220-mm, backwalls 325-mm, and roofs 220-mm thick. The inclined portion of the roof over the gun chamber was 180-mm thick. The greater backwall thickness was not only for protection, but it also shifted the center of gravity of the turret toward the after bulkhead for more favorable balance, operation, and construction of the turret. That procedure was followed in the turret design of World War I German battleships. In general, the armor thicknesses of the main turrets were similar to those used in the *Bismarck* and *Tirpitz*, which indicates that they were not derived from any altered protection requirements.

The barbette armor was in two courses, as in the *Bismarck* class. The lower armor ring, below the upper armor deck, was molded from 240-mm noncemented armor plate; the barbettes above the upper armor deck were protected by 365-mm face-hardened armor plate. The thickness of the armor around the periphery was constant, but it was reduced below the upper armor deck because that deck provided additional protection. The designers had increased the barbette armor thicknesses since they were convinced that barbettes were more vulnerable than turrets. The ammunition-handling system, powder cartridges, and flame-proof powder scuttles reduced the probability of a magazine explosion caused by a shell detonating in the turret. Such an explosion would be more likely within the barbette structure.

The secondary turrets had the same protection as was in the *Bismarck* and *Scharnhorst*. German designers saw no need to increase this protection, as it would further detract from the side armor belt, which was already as thin as practicable. The barbette armor for these turrets was 80-mm thick around the entire periphery. The height of the barbette armor varied with the turret location and ranged from 1.37 meters for the forward turrets to 1.61 meters for the middle turrets to 1.64 meters for the after turrets. (These distances represent heights above the upper armor deck.)

Extensive splinter protection, generally concentrated in the armored citadel between the armored decks, ranged from 15 mm to 20 mm in thickness, with heavier armor on the main divisional bulkheads, such as the longitudinal bulkheads. There was splinter protection for the 37-mm ammunition-handling rooms, the ammunition hoists to the 150-mm and 105-mm guns, the bulwarks, the open navigation bridge, charthouse, cable trunks, fire-control tower, antiaircraft directors, communication control, and the after side of the flag plot. The uptakes had 16-mm splinter protection.

Although these ships had less armor protection in certain areas than the *Bismarck* and *Tirpitz*, they had more armor at the bow and stern and in other important areas, such as the steering-gear room and the conning tower. The armor weight (not including light splinter protection of 15 mm or less) was 39.2 percent of the standard displacement in the "H"-class ships and 40.4 percent for the *Bismarck* class, but the armored citadel was the same percentage of the design waterline in both classes.* The only outstanding difference was that the armor belt of the *Bismarck*-class ships had a

---

*The propulsion plant was heavier in the "H"-class ships, and thus the armor percentage would decrease as a percentage of the light ship.

slightly greater resistance to shellfire than did that of the "H"-class ships. Nevertheless, the magazines and propulsion plant of the "H"-class ships were still well protected against 406-mm shellfire. The armored slopes, 105-mm thick in the *Bismarck*, had been increased to 120 mm.

Side protection. The torpedo-defense system was an important feature. The relatively large beam permitted a system similar to that used in the *Bismarck* class. Specifically, the inner torpedo bulkhead was to be held at a minimum distance of 5.5 meters from the shell at half draft. This dimension could not be maintained at the ends of the armored citadel, but it was intended to be not less than 3.25 meters. The torpedo bulkhead and inner structure were increased in thickness at these points to compensate for the reduced depth of the torpedo-defense system. The torpedo protection aft in way of the two main-battery turrets, however, presented some difficulties. Since the outer shafts emerged from the hull at this point and there was a bossing, the torpedo protection at half draft was less than the 3.25 meters desired. At this point a triple bottom was used under the turrets to compensate for the lack of adequate breadth. The main features of the side-protection system for the "H"-class battleships are shown in table 5-12.

TABLE 5-12
Characteristics of the Side-Protection System

Location	*Depth at Half Draft*	*Total Bulkhead Thickness in Side-Protection System*	*Thickness of Torpedo Bulkhead*
Turret Anton	10.10′ (3.08 m)	2.76″ (72 mm)	2.36″ (60 mm)
Turret Bruno	10.10′ (3.08 m)	2.76″ (72 mm)	2.36″ (60 mm)
Amidships	18.04′ (5.50 m)	2.08″ (55 mm)	1.77″ (45 mm)
Turret Caesar	6.75′ (2.06 m)	2.76″ (72 mm)	2.36″ (60 mm)
Turret Dora	4.21′ (1.28 m)	2.76″ (72 mm)	2.36″ (60 mm)

The side-protection system had slightly more breadth at the ends of the armored citadel than the *Bismarck* and *Tirpitz*. Such breadths, however, were less than those used in most foreign designs. A deeper torpedo-defense system could not be attained in these ships because the diesel propulsion plant required more volume, and the heavier-caliber main battery required larger magazines. To increase the breadth or length was not possible, because there would have been problems in docking these ships for routine overhauls.

It was decided to retain the double-bottom arrangement of the *Scharnhorst* and *Bismarck* classes in the H-39 design and to rely on subdivision to limit flooding resulting from an underbottom explosion. The depth of the double bottom, however, was increased to 2 meters under the propulsion plant and 2.75 meters under the turrets.

The ships were extensively subdivided, with 21 major watertight compartments and fairly extensive transverse compartmentation within the armored citadel. Much care was given to the side-protection system design to prevent heavy lists caused by flooding in magazines, machinery spaces, or wing compartments. Some of the important stability characteristics are summarized in table 5-13.

Although these ships had a greater beam, their metacentric heights were similar to those of the *Bismarck* class. This is attributed to the heavier main armament and more extensive superstructure.

TABLE 5-13
Stability Data

Displacement Condition	*Displacement*	*Metacentric Height (GM)*
Light condition*	52,566 tons (53,400 mt)	10.50′ (3.2 m)
Design displacement	57,618 tons (58,543 mt)	13.78′ (4.2 m)
Full load	62,591 tons (63,596 mt)	14.76′ (4.5 m)

*Corresponds to the German "Type Displacement."

Radar. Radar was not specified, but undoubtedly search radar and some fire-control radar would have been installed. The four German capital ships completed during that period carried radar, and the "H"-class ships, by the time they were completed in 1944, would have had even more sophisticated equipment, with the resulting reduction or elimination of the aircraft complement. In fact, an "H"-class battleship would probably have carried the same complement of electronics equipment that was provided on the *Tirpitz* in 1943 or 1944.

Propulsion. The propulsion plant of the "H"-class ships was most unusual. These vessels were to have three 4.8-meter-diameter propellers, each powered by four diesel engines with a maximum rated output of 13,750 metric horsepower and a normal rating of 12,500 metric horsepower per unit. Vibrations in the diesel installations of the *Deutschland*-class armored ships had been carefully investigated, and a new type of engine foundation, four meters deep from the top plate to its intersection with the bottom plating of the ship, had been designed. In between, a rather deep double-bottom structure also provided support, and the diesel-engine foundation was rigidly connected to it. Connecting this foundation to as much structure in the ship as possible as well as providing a massive foundation was considered a good measure against heavy vibrations, as more mass was involved. Construction of these foundations also required special procedures. The ships would have to be built with the stern section completed first. Welding would have created some deflection in the hull girder, and this had to be minimized because of the precise tolerances required in the mounting and alignment of the twelve propulsion diesels. The alignment between the engines and the reduction gears was critical.

Diesel engines. The diesel engines were a double-acting two-stroke, 9-cylinder type (MZ 65/95) designed and produced by Maschinenfabrik Augsburg-Nürnberg A.G. (MAN). They had a 65-cm bore, 95-cm stroke, and a maximum speed of 265 rpm. Four engines on each shaft were connected through a toothed coupling located between the engines and reduction gears. Each pair of diesels in line was connected to a common pinion by means of the synchronized toothed coupling. The pinions were a special flexible design, and the pinion rim was not rigidly connected to the hub. Instead, it was allowed to slide on it. Through the two pinions at the circumference of the joint, cylindrical holes were machined from pinion hub to pinion rim so that springs of a cylindrical shape, consisting of a number of split-up steel bushings—one inside the other—could be inserted. Under the varying torque of the diesel engines, the bushing

338

The battleships of the "H" class would have been propelled by twelve 9-cylinder diesel engines of the MZ 65/95 type, with four engines driving each of three propeller shafts. Each piston, such as that shown here, was designed to have a diameter of 650 mm and a stroke of 950 mm. (Courtesy of M.A.N.)

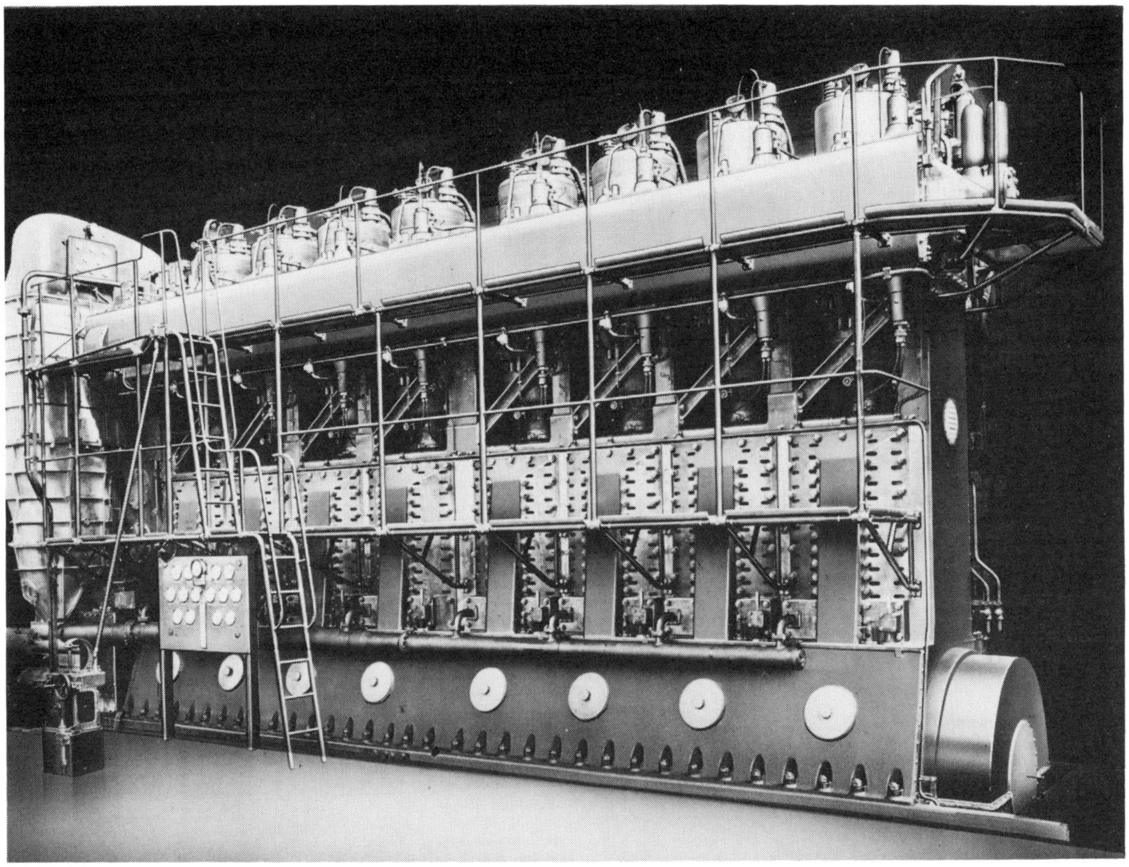

The Type MZ 65/95 diesel engine is shown on its test stand at M.A.N. in Augsburg, Germany, during World War II. A nine-cylinder version of this engine was to be used in the diesel propulsion plants of the battleship "H" series with the sole exception of Scheme B. The engine was approximately 7 meters high, and four of these were used on a shaft. (Courtesy M.A.N.)

springs would be shrunk and loosened alternately, thus decreasing the fluctuations in torque before reaching the gear teeth.

On the shaft line behind the main gear wheel, which was driven at a higher speed than the pinions, a hydraulic coupling was inserted to prevent the transmission of torsional vibration to the shaft and to disconnect the gear from the propeller when the diesel engine was idling. Two reduction gears were provided for each shaft, and the transmission ratio between the gears and the shafts was nearly 1:1 (shaft rpm—260), which indicates that the Germans were satisfied with their high-speed propeller designs.

After deducting the power loss of some 7–8 percent per shaft, which resulted from friction or slip in the gearing, the couplings, and the propeller-shaft supports, some 50,000 metric horsepower would have been available to each propeller. The rated output of the propulsion plant was 150,000 metric horsepower at the normal power load. At the overload power of 165,000 metric horsepower, the maximum speed of 30.4 knots would have been attained at the designed displacement of 57,617 tons. Speed and endurance capabilities of these ships are summarized in table 5-14.

The efficient propulsion plant and large fuel capacity would have provided an excellent endurance of 16,000 nautical miles at a cruising speed of 19 knots. Reserve capacity for battle operations would have raised the cruising endurance to 20,250 miles.

TABLE 5-14
Speed and Endurance Properties

Displacement	*Maximum Speed*	*Maximum Continuous Speed*	*Endurance at 19 Knots*
57,618 tons (58,543 mt)	30.4 knots	29.6 knots	8,100 miles
60,104 tons (61,070 mt)	30.1 knots	29.3 knots	15,200 miles
62,591 tons (63,596 mt)	29.8 knots	28.9 knots	20,250 miles

Electric plant. The electric plant of battleship "H" followed previous German practice in the *Bismarck* and *Scharnhorst* classes in that the electric current was 230-volt DC for the main circuit and 110-volt AC for equipment requiring alternating current. The electric plant, however, featured diesel units divided into four power stations. Within each station, there were two 920-kw diesel generators and one 460-kw AC generator. Although the main electrical circuit was direct current, the amount of equipment requiring AC power was increased over that in the *Bismarck* and *Tirpitz.*

Hull Characteristics.

Welding. There had been a gradual increase in the use of welding in German major warship construction, beginning with the light cruiser *Emden* in 1923, and the "H"-class ships would have been extensively welded. Full-scale tests on the *Preussen* had shown that riveting and welding both yielded connections of sufficient strength, but welding produced a slightly higher joint efficiency with less material. The structure of the side-protection system would have been all welded construction, as would the main-deck and splinter protection. Torpedo bulkheads, to be entirely welded, were arrayed like a membrane between the lower armor deck and the innerbottom. Underwater-explosion tests had indicated that welded torpedo bulkheads could bend more easily without rupturing joints or springing leaks. It was hoped that a flexible bulkhead design, as in the *Bismarck*-class ships but with improved construction, would dissipate a major portion of the explosive energy by elastic deformation. Furthermore, this type of construction would save considerable weight, which could be employed in additional armor to protect other parts of the ship.

The torpedo-bulkhead structure was carefully designed. Since it would be highly stressed by a torpedo explosion, it was essential that bulkhead end connections be well designed. It was decided to extend the torpedo bulkhead the full depth of the ship so that the armor decks and the double bottom would form a rigid supporting structure. The upper extension of the torpedo bulkhead between the two armor decks formed the main splinter defense in that area.

Rudders. The "H"-class battleships had three equal-sized rudders, each in a propeller race. They were espected to offer better maneuvering qualities while reducing the possibility of loss of all maneuvering control due to a single hit. The arrangement was similar to that used in the Italian *Vittorio Veneto* class.

The upper deck and the splinter protection were of 50-mm armor-type steel and were to be all-welded construction. The joint between the side armor, which tapered to 80 mm, and the side shell was also to be welded. Full-scale tests of welded and riveted construction had convinced the German naval constructors that both methods yielded joints of equal strength. For that reason welding was adopted in order to save weight.

Hull depth. A series of longitudinal bulkheads extending the length of the armored citadel and a double bottom 2 meters deep gave sufficient structural strength to allow the ship's depth to be reduced to 16 meters. Balancing the needs of protection, weight, and draft caused a problem in the design of these ships. Increased beam resulted in a better torpedo-defense system, but caused an undesirable increase in the weight of heavy deck protection. This would lead to more displacement and a deeper draft, which was intolerable. German naval constructors sought to arrange the structure in the most efficient manner, so that by increasing the beam and lengthening the ship, while decreasing the depth, the greater displacement would not result in deeper drafts. Such a savings in depth also produced a lower silhouette—ideal in battle but poor for seakeeping qualities. Because of the limited depth, the freeboard was only 5.8 meters amidships, increasing to 6.9 meters at the stem, which meant that the ships probably would have been wet gun platforms in moderate sea states.

Four docking keels were provided to support the immense beam and to reduce the stresses and loads when the ships were dry-docked.

Summary. The "H"-class design indicated that German naval constructors had clearly established the protection a battleship would require against air and underwater attack in World War II. In 1938 the German Navy recognized the need for a very strong antiaircraft defense against high-speed bombers at long ranges, or low-flying torpedo bombers at shorter ranges. There was a certain group of individuals within the German Navy who argued for separate batteries of antiaircraft and antiship guns, as they still thought in terms of the Battle of Jutland in World War I. The "H"-class ships were not suited for the type of combat operations that developed in the Pacific theater of World War II—a serious shortcoming, since if they had been completed, they would have been subjected to more intensive air attacks than had been anticipated during their design period. These ships had the requisite deck area and displacement for an antiaircraft battery larger than on any battleship designed, constructed, or operated in the European theater during World War II. The fact that the German Navy did not operate its own aircraft severely limited its opportunity to study the future role of air power in naval warfare and indirectly denied its ships the merits of a dual-purpose armament. By 1941, the Germans had developed an excellent 128-mm/61-caliber gun that had a maximum range of 20,000 meters, a muzzle velocity of 880 meters per second, and a rate of fire of 21 rounds per minute, but the opportunity to use such a gun on their naval ships had been lost. This gun was successful in shore-based defense against aircraft attack. It was to be mounted on the destroyers of the Z-52 class. The twin mounts were to be completely automatic. Studies had shown by 1942 that a standardized armament for battleships, cruisers, and destroyers was necessary and that this quick-firing gun would fulfill all requirements.

These ships had a different gunfire direction configuration than was used in previous ships. Four independent batteries could instantly shift to a common control, each unit having fire-control and gun operations gyro-stabilized. The 105-mm guns would have had radar and high-grade optics for spotting targets for each mount, and the centralized fire control was to be eliminated. The Germans felt that a high degree of mobility and rapidity of training was not possible in guns of 150-mm caliber or larger. If the 150-mm guns had been designed for antiaircraft defense, German ordnance designers have claimed that their effectiveness in surface gunnery would have been impaired. However, it is interesting to speculate what would have happened to such a battery if these ships had been completed in 1944 as planned. The *Tirpitz* had to use her 150-mm and 380-mm guns in antiaircraft defense by that time, and it is possible, though not probable, that the new 128-mm gun might have replaced the 150-mm guns, or the 150-mm turrets would have been provided with greater elevation.

These battleships would have had the most powerful main battery of any European-designed battleship in the World War II era, and their gunpower and protection would have made them very formidable opponents. Their armor protection was the most extensive of any battleships designed or constructed, although it was not the thickest. Fore-and-aft armor distribution was based on the experience of the battlecruiser *Lützow* in the Battle of Jutland and later appeared justified by the bow damage sustained by the *Bismarck* in her engagement with the *Hood* and *Prince of Wales*. The armoring of ship sections outside the citadel required thinning the main side belt. Despite tests indicating the low probability of hits on the main side belt, it remained a fact that a heavy-caliber hit there might penetrate and seriously damage the vitals of the ship.

The German naval constructors believed that the 300-mm belt armor, combined with 150-mm slopes, gave an equivalent belt thickness of 500 mm over the magazines. Above the lower armor deck, only the 141-mm upper citadel belt protected the spaces between the main armored deck and the upper armored deck. Upper armor belts of this type were not provided in other modern foreign battleships and appear to be of little value for the amount of weight involved. Such an arrangement in the *Bismarck* proved inadequate when she was shelled by the *Rodney* and *King George V*. The upper citadel belt was repeatedly penetrated by shells of 152-406 mm at close ranges. Although the sloping armor afforded traditional protection for the vitals and provided an excellent thrust block for the lower edge of the main side belt, it would permit water to enter above the lower armor deck if the upper portion of the belt were holed and the ship listed to that side. Thicker belt armor would have precluded this, but the Germans reduced the thickness since they combined belt, armor deck, and armor bulkheads to protect the vital sections of the ship. Whether this was good practice is debatable; the *Bismarck* action demonstrated some limitations in an upper vertical side belt.

The horizontal armor protection was also inadequate. The Germans conducted no aircraft bombing tests to determine practical armor thickness for deck protection. However, they extrapolated results of gunfire tests on armor plate to simulate aircraft bomb attacks, and believed that a 50-mm armor deck would be sufficient to initiate fuze action to explode a bomb on the lower armor deck. The inclusion of a third armor deck in battleship studies after H-42 and the increased upper armor deck thickness (60–80 mm) reflect their awareness of weaknesses in their concept of horizontal protection. This had been underscored when the *Scharnhorst* was bombed at La Pallice, where several bombs penetrated the upper and lower armor decks. The additional armor would have been more effective if it had been used in thickening the upper or lower armor decks instead of using a third deck. Because of displacement limitations, the increases in the horizontal armor were not enough to repel all bombs then in use. Considering the large surfaces involved in the horizontal armor system, increased deck armor would have required the removal of compensating weights saved from other areas of the ship to allow its inclusion, and the entire design could have been substantially weakened.

A strong point in the later designs of the "H"-class ships was the torpedo-defense system, which was increased from 5.5 meters to 16.5 meters. It was believed that the *Bismarck* and the H-39 lacked sufficient protection against torpedo attack or underwater shell hits, as evidenced by the fact that a 356-mm shell hit on the *Bismarck* by the *Prince of Wales* seriously damaged the torpedo bulkhead outside the turbo-generator room and the transverse bulkhead in the adjoining boiler room. The inability to control the flooding in these spaces caused their loss to the ship, which then had her electric power reserve cut and the total power reduced through the loss of a boiler.

The use of floatplanes on battleships was viewed as an obsolete practice by most naval powers by 1942. In Germany, however, the Luftwaffe did not appreciate the airplane's potential as a fighting extension of sea power. This also led to a difference of opinion over the use of the aircraft complement on the *Graf Zeppelin* and ultimately was a factor in the decision not to complete the ship. German battleships were greatly dependent on air reconnaissance, and developments in radar did not limit this dependence. This required the use of valuable space that could have been employed for additional antiaircraft weapons.

After 1942 the Germans did not seriously consider building battleships with very large dimensions and displacements. None of the staff from the Construction Office wanted to be involved with the H-42, a project that they deemed militarily of doubtful merit, because of German shipyard and harbor restrictions, and technically unwarranted for a "final victory." Only the 1939 and 1941 designs could have been erected on a launching ways—the later giants would have had to be constructed in a dry dock. All the designs from H-42 through H-44 deserve at best to be called conceptual study projects. Since aircraft had changed naval strategy completely, the Germans felt it was necessary to study their effects on ship-protection requirements.

Whether these later studies could have produced a successful battleship design is a matter of speculation. The professional approach, calculations, and plans were not conclusive, as all work was based on extrapolated data from previous designs—the *Bismarck, Tirpitz,* and H-39. It is pure whimsy to believe that the "H"-class giants could have been built in Germany in 1943 and 1944. From a technical viewpoint, the construction of such super warships was entirely feasible, as evidenced by the supertankers built in Japan 20 and 30 years later, but the geography of Germany—the shallow seas that lay between the German ports and the Atlantic—made it completely impractical and illogical. H-44 was such an enormous ship that it would have been unable to enter any German port.

The 37-mm machine gun was a standard weapon in all German capital ships' antiaircraft batteries. With a maximum range of 6,750 meters and a shell weight of 0.75 kg, these guns had limited effectiveness against high-speed aircraft, despite their ability to elevate to 80 degrees. (Albrecht Schnarke Collection)

Name	Battleship "H"	Battleship "J"
Builder	Blohm and Voss, Hamburg (Hull No. BV525)	Deschimag, Bremen (Hull No. W981)
Laid down	15 July 1939	1 September 1939
Disposition	Steel work dismantled on 25 November 1941 and contract canceled on 29 August 1942.	Steel work dismantled on 25 November 1941 and contract canceled on 29 August 1942

Name	Battleship "K"	Battleship "L"
Builder	Deutsche Werke, Kiel (Hull No. K264)	Kriegsmarine Werft, Wilhelmshaven
Disposition	Canceled	Canceled

Name	Battleship "M"	Battleship "N"
Builder	Blohm and Voss, Hamburg	Deschimag, Bremen
Disposition	Canceled	Canceled

Displacement

Battleship "H"

49,704 tons (50,502 mt) Light ship
57,617 tons (58,543 mt) Designed
62,592 tons (63,596 mt) Full load

Dimensions

Battleship "H"

911.25′	(277.75 m)	Length overall
872.7′	(266.0 m)	Waterline length
121.4′	(37.0 m)	Beam
51.5′	(15.7 m)	Depth at side
32.9′	(10.024 m)	Draft at design waterline
36.6′	(11.2 m)	Draft at full load

Hull characteristics at design draft—Battleship "H"

Displacement	57,617 tons (58,543 mt)
Draft	32.9′ (10.024 m)
Freeboard (amidships)	18.6′ (5.676 m)
Depth (amidships)	51.5′ (15.7 m)
Block coefficient	0.561
Prismatic coefficient	0.576
Waterplane coefficient	0.666
Midship coefficient	0.973
Tons-per-inch immersion	168.7 (64.6 mt/cm)
Moment to trim one inch	7027.8 (847.02 mt-m/cm)
Metacentric height	10.50′ (3.2 m) @ 52,566 tons (53,400 mt) 13.7′ (4.2 m) at 57,617 tons (58,543 mt) 14.8′ (4.55 m) at 62,591 tons (63,596 mt)

Frames

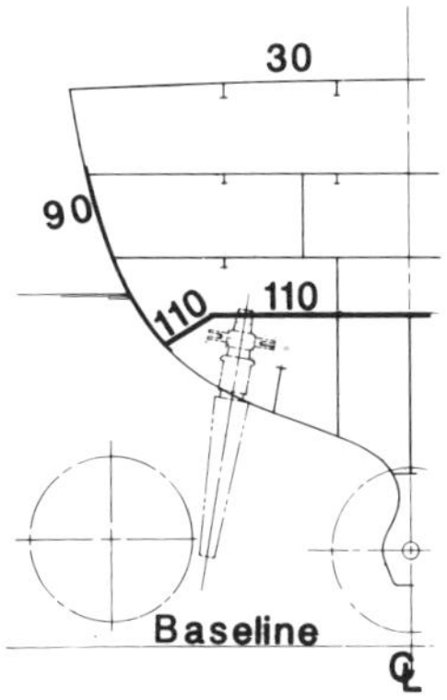

Frame 22.15

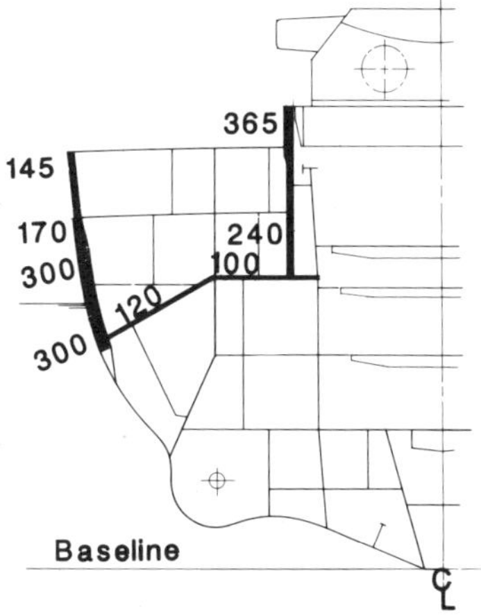

Frame 54.20

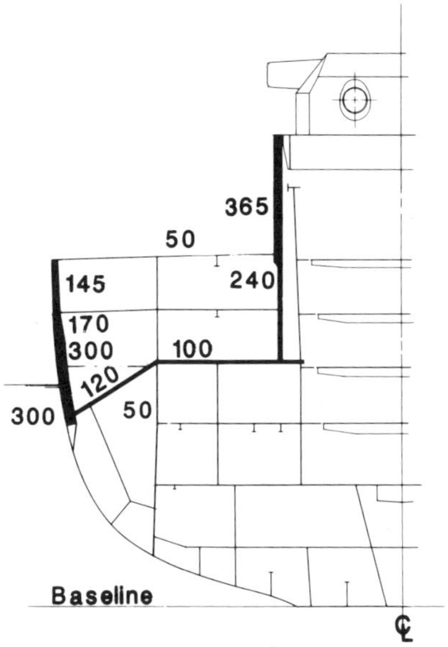

Frame 74.25

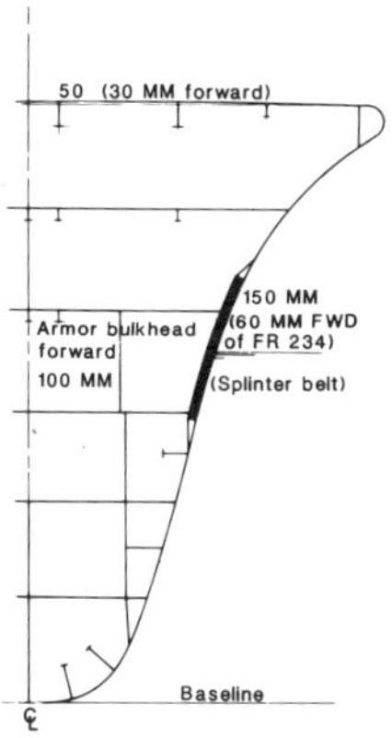

Frame 234

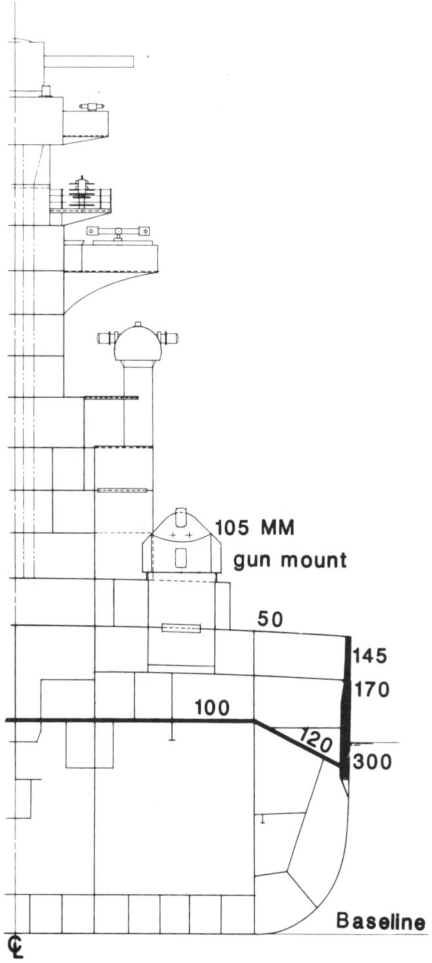

Frame 146.39

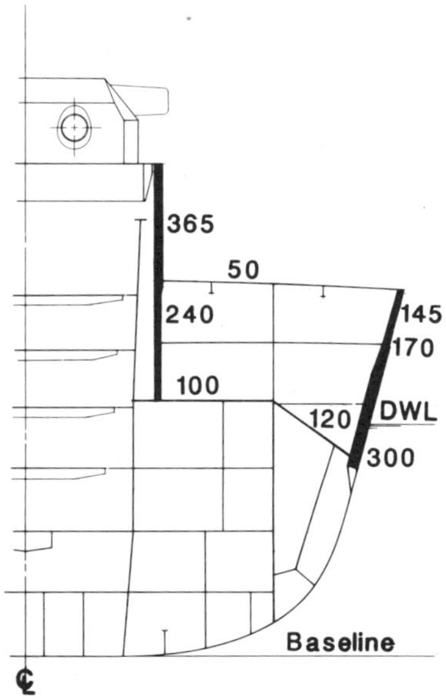

Frame 181.75

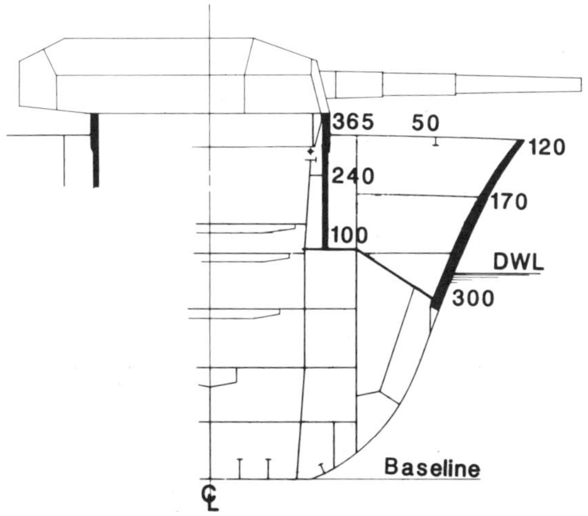

Frame 201.8

Midship Section

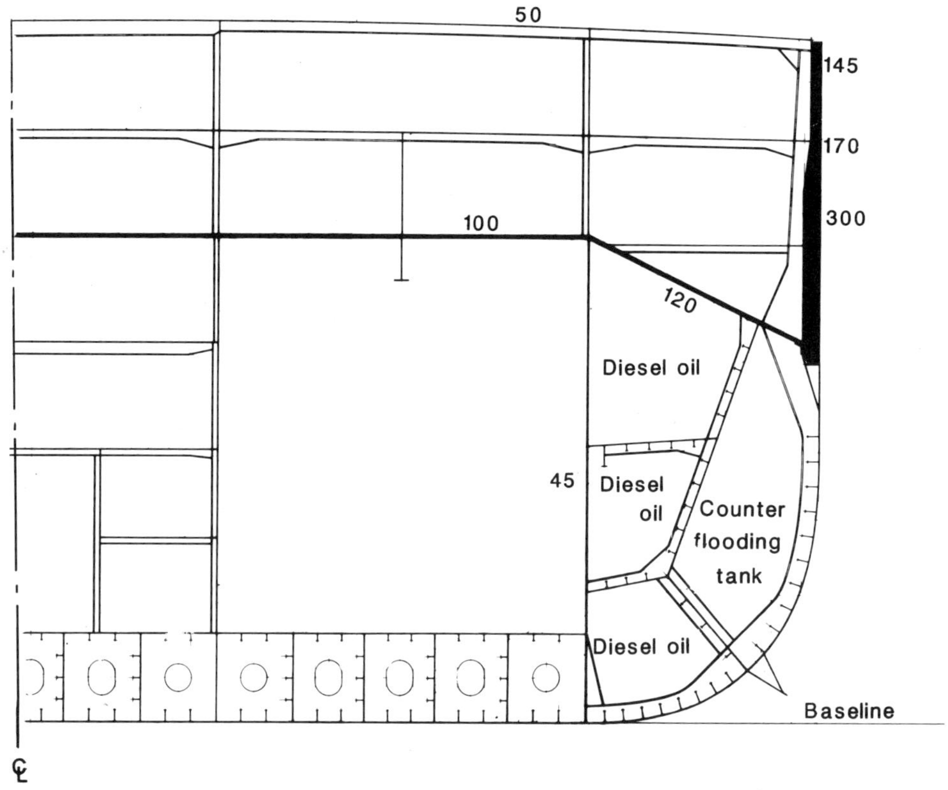

Armament
- Eight 16.0″/50-caliber (Mark 1934) (406 mm)
- Twelve 5.91″/55-caliber (Mark 1928) (150 mm)
- Sixteen 4.13″/65-caliber (Mark 1938) (105 mm)
- Sixteen 1.46″/83-caliber (Mark 1930) (37 mm)
- Twenty-four 0.79″ (20 mm)
- Six 21.0″ (533-mm) underwater torpedo tubes
- Four Arado-196 floatplanes

Armor protection		
Immunity zone	13,500 to 25,000 yards (12,434 to 22,860 m) against 16″/45-caliber gun (Mark 6) firing a 2,240-lb (1,016 kg) shell	
Amidships		
Belt armor		
Upper belt	5.91″ (150 mm)	
Lower belt	11.81″ (300 mm)	
Deck Armor	*Machinery Centerline*	*Magazine Centerline*
Upper deck	1.97″ (50 mm)	3.15″ (80 mm)
Lower armor deck	3.94″ (100 mm)	4.72″ (120 mm)
	5.91″ (150 mm)	7.87″ (200 mm)
Armor deck slopes—outboard	4.72″ (120 mm)	5.91″ (150 mm)

Machinery Schematic

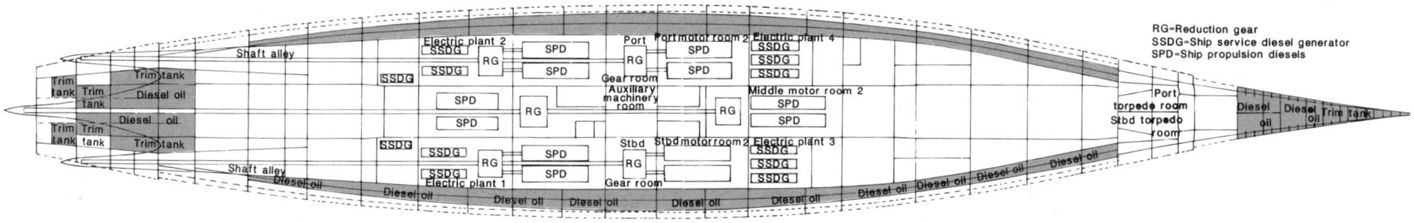

Turret armor	
face plates	15.75″ (400 mm)
sides	8.66″ (220 mm)
back plates	12.76″ (325 mm)
roof plates	
gun chamber	8.66″ (220 mm)
other	7.09″ (180 mm)
Barbette armor	
Centerline fwd	14.37″ (365 mm)
Sides	14.37″ (365 mm)
Centerline aft	14.37″ (365 mm)
Secondary gun protection	
face plates	3.94″ (100 mm)
sides	1.57″ (40 mm)
back plates	1.57″ (40 mm)
roof plates	1.57″ (40 mm)
barbettes	3.15″ (80 mm)
Conning tower armor	
centerline fwd	13.78″ (350 mm)
sides	13.78″ (350 mm)
centerline aft	13.78″ (350 mm)
roof plates	7.87″ (200 mm)
deck plates	3.15″ (80 mm)
communications tube	5.91″ (150 mm)
Underwater protection	
Designed resistance	550 pounds (250 kgs) of TNT
Side-protective system depth	16.89′ (5.15 m)
Side-protective system loading	Void-Liquid
Total bulkhead thickness	2.08″ (53 mm)
Underbottom protection	
Depth, one layer	5.58′ (1,700 mm)
Tank capacities	
Diesel oil	9,548 tons (9,700 mt)
Gasoline	6 tons (6 mt)
Reserve feed water	401 tons (407 mt)
Potable water	148 tons (150 mt)
Wash water	180 tons (183 mt)

Body Plan

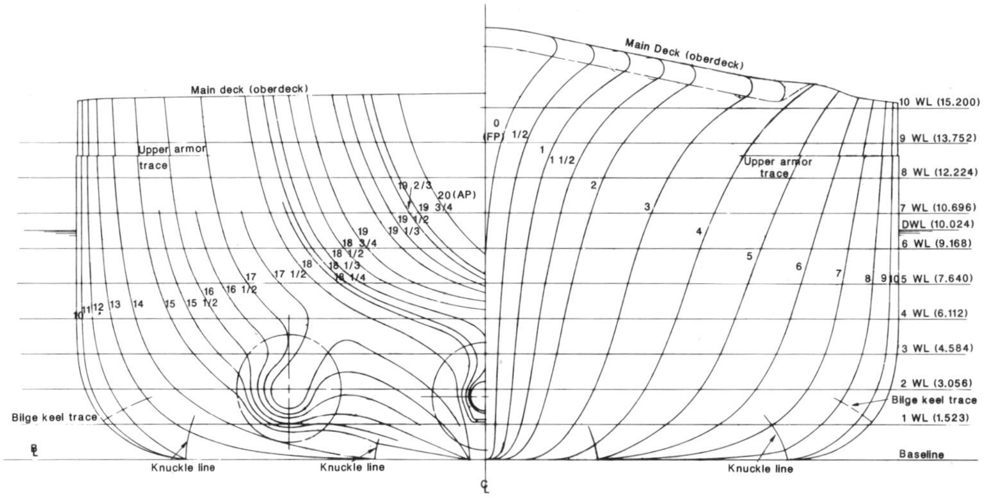

Machinery	
Engines	twelve M.A.N. MZ65/95 double-acting, two-stroke diesel engines with nine cylinders—maximum bhp 16,500 at 256 rpm
Shaft horsepower	162,743 (165,000 mhp)
Speed	30.4 knots
Nominal endurance	16,000 nautical miles at 19 knots
Generators	eight diesel generators at 920 kw each four diesel generators at 460 kw each total electric power—9,200 kw at 200 volts, DC
Propellers	three 15.8- (4.8-m-)diameter
Rudders	three
Miscellaneous	Weight Summary
Hull structure	11,979 tons (12,171 mt)—24.1%
Armor	20,619 tons (20,950 mt)—41.5%
Propulsion plant	4,828 tons (4,906 mt)— 9.7%
Electric plant	1,038 tons (1,055 mt)— 2.1%
Command & control	1,060 tons (1,077 mt)— 2.1%
Auxiliary systems	1,062 tons (1,079 mt)— 2.2%
Outfit & furnishings	1,053 tons (1,070 mt)— 2.1%
Armament	7,064 tons (7,177 mt)—14.2%
Margins	1,001 tons (1,017 mt)— 2.0%
Light ship	49,704 tons (50,502 mt)— 100%
Loads—complement	293 tons (298 mt)
stores	331 tons (336 mt)
ammunition	1,952 tons (1,984 mt)
aircraft	33 tons (183 mt)
potable water	150 tons (152 mt)
wash water	180 tons (183 mt)
reserve feed water	150 tons (152 mt)
diesel oil	9,548 tons (9,700 mt)
full-load	62,592 tons (63, 596 mt)

Outboard Profile (H-39 Des**position** (H-39 Design)

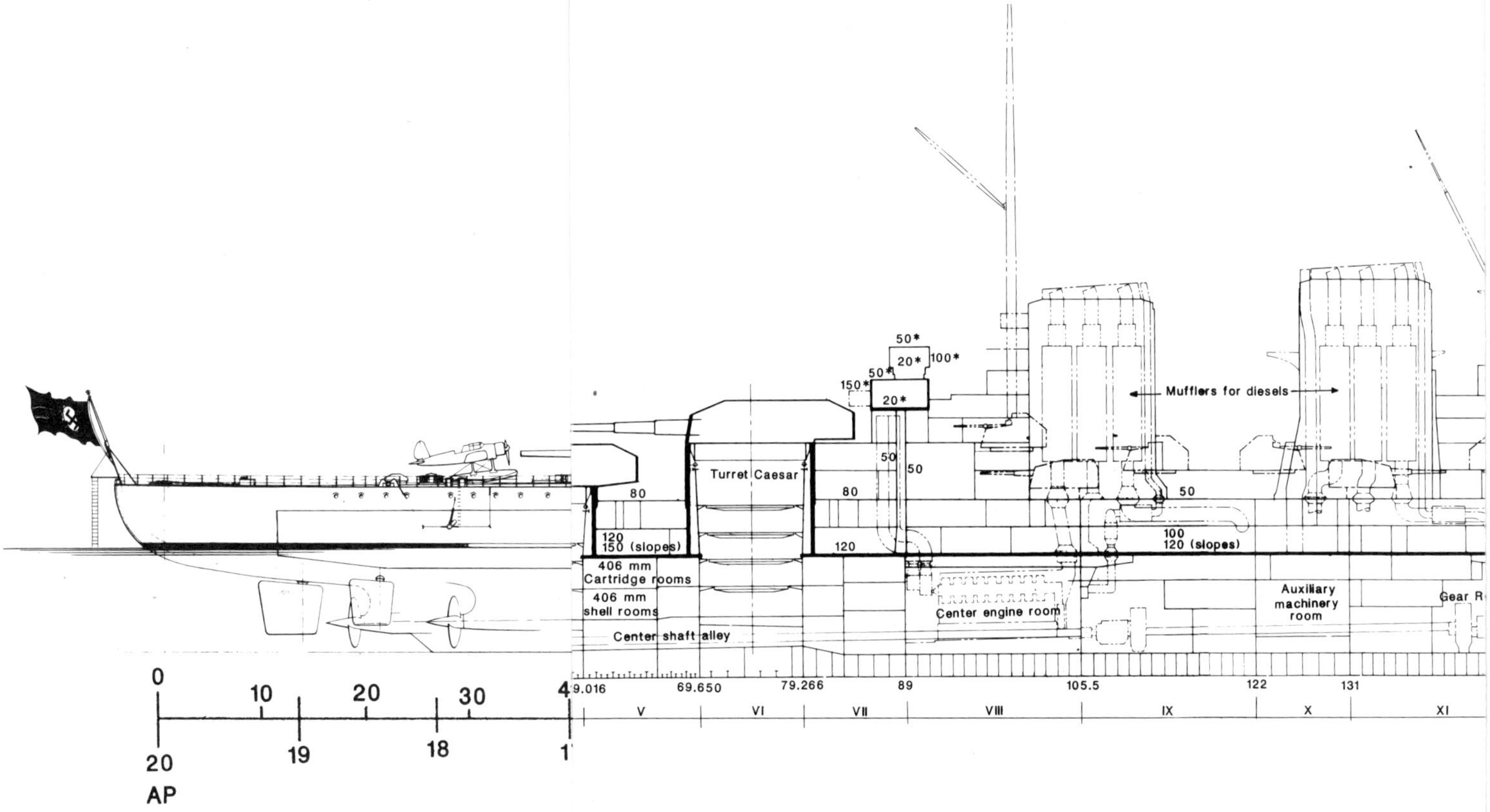

Topside View (1939 Desig

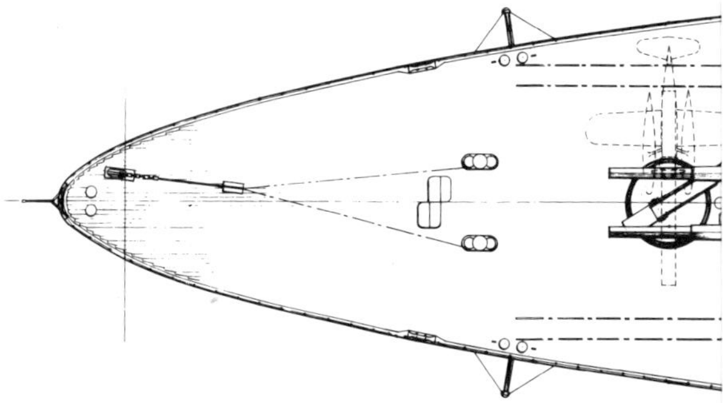

CHAPTER SIX

The "O"-Class Battlecruisers

During 1935–1937 the German Navy had an obvious numerical inferiority compared with the strength of other European navies, despite the construction of the *Scharnhorst*- and *Bismarck*-class ships. In 1937 the Naval Staff decided to increase warship strength by constructing fast armored ships of an improved *Deutschland*-type with a powerful armament and additional armor protection. German naval constructors had been pleased with the design of the armored ships, but two bomb hits on the *Deutschland* by Spanish Loyalist planes had indicated that her armor protection required strengthening.

Cruiser "P."

A new design featuring long endurance, higher speed, and more protection was sought. Twelve ships were to be built to this highly mobile armored ship design that would be suited for missions at sea and in coastal waters. Considering the numerical inferiority of the German Navy, the OKM expected only a limited measure of military success from such ships and only in diplomatic or commerce warfare situations. During the preliminary design, the new armored ship was designated as cruiser "P" (the "P" representing Panzer—armor). This design was derived from the same tactical considerations that determined the characteristics of the *Deutschland*-class armored ships. With the ability to outfight only light and heavy cruisers, it was mandatory that they have sufficient speed to avoid a possibly fatal gunnery duel with a battleship or battlecruiser. Also, high speed would enable them to select battle ranges when engaging slower and possibly more powerful ships. Since displacement was to be rigidly limited, high speed could only be realized by ruthlessly reducing armor protection, secondary armament, reserve buoyancy, and auxiliary systems. The requirements for cruiser "P" are tabulated in table 6-1.

Preliminary designs. Attempts to meet the above requirements led to 20 different designs during the ship-definition stage of the preliminary design, with displacements ranging from 20,000 to 30,000 metric tons and waterline lengths varying from 206 to 235 meters. The minimum length, using six 283-mm guns in triple turrets, was established at 217.5 meters, but to attain 34 knots it had to be increased to 229 meters to satisfy propulsion-plant requirements. The maximum displacement of 30,000 tons featured a ship with six 380-mm guns with light armor protection. Studies of the

The distinct familial resemblance of all the German capital-ship designs of the World War II era is apparent in this view of the "O"-class battlecruiser design concept. These ships, with their powerful armament of six 381-mm guns, high speed, and modest armor protection would have been remarkably similar to the handsome but fatally flawed British battlecruisers *Repulse* and *Renown*. (Painting by Richard Allison.)

TABLE 6-1

Design Requirements for Cruiser "P"

Displacement (designed)	19,679 tons (20,000 mt)
Armament	
Main	six 11.1" (283-mm) guns in either twin or triple turrets
Secondary	four 5.9" (150-mm) guns in twin turrets
Antiaircraft	eight 4.1" (105-mm) guns in twin mounts
Speed	34–35 knots
Endurance	15,000 miles at 19 knots
Protection	4.72" (120-mm) side belt with a 2.36" (60-mm) belt carried to the bow and a 4.33" (110-mm) sloping belt carried to the stern. See figure 6-1.

propulsion-plant arrangement revealed that the beam had to be a minimum of 25 meters, either for three or four shafts. If diesels were used, it had to be increased to 27 meters, and this would have required a plant 91.5 meters long and a side-protection system of 4 meters maximum width at half the designed draft amidships. The increased beam necessitated a greater length in order to maintain the 34-knot speed. However, this also required a longer armor box and larger displacement, which still resulted in some loss of speed. It was concluded that the speed requirements could not be met with an all-diesel drive with a displacement of 20,000 tons. To satisfy the endurance and speed requirements would lead to designs of greater size.

Other difficulties were encountered in the armor protection of these ships. With the 203-mm guns of Allied heavy cruisers in mind, the following armor requirements were laid down by the Naval Weapons Office:

• Side belt to armor deck	5.71"	(145-mm)	face-hardened armor
• Slopes	4.33"	(110-mm)	noncemented armor
• Side belt above armor deck	2.36"	(60-mm)	noncemented armor
• Armor deck over magazines	3.94"	(100-mm)	noncemented armor
• Other areas of armor deck	2.36"	(60-mm)	noncemented armor

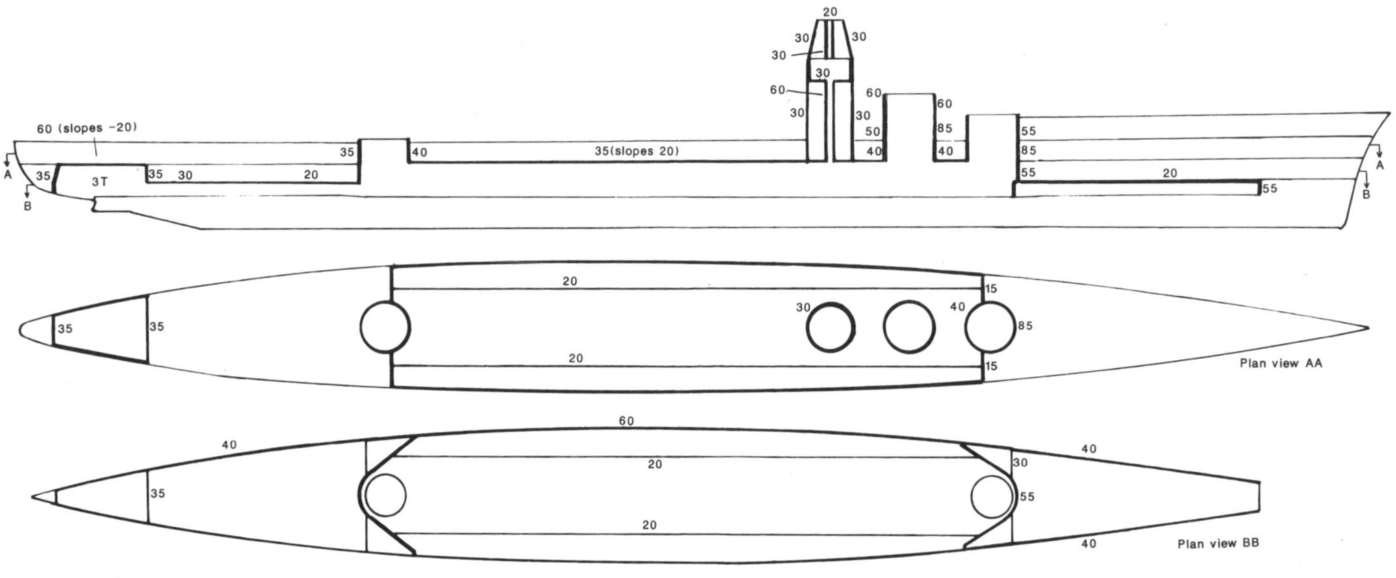

Figure 6-1. Cruiser "P" protection.

Such armor thicknesses were not considered realistic because in any engagement against a convoy, battle with heavy cruisers would be inevitable. It was believed that these proposed thicknesses were insufficient, and this was later substantiated by the damage the *Graf Spee* sustained in the River Plate Battle when she sustained 41 hits from shells of 203 mm and 152 mm. Also, as the vessel size increased, the required armor protection increased.

There were other difficulties involved in this design. To build 12 ships in three years would have required the full large-ship capacity of shipyards already heavily engaged in rebuilding the navy. There were problems with the projected main battery; foreign navies were not mounting 283-mm guns in any ships under order or projected for design; the smallest guns in use were the 305-mm weapons on Italian and American battleships of the World War I era. Tests had conclusively shown that the 283-mm/55-caliber gun, Model 1934, was far less effective than the 380-mm/47-caliber gun. Development of a 305-mm gun was considered, but the design and testing of such a gun would have delayed the project. The decisive factor in the choice of a main armament came in 1939 when Hitler abrogated the terms of the Anglo-German Naval Treaty of 1935. The battlecruisers *Gneisenau* and *Scharnhorst* had a number of modifications made during their construction, including one that would permit the replacement of the 283-mm triple turrets by 380-mm twin turrets. These 380-mm guns and their turrets had already been contracted for. Since the modification of the *Scharnhorst*-class battlecruisers would have required a prolonged layup in a shipyard, the strained political situation in 1939 made such an occurrence unlikely. The turrets and guns were now available, however, and an effort was made to find some use for them. Therefore, it was proposed that the main armament of cruiser "P" be changed to 380 mm, and this was the hour of birth of battlecruiser "O."

Design of Battlecruiser "O."

There were other mitigating circumstances for a change to a larger ship, as the requirements for cruiser "P" led to unsatisfactory solutions. The fuel consumption for the speeds projected was excessive, and the propeller loadings were too high. Propeller cavitation was feared. The weapons and the armor protection were out of proportion to the size of the ship desired. The reworking of the design had dragged on with little enthusiasm in 1937–38. Design work for battleship "H" began during the same period, and this took precedence over cruiser "P." It was not until 28 April 1938, when Hitler made known his views on the 1935 Anglo-German Naval Treaty and the danger of war with the United Kingdom increased, that interest in the battlecruiser project was revived. It was at this time that the OKM decided it wanted a larger ship with a main battery of 380-mm/47-caliber guns.

Hitler, feeling that all attempts at a mutual understanding with the United Kingdom had failed, called a meeting with the Naval Staff and Admiral Raeder. He desired a strong navy to menace the maritime trade of Great Britain, hopefully inducing the British to prefer an alliance to war. Hence, the concept of an armored ship with battleship guns came into being. The basic strategy was that such ships could attack convoys and force Britain and her allies to augment their convoy escorts with capital ships. This would seriously drain the resources of the Royal Navy. In the meantime, two German task forces, each consisting of three "H"-class battleships, a carrier, cruisers, and destroyers, would seek to engage convoy screens while a fast battleship with little armor protection (a battlecruiser in every sense of the word) would work over the convoy, assisted by U-boats. This was the basic rationale of the "Z"-plan, in which

TABLE 6-2
Design Requirements for Battlecruiser "O"

Displacement (designed)	29,518 tons (30,000 mt)
Armament	
Main	six 15" (380-mm)/47-caliber guns in three twin turrets
Secondary and antiaircraft	dual-purpose type
Speed	34 knots
Endurance	15,000 miles at 19 knots
Armor protection	to be based on the 203-mm guns of Allied heavy cruisers

three of these battlecruisers were projected. The characteristics of battlecruiser "O" are tabulated in table 6-2. All work on cruiser "P" terminated, and a complete changeover to battlecruiser "O" was made in mid-1939.

The armor-protection requirement, the use of 380-mm guns, and the requirement for high speed resulted in a very unfavorable relationship between the caliber of the main battery, the displacement, and the armor protection. With a displacement as large as 30,000 tons, a number of critics suggested that it would have been much better to concentrate on armor protection and particularly on deck protection. The displacement was limited to 30,000 tons to give the shortest possible construction time commensurate with the maximum thickness of armor specified. The estimated building period was three years instead of the four or more years required for a battleship. Model tests at the Hamburg Model Basin determined that a hull length of 250 meters and a beam of 30 meters would satisfy the requirement for a speed of 34 knots. Studies of the propulsion plant confirmed that an all-diesel drive was not feasible. Use of steam power for the centerline shaft shortened the armored citadel by 9 meters and permitted a beam reduction of 3.5 meters in the after portion of the armor box, where the hull volume began to decrease rapidly. Therefore, a mixed propulsion plant of diesels and a steam turbine were finally agreed upon.

Unfortunately for the German Navy, no successful dual-purpose gun had yet been developed. German ordnance experts also believed that giving a gun both air- and surface-defense capabilites would substantially diminish its effectiveness in surface actions. Thus, the medium-caliber battery was once again divided between 150-mm and 105-mm guns. This type of armament had been used in all German battleship designs for that period, but it was an unfortunate choice for the "O"-class battlecruisers because of their weight problem, lack of space, and limited armor protection. Only the mixed propulsion plant made it possible to offset the weight increase caused by the split-battery requirements.

By September 1939 a revised contract design had been completed, and the characteristics for these battlecruisers are tabulated in table 6-3.

The three ships, designated as "O," "P," and "Q," were to be built at the Deutsche Werke, Kiel; the Naval Dockyard, Wilhelmshaven; and Germania Werft, Kiel. In 1940, after the lines and ship arrangements had been completed, project drawings were approved by Adolf Hitler and Admiral Erich Raeder. Working drawings had not been prepared, but all work was accomplished under the direction of Admiral Werner Fuchs through the SN Service Division at the Deutsche Werke Shipyard in Kiel. No actual construction work was ever accomplished, aside from the initial procurement of materials and the issuance of some procurement orders.

TABLE 6-3

Design Requirements for Battlecruiser "O" (Revised September 1939)

Design displacement	31,142 tons (31,652 mt)
Full-load displacement	35,367 tons (35,945 mt)
Waterline length	807.1′ (246.0 m)
Waterline beam	98.43′ (30.0 m)
Depth at side	47.08′ (14.4 m)
Draft (design)	26.25′ (8.0 m)
Draft (full load)	28.87′ (8.8 m)
Armament	six 15″/47 (380 mm) paired
	six 5.9″/48 (150 mm) paired
	eight 4.1″/65 (105 mm) paired
	eight 1.46″/83 (37 mm) paired
	twenty 0.79″ (20 mm) machine guns in five quadruple mounts
	twelve 21.0″ (533-mm) torpedo tubes in four fixed triple mounts
	four Arado-196 floatplanes
Speed	33.5 knots at designed displacement
Shaft horsepower	173,600 (176,000)
Protection	3.54″ (90 mm) upper belt
	7.48″ (190 mm) main side belt
	1.18″ (30 mm) upper deck
	2.36″ (60 mm) armor deck

Armament. The principal characteristics of the guns for the "O"-class battlecruisers are tabulated in table 6-4.

Main Battery. The 380-mm guns, Model 1934, were to be the same as those used on the *Bismarck* and *Tirpitz.* This heavy armament meant that only capital ships could successfully engage these ships, and they would have to be of superior speed. The only possible Allied opponents, barring speed-reducing damage, would have been the French battlecruisers *Dunkerque* and *Strasbourg,* French battleships of the *Richelieu* class, and the British battlecruisers *Hood, Renown,* and *Repulse.* The *Bismarck-Hood* battle proved that the British battlecruiser's weak armor protection was vulnerable to the

TABLE 6-4

Gun Characteristics

Gun	15″ (380 mm)/47	5.9″ (150 mm)/48	
Shell weight (lbs)	1,764 (800 kgs)	99.8 (45.3 kgs)	
Muzzle velocity (fps)	2,690 (820 mps)	2,871 (875 mps)	
Maximum range (yds)	39,590 (36,520 m)	25,153 (23,000 m)	
Maximum elevation	30°	65°	
Gun	4.1″ (105 mm)/65	1.46″ (37 mm)/65	0.79″ (20 mm)/65
Shell weight (lbs)	33.3 (15.1 kgs)	1.64 (0.745 kgs)	0.291 (0.132 kg)
Muzzle velocity (fps)	2,952 (900 mps)	3,281 (1,000 mps)	2,952 (900 mps)
Maximum range (yds)	19,357 (17,700 m)	7,382 (6,750 m)	5,249 (4,800 m)
Maximum elevation	80°	80°	90°

The battlecruisers "O," "P," and "Q" were never laid down. The ships were to have been armed with 380-mm guns of the type mounted on the *Bismarck* and *Tirpitz*. This photo shows one of the 380-mm guns being lifted from the dock to the ship during the outfitting of the *Tirpitz* in Wilhelmshaven. (Courtesy Albrecht Schnarke.)

German 380-mm gun. The German ships also had a slightly more powerful main-battery gun than the two French battlecruisers, although the armor protection of the *Strasbourg* would have been sufficient against the German 380-mm shell.

Secondary battery. As the Rheinmetall concern and the Naval Weapons Office were unable to agree upon a satisfactory dual-purpose gun, and since the German Navy was reluctant to give up the 150-mm gun that was deemed necessary for defense against destroyers, the "O"-class battlecruisers had to do without a good dual-purpose gun. The elevation of the 150-mm guns, however, was to be increased from 40 degrees in the mounts used in the *Bismarck*, *Scharnhorst*, and "H" classes to 65 degrees in the mounts to be used in these battlecruisers. This meant that the 150-mm gun would be able to fire a shell against aircraft, and inevitably such a shell would have been provided some sort of fuze-setting capability. The rate of fire of the turret would still have been 6 rounds per minute. Two 150-mm twin turrets were located on the sides between the bridge and the forward funnel; the third was to be located aft, superimposed above turret Caesar. The forward turrets had a limited arc of fire of 170 degrees, compared to 330 degrees for the after turret. The use of only six such guns in the secondary battery, four guns bearing on either side, raises some doubts as to whether these ships could have defended themselves adequately against cruiser or destroyer attacks.

Long-range air defense was to be provided by eight 105-mm/65-caliber guns, stabilized as in other such mounts on German capital ships, but with limited training arcs, which reduced their effectiveness. Furthermore, these guns, and two high-angle directors to control them, were inadequate to oppose the type of air attacks made on the *Bismarck* and *Tirpitz*. The forward mounts, abreast of the forward funnel, had a training arc of 180 degrees with five degrees of the stern blocked out by superstructure. The two aft mounts, just slightly abaft the after funnel, had a training arc of 170 degrees, with 20 degrees of the forward quadrant blocked out by top hamper. A greater number of guns was not possible because of space and weight limitations.

Antiaircraft battery. Short- to medium-range antiaircraft protection was to be provided by eight 37-mm/83-caliber and twenty 20-mm machine guns, grouped around the funnels in much the same manner as the 105-mm guns. This light antiaircraft battery was totally inadequate, as war experiences on German capital ships would prove. Even the British *Renown* and *Repulse*, which were comparable to these ships in size, but built 20 years earlier, had superior antiaircraft protection. The designers of the light antiaircraft armament were handicapped by the availability of only single- and twin-mounted machine guns in 1939–1940. A quadruple-mounted 20-mm machine gun with a high rate of fire was under development at the time of design of this class, but its mounting was not stabilized. Another handicap to German capital ships' antiaircraft batteries was the competition for space in the superstructure between the guns, boats, and aircraft. Both the latter used up a considerable amount of space in these battlecruisers—one reason why the gun power of these ships was so limited.

Torpedo tubes. Although these ships were primarily commerce raiders, their initial design requirements did not include a torpedo armament. There was much discussion about arming these ships after the initial design was reviewed by the German Naval Staff. If they were to be raiders of merchant ships, torpedoes would give them more offensive power against destroyers and cruisers, especially since the secondary battery was rather weak. More importantly, torpedoes would sink a merchant ship quicker and save precious gun ammunition. The final "O"-class battlecruiser design featured twelve 533-mm triple torpedo tubes in fixed amidships deck mounts, with a torpedo magazine located between the port and starboard torpedo rooms. One of the exploratory preliminary designs had four fixed torpedo tubes on the main deck; however, the need for additional boat stowage forced the final torpedo-tube arrangement. Underwater torpedo tubes of the type used in battleship "H" were considered, but this arrangement would have required more space in the hull and a certain amount of additional protection, which was at a premium in these ships. For example, because space in them was so limited, the auxiliary boilers had to be located on the lower armor deck since there was not sufficient space for them in the machinery spaces. In such a volume- and weight-constrained design, below-deck torpedo tubes such as those in battleship "H" were simply impractical.

Aircraft. Each ship was to carry four Arado-196 seaplanes stowed in hangars amidships. The hangars had splinter protection of 14-mm armor against strafing attacks. A fixed catapult of the type used on the *Bismarck* and *Tirpitz* was mounted between the three hangars, and the aircraft were hoisted onto the catapult by large cranes. These cranes served both the aircraft and boats. In the final design, a boat platform was mounted over the catapult with sufficient headroom for flight operations. Two aircraft were stowed in a large hangar aft of the forward funnel, while single-plane hangars were provided abaft the catapult. The use of aircraft on battleships or battlecruisers has been questioned, but because of the lack of German aircraft carriers to search for convoys and provide information on convoy composition and escort strength, they were mandatory in these ships. Several arrangements of the aircraft facility were studied, but the boat complement finally dictated the arrangement shown on the outboard profile and overhead view.

Protection. The most noteworthy feature of these ships was the lack of adequate armor protection.Since they were designed to engage heavy cruisers screening convoys, it was felt that the armor protection should only be slightly in excess of that indicated in the original requirements for cruiser "P." Thus, normal capital-ship armor protection was not provided so that speed requirements and displacement limitations could be met.

The main side belt was vertical and divided into two thicknesses, as in previous German designs. The lower belt was to be a uniform 190-mm thickness throughout the citadel. The upper, or citadel, belt was to have been 90-mm thick.

Deck armor. Horizontal armor protection was less than that used in previous types. Deck-armor thicknesses are tabulated in table 6-5. Outboard of the torpedo bulkhead, the lower armor deck was sloped to provide reinforcement against shell hits near the waterline. Armor 80-mm thick was used on this part of the armor deck throughout the citadel. Aft, the lower armor deck was continued to protect the steering-gear room and shafts. As armor protection was limited, the outboard portion of the after armor deck was knuckled and made 110-mm thick.

TABLE 6-5
Deck-Armor Thicknesses

	Magazines	*Machinery*
Upper deck	1.18" (30 mm)	1.18" (30 mm)
Armor deck	2.36" (60 mm)	2.36" (60 mm)
	3.54" (90 mm)	3.54" (90 mm)

Conning tower. German designers still favored the concept of heavily protecting the main and secondary conning positions, even though armor on these ships, in general, was to be very limited. The forward or main conning tower was to have armor 200-mm-thick on the sides, 80-mm thick on the roof, and 30-mm thick on the deck. The main access tube was of 80-mm-thick armor. The only other armor in the superstructure was in the fire-control station on the bridge tower, with 30-mm plate on the sides and decks and 20-mm plate on the roof. The secondary conn had 50-mm sides and a 30-mm roof.

Armored bulkheads. There were to be three transverse armored bulkheads, as in the *Bismarck* and *Scharnhorst* classes, and a longitudinal 14-mm armored bulkhead between the diesel propulsion plants to provide additional protection for the machinery spaces. Such bulkheads would contain fragment damage from a detonating shell within the spaces. Transverse bulkhead armor at the extremities of the citadel ranged from 25 mm to 100 mm in thickness. Lesser thicknesses were used between the upper and lower armor decks where only splinter protection was required. Below the main armor deck there was a different arrangement. The forward bulkhead was 100-mm thick over three decks adjacent to the main-battery magazines; outboard of the torpedo bulkhead it was 80-mm thick, and because of the need to conserve weight and the decreased risk of penetration there, the lowest sections were 30-mm thick. The after bulkhead had 100-mm armor over one deck height, concentrated inboard between the torpedo bulkheads. The armor was 25-mm thick at the lower armor deck and 10-mm thick outboard of the torpedo bulkhead. This thin protection was reinforced by a local bulkhead, 30-mm thick, located behind the two main bulkheads. The third transverse armored bulkhead, aft of the steering-gear room, was 30-mm thick for one deck height above the lower armor deck and 80-mm thick below that deck.

With the limited displacement and a considerable unprotected area beyond the armored bulkheads, the main transverse watertight bulkheads were extended to the upper deck. This would protect the ship against excessive trim by the bow and consequent reduction of speed caused by shell hits forward. For the same reasons, 60-mm non-cemented armor plate was substituted for ordinary steel shell plating to form a splinter belt forward of the citadel. Such an arrangement, excellent protection against splinter damage, would only localize the effect of a large-caliber shell hit.

Turret armor. The heaviest armor on the ship was on the main turrets and barbettes, but it was far less than that used in the *Bismarck* and *Tirpitz.* In all German capital ships before World War II, main-battery turrets were heavily protected. The "O"-class battlecruisers were to have turret face plates 220-mm thick, sides and backwalls 180-mm thick, and a roof 50-mm thick—indications that they were highly vulnerable to guns of caliber equal to their own or to armor-piercing bombs dropped in a level-bombing attack from an altitude much lower than that to which either the *Scharnhorst-* or *Bismarck-*class ships were vulnerable.

Barbette armor was arranged in two courses, as in other German battleships of the period. Armor exposed above the upper armor deck was 180-mm thick; between the two armor decks it was 145-mm thick. The one exception was the after periphery of turret Caesar where the armor was only 80-mm thick—this because of the turret's proximity to the after armored bulkhead and also because of the low probability of sustaining a direct hit in the area. All exposed surfaces were face-hardened armor steel; behind the main armor belt, non-cemented armor of reduced thickness was used. In an effort to prevent magazine fires such as the one on the battlecruiser *Seydlitz* during the Battle of Dogger Bank in World War I, the Germans had done much research into the design of barbette structures. The ammunition-handling system in the "O"-class ships had to be carefully designed to prevent a magazine explosion resulting from a shell hit in the thinly protected barbette or turret.

The secondary turrets had 14-mm face plates, sides, and roofs, adequate only for splinter protection. Turret protection for the 150-mm guns was reduced drastically in order to provide armor for other parts of the ship. There was no armor protection for the barbettes and shell-handling equipment. Other turret machinery was exposed to the possibility of direct hits from shells of all calibers and damage by splinters.

Side-protection system. One important feature of these ships—the torpedo-protection system—was given priority consideration. Although it was similar in arrangment to systems employed in earlier ships, it was not as deep. The difficulty sprung from the fine lines, particularly those at the stern end of the system. With an emphasis on speed and a block coefficient of 0.512 (compared to the 0.554 of *Scharnhorst*), the protection at the ends of the citadel was less than the battleships of the *Bismarck* and "H" classes. Thus, it was possible for a British aerial torpedo to seriously damage the ship, as happened with the *Gneisenau* at Brest. The important features of the underwater-protection system in the "O"-class battlecruisers are summarized in table 6-6.

The torpedo bulkhead, constructed of 45-mm non-cemented armor plate (60 mm in the area of the main-battery magazines), was so arranged to deflect elastically and plastically to absorb some of the explosion's energy. The torpedo bulkhead would have featured the all-welded construction that was introduced in battleship "H".

TABLE 6-6

Characteristics of the Side-Protection System

Location	*Depth at Half Draft*	*Total bulkhead thickness*	*Torpedo bulkhead thickness*
Turret Anton	5.91′ (1.8 m)	2.68″ (68 mm)	2.36″ (60 mm)
Turret Bruno	8.53′ (2.6 m)	2.68″ (68 mm)	2.36″ (60 mm)
Amidships	13.78′ (4.2 m)	2.01″ (51 mm)	1.77″ (45 mm)
Turret Caesar	3.28′ (1.0 m)	2.68″ (68 mm)	2.36″ (60 mm)

Stability and subdivision. As these ships lacked good armor protection and a torpedo-defense system, it was important that careful attention be given to their stability and subdivision. Like other German capital ships of that period, they were designed to sustain the loss of buoyancy in two compartments amidships and in three at the ends without sinking. Watertight bulkheads, located in the orthodox manner and determined by the placement of the main-battery magazines, divided the ship into 18 watertight compartments. There were 21 compartments in the *Scharnhorst*, 22 in the *Bismarck*, and 21 in battleship "H." Since these vessels had a smaller number of watertight compartments, the Germans wanted to avoid large lists and trims. To increase the reserve buoyancy, German naval constructors introduced the use of "Ebonite Mousse" for the first time in their warship designs. This material was a French-developed vulcanized foam rubber that had a specific gravity of 0.02 to 0.05 and was to be installed in the void spaces within the region of the armor box. Because of weight problems, no installation of "Ebonite Mousse" was planned for the bow and stern portions outside the armored citadel. The effect upon any undesirable trim change was comparatively small, due to the large longitudinal stability of the exceedingly long vessel. If the ship was intact, the use of this water-exclusion material meant an additional weight, which in the case of battlecruiser "O" amounted to 150 to 200 tons. In a damaged ship, as soon as these void spaces filled with "Ebonite Mousse" were flooded, the water-exclusion material offset the loss in displacement that results from the damage sustained. The use of this water-exclusion material also improved the resistance of the underwater-protection system, so that these battlecruisers had a slightly more effective torpedo-defense system than the *Scharnhorst* and *Genisenau.*

The stability characteristics of the "O"-class battlecruisers, as described by their metacentric heights, are tabulated in table 6-7.

TABLE 6-7

Stability Characteristics

Condition	*Displacement Tons*	*(mt)*	*Metacentric height (GM) Feet*	*(m)*
Designed displacement	31,143	(31,652)	10.47	(3.19)
Full-load displacement	35,367	(35,945)	11.91	(3.63)

Radar. Radar was not specified in the contract specifications; however, there is little doubt that some search and fire-control radar systems would have been fitted when these ships went into service. Their radar equipment would have been similar to that of the *Tirpitz* in 1943 or 1944.

Propulsion Plant. Speed was a major factor in the design of the "O"-class battlecruisers. To attain very high speeds, the armament and protection were limited to accommodate the propulsion plant. The hull form and propulsion equipment were developed to produce a top speed of 34 knots and a cruising speed of 19 knots. With the diesel plant at full power, the ship could make a speed of 25 knots. The diesel-steam propulsion plant had a low fuel-consumption rate; accordingly, a large cruising radius of 14,000 to 15,000 nautical miles was anticipated at the full displacement and speed of 19 knots. Such a large radius was obtained with only the diesels in operation.

Diesel engines. The diesels were double-acting 24-cylinder units arranged four to a shaft, with an auxiliary 5-cylinder engine of the same bore and stroke providing intake air for the main engines and driving the water, oil, and lube-oil pumps. The result was a V-type engine that combined features of both the main and auxiliary units. Eight such engines were required for each ship, and they were to be designed and built by Maschinenfabrik Augsburg Nürnberg, AG (M.A.N.). These engines were also known as the "pocket battleship engines" and had been in use as experimental engines as early as 1940. Based upon a continuous rating of 13,750 mhp, they had a weight of 12.5

The "O"-class battlecruisers would have had a mixed type of propulsion plant, with a centerline shaft driven by a steam-turbine installation and each wing propeller shaft powered by four M.A.N. double-acting 24-cylinder V-type diesel engines, one of which is shown in this view (Courtesy M.A.N.)

kilograms per horsepower and had a maximum rating of 15,300 mhp, with an engine speed of 450 rpm. An interesting feature of the engines was that they could develop 70 percent of their full power in reverse. A group of four were to be connected to a reduction gear for the outboard shafts, and Fottinger hydraulic couplings were arranged between the gear and each of the engines. This would counter the fluctuations in the rpms of the four diesels when more than one were connected to the shaft. Although the entire project was suspended when war was declared, M.A.N. made a successful test run of the prototype engine for almost 300 hours.

Steam plant. The centerline shaft was powered by a steam turbine and four high-temperature-and-pressure boilers. All four boilers were contained in a single fireroom similar to the arrangement used in the *Scharnhorst*-class battlecruisers. The use of a single fireroom was a weight-saving measure, but a risky arrangement in a ship with very limited protection. The boilers were to be produced by the Wagner Hochdruck Dampfturbinen-Gesselschaft and operated at a working pressure of 55 kilograms per square centimeter with superheat at 920 degrees centigrade. The shaft was rated at 55,000 metric horsepower normal and 60,000 metric horsepower maximum. The turbine arrangement was one used in the *Scharnhorst* and *Bismarck* classes. There were to be three cylinders, with the high pressure and intermediate pressure located aft of the single reduction gear that drove the side pinions, and the double-flow low pressure located ahead of the gear driving the center or top pinion. Both the high pressure and the intermediate pressure were to consist of two-row Curtiss first stages, but the rest of the steam path was reaction. The backing turbine consisted of a two-row Curtiss high-pressure element in a separate casing located abaft the reduction gear driving the center pinion, and exhausting to a four-row reaction element in the after end of the low-pressure ahead casing. The turbines were to be built by Satz, Brown, and Boveri. To expedite the construction, the turbines for the three vessels would have come from those specified for aircraft carrier "B," because these were in production. The aircraft carrier was not considered as important to the war effort as were these three battlecruisers.

Electric plant. The maximum electrical capacity was established at 7,360 kw, to be produced by two turbo-generators of 460 kw and six diesel generators of 920 kw. There were four electric plants, two forward of the diesel plant and two abreast of the boiler room. The ships' light and power was to be 220-volt direct current; however, a 110-volt AC circuit was also to be installed for that equipment requiring alternating current. The capacity of the AC system was 460 kw supplied by a diesel generator forward of the diesel plant. There was no casualty power system, as German naval practice required 100 percent backup power supply. The normal shipboard current was 2,760 kw during battle conditions.

Hull Characteristics.

There were three rudders in the "O"-class battlecruisers, with each rudder located in the propeller race and a separation of 9.45 meters between the outboard and centerline rudders. The outboard rudders were slightly smaller and located slightly forward of the centerline rudder. The desirability of a three-rudder arrangement was later shown in the loss of the *Bismarck*, although there is a possibility that the "O"-class rudder arrangement may have been derived from the Italian *Vittorio Veneto*-class ships, which also had three rudders.

Welding. Welding would have been extensively used in these ships, as German naval constructors recognized that welded structure saved considerable weight, yet yielded connections that were more efficient and stronger than a comparable riveted joint. The hull was to have been 90 to 95 percent welded, and because of the short building period projected, the struts, bossings, and stern post were also to have been fabricated weldments.

Structure. Because of the reduced plating thicknesses in the forecastle and stern, special supports had to be provided under the upper deck so that it could effectively resist the blast pressure of the main-battery guns.

Summary. The "O"-class battlecruisers would have had a powerful main battery, and a rather unusual propulsion plant of great power, but their weak protection made them obsolete even during the design period. The poor choice of features was based on concepts that would have been justified 25 years earlier; fortunately, the ships were never built.

It seems that the decision to design and build such ships ultimately rested on Adolf Hitler, who was always enthusiastic about large ships. He saw them as a formidable foe to surface escorts of Allied convoys to Europe. Even in Germany, however, the "OPQ" battlecruisers were considered extremely poor concepts. In certain professional circles this project became known as "*Ohne Panzer Quatsch*"—colloquial German literally translated as "without armor nonsense"—because of their insufficient armor protection, shallow torpedo-defense system, and limited armament. Their subdivision was more carefully designed than in most German warships of the period as a result of the limited protection. The additional use of a water-exclusion material in the wing voids was a positive approach in German warship design and diminished the effect, to some extent, of the loss of armor and underwater protection.

The primary mission of these ships was to seek out and destroy merchant ships. To accomplish this they would have to engage heavy cruisers, for which it was believed that at least 283-mm guns were necessary. The decision to use 380-mm guns is considered sound, but the ships approached the size and cost of battleships without the protection expected in such large ships. It would seem that a ship armed with 380-mm guns and with protection similar to that of the *Gneisenau* and *Scharnhorst* would have been far better suited for commerce raiding. The fairly successful war cruise of those two battlecruisers in 1941 indicated that six 380-mm guns in three turrets would have been a better solution, due to the fact that convoys were being protected by battleships armed with 381-mm guns.

Accepting the wisdom of the decision to use the bigger guns, the fact remains that the secondary battery and the antiaircraft armament were unsatisfactory. A dual-purpose battery would have permitted a considerable saving in weight and improvement in firepower to be realized. It was not possible to fulfill the demand for a dual-purpose gun, an unfortunate drawback in German naval ordnance, and thus deprived these displacement-limited warships of a more suitable type of armament. The *Bismarck*- and *Scharnhorst*-class ships had more numerous 105-mm and 150-mm batteries, but their antiaircraft protection was inadequate against high-speed aircraft. This leads to the conclusion that the "O"-class battlecruisers could have been quickly disabled and eventually destroyed by aircraft.

The greatest weakness in these ships was their lack of armor protection and the almost total absence of horizontal protection in an era when aerial bombs and long-range projectiles were increasing in destructive power. These battlecruisers would have been extremely vulnerable to heavy-caliber shell hits and unquestionably could never have served in a battle line.

The lack of aircraft carriers in the German Navy resulted from a German lack of appreciation of the airplane as a factor in sea power. World War II would prove that the battleship and battlecruiser had become obsolete and that the construction of any large capital ships other than aircraft carriers was ill-advised. This was particularly true for this totally obsolete battlecruiser type, which had the great size and expense of the battleship yet had completely inadequate protection.

These are the 105-mm dual-purpose guns on the battleship *Tirpitz,* which was stationed in Norway when this photo was taken. Note the camouflage on the main deck against aircraft detection and the torpedo nets arrayed around the ship. The 105-mm gun was the standard heavy antiaircraft weapon on all German capital ships built or projected in the World War II era. The problem with the mount on the *Scharnhorst-* and *Bismarck*-class ships was that it did not have its own central ammunition supply, and the gun crew could be exposed to strafing attacks by aircraft. Only in the "H"-class battleships and the "O"-class battlecruisers were these mounts to be enclosed, making them more useful in a dual-purpose role. With a maximum range of 17,700 meters and a 15-kg shell, these guns were effective against the Albacore or Swordfish biplanes, but not against the Martlets or Barracudas that the Royal Navy used later in the war. Important in air defense was the need to destroy aircraft before they released their weapons.

Name	Battlecruiser "O"	Battlecruiser "P"	Battlecruiser "Q"
Builder	Deutsche Werke, Kiel	Kriegsmarine Werft, Wilhelmshaven	Germania Werft Kiel
Disposition	Canceled	Canceled	Canceled

Displacement
31,152 tons (31,625 mt) Design
35,155 tons (35,720 mt) Full load

Dimensions
814′ 2″ (248.15 m) Length overall
807′ 1″ (246.0 m) Waterline length
98′ 5″ (30.0 m) Beam
50′ 10″ (15.5 m) Depth
28′ 10″ (8.8 m) Draft at 35,377 tons (35,945 mt)
26′ 3″ (8.0 m) Draft at 31,152 tons (31,625 mt)

Armament
Six 15″/47-caliber (Mark 1934) (380 mm)
Six 5.9″/48-caliber (Mark 1938) (150 mm)
Eight 4.1″ /65-caliber (Mark 1933) (105 mm)
Eight 1.46″/83-caliber (Mark 1930) (37 mm)
Eight 0.79″ (20 mm)
Twelve 21.0″ (533 mm) deck-mounted torpedo tubes
Four Arado-196 floatplanes with one double catapult

Armor Protection
Amidships

Upper belt	3.54″ (90 mm)	
Main side belt	7.48″ (190 mm)	
Deck armor	(centerline)	(outboard)
upper deck	1.18″ (30 mm)	1.18″ (30 mm)
armor deck	2.36″ (60 mm)	3.15 (80 m)—slopes
total	3.54″ (90 mm)	4.33″ (110 m)

Turret armor

face plates	8.67″ (220 mm)
sides	7.09″ (180 mm)
back plates	7.09″ (180 mm)
roof plates	1.97″ (50 mm)

Barbette armor

centerline fwd	7.09″ (180 mm)
sides	7.09″ (180 mm)
centerline aft*	7.09″ (180 mm)

Secondary gun protection
face plates, sides, back plates, roof plates } 0.55″ (14 mm)

Conning tower armor

centerline fwd	7.87″ (200 mm)
sides	7.87″ (200 mm)
centerline aft	7.87″ (200 mm)
roof plates	3.15″ (80 mm)
deck plates	1.18″ (30 mm)
communication tube	3.15″ (80 mm)

*Note that aft barbette for turret Caesar had only 3.15″ (80 mm).

Frames

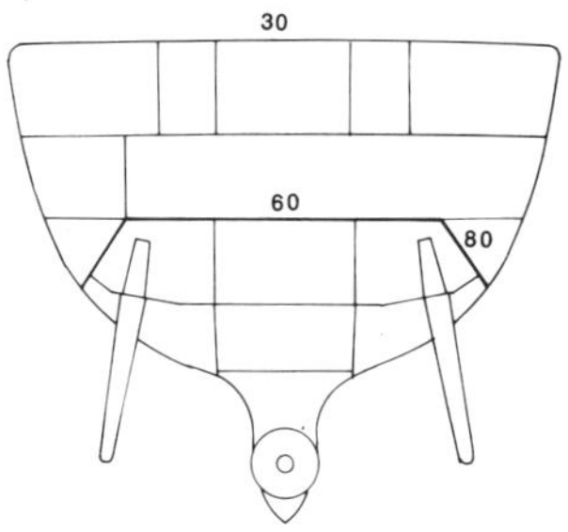

Frame 19.75—Steering-gear room

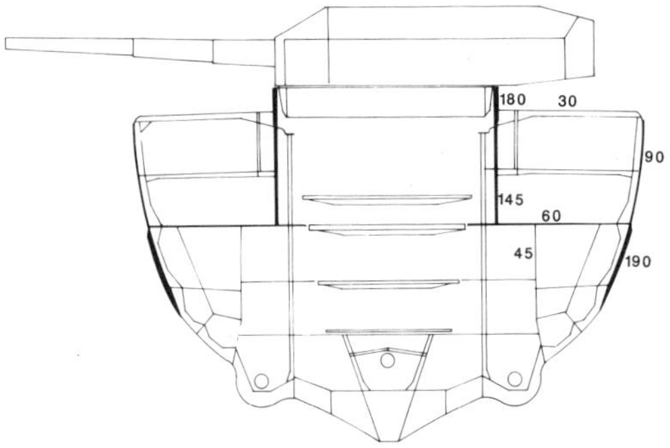

Frame 49.2—Turret Caesar

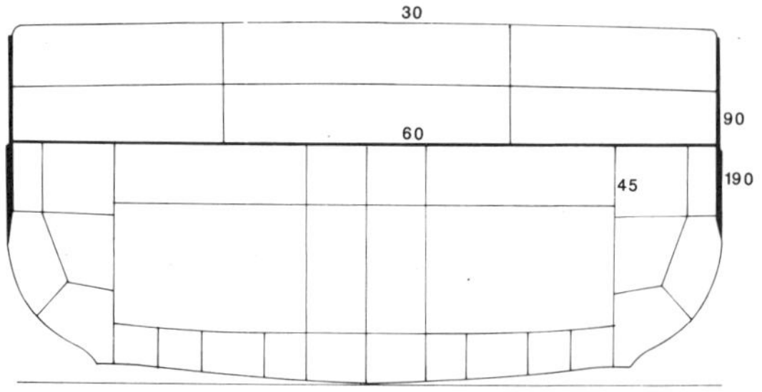

Frame 119.4

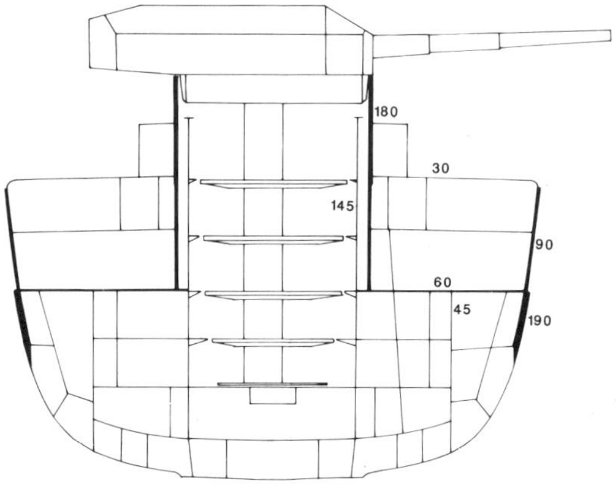

Frame 160.5—Turret Bruno

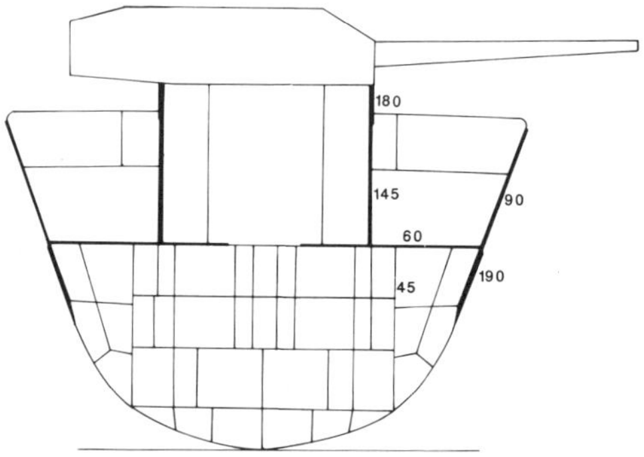

Frame 184.65—Turret Anton

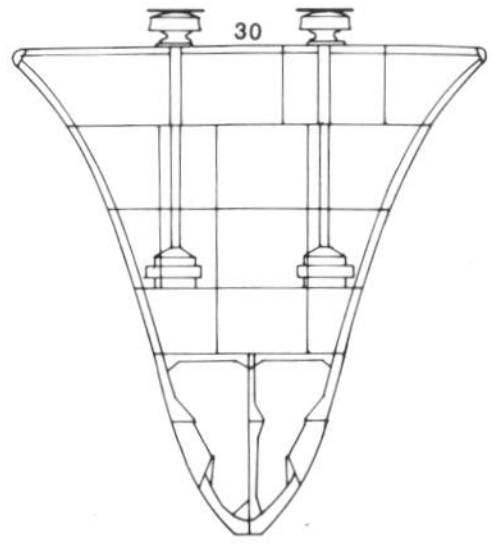

Frame 221.5

Underwater protection	
Side-protective system depth	13.78′ (4.2 m)
Side-protective system loading	Void-Liquid
Total bulkhead thickness	2.01″ (51 mm)
Bottom protective system depth:	5.58′ (1.70 m) One layer

CHAPTER SEVEN

The Vittorio Veneto Class

The largest and most powerful warships ever built in Italy, the battleships of the *Vittorio Veneto* class were the first "35,000 tonners" to come under the provisions of the Washington Treaty. Although the Washington Naval Conference of 1922 gave Italy authority to build up to 182,800 tons of new capital ships (including 70,000 tons prior to 1932), the Italians waited more than a decade before beginning construction.*

During World War I, the Italian Navy did not commence the construction of new battleships, with the exception of the *Francesco Carraciolo.* Instead, it concentrated on smaller ships suitable for operations in the Adriatic, as their decision to go to war on the side of Great Britain and France made the construction of heavy oceangoing ships unnecessary. Four battleships of the *Francesco Carraciolo* class, the first of which was laid down on 12 October 1914, were to be the backbone of the Italian battle fleet. These ships were designed to a normal displacement of 34,000 metric tons, a main battery of eight 381-mm/40-caliber guns, and a speed of 28 knots. Only the hull of the name ship of the class was completed. It was launched on 12 May 1920, only to be scrapped three years later. Three ships of the *Conte Di Cavour* class (laid down in 1910) and two ships of the *Caio Duilio* class (laid down in 1912) were completed during the war. The *Leonardo Da Vinci,* one of the *Conte Di Cavour* class, was seriously damaged by a magazine explosion and fire on 2 August 1916. She was a total loss, and the hulk was scrapped in 1923.

When the Washington Conference was convened, Italy was experiencing a great financial crisis, and a naval armaments race was something that this country could ill afford. The Italians accepted naval parity with France without too much opposition, but compromised this acceptance with the insistence that the Italian Navy be allowed to complete new battleships, since none had been completed from 1916–1921. A limit of 70,000 tons on new construction was agreed upon for both France and Italy.

By 1928, Italy's battle fleet was obsolescent and rapidly becoming overaged. It was decided to let attrition reduce its size, while design schemes for conversions and new ships were studied. The *Dante Alighieri,*† *Conte Di Cavour,* and *Guilio Cesare* were demobilized and placed in reserve. The two battleships of the *Duilio* class were left in service.

---

*The Washington Treaty defined limits of standard displacement in long tons (2,240 lbs = 1 ton)
†This battleship, laid down in 1907, was the Italian Navy's first dreadnought.

The *Roma* explodes spectacularly after a direct hit by a German radio-controlled bomb, the FX-1400. This is the first instance of a major warship being destroyed by a guided missile. The ship was destroyed on 9 September 1943 while en route to Malta to be surrendered. (Imperial War Museum)

The Italians began the construction of several battleships to be armed with 381-mm guns during the World War I era. The launch of the *Francesco Caracciolo* on 12 May 1920 is shown here. The vessel was later scrapped under the terms of the 1922 Washington Treaty. (Aldo Fraccaroli)

The Italian Navy pioneered the triple turret. The *Dante Alighieri* was the first capital ship in the world to mount such an armament. (Aldo Fraccaroli)

Italian design teams began studies of a battleship type that could best use the 70,000 tons allotted to Italy under the terms of the Washington Treaty. Considerable pressure was placed on the Italian and French governments to build battleships of smaller displacement and smaller gun calibers than those completed by the major naval powers during the 1920s. With the climate favorable towards smaller-displacement capital ships, the Italian Navy favored a ship of 25,000 tons standard displacement, as such small ships were feasible if given limited endurance. Such a limitation would have been of enormous advantage to the Italian ships, which could be constructed for service only within the Mediterranean, while likely opponents would have been required to sacrifice other military features to attain adequate endurance with such a limited displacement.

The first design study seriously considered in 1928 (figure 7-1) was a small-displacement capital ship with the principal characteristics summarized in table 7-1. The heavy armament and high speed of these ships was achieved at the expense of protection and endurance. The Italian Navy did not need great endurance for operations in the Mediterranean.

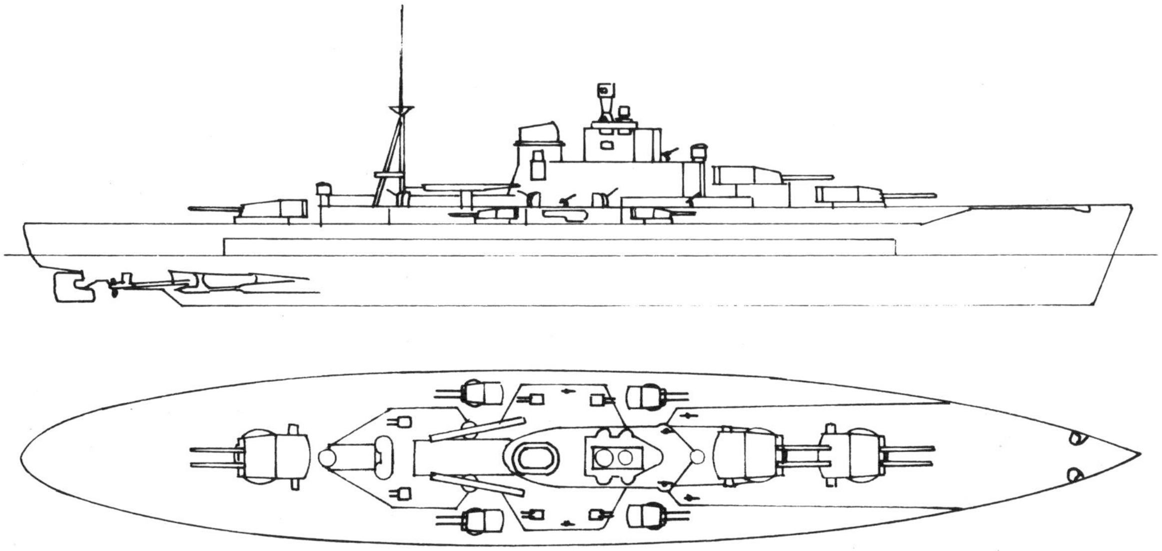

Figure 7-1. 1928 battleship study.

TABLE 7-1

Characteristics of 23,000-Ton Design (1928)

Displacement (standard)	23,000 tons (23,369 mt)
Dimensions	
Length	639.76′ (195 m)
Beam	95.14′ (29 m)
Draft	26.25′ (8 m)
Armament	
	six 15″ (381 mm) paired
	eight 6″ (152 mm) paired
	twelve 4.1″ (100 mm) paired
	two aircraft
Speed	28–29 knots

A June 1928 design of the Committee on Naval Projects (CPN) was tested at the model basin at La Spezia. Design number 45367 was assigned to a ship of 26,200 metric tons normal displacement. No further details are known concerning this study, although it is possible that the model tests were for the 23,000-ton ship (standard displacement) at a more heavily loaded condition.

The Italian design teams also carried out preliminary design studies of "treaty battleships" in 1928. Based on a standard displacement of 35,000 tons, it was determined that a main armament of six 406-mm guns would be possible with a speed of 29–30 knots and protection against the 406-mm gun. These heavier ships could be given better armor and underwater protection than the smaller ships. Insofar as speed was concerned, the Italian Navy wanted its new battleships to have the same speed as the heavy cruisers of the *Trento* class. No construction funds were appropriated. For political reasons, especially with regard to France, the Italians did not want to be the first to initiate the construction of new battleships.

The London Naval Conference of 1930 did not resolve the question of the size of the future battleship and its gun power. In fact, the conference rejected the British proposal that the standard displacement be further limited to 25,000 tons and the main armament be limited to 305-mm guns, but the naval building holiday was advanced to 31 December 1936. However, Italy and France were still permitted to construct new ships totaling no more than 70,000 tons standard. This brought increased pressure by Great Britain on the Italian and French governments to build smaller ships with 305-mm guns as their main armament. The Italians abandoned the studies of small battleships after the 1930 conference because such small ships were attractive only if foreign contemporaries were of similar limited size.

By 1932 the Italian government was confronted with an entirely different naval situation. The Germans had begun the first of their three "pocket battleships" in 1928, and the French had laid down the new battlecruiser *Dunkerque* just before the end of 1931. The displacement of this ship was 26,500 tons and the main armament featured 330-mm guns in two quadruple turrets.

During September–December 1932, the Ansaldo firm of Genoa, under contract to the Italian Navy, prepared a series of alternate versions of a design similar to the German pocket battleships. No final definition of Design 770 (figure 7-2) was ever reached, the propulsion system and protection not being resolved from among several alternate schemes. However, the general characteristics of the ships at the time the design work was abandoned are tabulated in table 7-2.

The armament of these "pocket battleships" would have been extraordinary. The Italian Navy was not willing to accept a smaller gun caliber than on the *Dunkerque*, and it was decided to arm these ships with 343-mm guns. It was impossible to mount eight guns and achieve an 18,000-ton displacement, unless the navy was willing to make major sacrifices, either in the armament, protection, or speed. A six-gun main battery was possible, however, and the Italians had workable triple-turret, heavy ordnance designs that had been constructed for the *Conte di Cavour* and *Dante Alighieri*. Thus, triple turrets with 343-mm guns were chosen for the main battery, one forward and one aft. Each main-battery turret was given as large an arc of fire as possible—300 degrees. This complicated the placement of the secondary and antiaircraft guns because of the blast from the main guns and ultimately forced a concentration of firepower. The 152-mm twin turrets were arranged two abreast, superfiring over the 343-mm main-battery turrets. The 152-mm caliber of the secondary battery was considered the most suitable gun size to stop a destroyer. Three 100-mm dual-purpose twin mounts were grouped amidships, to port and starboard, relatively clear of the blast effects of the 152-mm and 343-mm guns. The projected sextuple 37-mm machine-gun enclosed

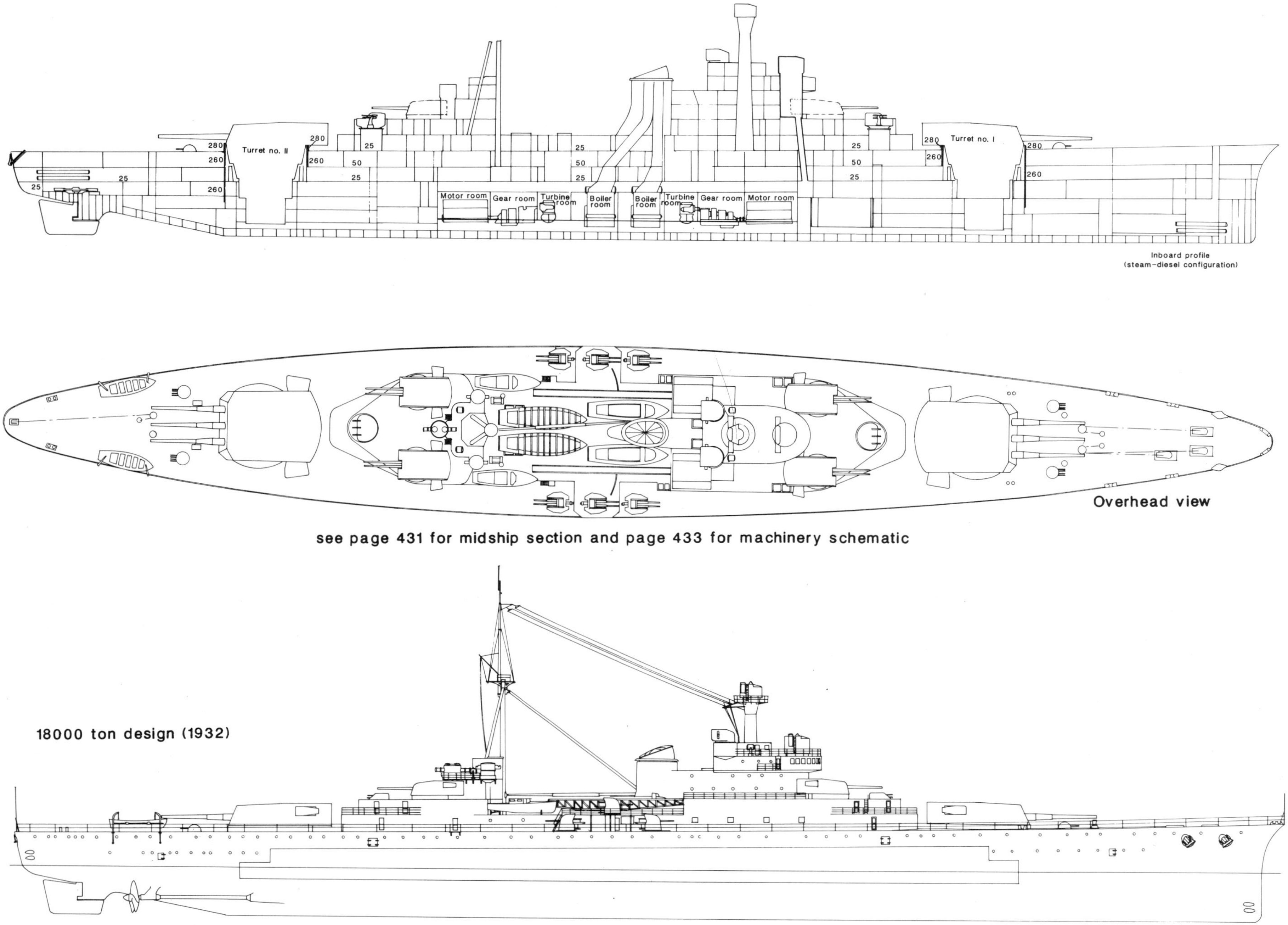

Figure 7-2. Design 770—profile and overhead views.

TABLE 7-2
Characteristics of 18,000-Ton Design

Displacement (standard)	18,000 tons (18,289 mt)
(normal)	19,192 tons (19,500 mt)
Dimensions	
Length	606.96 (185.00 m)
Beam	86.12′ (26.25 m)
Draft	25.10′ (7.65 m)
Depth (amidships)	43.64′ (13.30 m)
Armament	six 13.5″ (343 mm) tripled
	eight 6.0″ (152 mm) paired
	twelve 3.9″ (100 mm) paired
	twelve 1.46″ (37 mm) sextupled
	sixteen 0.51″ (13 mm) quadrupled
	torpedo tubes
Shaft horsepower	78,906 (80,000 mhp)
Maximum speed	26 knots

mounts were to be positioned between the 343-mm and 152-mm turrets, forward and aft. Such a mount did not then exist in the Italian Navy.

The compact superstructure arrangement and the placement of the necessary gun positions did not permit deck-mounted torpedo tubes. The Italians decided to use six submerged torpedo tubes forward and four above-water torpedo tubes aft, just above the deep-draft waterline. The torpedo-handling rooms aft had minimal protection, being protected by only 25-mm armor. No protection was provided the forward spaces.

Four aircraft and two trainable catapults were to be carried. Because of blast from the 343-mm guns and the limited space at the stern, the aircraft could not be positioned aft. Instead, a port and starboard hangar in the midship superstructure was planned.

The protection provided these vessels was a compromise. On a ship of such limited displacement whose emphasis was on offensive power, heavy protection was not possible. This was further complicated by the need for a speed of 26 knots. To best protect these ships, two schemes were proposed:

- A *Nelson*-style protection modified to Italian standards. The main side belt would be 210-mm outward-sloping armor with a 70-mm vertical side belt.*
- The classical turtle-deck armor system with 280-mm vertical armor on the side shell supplemented by 25-mm armor slopes.

Both schemes featured a very lightly armored deck structure. The horizontal armor would have been distributed over three decks and would have been 25-mm, 50-mm, and 25-mm thick respectively, except in way of the main-battery magazines where the 50-mm armor thickness would be increased to 75 mm.

The underwater protection was severely limited because of the ship size; there is some question whether it would have been effective against contemporary torpedo warheads. The much-larger-displacement USS *Alaska*-class ships also lacked adequate underwater protection. The characteristics of the two underwater protection systems proposed for the 18,000-ton battleship study are summarized in table 7-3.

---

*Outward-sloping armor—the top edge of the armor would be farther outboard than the lower edge.

TABLE 7-3

Characteristics of Underwater Protection—18,000-Ton Design*

	Mean Vertical Extent	*Breadth at Half Draft*	*Torpedo Bulkhead Thickness*	*Total Bulkhead Thickness*
Design A (Turtle Deck)	22.3′ (6.8 m)	14.27′ (4.35 m)	0.28″ (7 mm)	0.67″ (17 mm)
Design B (Modified *Nelson* Type)	25.9′ (7.9 m)	14.27′ (4.35 m)	0.28″ (7 mm)	0.63″ (16 mm)

*See page 431 for midship sections.

The propulsion plant was thoroughly studied, although a final arrangement was never decided upon. The choice of a propulsion plant was complicated by the desired 26-knot speed and the sought-after balance between the armament and protection. To achieve the limited displacement, the endurance was reduced, allowing the ship to operate economically in the Mediterranean, but not outside of it. At the time the design was abandoned, the Ansaldo firm had considered four different propulsion plant schemes:

- Steam electric drive—two shafts
- Diesel electric drive—four shafts
- Steam and diesel electric drive—four shafts (see page 433, figure 7-11)
- Steam geared drive—two shafts

The steam electric drive was preferred. The use of diesels was an attempt to gain the fuel efficiency of the diesel engine in a ship severely limited in displacement. The difficulty in the use of this engine was the engine size available from the manufacturers; it was very desirable to avoid the need to build and test special engines for these ships. If diesels were to be used, four shafts were required to attain 80,000 mhp. (There was some question whether the technology of that period would permit 40,000 mhp to be developed per shaft.) The propulsion plant of the 18,000-ton pocket battleship was an unresolved question when work on the design of the ships was stopped in December of 1932.

The Italian Navy decided that such small capital ships were impractical and that better results could be achieved by increasing displacement and speed. The 18,000-ton battleship study did highlight the problems in the choice of an adequate propulsion plant and later served as a basis for the preliminary design of the *Vittorio Veneto*-class battleships in 1934. Although the 18,000-ton battleship could stand up to 203-mm and possibly 280-mm guns, it lacked the protection to withstand 381-mm gunfire.

In the meantime, France had begun the construction of the *Dunkerque.* This meant that existing Italian battleships would be outclassed in the Mediterranean by a fast and modern battlecruiser. The Committee on Naval Projects decided to investigate the design of a ship that would be comparable in power and speed to this French battlecruiser. A 1933 design study produced a ship with the characteristics summarized in table 7-4.

The arrangement of the 26,500-ton battlecruiser was strikingly similar to the 18,000-ton "pocket battleship" plan prepared by the Ansaldo firm. The primary difference was in the arrangement of the main battery, where there were four twin turrets, equally disposed forward and aft. Although two quadruple turrets would have been more readily accommodated on such a limited displacement, Italian industry would have experienced considerable difficulty with the design and construction of such

TABLE 7-4
1933 Battlecruiser Design

Displacement (standard)	26,500 tons (29,925 mt)
Dimensions	
Length	656.17′ (200.00 m)
Beam	89.40′ (27.25 m)
Armament	eight 13.4″ (343 mm) paired
	twelve 6.0″ (152 mm) paired
	sixteen 1.46″ (37 mm) quadrupled
	four aircraft
Speed	29 knots
Protection	
Armor belt	9.84″ (250 mm)—main side belt
Deck armor	5.91″ (50 mm)—three decks, total

turrets, having had no previous design experience. Therefore, a twin-turret arrangement was sought from the very earliest of preliminary designs.

The secondary battery had an arrangement similar to that of the 18,000-ton design, with 152-mm wing turrets deployed fore and aft. However, the added displacement permitted an additional amidships 152-mm turret on either beam. This battery was strictly for surface targets and had very limited air-defense capabilities. Air defense was to be provided by a total of four 37-mm quadruple machine-gun mounts evenly distributed fore and aft, to port and starboard. This antiaircraft defense was weak in comparison to that of the *Dunkerque*, which had a dual-purpose secondary armament and more numerous machine guns.

The proposed 26,500-ton battlecruiser would have been equipped with four aircraft and two trainable catapults. The lack of sufficient space amidships prevented the provision of hangars.

One of the first studies in the progression of design for the *Vittorio Veneto*-class battleships was this 26,500-ton battlecruiser type with 343-mm guns in four twin turrets. This design was prompted by the French *Dunkerque* and would be later canceled. Instead, the Italians decided to modernize four older battleships. (Italian Navy)

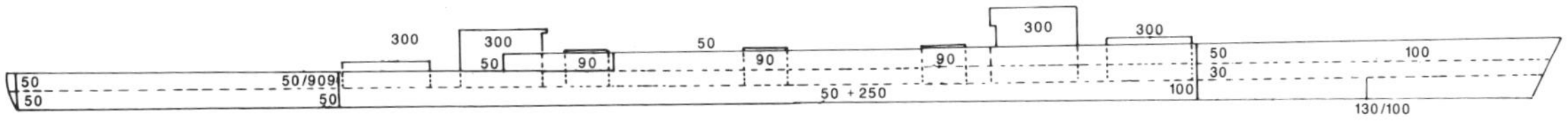

Figure 7-3. 26,500-ton battlecruiser armor profile.

The armor protection would have been of the classic type with a 250-mm side belt and a 50-mm armor slope. The deck protection was weak, featuring only three decks with 50-mm armor plate each. The scheme of protection is shown in figure 7-3.

The propulsion plant would have been a steam geared-turbine drive, as it was clear from the previous 18,000-ton ship design that even a steam electric drive would not be suitable for this battlecruiser design. In fact, the steam power plant provided this design was very similar to that adopted in the *Vittorio Veneto* class.

Finally, the Italian Navy decided to abandon work on the rather advanced design of the 26,500-ton battlecruiser and to revert to a 35,000-ton "treaty battleship." As the design and construction of these new ships would inevitably be a time-consuming process, the naval authorities decided to extensively modernize the four obsolete battleships of the *Conte de Cavour* and *Andrea Doria* classes as a stop-gap measure to counter French new construction programs. As events were to prove, this was an unfortunate decision. The reconstructed ships were soon shown to be extremely vulnerable to modern ordnance, and they proved of very little value during World War II. The Italians may well have been better advised to build one or two of the 26,500-ton battlecruisers rather than reconstruct the four old battleships. In any event, this abandonment of the 26,500-ton battlecruiser design set the stage for the development of the *Vittorio Veneto* class.

Outline requirements set by the Italian Navy for the new battleship design are summarized in table 7-5.

TABLE 7-5

Design Requirements—Vittorio Veneto

Displacement (standard)	35,000 tons (35,562 mt)
Armament	The maximum number of 381-mm or 406-mm guns suitable to the displacement
Maximum speed	30 knots
Endurance	For a round trip to any point in the Mediterranean Sea with adequate reserves for combat
Armor protection	Immunity zone against the main battery between ranges of 20,232 to 30,075 yards (18,500 to 27,500 meters)
Side protection system	To withstand a contact detonation of a 770-pound (350-kg) TNT charge

By early 1934, several hull forms had been chosen, and a series of model tests were conducted at the model basin in La Spezia.

- CPN Design 54265 of February 1934—displacement 37,000 metric tons
- La Spezia model basin design—displacement 35,900 metric tons
- Hulls B191, B192, and B193 (preliminary hulls for design of hull B194)—displacement 40,000 metric tons, speed 34 knots
- Hull B194, design of CRDA Trieste and Ansaldo—displacement from 36,363 to 48,000 metric tons

The decision to construct new battleships of the maximum displacement permitted by the Washington Treaty involved the complete revision of earlier plans. Such ships, fully half again the displacement of any Italian naval ships in service, represented a major technical challenge to the Italian design teams. Traditionally, naval ship design relies heavily on earlier ships as a baseline for design development. Such a substantial increase in ship displacement, in addition to the pre-World War I design vintage of existing ships, greatly inhibited the utility of this design technique during the development of the *Vittorio Veneto* design.

It was decided to revert to 381-mm guns for the main battery, as such guns were considered to be within Italian industrial capabilities, while heavier guns promised serious difficulties. Additionally, the Italians believed that the smaller caliber would permit a superior compromise between the desired offensive and defensive capabilities.

The design of the new battleships, by teams headed by Umberto Pugliese, General Inspector of the Engineer Corps, responded to problems posed by progress in ordnance technology with imaginative design approaches, the most unique of which was the underwater side-protection system commonly referred to as the "Pugliese system."

The Pugliese system featured a large hollow cylinder installed along the length of the citadel on both sides of the ships. It was to be kept void at all times, and completely immersed in fuel oil, water, or a mixture of the two. The supporting structure bounding this protective system was more heavily constructed than the hollow cylinder, which was designed to absorb explosive energy by being hydraulically crushed. The resulting turbulence would further dissipate the destructive forces of the underwater detonation. Venting plates, fitted to foreign battleships, were not included in the design of this system so that the full force of an underwater explosion would be concentrated on the destruction of the interior cylinder. The concentric inboard bulkhead was made of very heavy structure and high-strength (Elevata Resistenza—ER) steel. Unfortunately, because welding technology was not sufficiently advanced, the joints of this bulkhead had to be riveted. The Italians considered the system to be effective and light in weight relative to more conventional torpedo-protection systems.

The *Vittorio Veneto* was a strikingly handsome ship. She and her sister ships were the first of the 35,000-ton "treaty" battleships to be built and the first Italian battleships to mount an armament of 381-mm guns.

This novel system was developed in 1930, tested on a model, and then tested full-scale on the old tanker *Brennero.* The results were so promising that the system was fitted on the reconstructed battleships of the *Conte Di Cavour* and *Duilio* classes. The limited space available and modifications to the system necessitated by the old hull design of these battleships inhibited the effectiveness of the new system. In order to increase its effectiveness, it would have been necessary to increase the beam of these older ships, and this would have resulted in a more costly and extensive reconstruction than was economically feasible. The *Vittorio Veneto*-class battleships were the only Italian ships to be built from the keel up to make use of this system.

Final characteristics of the *Vittorio Veneto* were approved by the CPN on 12 July 1935, implementing the outline design established in 1934. In October these ships were officially ordered.

The Italians were successful in developing a well-balanced design with good protection, high speed, and adequate armament. These were the first capital ships with the "classic" main-battery arrangement—three triple turrets, two forward and one aft. Technological progress subsequent to the design of these ships revealed some defects, particularly in the horizontal protection and antiaircraft armament.

As built, the *Vittorio Veneto* had the basic characteristics outlined in table 7-6.

As the *Veneto* design evolved, it became apparent that a main armament of 381-mm guns and a secondary battery of 152-mm guns—in combination with the protection requirements, a maximum speed of almost 30 knots, and the use of the Pugliese systems—would result in a battleship of more than 35,000 tons standard displacement.

TABLE 7-6

Characteristics—Vittorio Veneto

Displacement	
Light ship	37,613 tons (38,216 mt)
Standard	40,516 tons (41,167 mt)
Design	42,935 tons (43,624 mt)
Full load	45,029 tons (45,752 mt)
Dimensions	
Length overall	780.05′ (237.76 m)
Length between perpendiculars	734.07′ (224.05 m)
Maximum beam	108.10′ (32.95 m)
Waterline beam	106.41′ (32.43 m)
Mean draft at full load	34.25′ (10.44 m)
Armament	nine 15″ (381-mm)/50-caliber guns—tripled
	twelve 6″ (152-mm)/55-caliber guns—tripled
	twelve 3.5″ (90-mm)/50 guns—single
	twenty 1.46″ (37-mm)/54 guns—paired
	twenty-four 0.79″ (20-mm)/65 machine guns
	three aircraft, one catapult (no hangar)
Shaft horsepower	128,322 (130,000 mhp)—normal maximum
	138,085 (140,000 mhp)—overload
Maximum speed	30 knots at design displacement
	29 knots at full-load displacement
Endurance	4,700 nautical miles at 14 knots
	3,900 nautical miles at 20 knots
Protection	14.17″ (360 mm) total thickness—main side belt
	6.65″ (169 mm) maximum—total deck thickness

Any reduction in displacement would have meant a ship of lesser speed, protection, or armament. The Italians decided to design to a larger standard displacement rather than accept lesser military characteristics.

The standard displacement of these ships substantially exceeded the limits set by the Washington Treaty. This was characteristic of all the "treaty battleships" of the 35,000-ton class by the time they entered service. It is important to note that the 1922 treaty limitations on displacement remained in effect until 31 December 1936. The Italians were not signatories to the subsequent naval limitations treaties.

Vittorio Veneto—Operational History. The *Vittorio Veneto* was laid down by Cantieri Riuniti dell 'Adriatico, in Trieste, on 28 October 1934, launched on 25 July 1937, and completed on 28 April 1940. The first new Italian battleship commissioned since 1916, she was destined to have an extremely active career.

The ship entered service on 1 May 1940, but it was not until 2 August 1940 that she was ready for action. Her first mission of World War II came on 31 August as part of the largest concentration of Italian battleships to put to sea during the war. In addition to the *Vittorio Veneto*, the battleships *Littorio*, *Cesare*, *Cavour*, and *Duilio*, accompanied by 10 cruisers and 34 destroyers, participated in a mission to challenge the British Mediterranean Fleet, based at Alexandria, Egypt. However, British air reconnaissance spotted this large concentration of warships, and the British ships turned away. A severe storm intervened during the pursuit, and this made it difficult for the destroyers to maintain their station; the Italian aircraft failed to locate the British ships. The Italian fleet returned to base, but was sent out again on 6 September to engage a British force from Gibraltar. The British ships steamed south into the Atlantic Ocean, however, and the Italian ships were recalled.

The *Vittorio Veneto*, stationed at Taranto with four other battleships, was inactive until November 1940. Also in the anchorage were nine cruisers and a large number of

The *Vittorio Veneto* at a mooring buoy, about 1940. An interesting variety of small craft are arrayed along her port side. (Aldo Fraccaroli)

The *Vittorio Veneto* rides at anchor as she nears completion. The outfitting of the ship was not yet complete. The 90-mm antiaircraft mounts had not been installed at the time this photo was taken.

destroyers and auxiliaries. The port was defended by 21 90-mm gun positions, plus a number of 37-mm and 20-mm automatic close-range weapons. There were a large number of searchlights and 27 barrage balloons, but the absence of radar and the lack of adequate torpedo netting round the ships were very significant factors. Italian industry, hard pressed by the war, could not meet the order requested by the navy; thus, torpedo nets could not be drawn completely around the battleships.

The Italian invasion of Greece in October 1940 and the Italian battleship strength concentrated at Taranto brought a response from the Royal Navy. During the night of 11 November, the British carrier *Illustrious* approached to within 170 miles of the Italian naval base at Taranto, undetected. At 2035 a first wave of 21 Swordfish torpedo planes attacked the Italian fleet based there. A second wave of 21 planes attacked approximately one hour later. During these two attacks, a total of two torpedoes were launched at the *Vittorio Veneto*, but they missed. However, three other battleships in the anchorage at the time were hit by torpedoes and sank at or near their moorings. These ships were the *Littorio*, *Duilio*, and *Cavour*. This stunning attack crippled the Italian battle fleet and reversed the relative naval strength in the Mediterranean theater in favor of the Royal Navy. Undamaged, the *Vittorio Veneto* was ready for action and later was moved north to Naples to be safe from any other British air attacks.

Cape Teulada engagement. The *Vittorio Veneto*, serving as the flagship of Admiral Inigo Campioni, left Naples in company with the *Cesare*, six cruisers, and 14 destroyers to take part in the rather inconclusive engagement off Cape Teulada on 27 November. The Italian fleet attempted to intercept a Malta-bound convoy that was being covered by the *Renown*, *Ark Royal*, two cruisers, and nine destroyers. Another covering force of the *Ramillies*, three cruisers, and five destroyers left Alexandria to strengthen the

The *Vittorio Veneto* is shown here on 28 March 1941 on her way to Taranto after the Battle of Cape Matapan. The battleship's stern is low in the water due to the flooding—almost 4,000 tons—caused by a torpedo and a bomb hit. The floatplane perched precariously on the stern had been damaged by the blast from the after 381-mm gun turret. (Italian Navy)

Gibralter group during the last part of the convoy's passage to Malta. Inept air reconnaissance foiled the efforts of the superior Italian force, and exaggeration of the British forces' strength contributed to Admiral Campioni's decision to break off contact. The Italians, after their disastrous losses at Taranto, could not afford a fleet engagement with equal or superior opponents. Admiral Campioni's orders were to seek an engagement only if he had clear superiority. The *Vittorio Veneto* did briefly engage British cruisers at long range and bracketed them on her fourth salvo. The cruiser *Berwick* was hit in the exchange, but a signal to all Italian ships to head for Naples ended this action. None of the Italian battle fleet was damaged in this engagement.

British long-range bombers attacked the Italian fleet based at Naples on 14 December and damaged the heavy cruiser *Pola.* This raid, the culmination of attacks on southern Italian ports, forced the transfer of heavy fleet units to Sardinia. This was short-lived, however, because the *Vittorio Veneto* returned to Naples on 20 December. Despite the risk of air attack, the Italian fleet had to return because the Italian Naval Staff reasoned that Alexandria-based convoys would reach Malta with relative ease while the Italian fleet was in Sardinia.

Another British air raid on Naples during the evening of 8 January 1941 damaged the reconstructed battleship *Giulio Cesare.* The following day, the *Cesare* departed for La Spezia with the *Vittorio Veneto* providing cover. The *Cesare* required significant battle-damage repairs, leaving the *Vittorio Veneto* the only Italian battleship in service. From January through March the Italian Navy waited for an opportunity to use its only operational battleship in conjunction with their large cruiser force. An ambitious operation was planned against British convoys to Greece on 24 March, with extensive air coverage to be provided by the Italian Air Force and Fliegercorps X of the Luftwaffe. The operation was delayed two days at the request of the Germans, who wished to arrange for effective coordination of their air activities in support of the operation. The Italian Navy's decision to take part in this combined operation stemmed from an inaccurate German report that only one battleship, the *Valiant,* was present in the eastern Mediterranean.

In early April of 1941, the badly damaged *Vittorio Veneto* was towed to a shipyard in Taranto for repairs. In these two views the ship is shown with two large cylinders placed on her bow. These were filled with water in order to reduce the ship's trim by the stern due to her damage. (Italian Navy)

The *Vittorio Veneto* sortied from Naples under the command of Admiral Angelo Iachino during the evening of 26 March, accompanied by the heavy cruisers *Zara, Fiume, Pola, Trento, Trieste,* and *Bolzano,* light cruisers *Luigi Di Savoia Duca Degli Abruzzi* and *Giuseppe Garibaldi,* and 13 destroyers. They were to attack supply ships and troop transports from Alexandria, but the Italian force was detected by a British Sunderland reconnaissance plane at 1220 on 27 March. The element of surprise had been lost.

Encounter off Guado. On the morning of 28 March, three Italian cruisers encountered the outer screen of a British battle fleet consisting of the *Valiant, Barham, Warspite, Formidable,* and nine destroyers. The British cruisers had been detached from a troop convoy to Greece when it was learned of the Italian cruisers' presence in this part of the Mediterranean. Four British cruisers (*Gloucester, Orion, Ajax,* and *Perth*) and four destroyers were cruising in a position 30 miles south of the rocky islet called Guado Island. The Italian cruisers followed the British for approximately one hour, then deliberately reversed course to lure these British cruisers towards the main battle fleet. The British ships fell for the ruse and began to follow the Italians until the *Vittorio Veneto* suddenly appeared at 1058.

When the *Vittorio Veneto* came within sight of the British cruisers, she instantly opened fire at an initial range of some 23,000 meters. The Italian fire was accurate, but with too large a spread. Fire was directed on the *Orion* and a near miss was scored, which caused some minor damage. The British cruisers and destroyers quickly made smoke and retreated from the battle. Only the *Gloucester* remained visible to the Italians, and the *Vittorio Veneto* switched fire to her and repeatedly straddled the cruiser with 381-mm shells. However, a British destroyer raced between the *Gloucester* and the battleship and made smoke to conceal her from further action.

Around 1115, when the *Gloucester* was in very grave danger, six British Albacore torpedo bombers and two Fulmar fighters arrived on the scene from the *Formidable.* Two German JU88 fighter bombers tried to intercept; one was shot down while the other escaped. It took some 12 minutes for the British planes to gain an attack position on the *Vittorio Veneto.* At 1127 the battleship ceased fire and maneuvered to defend herself against the new threat.

The subsequent air attack was unsuccessful, but did enable the British cruisers to disengage without further mishap. The appearance of British air support resulted in a decision by Admiral Iachino to break off the pursuit and return to port. The combination of Fulmars and Albacores meant that a British carrier was at sea, and there could be battleships escorting her. The Italians turned homeward at 1130, being in an exposed position without the promised Italo-German air support, while British aircraft began to continuously track the formation. Commencing at 1207 and for the next five hours, the Italian force would be subjected to eleven separate air attacks.

The *Vittorio Veneto* underwent two high-level bombing attacks by Royal Air Force Blenheims from bases in Greece at 1420 and 1450. Although some bombs fell close to the Italian battleship, no damage was sustained. At 1519 that afternoon, the *Vittorio Veneto* was attacked by a Fleet Air Arm force of five torpedo bombers, three Albacores and two Swordfish, and two Fulmar fighters. The combined attack was executed to perfection, as the Albacores approached the battleship from ahead, supported by the Fulmars. The battleship began a 180° turn to starboard to avoid this torpedo attack; however, the two Fulmars dived on the battleship and machine-gunned her decks. This maneuver distracted the Italian gunners. As the Albacores closed with the *Vittorio Veneto,* the ship began her starboard turn, which in hindsight was a mistake because she should have continued to head towards them and attempt to comb the torpedo wakes. Meanwhile, the lead plane was brought under heavy fire and was brought down some 1,000 meters off the battleship's starboard bow, but its torpedo had been launched.

Torpedo damage. The torpedo hit seconds after its launching aircraft crashed into the sea. It detonated on the port quarter above the starboard propeller, some 6.1 meters below the waterline. The torpedo struck a particularly vulnerable part of the ship, and the damage sustained was extensive. (See figure 7-4.) The port outboard shaft broke within the thrust-bearing housing, as the torpedo explosion sheared off the port outboard propeller. The whipping forces of the explosion caused salt water to enter the lubrication system for the inboard shaft. This jammed the line bearings and forced this shaft to be shut down. The auxiliary port rudder was jammed, and the pump for the steering-gear motor of the starboard auxiliary rudder was temporarily put out of action. The size of the torpedo warhead (227 kilograms) and the circumstances were very similar to those experienced by the British *Prince of Wales* when she was torpedoed in the South China Sea on 10 December 1941. (See chapter 5 of Volume II for further details.)

Heavy flooding occurred within minutes after the explosion. Some watertight hatches and doors sprung open in the area of the torpedo hit. Their closing by repair parties was later complicated by the flooding and structural damage from the explosion's blast and the resultant shock response. The port longitudinal bulkhead outside and the transverse bulkhead between the diesel generator room and the aft pump room were demolished. Extensive damage to the structure of the longitudinal bulkhead took place over a distance estimated to be some 10 to 15 meters. There was also a large vertical fracture in this bulkhead in the area just below the weather deck, and a smaller one on the starboard side. Instant flooding took place in the bounding spaces inboard and wing compartments outboard. By 1530 the *Vittorio Veneto* had almost 3,500 tons of flooding water in her stern. The battleship went down by the stern, listed some 4–4.5 degrees to port, and lost all her motive power.

At 1530, while the battleship was motionless, a Royal Air Force Blenheim dropped a bomb that exploded in the water near the stern end. A great column of oil and water fell on the quarterdeck. The bomb fell close enough to the ship so that the side shell was

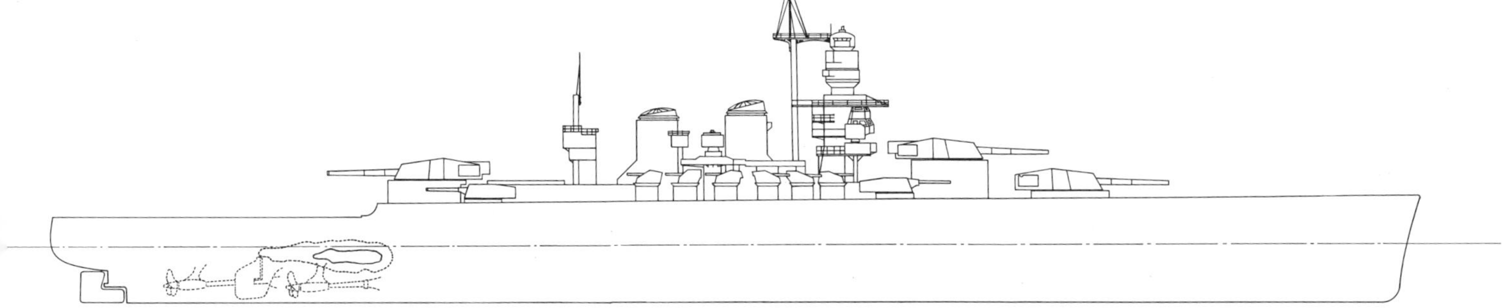

Figure 7-4. Vittorio Veneto—*torpedo damage, 28 March 1941.*

riddled with bomb splinters and the plating was deformed by the explosion's shock wave. The resultant whipping of the ship disabled the main steering gear, and some flooding took place in the steering room. At this point, there were some 3,750 tons of flooding water in the stern.

Immediately after the torpedo attack, damage-repair parties were sent to the stern. There was much damage on the second deck (1° Corridoio). Furniture in the cabins was thrown around, and there was some minor flooding in the compartments over the point of the torpedo's impact. These spaces were made tight and the water pumped out. Due to the loss of the aft pump room, the repair parties' efforts were complicated on the decks below the second deck. This meant that emergency hand pumps had to be used in this section of the ship. They lacked the capacity to dewater those spaces made tight by shoring and other emergency measures. Therefore, the containment and reduction of flooding in the stern of the *Vittorio Veneto* was very difficult and time consuming. Counterflooding of forward and starboard voids reduced the trim and list. However, the effort to regain lost buoyancy was to continue into the night.

The engineering repair parties had more success. Emergency hand steering was used until the hull repair parties had the starboard auxiliary steering-gear engine back in operation around 1642. In a remarkably short time, the two starboard shafts were back in operation, and the ship gradually built up speed. First, a speed of 10 knots was attained, and then 16 knots by the time a British reconnaissance plane spotted her at 1820.

Admiral Iachino radioed the Italian Naval Command of his ship's condition and requested immediate fighter protection. The latter could not be provided, but he was informed that a British fleet of one battleship and one carrier was approximately 175 miles from his position. This report was based on earlier reconnaissance and led Admiral Iachino to make some incorrect decisions. Actually, a British battle fleet consisting of the *Warspite*, *Valiant*, and *Barham* was only 50 miles behind the Italian fleet and closing at a relative speed of 7 knots.

The Italian admiral decided to regroup his forces, and around 1900 the *Vittorio Veneto* was able to increase speed to 19 knots. The battleship took the center position in the fleet formation with two destroyers ahead and two behind. There was a column of three cruisers on each flank, with other destroyers stationed outboard of them. This formidable battle group was to be subjected to one more torpedo attack in a final British effort to slow down the *Vittorio Veneto* and enable the British battleships to overtake her.

The British were aware from air reconnaissance that the *Vittorio Veneto* had been damaged. Eight torpedo bombers were launched from the *Formidable*, and these gained contact with the Italian fleet around 1900. Just after sundown they commenced their attack and hit the heavy cruiser *Pola*, leaving her dead in the water. The British thought they had again slowed the *Vittorio Veneto.*

The Battle off Cape Matapan. British efforts to overtake the fleeing battleship led to a violent night encounter in which point-blank (*maximum* range was less than 4,000 meters), radar-directed fire from the three British battleships sank the cruisers *Pola*, *Zara*, and *Fiume* and two destroyers in a matter of minutes. The latter two ill-fated Italian cruisers had been detached to assist the *Pola.* The Italian high command was once again frustrated by the lack of air reconnaissance and radar, with which the Italian ships were not equipped. Admiral Iachino, based on the reconnaissance data provided to him, had been under the mistaken impression that the British units had turned homeward.

During the voyage to Taranto, progressive flooding threatened all efforts to save the *Vittorio Veneto.* The starboard shafts operated normally, but speed was later reduced due to fear of water contamination of the lubrication system. British carrier-based aircraft launched for a night attack failed to find the ship.

Four months were required to repair the battle damage. In August 1941 the *Vittorio Veneto* was finally able to resume operations. On 23 August the *Vittorio Veneto, Littorio,* four cruisers, and a destroyer screen put to sea to intercept a British force from Gibraltar consisting of the *Nelson, Ark Royal,* a cruiser, and destroyers. After British reconnaissance planes sighted the Italian force, the British fleet returned to base and the Italians returned to port.

First battle of Sirte. On 24 September 1941, nine fast British cargo ships left Gibraltar for Malta. Every means was taken in the dispatch of this convoy so that its sailing would be concealed from Italian intelligence. The true strength of the covering force of this convoy was not known by the Italian naval command, and based on their best intelligence information, they thought that it was almost the same force that had sailed into the Mediterranean a month earlier. However, unknown to the Italians, the battleships *Rodney* and *Prince of Wales* were also included. The British had hoped to lure the Italian fleet into decisive battle with three of their most powerful battleships. An Italian fleet, consisting of the *Vittorio Veneto* and *Littorio,* five cruisers, and 14 destroyers left Naples on 26 September under the command of Admiral Iachino to intercept the British force. During the afternoon of 27 September, the *Nelson* was torpedoed by Italian aircraft and was ordered to stay with the convoy while the remainder of Force H would head north to battle the Italian fleet. Neither side sighted the other, although at one point the two fleets sailed on parallel courses, some 70 miles apart, for several

The *Vittorio Veneto* is shown at anchor some time during 1941–42. (Aldo Fraccaroli)

This starboard view of the *Vittorio Veneto* was taken during the period 1941–42. Note the two Ro.43 floatplanes on the fantail. The ship was equipped with a single catapult, but additional aircraft could be carried on skids in the fantail area. There were no hangar facilities aboard these ships.

hours. The Italian fleet cruised east of Sardinia until 1400 when its mission was terminated by an order from Admiral Iachino to return to base. Once again, the lack of adequate reconnaissance had frustrated an Italian admiral in determining the British fleet's location.

Torpedo damage. On 13 December 1941, the *Littorio* and *Vittorio Veneto,* with a screen of four destroyers, provided distant coverage for an important five-ship convoy to Tripoli. Close support was provided by a *Duilio*-class battleship, three cruisers, and three destroyers. However, a radio diversion by British light units gave the impression that the British battle fleet in Alexandria was at sea, resulting in nervous Italian admirals recalling the warships and the convoy. While returning to her base at 0859 on 14 December 1941, the *Vittorio Veneto* was torpedoed by the British submarine *Urge* off Cape dell Armi in the Straits of Messina. A spread of three torpedoes was fired at the Italian battleship, and one exploded on the port side abreast of the after turret (see figure 7-5). The force of the explosion of 340 kilograms of TNT was successfully absorbed by the side-protection system. A large hole some 13 meters in length was ripped open in the hull. About 2,032 metric tons of sea water flooded this system over three main compartments. The ship rapidly listed 3.5 degrees to port, and the stern went down 2.20 meters. The port list was reduced by one degree by flooding bilge compartments on the starboard side outboard of the forward turret. There were small cracks in the hull structure that affected other spaces in the ship. However, the stern trim was maintained until she reached port. The ship reached Taranto under her own power, but with a heavy loss of life. The extensive damage sustained put this battleship out of service for many months of repair.

Convoy battle of June 1942. The *Vittorio Veneto* was again ready for action in June 1942. During mid-June the Royal Navy made a two-pronged attempt to resupply Malta. A convoy of five ships was dispatched from Gibraltar with coverage by a British force consisting of the *Malaya,* two carriers, 4 cruisers, and 17 destroyers. This force would cover the convoy as far as the Skerki Channel between the African coast and Sicily. The

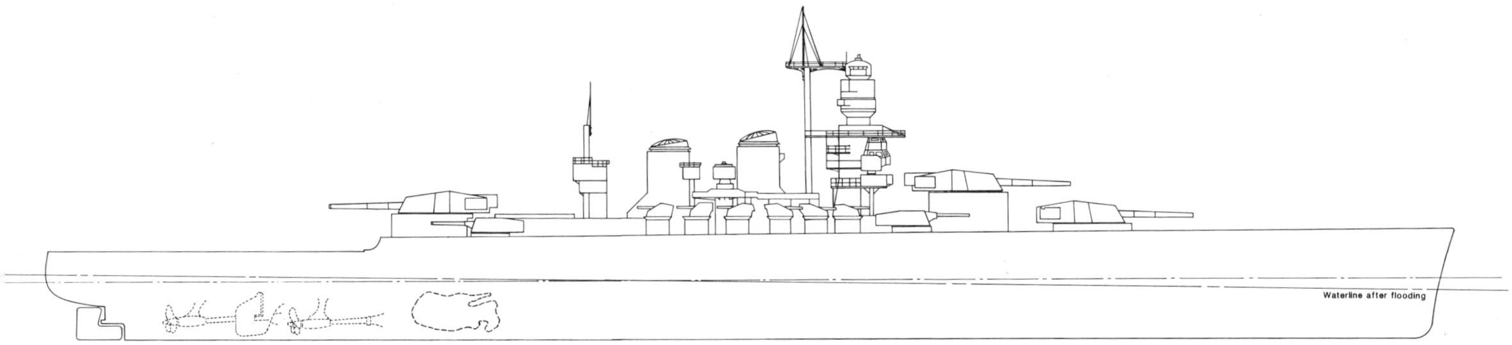

Figure 7-5. Vittorio Veneto—*torpedo damage, 14 December 1941.*

final escort would be one cruiser and nine destroyers from this point to Malta. Coming simultaneously from the east would be eleven merchant ships, covered by the British Mediterranean Fleet. The *Vittorio Veneto* and *Littorio* left Taranto at 1430 on 14 June to intercept the westbound convoy, while an Italian cruiser division was poised at Palermo, Sicily, to intercept the eastbound convoy.

The Italian battleships were to intercept the westbound convoy and its covering force at approximately 0930 on 15 June. While en route to their interception point, they were attacked by British Wellingtons at 0340 and at 0610 by nine Beauforts. However, only the heavy cruiser *Trento* was torpedoed. By 0828 the Italian battle fleet was still heading south, some 150 miles from the convoy. At 0905 American B-24 bombers, using the Norden bombsight, conducted a high-level attack from an altitude of 12,800 meters. A number of near misses were made on the two battleships, and one bomb hit the *Littorio* on her forward turret.* The attack did not slow down the Italian fleet or the *Littorio.* Some forty minutes later a flight of Beauforts equipped with torpedoes tried to attack the Italian fleet. By this time some air coverage had been provided for the Italians, and German Me-109s shot down two planes and damaged five others, but the remaining British aircraft pressed home their attack in the face of heavy antiaircraft fire. No hits were made.

Admiral Iachino continued on a southeast course to intercept the British force, but by 1500 he knew he would have to fight a night action for which his crews were not trained. He altered course towards Navarino where he felt that he could make a fresh start or intercept on the next day. During the withdrawal, the Italian fleet was attacked again by torpedo-bearing Wellingtons. This time they were successful in torpedoing the *Littorio.* In the meantime, the British force the Italian fleet was searching for had been weakened by German and Italian aerial assaults. Around 1830 the decision was made by the British to return the convoy and warships to Alexandria. The Italian fleet returned to base, but minus the cruiser *Trento,* which was left dead in the water after an aerial torpedo hit and was sunk by a torpedo from a British submarine. The Gibraltar

*This was one of the few instances during World War II of the successful high-level bombing of ships under way in the open sea.

convoy was also decimated by the Italians. Out of 16 vessels dispatched to reach Malta, only one ship got through.

This was to be the last combat operation for the *Vittorio Veneto*, as the serious fuel shortage prevented any large-scale activity by Italian naval units. On 12 November 1942 the *Vittorio Veneto* and *Littorio* were moved from Taranto to Naples. The *Vittorio Veneto*, *Littorio*, and *Roma* departed for La Spezia on 6 December. By the end of 1942 the only active Italian battleships were the three of the *Vittorio Veneto* class; all otherwise operational, reconstructed battleships were out of service due to the lack of fuel. That, plus a shortage of escort craft, reduced the *Vittorio Veneto*-class ships to a very limited combat potential. The *Vittorio Veneto* was used to supplement antiaircraft defense of some Italian cities like La Spezia. The 152-mm and 381-mm guns had time-fuzed shells so that these ships could provide barrage fire.

Bomb Damage. La Spezia was heavily bombed on 19 April 1943. The *Vittorio Veneto* escaped damage then, but was not so fortunate during another raid at 1345 on 5 June when U.S. B-17 bombers carrying 908-kilogram armor-piercing bombs made two hits on the port side near the forward turret (figure 7-6). The first bomb, a dud, hit the port side near frames 160-161 and penetrated the armor-deck system before passing through the side-protection system and out the bottom. Fragments penetrated the watertight bulkhead at frame 159 and the absorbing cylinder immediately aft.

The bomb penetrated a total thickness of 216-mm of steel plating (including a 100-mm armor deck), and made a hole 700-mm in diameter in the upper and lower absorbing cylinder in the side-protection system before finally passing through the ship's lower structure near bulkhead 159. The bomb left a 390-mm-diameter hole at its point of entry on the forecastle deck (USN 01-level) and the main deck below, whereupon it entered the cofferdam side-armor structure. At the second-deck level it left a 385-mm hole tangent to the 7-mm armor belt. The latter was bulged outboard with a maximum distortion of 50 mm. The covering of the bulge was pierced, and a 400-mm hole was made in the doubler plate. The bomb traversed the absorbing cylinder, the interior plates of the double bottom, and the bottom plating, where a 500-mm hole was

The *Vittorio Veneto* is at anchor under the guns of Fortress Malta. Note the single-wing reconnaissance floatplane and "Gufo" radar apparatus on top of the forward command tower. (Imperial War Museum)

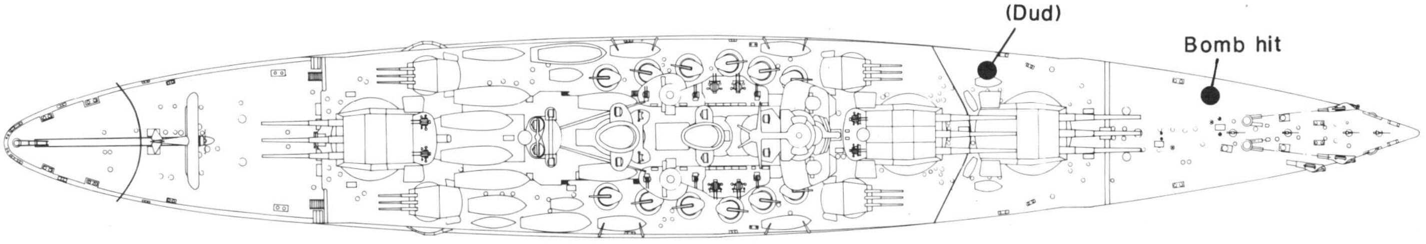

Figure 7-6. Bomb damage (Vittorio Veneto), *5 June 1943.*

made. The transverse bulkhead (frame 159) adjacent to the point of trajectory had fragmented the holes over a height of 4.60 m. The bulkhead was also pierced by an armor splinter that produced a tear in the absorbing cylinder of the bulge aft.

The second bomb struck the port side near frame 197 and went through the side of the ship near the second platform before exploding in the water. Plating from frames 188 to 203 was deformed to a maximum deflection of 1.5 meters by the water hammer of the bomb's explosion in the water. This deflection was accompanied by six large tears in the plating over an area of 60 square meters. The small frames were also buckled, which required extensive repairs.

Further damage was sustained in the area of frame 197:

- *Weather deck*—a circular opening 390 mm in diameter
- *Forecastle deck*—a circular opening approximately 390 mm in diameter and small fractures in the bulkhead adjacent to the hole
- *2nd deck*—a 390-mm hole in the 70-mm splinter deck
- *3rd deck*—a single 390-mm hole with five smaller irregular holes caused by the fragments from the deck over

The ship was sent to Genoa for repairs, since the dockyard facilities in La Spezia had sustained such serious damage from aircraft bombs that they could not accommodate her. Once the repairs had been completed, the *Vittorio Veneto* returned to La Spezia. There she was to remain inactive until the Italian surrender in September 1943.

The *Vittorio Veneto* is seen from an RAF Baltimore escort plane during her approach to Malta. Note the camouflage scheme, the lone reconnaissance aircraft on the stern catapult, and the "Gufo" radar on top of the forward command tower. (Imperial War Museum)

"Be pleased to inform Their Lordships that the Italian Battle Fleet now lies at anchor under the guns of the fortress of Malta" was signaled by the Royal Navy's Admiral Cunningham on 11 September 1943. The *Italia* steams to surrender to the Allies. Crew members of the Royal Navy battleship *Valiant* line the rail to view this historic event. (Imperial War Museum)

This famous view shows the *Littorio* in 1940 shortly after completion.

The *Vittorio Veneto* is shown leading the cruiser *Eugenio de Savoia* and the *Italia* (right) to Alexandria. The photo was taken from a floatplane from the *King George V*, which, with the *Howe* and other British warships, provided an escort. The two Italian battleships were interned for the duration of the war in the Great Bitter Lake south of Suez, and they did not return to Italy until 1947. (Imperial War Museum)

At 0300 on 9 September 1943 the *Vittorio Veneto* left La Spezia for Sardinia, where the Italians had hoped to establish the king and his government in an effort to retain some influence on their destiny. This scheme was thwarted by the German occupation of Sardinia during the daylight hours of 9 September 1943, and after the formation was attacked by German planes and the *Roma* sunk, the *Vittorio Veneto* went on to Malta with the remaining units of the Italian fleet, arriving there on 11 September.

The armistice ended the combat career of the *Vittorio Veneto*, the most active Italian capital ship in World War II. She sortied on a total of 56 missions, took part in 11 offensive actions, changed bases 12 times, and conducted 33 exercises. She steamed 17,970 miles during 1,056 hours under way and burned 20,288 tons of fuel. She was inactive for repairs and other causes for a total of 199 days.

Final Disposition. The ship sailed on 14 September 1943 for Alexandria, arriving there on 17 September. A month later she went to Lake Amaro (Great Bitter Lake) and remained there under virtual internment until 6 February 1946 when she returned to Italy, arriving at Augusta, Sicily, three days later. She returned to La Spezia on 14 October 1947, was decommissioned on 3 January 1948, and was formally stricken on 1 February 1948. Although the Paris Peace Treaty had ceded the ship to Great Britain, the British government permitted the Italians to scrap her. This was done in La Spezia during 1948–1950.

Littorio—Operational History. The *Littorio* (later the *Italia*), second ship of the *Vittorio Veneto* class, was laid down on 28 October 1934 by Cantieri Ansaldo of Genoa. She was launched on 22 August 1937, completed on 6 May 1940, and was commissioned on 24 June 1940. She joined the 1st Naval Squadron at Taranto under Admiral Bergamini's flag.

The *Littorio* was considered to be a high-prestige ship in the Royal Italian Navy and her crew was carefully selected. Even so, the *Littorio* was rushed into service with the crew not completely trained. Civilian engineers were on board the first few months to check and adjust equipment.

After two brief sorties of no consequence, the *Littorio* was anchored at Taranto (Mar Grande) with five other battleships, about five cruisers, and 20 destroyers during the night of 10–11 November 1940. A total of 42 Swordfish aircraft, launched by the carrier *Illustrious*, made two raids on the anchorage. Eleven of the planes carried torpedoes, the others were armed with bombs. The raid was highly successful, resulting in damage to the old battleships *Conte di Cavour* and *Caio Duilio* as well as the *Littorio*, two cruisers, and several fleet auxiliaries. The *Cavour* was so badly damaged that she never saw action during World War II again; both the other battleships had to undergo extensive repairs. The cost to the British was only two aircraft shot down.

Torpedo damage. The *Littorio* was hit almost simultaneously by two torpedoes at 2315. One detonated on the starboard side slightly abaft the forward turret in the vicinity of frame 163. A 7.5-meter hole was punched in the ship, plating was dished in, and the plating beneath the bulge structure was caved in. Several double-bottom spaces in way of the protection system were flooded from frames 159–165.

A second torpedo exploded under the stern on the port quarter at frame 9, tearing holes in the port and starboard sides just abaft the steering-gear room. The rudder was partially destroyed, and there was considerable structural damage to the foundations of the steering gear from the shock response from the torpedo explosion. Despite all damage, the ship was never in immediate danger of sinking.

At 0001 on 12 November, a third torpedo hit the starboard bow at frame 192, tearing a hole between frames 187 and 192 and distorting plating from frames 183–199. There was extensive flooding (figure 7-7). The ship was deliberately grounded in shallow water to facilitate salvage; she settled into the mud bottom with a starboard list, the bow awash back to the forward turret.

Salvage was made hazardous by an unexploded torpedo found in the mud beneath the ship's keel. Finally after much careful work, because sudden shifts in the magnetic field might detonate the torpedo warhead, the *Littorio* was dry-docked on 11 December

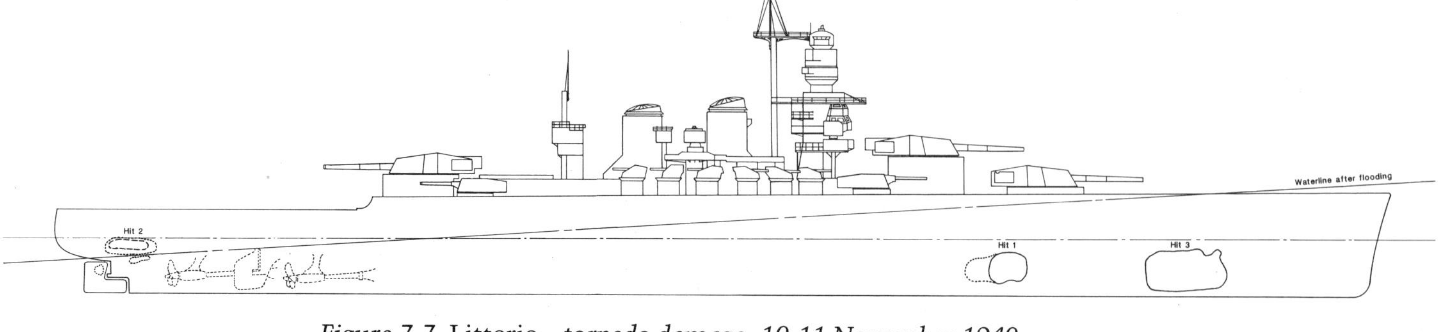

Figure 7-7. Littorio—*torpedo damage, 10-11 November 1940.*

1940; repairs were completed on 11 March 1941, and she was back in service by the end of that month. On 1 April Admiral Iachino's flag was again hoisted on *Littorio*'s foremast. The ship was inactive for the next four months, but in August and September some inconclusive sorties were carried out with the *Vittorio Veneto*, as previously described. On 17 December 1941 the *Littorio* participated in the inconclusive First Battle of Sirte Gulf; this marked the first time in two months that an entire Italian convoy reached Libya.

First battle of Sirte Gulf. The *Littorio, Doria,* and *Cesare,* two cruisers, and ten destroyers were the distant covering force for the convoy, which had as its close escort the *Duilio,* three cruisers, and four destroyers. In the meantime, a heavily escorted tanker had left Alexandria in company with three British cruisers and a dozen destroyers. A German reconnaissance plane spotted this force and erroneously reported the tanker as a battleship. With the two forces 250 miles apart, Admiral Iachino headed his distant covering force eastward, toward the British. By 1730 the Italian ships observed antiaircraft fire to the east, and shortly thereafter the superstructures of the British ships hove into view. The *Littorio* opened fire at 1753 at the extreme range of 32,000 meters. Despite such a range, however, the Italian 381-mm gun was capable of accurate fire. As the range closed, the *Andrea Doria* and the *Guilio Cesare* opened fire with their 320-mm guns, and the heavy concentrated fire from the three Italian battleships covered the Italian destroyers in their dash towards the British ships. At 1800 the action

The *Littorio* (later renamed *Italia*) displays the modest bulbous bow given the class. Such bows markedly improved the hydrodynamic properties of the ships.

Royal Navy Swordfish torpedo bombers attacked units of the Italian fleet at Taranto on the evening of 11 November 1940. The *Littorio* was struck by three torpedoes forward that caused heavy flooding and extensive structural damage from the whipping response to the detonations. This photo shows the ship's stern high out of the water with a Ro.43 floatplane upended by the whipping action of the hull girder. (Italian Navy)

After the *Littorio* had been made tight and an unexploded torpedo removed from under her bow, the ship was repaired at Taranto. (Italian Navy)

ended with the British ships retreating behind a smokescreen, and then the battle area was plunged into darkness. No further action occurred, but the battleship force steamed south to provide distant cover for the Italian convoy, which reached Tripoli on 19 December.

Second battle of Sirte Gulf. On 21 March 1942, two Italian submarines patrolling in the eastern Mediterranean radioed that a British convoy was headed west with cruiser and destroyer protection. The Italian Naval Command ordered the *Littorio* to sail from Taranto with four destroyers, and three heavy cruisers with four destroyers sortied from Messina. By the morning of the 22nd, the British force of five cruisers and eighteen destroyers had closed up to cover a vital Malta-bound convoy, which was slowed by a severe storm out of the southeast. In the heavy weather, one of the *Littorio*'s escorting destroyers had an engine malfunction and had to return to port. At 1424 the two opposing forces sighted each other. The Italian cruiser-destroyer force, believing that three battleships were present in the British formation, laid down a smokescreen and retreated to the northwest. The British followed the Italian cruisers, coming under their fire at 1435. In the meantime, the British convoy was diverted south, screened by a cruiser and six destroyers.

At 1618 the *Littorio* and her escorts joined with the Italian cruisers and destroyers. By this time the storm had intensified, and dense spray caused by gale-force winds whipped over the waves and lessened visibility. Admiral Iachino decided to head west towards Malta to cut off the British forces since his heavier ships would be able to fare much better than his adversaries in terms of ship speed and seakeeping. The British laid down a smokescreen to make rangefinding even more difficult for the Italians. However, the Italian ships continued to fire their guns until night fell. Shortly after 1830, British destroyers made a valiant close-range torpedo attack against the *Littorio*, but blistering fire from her 381-mm and 152-mm guns repulsed the attack.

During the battle, the muzzle blast from the after 381-mm turret set fire to an Ro.43 floatplane on the fantail, again demonstrating that aircraft on battleships presented a dangerous fire hazard. The fire that blazed up from the damaged aircraft led the British to believe they had scored a torpedo hit on the battleship. As the *Littorio* withdrew from the action at 1851, a 120-mm shell exploded above the fantail to starboard, near frame 27. Fragments sprayed over a radius of some 20 meters, several plates were damaged,

The *Littorio*, which was renamed *Italia* on 30 June 1943, is at anchor in this May 1942 photo taken in Taranto. The ship's anchorage is protected by a system of anti-torpedo nets, whose buoys can be seen. Such measures were mandatory after the devastating attack on this port by British carrier aircraft in November 1940. (Italian Navy)

and approximately a square meter of wood planking was destroyed. A near-miss to port caused only minor damage to the side shell plating aft. The *Littorio* also sustained some storm damage, and two Italian destroyers were lost on the way back to port. During this action, 181 rounds were fired by her 381-mm guns.

Battle of mid-June (12–16 June 1942). Another attempt in June against an Alexandria-to-Malta convoy by the *Littorio* and *Vittorio Veneto* was successful in forcing the British to turn back. The *Littorio* and *Vittorio Veneto,* four cruisers, and 10 destroyers sortied from Taranto on 14 June. British reconnaissance aircraft promptly sighted the Italian units, and several night air raids were unsuccessfully launched in an effort to turn the Italian force away from the convoy.

This photograph was taken with a long-focus camera from one of the escorting RAF aircraft. Officers and crew of the *Italia* line the decks and even sit on the gun mounts while Captain Calamai in his white uniform on the bridge looks at the British aircraft through his binoculars. Some of the bridge arrangements and details not in the foldout drawing can be observed here. (Imperial War Museum)

Italian sailors on the *Italia* watch as a launch containing Vice Admiral W. G. Tennant (RN) pulls alongside. Port side details abreast of the funnel can be seen here, including one of the 90-mm gun mounts. (Imperial War Museum)

Italian sailors line the deck as the *Italia* arrives in Malta. The destroyer *Artigliere* is seen to the starboard bow of the Italian battleship. Note how the Pugliese underwater hull form is integrated into the upper ship form. The "Gufo" radar apparatus is visible on top of the forward command tower. (Imperial War Museum)

On 9 September 1943, the three *Vittorio Veneto*-class battleships sailed from La Spezia, presumably to join their German allies to attack American and British invasion forces at Salerno. The Italian Naval Command made this agreement with the German Naval Command in Italy to disguise their real destination—Sardinia. However, German forces occupied this Italian island before these ships could reach there, so the Italian naval force headed for Malta to surrender. Here are views of the *Vittorio Veneto* (top) and of the *Italia* (bottom) on 10 September 1943 as they neared Malta. (Imperial War Museum)

Bomb damage. Two air attacks on 15 June resulted in damage to the *Littorio.* At 0905 that morning, American B-24 bombers in a high-level attack made a single hit on the forward turret. A rangefinder hood was damaged, as was the barbette, and several holes were punched in the forecastle deck from bomb splinters. However, the turret remained operational, and the ship continued on her mission, participating in the pursuit of the fleeing convoy.

Torpedo damage. At 1400 on the 15th of June, the pursuit was ordered discontinued. That evening, as the *Littorio* returned to her base, a particularly violent air attack from Malta-based Wellington torpedo bombers resulted in one torpedo hit at 2339. The torpedo detonated near frame 194 on the starboard bow and tore a large hole in the shell plating. About 1,600 metric tons of water flooded the double bottom and lower compartments in the bow (figure 7-8). Some 350 tons of counterflooding water aft helped to improve the trim and remove the list.

Repairs were completed by 27 August 1942, but the ship was destined to carry out no more combat missions. For the Italian Navy, the fuel situation had become so critical that the reconstructed battleships *Doria* and *Duilio* were transferred to reserve

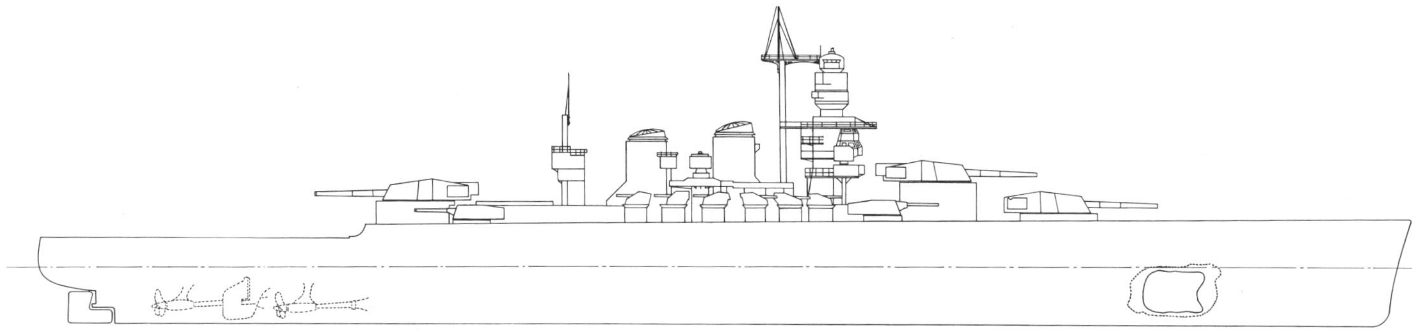

Figure 7-8. Littorio—*torpedo damage, 15 June 1942.*

status in December 1942. Only the *Littorio* and her sisters were to remain operational, but only for specific missions.

On 13 November 1942 the *Littorio* was transferred to Naples, but returned to Taranto on 4 December. From December 1942 to June 1943 the *Littorio* was used to defend the cities of Taranto, La Spezia, and Genoa from Allied air attacks. The 152-mm and 381-mm guns had special time-fuzed shells so that these heavy guns could provide barrage antiaircraft fire.

A bombing raid on La Spezia on 14 April 1943 resulted in one hit on the port side of number 2 main-battery turret; armor plate was ruptured and distorted, and fragments caused minor damage to deck guns and plating in the area. This damage was repaired at La Spezia. A particularly severe raid on 5 June, which damaged the *Vittorio Veneto*, caused only minor damage to the *Littorio*. Again, despite the critical condition of the shipyard and the naval base, repairs could be made at La Spezia.

On 25 July 1943, subsequent to the fall of Premier Mussolini and the Italian fascist regime, the *Littorio* was renamed the *Italia*. On 9 September, while the *Italia* was underway with other units of the Italian fleet off La Maddalena, Sardinia, she was hit by a heavy German radio-controlled bomb of the type known as "Fritz-X." The bomb hit the forecastle deck to starboard at frame 162½, penetrated the side of the ship, and detonated in the sea close to the shell plating. About 190 square meters of plating were damaged, and eight double-bottom compartments, three bilge compartments, and two interconnecting sections were flooded. About 830 metric tons of water flooded in forward, and about half that much counterflooding water was taken on aft. A second bomb exploded close aboard the port quarter, causing fuel oil from a damaged tank to flow into the refrigerated spaces (figure 7-9). The damage later was temporarily repaired by installing a patch over the area torn open by the bomb.

Final Disposition. There was a proposal to use the *Italia* as a bombardment ship in the Allied landings in Southern France, but it was finally abandoned for political considerations. After brief stays at Malta and Alexandria, the *Italia* was interned in the Suez Canal area from 18 September 1943 until 4 February 1947. She

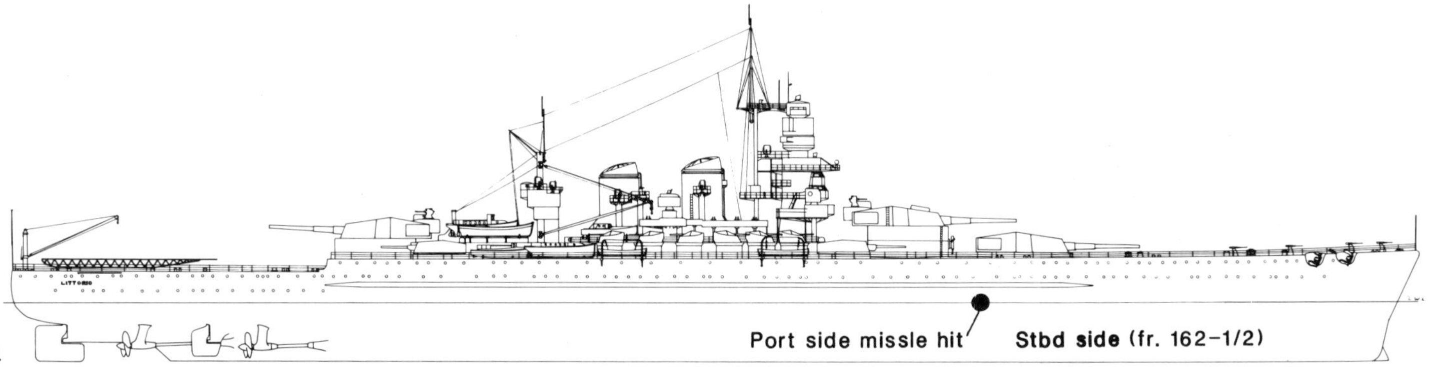

Figure 7-9. Italia*—guided-missile bomb damage, 9 September1943.*

then sailed to Augusta, Sicily, where she arrived on 9 February. Ceded to the United States by the Paris Peace Treaty, the ship was turned back to Italy for scrapping. She was decommissioned and stricken from the Navy List on 1 June 1948 at La Spezia, and scrapped there.

The *Italia* (ex-*Littorio*) conducted a total of 46 missions during her wartime career, nine of which were combat operations. She steamed a total of 13,583 miles during 755 hours underway, burning some 17,740 metric tons of fuel. She was inactive for repairs and other causes for 251 days.

Roma—Operational History. The *Roma,* the first of two slightly modified versions of the *Vittorio Veneto* class, was laid down by Cantieri Riuniti dell'Adriatico, in Trieste, on 18 September 1938. While the *Vittorio Veneto* and *Littorio* were under construction, the decision to reinforce the Italian fleet with two more battleships was made because of increasing international tensions, Italy's withdrawal from the League of Nations, and the abandonment of negotiations on arms limitation. The Italians considered that even if the four old battleships were modernized, two new *Vittorio Veneto*-class ships would not give them sufficient strength to confront a Franco-British naval alliance in the Mediterranean; hence, the determination to build two more ships, named *Roma* and *Impero.*

The *Roma* was launched on 9 June 1940 and was commissioned on 14 June 1942. On 21 August she arrived in Taranto, where she joined the 9th Naval Division. She was the last Italian capital ship to enter service, and had a very brief and tragic career. She conducted training exercises and changed bases several times. In June 1943, while anchored at La Spezia, she was damaged by American air raids. Her movements from port were also restricted because of the severe fuel shortage.

Bomb damage. At 1359 on 5 June, hits by two 908-kilogram armor-piercing bombs caused extensive flooding in the bow. One bomb struck about one meter from the starboard side near frame 222, pierced the bilge keel, exploded in the sea, and caused leakage between frames 221–226 over an area of 32 square meters, with flooding from frame 212 forward to the stem. A second bomb detonated in the water close aboard to port, causing leakage over an area of about 30 square meters between frames 198 and 207, below the second platform. The chain locker, bilge pumps, depth-finder room, and miscellaneous other compartments were flooded by a total of some 2,350 tons of water.

This August 1946 photograph of the *Italia* was taken in Lake Amaro at the southern end of the Suez Canal. The battleship was surrendered to the Allies in Malta on 11 September 1943 and was awarded to the United States under the terms of the peace treaty. She returned to Italy in 1947 and was scrapped at La Spezia in 1948–49. Note the awning over the forecastle and the temporary patch near the forward turret. The wartime camouflage has been removed. (Imperial War Museum)

Damage would have been more severe if the bombs had been high-explosive rather than armor-piercing types.

During the evening of 23-24 June, two more bombs damaged the *Roma*. The first struck the weather-deck scuppers, starboard side aft, penetrated the deck, and destroyed several staterooms below. These were flooded by wash water from broken pipes. The second detonated on the starboard side of the face plate of number 3 turret without penetrating the armor. There was limited local structural deformation to the backing structure, but no damage of consequence, and no flooding (figure 7-10). As a result of the bomb damage, the *Roma* went to Genoa for repairs, where she arrived on 1 July. She returned to La Spezia on 13 August.

Loss of the Roma. The *Roma* went to sea with other units of the *Vittorio Veneto* class and the Italian fleet on 9 September 1943. As stipulated in the Italian armistice with the Allies, she was the flagship of Admiral Carlos Bergamini, who commanded the fleet formation. The Italian fleet left La Spezia and was joined by three cruisers from Genoa. The fleet made a feint towards Salerno, ostensibly to attack the Allied invasion force heading there, and then altered course towards Malta. German intelligence quickly learned of the defection, and the Luftwaffe ordered planes armed with guided bombs to attack and destroy the Italian ships. En route to Malta, when in the Gulf of Sardinia, the *Roma* was attacked by German aircraft. The Allies had promised air cover for the passage to Malta as part of the surrender terms.

During the early afternoon, airplanes appeared on the horizon and followed the formation for some time. Too distant for identification, they were thought by the Italians to be the promised air cover. It was not until the *Italia* and *Roma* were attacked that these aircraft were recognized as German Dornier 217s. At that time Italian antiaircraft guns opened fire, and the ships began evasive maneuvering.

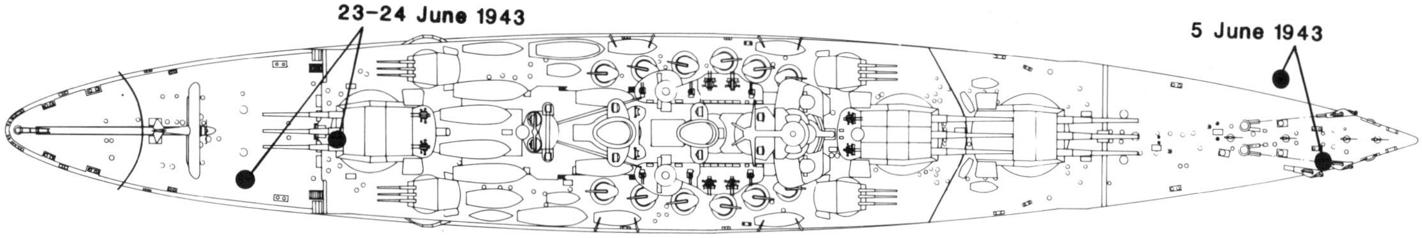

Figure 7-10. Roma—*bomb damage, June 1943.*

The *Roma* is shown with an interesting dazzle camouflage scheme in 1942. Such systems were intended to make it difficult to judge the ship's heading and distance. (Italian Navy)

The *Roma* is shown here in 1942. Note the camouflage scheme and the exposed machine-gun mounts on the forecastle and the Ro.43 floatplane aft. (Aldo Fraccaroli)

The *Roma* is being towed from Trieste, where she was built, to Venice for dry-docking and final completion of outfit in this November 1941 photo. (Aldo Fraccaroli)

The first attack at 1537 was unsuccessful, but a second attack some 15 minutes later produced hits that damaged the *Italia* and caused severe damage that later destroyed the *Roma*.

The *Roma* was hit by the "Fritz-X"-type radio-controlled bomb (PC-1400X), which, when dropped from a height of 6,000 meters, was capable of penetrating some 130 mm of deck armor. Its total weight was 1,570 kilograms, with an explosive charge of 300 kilograms. Fritz-X was the equivalent of a semi-armor-piercing bomb in destructive effect; when controlled by a skilled bombardier, it could be placed consistently within a 4.5-meter radius of the aiming point. Its operational record was good, with hits scored in approximately 30 percent of its launches. Similar bombs hit the battleship *Warspite* (three times), the cruiser USS *Savannah*, and numerous other ships. Later in the war, the Allies learned to jam the radio transmissions used to control these bombs, drastically reducing their effectiveness. Truly, the *Roma* was sunk by a powerful new type of guided missile, specifically designed for use against heavily armored targets.

One bomb struck the starboard side of the *Roma* between frames 100 and 108, passed through the Pugliese system, and exploded under the hull. The Pugliese system was bodily shifted inboard, as the riveted joint at the bottom failed under the stresses imposed by the detonation. The explosion of such a large charge caused severe damage to the shell and framing of the hull girder. The after engine room and boiler rooms 7 and 8 flooded rapidly, resulting in the loss of power to the inboard propellers. The large infusion of salt water in a short period of time and the rupture of electric cables caused much arcing. This caused severe fires in the after section of the ship. The *Roma* soon fell out of formation, paralyzed by the sudden loss of power and the massive flooding.

Around 1602 a second bomb struck the *Roma* between frames 123 and 136 slightly to starboard. The bomb continued into the ship, probably exploding in the forward engine room, causing additional fires and flooding the magazines for number 2 main-battery turret and the forward 152-mm turret on the port side. The explosion of this bomb caused massive flooding and excessive strain on an already weakened hull girder. The number 2 381-mm turret was blown overboard by the violent explosion of its magazines a few seconds later.

The *Roma* is shown here completing her outfitting. Torpedo nets to the stern protect the ship.

This is a view of the port quarter of the *Roma* taken in 1943 at La Spezia. Note the camouflage scheme. Although this ship was completed, she saw no active war service.

The *Roma* was the most handsome of the battleships of the *Vittorio Veneto* class, with a modified bow form that gave greater freeboard.

There was extreme flooding forward, and the forward bridge/command tower tilted to starboard and forward. The ship also began to heel to the starboard side and finally capsized, breaking in two parts. Of a total of 1,849 officers and men on board, only 596 survived.

An official inquiry by the Italian Navy was conducted into the loss of the *Roma*. It was concluded that the ship was lost to massive damage caused by two unusually large bombs, of which little was known. However, the ship was fought by her crew in the best traditions of the Italian Navy.

The *Roma*'s career lasted 15 months, during which she participated in no combat missions. She made 20 sorties, steamed 2,492 miles during 133 hours underway, and consumed 3,320 metric tons of fuel. She was out of service for repairs for 63 days.

The FX-1400 bomb was designed for use against heavily protected targets. The *Roma* was destroyed by two of these bombs, one of which hit the ship, while the second exploded in the water nearby.

Dornier Do-217K bombers of this type attacked the *Roma*. Each carried one FX-1400 bomb.

This sequence of photographs depicts the stunning destruction of the *Roma* on 9 September 1943. The two top views show the massive explosion that devastated the forward portion of the ship. The middle view shows the two halves of the ship moments later. Note that the ship has capsized and is plunging bottom up. The bottom view shows the bow just before it disappeared below the surface. Note the escorting ships closing the wreck to rescue survivors. (Two bottom photos: Imperial War Museum)

Impero - Operational History. The *Impero* was laid down by Cantieri Ansaldo in Genoa on 14 May 1938 and launched on 15 November 1939. In May 1940, the likelihood of war with France meant that the ship could be attacked by Allied bombers. This led to plans to have the battleship towed to Trieste for completion by the yard building the *Roma*, but that yard could not work on two battleships simultaneously. Therefore, on 8 June 1940 she was towed to Brindisi, where the propulsion machinery and some gun mounts were completed. While stationed there, the *Impero* was subjected to Allied bombing attacks, but no damage was sustained. The Italian Navy had decided to concentrate on destroyer and escort construction and the *Impero*'s completion had been deferred.

On 22 January 1942 the *Impero* moved to Venice under her own power. Still incomplete, she had a commanding officer and ship's company assigned, and sufficient

The *Impero* is photographed on the building ways at Ansaldo, Genoa, on 15 November 1939, the day of her launch. The ship was later moved to Brindisi, then to Venice, and finally to Trieste, where she eventually fell into German hands. She was never completed.

A floating crane is lowering the roller-path structure of the after main-battery turret in this 1941 view of the *Impero* at Brindisi. The *Impero* was never completed due to the shortage of steel and in January 1942 was towed to Venice. She was to remain a hulk for the duration of the war and was later moved to Trieste, where she was taken over by the Germans when Italy capitulated. (Aldo Fraccaroli)

armament for limited antiaircraft and surface defense. Later, the *Impero* went to Trieste, where she was later abandoned to the Germans. Construction was far behind schedule due to material shortages and a low priority of construction. The battleship had been laid down more than five years earlier, and was of limited military value.

Final Disposition. At the end of the war the *Impero* was found, half sunk—the Germans had used her as a target ship. She had been damaged by an Allied air raid on 20 February 1945. The hulk was scrapped at Venice, and the name *Impero* formally stricken from the Navy List on 27 March 1947.

Armament. Guns carried by the *Vittorio Veneto*-class ships had characteristics as summarized in table 7-7.

Main battery. The new 381-mm, Model 1934 gun, designed by Ansaldo, was built by both Ansaldo and Odero-Terni-Orlando. This gun was extremely powerful for its caliber, with a maximum range exceeding that of all other battleship guns despite the modest maximum elevation of 35 degrees. The armor-piercing shell could penetrate

TABLE 7-7
Gun Characteristics

Gun	15″/50 (381 mm)	6″/55 (152 mm)
Shell weight (lb)	1,951 (885 kg)	110 (50 kg)
Muzzle velocity (fps)	2,789 (850 mps)	3,035 (925 mps)
Maximum range (yd)	46,216 (42,260 m)	27,230 (24,900 m)
Maximum elevation	35°	45°
Gun Shell weight (lb)	3.5″/50 (90 mm)	37 mm/54
Shell weight (1 lb)	22.333 (10.13 kg)	1.83 (0.83 kg)
Muzzle velocity (fps)	2,723 (830 mps)	2,625 (800 mps)
Maximum range (yd)	18,097 (15,548 m)	7,688 (7,030 m)
Maximum elevation	75°	90°
Gun	20 mm/65	
Shell weight (lb)	0.30 (0.136 kg)	
Muzzle velocity	2,625 (800 mps)	
Maximum range (yd)	6,244 (6,110 m)	
Maximum elevation	100°	

Vice Admiral W. G. Tennant, former captain of the battlecruiser *Repulse* when she was lost on 10 December 1941, talks with Captain Calamai of the *Italia* in Malta. Behind them is one of the four triple 152-mm turrets. (Imperial War Museum)

A view of the after main-battery turret of the *Vittorio Veneto.* The photographer apparently stood on the floatplane catapult to get this view. Note the catapult structure in the center foreground. (Aldo Fraccaroli)

The *Littorio* operating at high speed in 1940. The limited arcs of fire for the various gun mounts and turrets amidships are apparent. (Italian Navy)

348 mm of vertical armor at a range of 23,774 meters. The designed rate of fire was 1.33 rounds per minute. Recoil was 1 meter.

The relative superiority of the gun over contemporary battleship guns of similar caliber is shown in table 7-8.

TABLE 7-8
Comparative Gun Performance

Gun	*Nation*	*Penetration @ Muzzle*
18.1″/45 (460 mm)	Japan	34.06″ (865 mm)
16″/50 (406 mm)	USA	32.62″ (828 mm)
16″/52 (406 mm)	Germany	31.72″ (806 mm)
16″/45 (406 mm)	USA	29.74″ (755 mm)
16″/45 (406 mm)	UK	29.03″ (737 mm)
15″/50 (381 mm)	Italy	32.07″ (814 mm)
15″/50 (380 mm)	France	30.93″ (786 mm)
15″/42 (381 mm)	UK	29.28″ (744 mm)
15″/47 (380 mm)	Germany	29.17″ (741 mm)
14″/45 (356 mm)	UK	26.29″ (668 mm)
13″/52 (330 mm)	France	27.81″ (706 mm)
12″/50 (305 mm)	USA	24.48″ (622 mm)
11.1″/54 (283 mm)	Germany	23.79″ (604 mm)

Note: Penetration data, obtained using empirical U.S. Navy equations, are probably not precisely correct, but relative merits are believed accurately represented.

A particularly dramatic view of the *Littorio* (foreground) and the *Vittorio Veneto* on gunnery exercises during the summer of 1940.

The billowing clouds of gunpowder from heavy battleship guns severely obscured vision from the firing ship, a crippling factor in the era before radar-directed gunfire. The *Littorio* is conducting a gunnery exercise in the summer of 1940.

Only the Japanese 460-mm guns and the U.S. 406-mm/50-caliber gun had superior penetration at the muzzle. As the Italian armor-piercing shell was relatively heavy for its diameter (more so than any other shells of equal bore) and had higher muzzle velocity, long-range penetration capabilities should have been similarly outstanding. This superb performance had its price; the barrel life, about 110–130 rounds, was less than half that of other nations' guns.

The rotating weight of the turrets was about 1,595 metric tons each. Maximum training and elevating rates were 6 degrees per second. The outside diameter of the roller path was 11.8 meters.

Secondary battery. As in all Axis battleships, the secondary armament was of mixed caliber.

Four triple turrets mounting 152-mm/55-caliber, Model 1934 guns were provided for surface defense. The maximum range of 24,900 meters was ample against light cruisers or destroyers, freeing the main battery to engage cruisers or capital ships. Maximum elevation, of only 45 degrees, severely limited the potential effectiveness of these guns in antiaircraft defense.

Primary defense against aircraft was provided by twelve 90-mm/50-caliber, Model 1939 guns in single mounts. These guns had good ballistic properties for their caliber, but were hampered by a multitude of technical difficulties, causing a crucial weakness in the defensive armament. The loading equipment, designed for a rate of fire of 12 rounds per minute, was failure-prone. This problem was compounded by faulty projectiles, defective gun-mount stablizing equipment, and lack of adequate training in fire control. The Italian 90-mm antiaircraft batteries were simply too unreliable.

Light antiaircraft battery. Light antiaircraft batteries of 20 37-mm and 24 20-mm machine guns were installed in the *Veneto* and *Littorio* as completed. The newer *Roma*

All but the main-battery armament of the *Vittorio Veneto*-class battleships is shown in this view. The 37-mm Breda twin antiaircraft machine guns are shown in the left foreground, with a 20-mm mount immediately behind it. Several single-mounted 90-mm antiaircraft guns can be seen on the main deck as well as the guns of the port after 152-mm turret. The crowded conditions amidships are vividly shown here.

This wartime photograph of the *Italia* at speed shows the amidships portion. Note that the 152-mm guns are trained forward and outboard and that 20-mm gun positions have been added to the two main-battery turrets. (Imperial War Museum)

This starboard stern view of the *Italia* shows the floatplane launching facility. The after turret had to be raised in these ships to clear this area of blast damage from astern 381-mm gunfire. Note the radar device mounted atop the rangefinders on the forward command tower. (Imperial War Museum)

carried 20 37-mm and 32 20-mm guns on completion. The antiaircraft armament of the *Littorio* and *Vittorio Veneto* was increased in 1942 with the addition of four 20-mm/65 twin mounts, two abreast of turret Number 2 and one on top of each forward 152-mm turret. Contemporary foreign battleships had much more numerous machine gun batteries.

Aircraft. A single catapult was installed on the fantail. Three Ro.43-type floatplanes were originally assigned to each ship. One or two of them were later replaced by Reggiane 2000-type fighter aircraft that could not be recovered at sea. No hangar facilities were provided.

Protection. The Italian Navy was firmly convinced that in the *Vittorio Veneto* class it had for the first time developed a battleship that was adequately protected against shell fire. Test firing of the new 381-mm guns confirmed the fact that the side armor could withstand the 885-kilogram armor-piercing shell when fired from a range of 16,000 meters.

Side protection. The main side-belt armor, inclined 8 degrees from the vertical, was a sandwich of steel plates with wood backing. Two outer plates, 70 mm superimposed over 10 mm, were located at an interval of 250 mm from the main 280-mm armor plating. A wooden cushion 50-mm thick was supported by 15-mm steel plates, which formed the inner boundary of the side-armor sandwich system. This system extended some 675 mm from the exterior shell plating to the inboard boundary of the backing plates, including 375 mm for steel armor and structural plating. Each composite element consisted of a panel about 3 meters square, framed by armor 60-mm thick. There was no lower side-belt armor.

The external armor plate was designed to tear off the false cap of an armor-piercing shell while the main armor structure broke up the shell. In tests, the system functioned as designed, when a 381-mm projectile was fired from a limiting range of 16,000 meters. The armored splinter-protection bulkheads contained fragments resulting from this projectile, but some shells did manage to penetrate the belt-armor system before breaking up. Similar tests indicated that the 70-mm outer plating was sufficient to withstand the effects of the detonation of a 203-mm high-explosive shell.

Splinter protection was provided by two inboard bulkheads, the first of which was parallel to the belt armor with a thickness of 36 mm. The second, with an inboard inclination, was of 24-mm plates.

Belt-armor protection extended beyond the armored citadel. Forward, 130-mm armor on 11-mm backing plates extended from frames 176 forward to 199 (the Italian Navy numbered frames from aft forward, with frame 0 located at the centerline rudder post), while 60-mm armor on 9-mm backing plates extended farther forward to the bow. There was no side armor aft, although limited protection was given by an internal side-armor system of 60-mm plates on 10-mm backing plates. This system extended from frame 24 forward to 51.

Deck armor. These ships were given good horizontal protection for the time of their design. The system was considered satisfactory by the designers, but World War II developed a need for particularly massive deck armor to counter increasingly powerful bombs and projectiles. Deck armor protection is tabulated in table 7-9.

Turret armor. The main-battery guns had good armor protection. Turret face plates were 380-mm thick, side plates were either 200 mm (forward) or 130 mm, and top plates were 200 mm. The backing-plate thicknesses varied considerably—380 mm on the after turret. The barbette armor was similar in scale, with equal thicknesses over the circumference in all cases. For all locations where the barbettes were not covered by deck plating, the armor was 350-mm thick. Between the main and the third decks, the barbette protection was 280 mm. There was no barbette armor below the third deck.

The secondary guns had remarkably heavy protection, with 280-mm turret face plates, 130-mm side plates, and 80-mm back plates. The top plate was 150-mm thick forward and 105-mm thick aft.

The barbettes, where not shielded by deck armor, were 150-mm thick. Below decks, the barbettes were given 100-mm plating. As was the case for the main-battery barbettes, there was no armor protection below the third deck.

TABLE 7-9
Deck Armor Thickness

	Over Machinery *Centerline*		*Outboard*	
Main deck	1.42″ + 0.35″	(36 mm + 9 mm)	1.42″ + 0.35″	(36 mm + 9 mm)
Second deck	0.47″	(12 mm)	0.47″	(12 mm)
Third deck	3.94″ + 0.47″	(100 mm + 12 mm)	3.54″ + 0.35″	(90 mm + 9 mm)
Total	6.65″	(169 mm)	6.14″	(156 mm)

	Over Magazines *Centerline*		*Outboard*	
Main deck	1.42″ + 0.35″	(36 mm + 9 mm)	1.42″ + 0.35″	(36 mm + 9 mm)
Second deck	0.47″	(12 mm)	0.47″	(12 mm)
Third deck	5.91″ + 0.47″	(150 mm + 12 mm)	3.94″ + 0.35″	(100 mm + 9 mm)
Total	8.62″	(219 mm)	6.52″	(166 mm)

The antiaircraft gun mounts were given limited protection against strafing attacks and fragmentation damage by shield plating ranging from a maximum of 40 mm to a minimum of 12 mm in thickness.

Conning tower. These ships lacked the conventional low-lying, heavily armored conning tower. Instead, the entire forward bridge tower was provided varying degrees of armor protection, with a two-level conning tower structure located relatively high up in the superstructure. The first six levels within the forward bridge tower, starting at the armor deck, were protected by 60-mm plating. An armored communications tube with 200-mm plating between the third and the seventh level, and 150-mm plating between the first and the third level, connected the citadel structure and the triple-level heavily armored structure above. The captain's and admiral's bridges were on the seventh and eighth levels respectively, and the second fire-control station was on the ninth level. All three had relatively heavy armor: Above the ninth level were two levels with optical rangefinders and the top level with the first fire-control station, all unprotected. Although there was a remarkable degree of medium armor protection, the conning-tower armor was inferior to that of most World War II battleships. The tower structure could have been riddled by heavy shellfire at relatively long ranges.

Armored bulkhead. The transverse armored bulkheads, which in conjunction with the side armor and the deck protection formed the boundaries of the armored citadel, were not as heavy as in some foreign designs.

Between the first and third decks, the forward transverse bulkhead was only 70-mm thick, while 200-mm plates superimposed on 10-mm backing plates comprised the

TABLE 7-10
Conning-Tower Armor

9th level	top:	4.72″ + 0.39″	(120 mm + 10 mm)	fore
		3.54″ + 0.39″	(90 mm + 10 mm)	aft
	sides:	8.86″ + 0.98″	(225 mm + 25 mm)	fore
		6.89″ + 0.98″	(175 mm + 25 mm)	aft
8th level & 7th level	sides:	9.84″ + 0.39″	(250 mm + 10 mm)	fore
		7.87″ + 0.39″	(200 mm + 10 mm)	aft
7th level	bottom:	3.54″ + 0.39″	(90 mm + 10 mm)	

main forward transverse armored bulkhead below the armor (third) deck. At a point 23 frames forward of this bulkhead, another armored bulkhead of 60-mm armor on 10-mm backing plates, below the armor deck, provided added resistance. Aft, the upper part of the transverse bulkhead was 70-mm armor, while the main bulkhead was 280-mm armor on a 10-mm backing bulkhead. Outboard, the armor thickness was reduced to 250 mm on 10-mm backing plates. This heavy protection aft extended only between the third and fourth decks. Forward, the lower transverse bulkhead had 70-mm armor, while the after bulkhead was of an equal thickness.

Aft of the citadel, a combination of deck armor and slopes to the shell plating protected the steering gear and other vital equipment. Between frames 24 and 54, third-deck armor consisted of 36-mm plates on an undercourse of 8-mm plates. Vertical armor down to the next deck was a sandwich of 60-mm plates over 10-mm backing plates. The fourth deck inboard was 100-mm plates on 8-mm plates, while the slopes outboard were arranged in a manner similar to that of the main side-belt plating, with successive courses of 50-mm, 9-mm, a small interval, followed by 200-mm over a backing structure of wood and 8-mm plates. This array was sloped down and outboard at an angle of about 30 degrees from the horizontal. Further aft, over the steering gear, protection was confined to the fourth-deck armor of 100-mm plates over 8-mm backing plates. The slopes were of the same thicknesses, inclined some 45 degrees from the horizontal. The after bulkhead of the steering-gear room was a sandwich of 200-mm and 10-mm plates. This distribution of protection was more generous in area than in later capital-ship designs, although the armor thicknesses in later ships were frequently much greater.

Underwater protection. The Pugliese system of underwater side protection was designed to withstand 350 kilograms of TNT detonating in contact with the side shell plating. The innermost cylinder was fabricated of 7-mm steel plates. The special steel plating forming the inboard boundary of the liquid-loaded chamber was from 28-mm to 40-mm thick.

In theory, the innermost cylinder was to collapse under hydraulic and shock loadings before the adjacent structure inboard failed. To accomplish this, the cylinder was made 3.8 meters in diameter and surrounded by oil or water so that substantial energy from the gas jet could be absorbed and dispersed before it could reach the holding bulkhead. To achieve this, no venting plates were fitted. The inboard bulkhead of the system was configured so that the gas jets would be directed towards the surface of the cylinder and holding bulkhead by hydraulic action. In this manner, the energy would be absorbed by the crushing of the cylinder, and this would cause an instant reduction in the pressure, since the gas jet would suddenly expand. This would relieve the forces on the holding bulkhead that would be subjected to high shear forces at the moment of the cylinder's destruction. Thus, it was important to use very thick and well-supported steel plates for the holding bulkhead. The Italians used a special high-strength silicon manganese steel.

Unfortunately, two fundamental defects prevented the system from attaining its potential performance. First and most importantly, the riveted joint between the system and the hull bottom structure was an area of excessive stress concentration and high shear loading in the rivets. It appears that the efficiency of the joint, assumed in design, was estimated too high. These conditions led to massive failures in the riveted connections, even when exposed to large noncontact explosive shock loadings. This weakness also meant that the innermost cylinder would not collapse as designed, but that the riveted seam of the holding bulkhead would fail, which would lead to massive flooding beyond the holding bulkhead. Second, space limitations prevented the reten-

TABLE 7-11

Characteristics of the Side Protection System*

Location	*Cylinder Diameter*	*System Breadth at Half Draft*
Turret one	7.48′ (2.28 m)	18.07′ (5.51 m)
Turret two	9.97′ (3.04 m)	21.19′ (6.46 m)
Amidships	11.58′ (3.80 m)	23.69′ (7.22 m)
Turret three	7.48′ (2.28 m)	19.32′ (5.89 m)

*For the purposes of this analysis, the holding bulkhead is considered to be the vertical plane bulkhead forming the outboard boundaries of the machinery and magazine spaces.

tion of the maximum diameter of the hollow cylinder (3.8 meters) throughout the length of the citadel, as shown in table 7-11. This was due to the hull fineness, a problem in all capital ships designed during this era of the high-speed battleship.

The design of the Pugliese system did not permit efficient utilization of the strength of the steel used in the system. In conventional multiple-bulkhead systems, presuming the detail design was carefully done, the longitudinal bulkheads acted as membranes under explosive loadings, deforming elastically and then plastically before rupturing, thereby taking advantage of the inherent strength of the steel bulkhead. Much of the strength of the steel in the innermost cylinder of the Pugliese system, on the other hand, was wasted. Once the external pressures on the hollow cylinder caused the onset of buckling, the subsequent collapse of the structure was the result of structural instability phenomena, with very little energy absorption. (Cylinders are much more efficient in resisting internal pressures than they are when externally stressed.)

The Pugliese system was a relatively lightweight but bulky side-protective system. Because of the curvature of all plating members, battle-damage repairs must have been appallingly difficult. The Italians were convinced of the merits of this system, but no other capital ships except those built in the Soviet Union departed so radically from the conventional type of multiple bulkhead side-protective systems.

The Italian Navy was extremely concerned that large underwater explosions beneath the underbottom could seriously cripple a battleship. Therefore, it was decided to employ a modest triple-bottom structure that would be confined to the vitals for the entire length of the armored citadel. The vertical extent of the system was about 2.1 meters, with the lower series of compartments having a height of about 1.25 meters. The system was designed to have the lower layer kept liquid-loaded, while the upper layer was void. It was never tested in combat.

Radar. Italian development of radar began in 1935 when Professor Tiberio of the Istituto Eletrotecnico e delle Communicazioni began experiments that resulted in the following models:

- a 2-meter wavelength set using linear modulation of frequency (1936–1938);
- a 2-meter wavelength set using coincidence techniques (1938–1939);
- 70-centimeter and 1.5-meter wavelength sets, both operating on impulse principle (end of 1939).

The 70-centimeter wavelength equipment gave good results, with 12,000-meter detection ranges for surface ships, and 30,000-meter ranges for aircraft. Industrial

Only a few radar sets were produced by Italian industry, despite pioneering research work begun as early as 1935. The antenna array for the E.C.3-type radar is shown on the forward bridge tower of a *Veneto*-class battleship. This radar, also known as "Gufo" (Owl) was capable of detecting surface ships at ranges of about 12,000 meters, and aircraft were detected at 30,000-meter ranges. The Italians, like the Germans and the Japanese, were substantially behind Allied technological progress in the development and operational use of radar systems.

production was not started despite the successful trials. Only after the Battle of Cape Matapan (28 March 1941) was the Tiberio radar revived and experiments conducted on board the torpedo boat *Carini*.

Fifty radar sets were ordered, but various technical problems resulted in the production of only a few sets. The *Littorio* had this equipment installed in October 1942, with a formal designation E.C.3 for Istituto Eletrotecnico e delle Communicazioni Number 3). The name "Gufo" (owl) was used by the Italians for this radar. The *Vittorio Veneto* and *Roma* were similarly equipped.

Propulsion Plant. These were the first Italian capital ships to be built with a designed maximum speed of 30 knots (reconstructed ships of the *Cavour* and *Duilio* classes were capable of maximum speeds of 28 and 27 knots, respectively). The relatively limited required endurance allowed a powerful propulsion plant without undue increase in displacement or length.

Boilers. Eight Yarrow-type boilers were arranged, two each, in four boiler rooms. The boilers were designed to operate at a pressure of 25 kilograms per square centimeter with steam superheated to 325 degrees centigrade. Boiler rooms were not subdivided longitudinally, lessening the chances of excessive lists resulting from flooding in a given space.

Turbines. Four sets of Belluzzo geared turbines were arranged in two machinery spaces, each containing two turbine units. One engine room was forward of the boilers, the other aft. Each turbine set consisted of a high-pressure turbine with a cruising turbine incorporated, an intermediate-pressure turbine with an astern turbine incorporated, and a low-pressure turbine likewise with an astern turbine incorporated. Each turbine set drove a three-bladed Scaglia-type propeller.

The geared turbines were rated at a normal maximum of 130,000 metric horsepower, good for a nominal maximum speed of 30 knots. In World War II actions, the ships made about 29 knots at displacements of about 46,000 metric tons. The overload rating of the turbines was 140,000 metric horsepower. The full-power trials of the first two ships of the class yielded the following results:

TABLE 7-12

Full-Power Trials

	Vittorio Veneto	*Littorio*
Displacement	41,473 tons (42,138)	41,122 (41,782)
SHP	132,775 (134,616)	137,652 (139,561)
Speed, knots	31.428 at 238 RPM	31.292 at 239 rpm*

*Slower with more power and lighter displacement possibly due to rougher sea, more hull fouling, or adverse trim.

The limited endurance permitted a proportionately greater weight of protection and armament than was possible in ships of other nations designed for open-ocean operation. The designed endurance was 4,580 miles at 16 knots; 3,920 at 20; and 1,770 at 30.

Electric plant. The ships had eight 220-volt DC turbo-generators, each rated at 450 kw, 2,020 amperes—four in the upper level of the forward engine room, and the others similarly placed in the after engine room, with a total output of 3,600 kw, 16,160 amperes.

Four Fiat MS 368-type emergency 220-volt DC diesel generators, each rated at 800 kw, 3,640 amperes, were located beyond the citadel, two forward and two aft. Their total output was 3,200 kw. Alternating current was provided by three Ansaldo A 625/4-type alternators, each rated at 62.5 kw, 164 amperes, 62.5 volts, and 50 cycles.

Hull Characteristics. When these ships were designed, model-basin tests indicated that very long hulls designed for high speed should have bulbous bows, and the *Vittorio Veneto* and the *Littorio* were so built. However, the first sea trials of the *Veneto* revealed excessive vibrations in certain parts of the hull, which were finally determined to be the result of too-blunt waterline endings at the bow. The ships were then given a modified bow of the same total displacement with the same bulb and improved form, but finer endings. This gave satisfactory results. High-speed vibrations were reduced by this modification, but at the price of more pronounced bow waves and reduced resistance to pitching. The hull forms of the last two ships were similarly modified during construction.

Especially large bilge keels were fitted in an effort to reduce rolling in rough seas. The cruiser-type stern was more rounded than on the reconstructed battleships.

The structure of most capital ships of this era, of necessity, was of the combination type, with transverse framing amidships to provide the added strength and rigidity necessary to support heavy armor, while the structurally more efficient longitudinal framing was employed at the bow and stern.

The battleships of the *Vittorio Veneto* class conformed to the general rule with regard to capital-ship structural arrangements. As was usually the case in the mid-1930s, welding was confined to non-structural members, although fillet welding was extensively used in the protective system to supplement the riveted structure. Even there, riveting was far more important in the overall structure of the hull girder.

Rudders. For the first time in the Italian Navy, three rudders were installed on these capital ships. The main rudder was located on the centerline well aft; two auxiliary rudders were situated between the inboard and outboard propeller shafts, forward of the inboard props. This arrangement, with considerable longitudinal separation of the main and the auxiliary rudders, was designed to ensure steering control in the event of damage to either the main or an auxiliary rudder. The wisdom of this design decision was proven in World War II, when loss of steering control contributed to the sinking of the *Bismarck* and *Prince of Wales*, but did allow the *Veneto* to escape her British pursuers.

Summary. The battleships of the *Vittorio Veneto* class represented a good combination of armament, speed, and protection, although they had several serious defects that detracted from an otherwise excellent design.

The most apparent weakness was in the 90-mm antiaircraft batteries. The disposition of twelve guns in single mounts was a decidedly inefficient arrangement, greatly increasing the size of target presented to enemy fire. Furthermore, numerous difficulties encountered with these guns meant that the ships frequently lacked fully operational batteries.

The horizontal protection proved to be inadequate in service. This vulnerability (shared in varying degrees by all warships) directly contributed to the loss of the *Roma*. The deck-armor disposition was relatively inefficient, with two primary armor decks, each a sandwich of two layers. As a result, for example, the effective single-plate armor thickness (about 130 mm) on the centerline amidships was much less than the aggregate thickness of 169 mm.

The damage-control equipment was gravely deficient. In particular, the lack of portable power-driven pumps at least once, in the torpedoing of the *Veneto* on 28 March

1941, almost led to the loss of the ship due to progressive flooding. The difficulty in controlling this flooding was due to the loss of the aft pump room. The forward and aft pump rooms in Italian battleships were located beyond the armor box in very vulnerable positions. When the pump room was lost in the *Vittorio Veneto*, there was a lack of portable power-driven pumps to handle the 3,750 tons of flooding in the stern. Moreover, some of this flooding could have been limited by better subdivision. The use of watertight trunks in major watertight compartments might have prevented the flooding of large shell-to-shell compartments on decks above the torpedo impact. The rupture of watertight hatches from the torpedo explosion led to progressive flooding of large spaces above.

The modest endurance requirements permitted a limited fuel capacity and machinery-space compactness intolerable in ships designed for long-range operations. The fact that these ships would not leave the Mediterranean made it possible for the Italians to use the weight saved for heavier protection, greater speed, or better redundancy of systems, such as the rudders.

The rudder arrangement was good from the standpoint of providing at least a limited residual steering capability after damage to the main rudder, but better maneuverability would have been provided by twin rudders located off the centerline in the propeller races.

The 381-mm guns were remarkably potent weapons, being ballistically superior to all contemporary capital-ship guns of equal bore. For that matter, many modern 406-mm guns were inferior to the Italian 381 mm in terms of armor penetration at given ranges. As might be expected, this superb performance was achieved at the price of a relatively short barrel life.

Perhaps the most controversial aspect of the design of the *Vittorio Veneto*-class battleships was the novel Pugliese system of side protection. Although it was never defeated in the several instances of torpedo contact detonations, the *Roma* experience confirmed the susceptibility of the system to failure under heavy loadings from noncontact detonations, due to the inadequacy of the riveted joint joining the protective system to the bottom structure of the hull girder. Despite the system's demonstrated adequacy in service against torpedo attack, the multiple bulkhead systems more often used were more efficient in terms of obtaining the maximum possible resistance from the structure within the system. This resulted from the fact that much of the strength inherent in the structure of the Pugliese system was wasted once the hollow cylinder within the system began to collapse. The membrane-type action within conventional systems, which use both elastic and plastic deformation to absorb explosive energy, is much more effective in this respect. Obviously, battle-damage repairs to such a complex structure would have been quite difficult. In fairness, Pugliese must be given credit for the development of a novel protective system, which performed well in service.

The *Vittorio Veneto*-class battleships were extremely good warships, exceptionally well designed for their time. The rapid evolution of ordnance technology that hastened the end of the battleship era soon outstripped the capabilities of these ships, although the *Veneto* and *Littorio* served well during Italy's Mediterranean campaigns against the Royal Navy.

Name	*Vittorio Veneto*	*Littorio* (*Italia*—30 July 1943)
Builder	Cantieri Riuniti dell'Adriatico, Trieste	Cantieri Ansaldo, Genoa
Laid down	28 October 1934	28 October 1934
Launched	25 July 1937	22 August 1937
Completed	28 April 1940	6 May 1940
Operational	1 May 1940	2 August 1940
Disposition	Stricken 1 Feb. 1948, scrapped at La Spezia 1948–50.	Stricken 1 June 1948, scrapped at La Spezia, 1948–50.

Name	*Roma*	*Impero*
Builder	Cantieri Riuniti dell'Adriatico, Trieste	Cantieri Ansaldo, Genoa
Laid down	18 September 1938	14 May 1938
Launched	9 June 1940	15 November 1939
Completed	14 June 1942	. . .
Operational	21 August 1942	. . .
Disposition	Sunk 9 Sept. 1943 in Gulf of Sardinia, 41°10′N., 8°40′E.	Stricken 27 March 1947, scrapped at Venice.

Displacement

Vittorio Veneto	
(17 Oct. 1939)	37,613 (38,216 mt) Light ship
	40,516 (41,167 mt) Standard
	42,935 (43,624 mt) Normal
	45,029 (45,752 mt) Full load
(1940)	37,142 (37,738 mt) Light ship
Littorio (1940)	
	37,820 (38,427 mt) Light ship
	40,723 (41,377 mt) Standard
	43,143 (43,835 mt) Normal
	45,237 (45,963 mt) Full load
Roma (1942)	
	37,794 (38,400 mt) Light ship
	40,990 (41,650 mt) Standard
	43,797 (44,050 mt) Normal
	45,473 (46,203 mt) Full load
Impero	
	similar to *Roma*

Frames

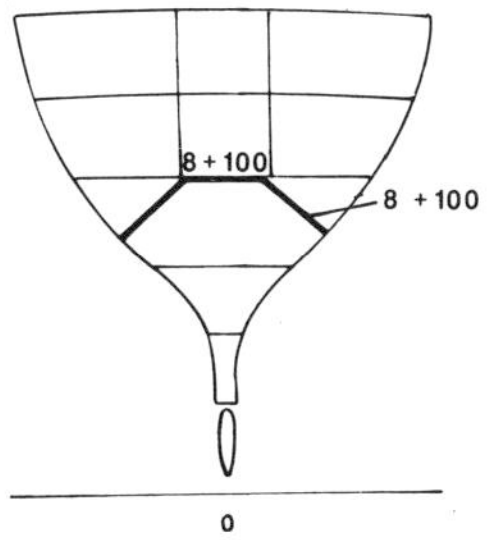

Frame 0

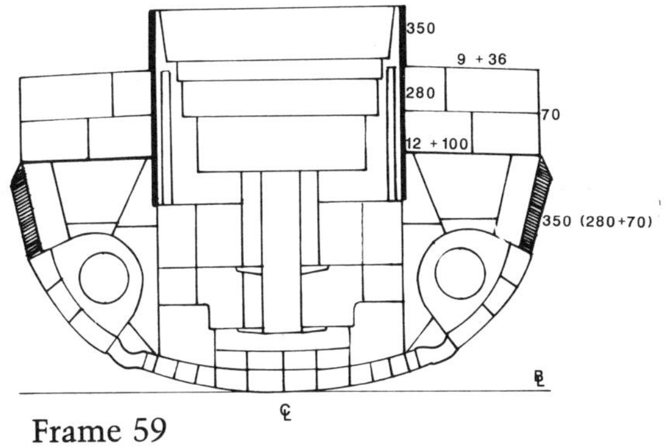

Frame 59

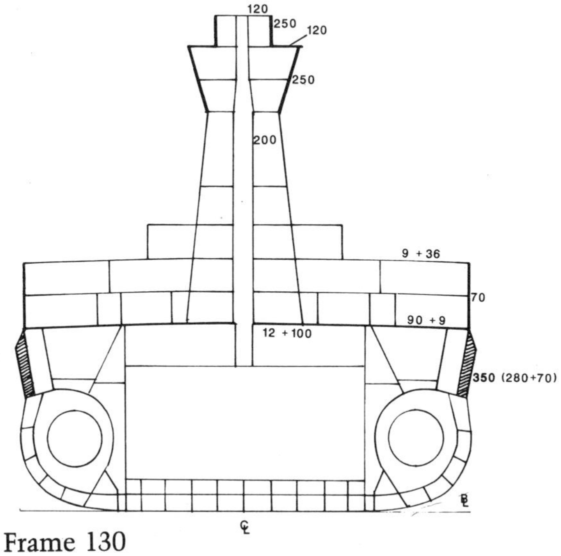

Frame 130

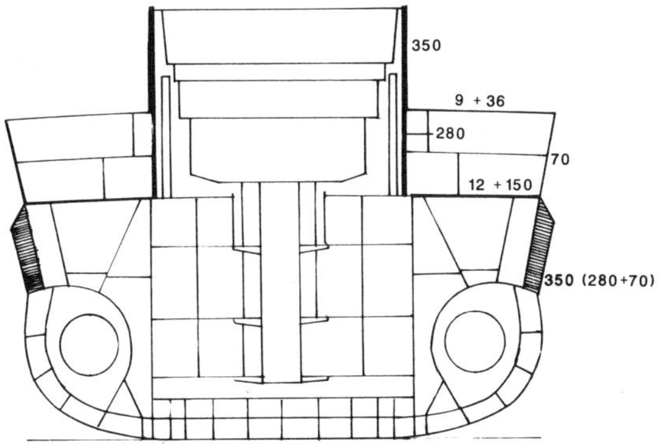

Frame 150

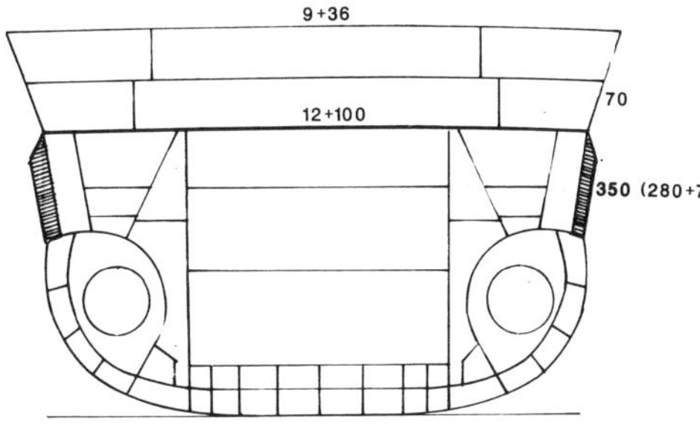

Frame 160

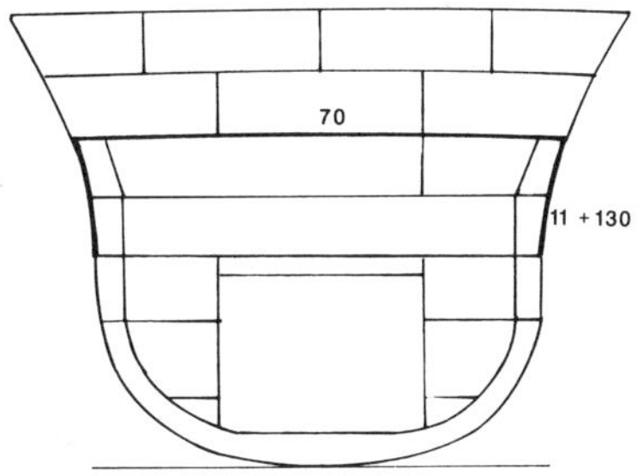

Frame 180

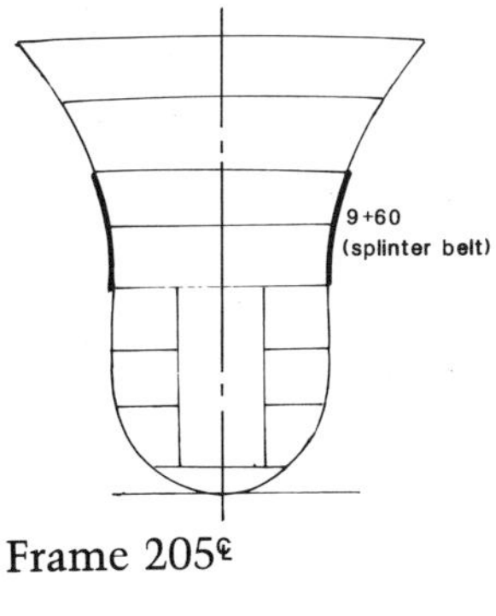

Frame 205℄

Midship Section

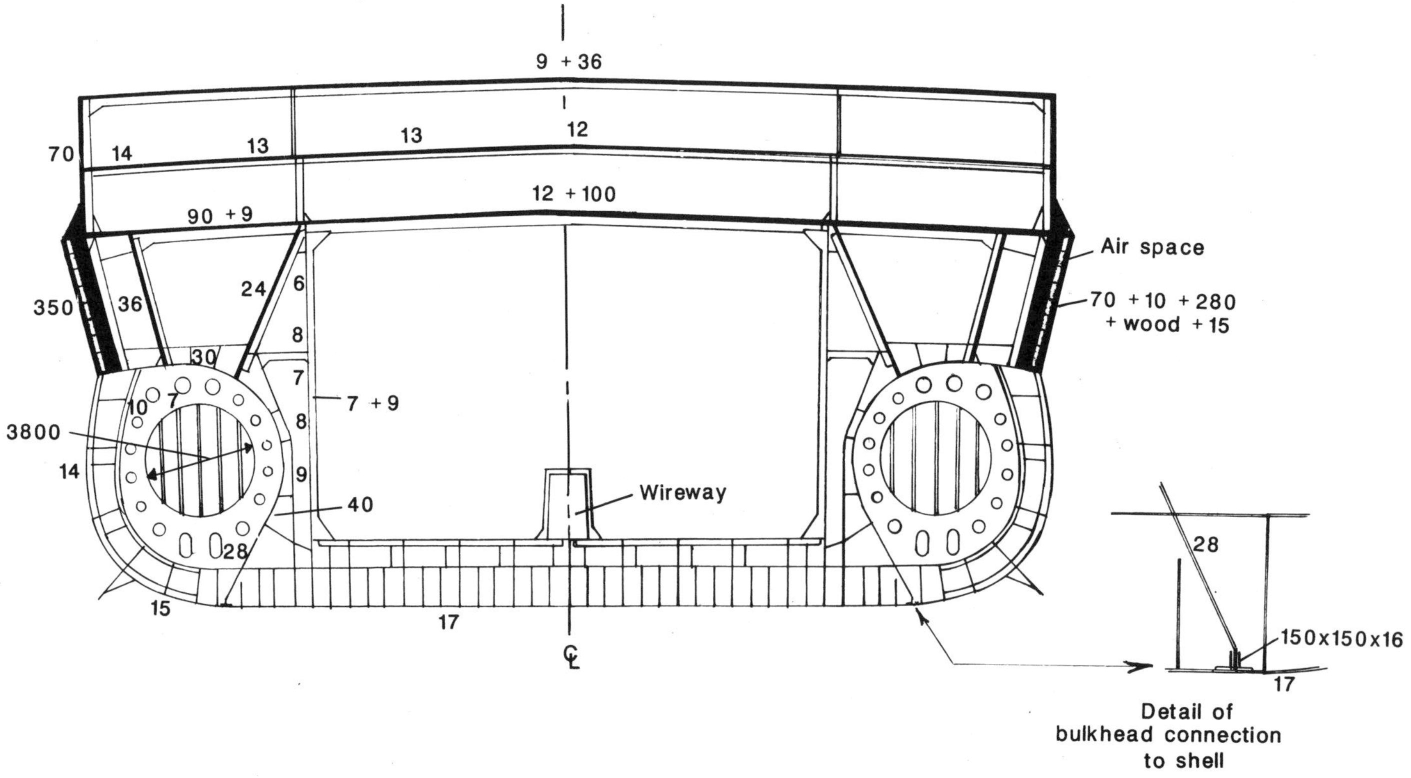

Dimensions

Vittorio Veneto and *Littorio*

780.05′	(237.76 m)	Length overall
734.07′	(224.05 m)	Length between perpendiculars
108.104′	(32.95 m)	Maximum beam
106.408′	(32.433 m)	Waterline beam
29.396′	(8.98 m)	Mean draft @ 37,613 tons (38,216 mt)
34.252′	(10.44 m)	Mean draft @ 45,029 tons (45,752 mt)

Roma and *Impero*

787.728′	(240.1 m)	Length overall
734.07′	(224.05 m)	Length between perpendiculars
108.104′	(32.95 m)	Maximum beam
106.408′	(32.433 m)	Waterline beam
34.252′	(10.44 m)	Mean draft @ 45,473 tons (46,203 mt)

Hull Characteristics at Designed Draft

Vittorio Veneto

Displacement	45,029 tons (45,752 mt)
Mean draft	34.252′ (10.44 m)
Hull depth amidships	47.244′ (14.40 m)
Block coefficient	0.565
Prismatic coefficient	0.765
Waterplane coefficient	0.679
Midship section coefficient	0.959
Tons per inch immersion	131.494 (52.6 mt/cm)
Actual metacentric height	17 October 1939
	2.228′ (0.679 m) @ 37,613 tons (38,216 mt)
	5.272′ (1.668 m) @ 45,029 tons (45,752 mt)

Midship Sections—18,000-Ton Pocket Battleship Study

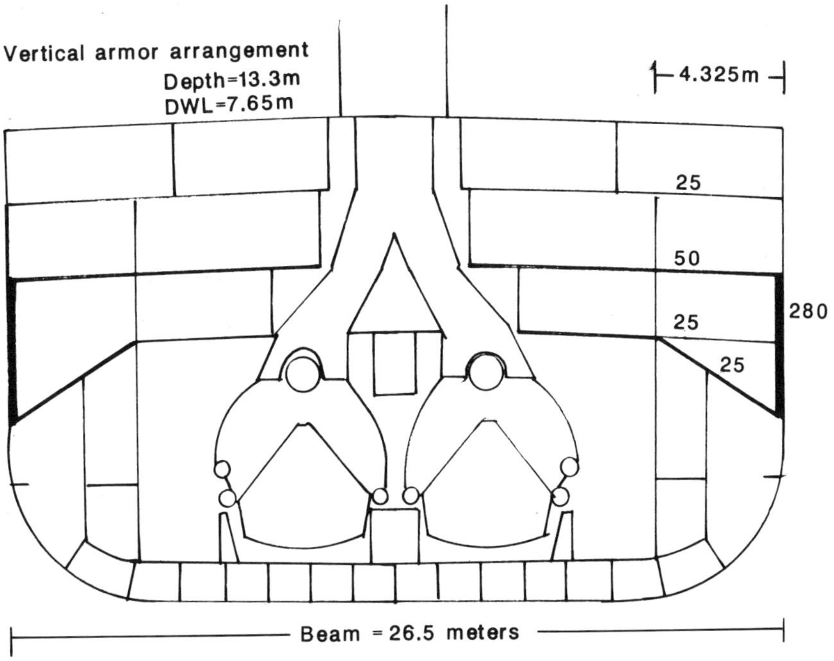

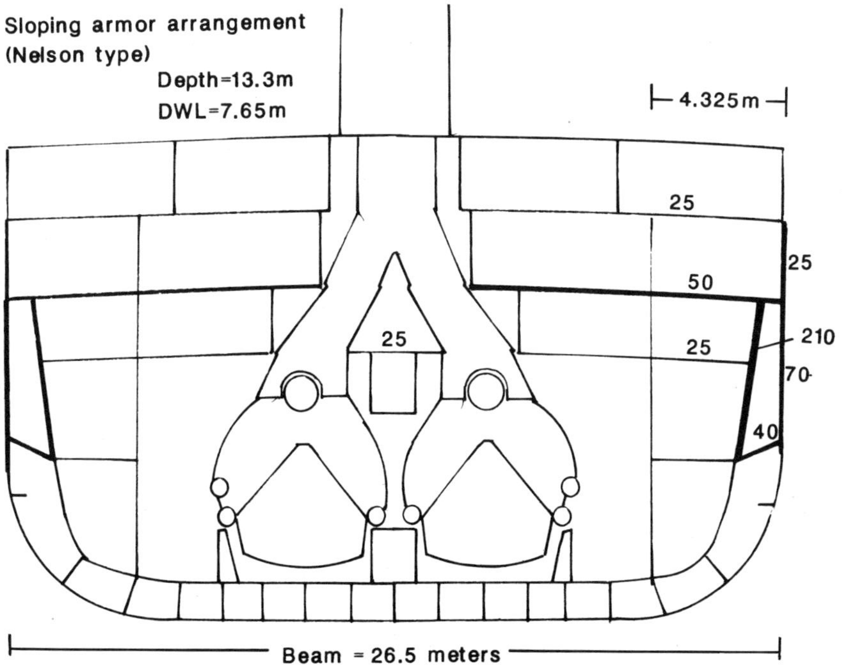

Machinery Schematic (*Vittorio Veneto* Class)

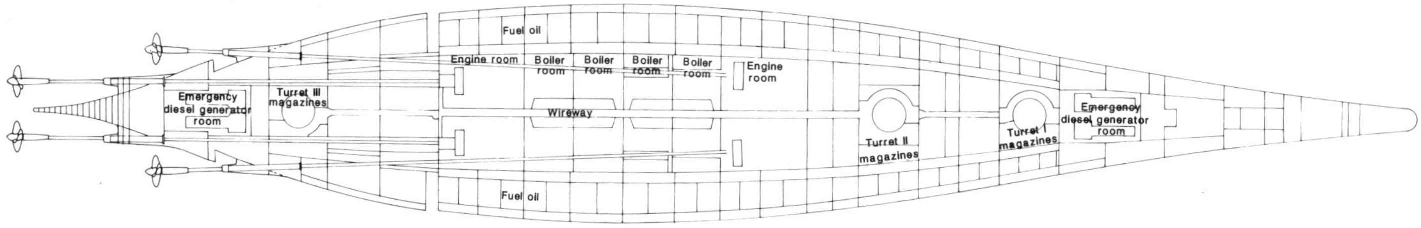

Armament

Nine	15″/50-caliber guns (Model 1934) (381 mm)	
Twelve	6″/55-caliber guns (Model 1934) (152mm)	
Twelve	3.5″/50-caliber guns (90 mm)	

Antiaircraft machine guns:

	37mm/54	20 mm/65	8 mm
Vittorio Veneto and *Littorio*			
(1940)	20	24	6
(1942)	20	32	6
Roma			
(1942)	20	32	6

Armor Protection (Refer to plans for arrangement details.)

Immunity zone From 17,498 to . . ,. . . yards (Citadel)
(16,000 to . . , . . . m.)
Italian 15″/50 (381 mm) firing 1,951-lb (885-kg) shell.

Amidships

Belt armor 2.76″ on 0.39″, 9.84″ interval, then 11.02″, inclined 8° (70 mm-10 mm, 250 mm, 280 mm)
Total thickness 14.17″ (360 mm)

Upper-belt armor 2.76″ (70 mm)

Deck armor

	Centerline	
main	1.42″ + 0.35″	(36 mm + 9 mm)
second	0.47″	(12 mm)
third	3.94″ + 0.47″	(100 mm + 12 mm)
	6.65″	(169 mm)
	Outboard	
main	1.42″ + 0.35″	(36 mm + 9 mm)
second	0.47″	(12 mm)
third	3.54″ + 0.35″	(90 mm + 9 mm)
	6.14″	(156 mm)

Turret armor

face plates	14.96″ (380 mm)
fore side plates	7.87″ (200 mm)
after side plates	5.12″ (130 mm)
back plates	14.96″; 8.27″; 11.81″ (I; II; III) (380 mm; 210 mm; 300 mm)
roof plates	7.87″ (200 mm)

Barbette armor

Sides & centerline 13.78″–11.02″ (350 mm–280 mm)

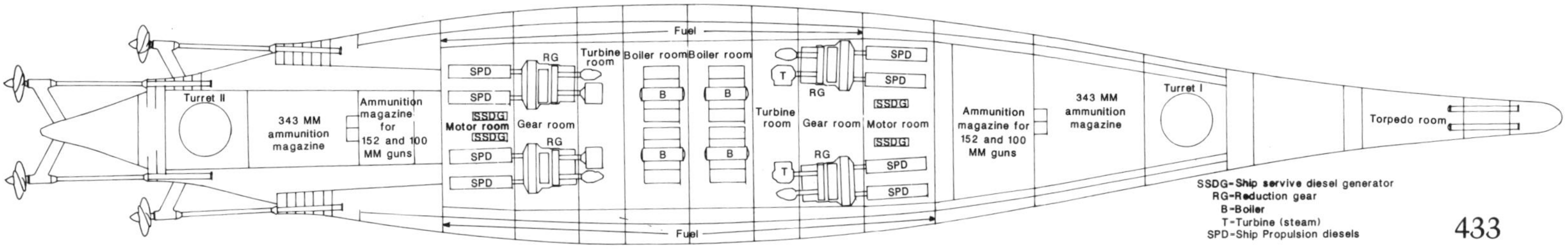

Figure 7-11. Machinery schematic—(18,000-ton pocket battleship study.)

Secondary gun armor		
face plates	11.02″	(280 mm)
fore side plates	5.12″	(130 mm)
after side plates	3.15″	(80 mm)
fore roof plates	5.91″	(150 mm)
after roof plates	4.13″	(105 mm)
back plates	3.15″	(80 mm)
barbettes	5.91″–3.94″	(150 mm–100 mm)
Antiaircraft gun armor		
gun shields	1.57″–0.47″	(40 mm–12 mm)

Conning-tower armor			
roof plates	4.72″ + .39″	(120 mm + 10 mm)	Fore
	3.54″ + .39″	(90 mm + 10 mm)	Aft
sides	8.86″ + .98″	(225 mm + 25 mm)	Fore—9th level
	9.84″ + .39″	(250 mm + 10 mm)	Fore—7th & 8th levels
	6.89″ + .98″	(175 mm + 25 mm)	Aft—9th level
	7.87″ + .39″	(200 mm + 10 mm)	Aft—7th & 8th levels
bottom plates	3.54″ + .39″	(90 mm + 10 mm)	

Underwater Protection	
Designed resistance	772 pounds (350-kg) TNT
Side protective system depth	23.69′ (7.22 m) @ half draft amidships
Hollow cylinder diameter	11.58′ (3.80 m) amidships
S.P.S. designed loading	cylinder void; remainder liquid
Bottom protective system depth	6.89′ (2.10 m)
upper layer	2.79′ (0.85 m) (void)
lower layer	4.10′ (1.25 m) (liquid loaded)

Body Plan

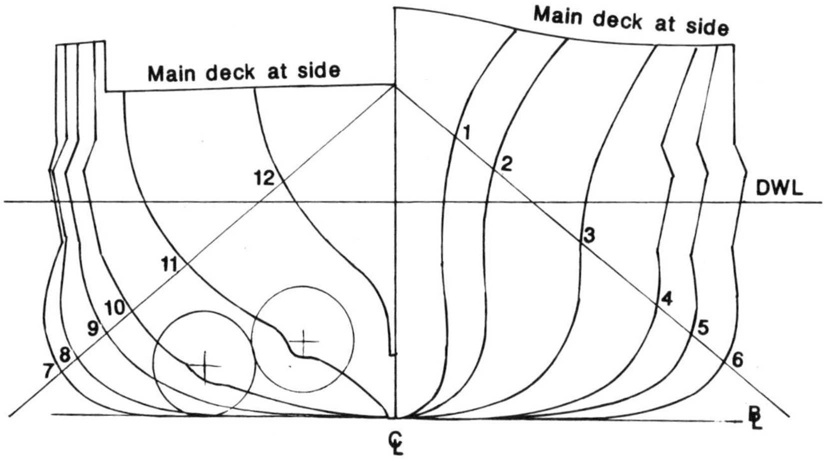

Tank Capacities

Vittorio Veneto	
fuel oil	3,966.4 tons (4,030 mt)
lube oil	129.0 tons (131.07 mt)
gasoline	17.3 tons (17.6 mt)
reserve feed water	369.1 tons (375 mt)
Littorio	
fuel oil	4,161.3 tons (4,228 mt)
lube oil	108.3 tons (110 mt)
gasoline	17.3 tons (17.6 mt)
reserve feed water	421.5 tons (428.3 mt)
potable water	329.3 tons (334.6 mt)
Roma and *Impero*	
and *Impero*	4,047.9 tons (4,113 mt)
fuel oil	145.2 tons (147.5 mt)
lube oil	17.3 tons (17.6 mt)
gasoline	329.3 tons (334.5 mt)
reserve feed water	333.1 tons (338.4 mt)
potable water	

Machinery

Boilers	eight Yarrow-type watertube boilers with superheaters
Pressure:	355.6 psi (25 kg/cm^2)
Temperature:	618° F. (325° C.)
Turbines	four sets Beluzzo geared turbines, each with 1 HP turbine with cruising turbine incorporated 1 IP turbine with astern turbine incorporated 1 LP turbine with astern turbine incorporated
Shaft horsepower	128,222 (130,000 mhp) normal maximum 138,085 (140,000 mhp) overload
Maximum speed	30 knots—normal displacement 29 knots—full-load displacement

Trials	
Vittorio Veneto	31.428 knots @ 238 rpm @ 132,775 horsepower (134,616 mhp) @ 41,473 tons (42,138 mt)
Littorio	31.293 knots @ 239 rpm @ 137,652 horsepower (139,561 mhp) @ 41,122 tons (41,782 mt)
Nominal endurance	4,580 miles @ 16 knots 3,920 miles @ 20 knots 1,770 miles @ 30 knots

	Vittorio Veneto	*Littorio*
Actual endurance	4,700 miles @ 14 knots	4,870 miles @ 14 knots
	3,900 miles @ 20 knots	4,050 miles @ 20 knots

Generators	eight ship's service turbogenerators (450 kw) four Fiat MS-368 emergency diesel generators (800 kw) three Ansaldo A625/4 alternators (62.5 kw) total ship's service capacity: 3,600 kw @ 220-volts, DC and 187.5 kw, 62.5-volts, AC
Propellers	four Scaglia three-bladed propellers, 15.748′ (4.80 m) diameter
Rudders	one main rudder on the centerline two auxiliary rudders aft

Miscellaneous	
Complement	1,920 (120 officers/1,800 enlisted)
Aircraft	3
Catapults	1

Vittorio Veneto Weight Summary—17 October 1939

Light ship displacement	37,613 tons (38,216 mt)
Supplies and stores	1,840 tons (1,869 mt)
Liquid loads	4,968 tons (5,049 mt)
Machinery/hull liquids	608 tons (618 mt)
Full-load displacement	45,029 tons (45,752 mt)

Vittorio Veneto Weight Summary—1940

Hull structure (excl. armor)	10,402 tons (10,569 mt)	28.01%
Armor	13,331 tons (13,545 mt)	35.89%
Fittings	4,577 tons (4,650 mt)	12.32%
Machinery	2,367 tons (2,405 mt)	6.37%
Armament	6,465 tons (6,569 mt)	17.41%
Light ship displacement	37,142 tons (37,738 mt)	100%

Littorio Weight Summary—1940

Hull structure (excl. armor)	10,441 tons (10,608 mt)	28.06%
Armor	13,451 tons (13,669 mt)	36.16%
Fittings	4,583 tons (4,656 mt)	12.32%
Machinery	2,267 tons (2,303 mt)	6.09%
Armament	6,462 tons (6,565 mt)	17.37%
Light ship displacement	37,204 tons (37,801 mt)	100%

1985 Allison

CHAPTER EIGHT

Spain

On 8 September 1939, a few months after the end of a three-year civil war of noteworthy violence and brutality, the victorious General Francisco Franco signed the secret Law of the Fleet Program ("Ley de Programa de Flota") at Burgos. This ambitious, if not grandiose, naval construction program projected an impressive expansion of the Spanish Navy over the following eleven years, as outlined in table 8-1.

The only ship types that were explicitly specified were the capital ships. The battleships were to be license-built versions of the Italian *Vittorio Veneto*-class ships, while the armored cruisers were to be based on Spanish designs for unusually powerful heavy cruisers, known as "Proyecto 138." The evolution and nature of Spanish plans for the construction of capital ships will be described in this chapter.

Although Spain was no longer a major naval power by the time the dreadnought era dawned in 1906, the Spanish did not hesitate to join in the construction of these revolutionary new warships. This interest in dreadnought-type capital ships persisted until the end of the battleship era during World War II.

The España Class.

The Maura-Ferrandiz Navy Law of 7 January 1908 authorized the construction of three small dreadnought-type battleships for the Royal Spanish Navy. Sociedad Española de Construcción Naval (SECN), El Ferrol, built the ships, with extensive technical and material assistance from the United Kingdom. The outbreak of World War I delayed the completion of the third ship until after the end of the war. The name ship of the class, the *España,* was wrecked off Cabo Tres Forcas, Morocco, in August of 1923. The second of the class, *Alfonso XIII,* was renamed *España* after the declaration of the Spanish Republic in 1931. The *Jaime I* took almost 10 years to complete, as a result of delays in material deliveries from the United Kingdom during the war.

These were the smallest dreadnoughts to be built, with a normal displacement of only 15,452 tons. The main battery was eight 305-mm/50-caliber guns in twin turrets, while the secondary armament consisted of twenty 102-mm guns. As one might expect with such heavily gunned, small-displacement ships, the scale of protection was quite modest, and the maximum speed was only some 19.5 knots. Although these ships were markedly inferior to most contemporary dreadnoughts of almost twice the displacement, they were quite sufficient for Spanish requirements. In effect, the *España* and her

This artist's conception by Richard Allison depicts the estimated characteristics of the Spanish version of the *Vittorio Veneto*-class ships. The Spanish would probably have substituted 120-mm guns for the unreliable 90-mm guns on the Italian ships, and improved horizontal protection would have been a very desirable design improvement. Overall, there would have been few apparent differences between the Spanish and the Italian ships.

TABLE 8-1

Ley de Programa de Flota

Artículo 1:	4 Battleships 2 Armored cruisers 12 Light cruisers 54 Destroyers 36 Torpedo boats 50 Submarines 100 Motor torpedo boats Auxiliaries Naval bases and schools Shops and yards Equipment and spares
Artículo 2:	Finances: a total of 5,000,000,000 pesetas to be disbursed over a period of 11 years. At 1939 exchange rates, 16 pesetas equaled one dollar.
Artículo 3:	Provided directives on how to develop the plan.
Artículo 4: *Artículo 5:*	Ordered the "complementary norms," authorizing some specified individual, normally the naval secretary, to deal with the details of implementing the plan. These latter articles were typical of Spanish laws.

two sisters were dreadnought-type coastal-defense ships with a relatively heavy armament, modest displacement, limited armor, and minimal speed.

These three small dreadnoughts, known as the "Primera Escuadra," were to be joined by a further three dreadnoughts, projected in the 1914 "Plan de la Segunda Escuadra." The *Reina Victoria Eugenia*-class ships would have displaced some 21,000 tons, with eight 340-mm guns and a speed of about 21 knots. The outbreak of the First World War aborted this program, which presumably would have relied heavily on British technical and material assistance.

The Postwar Era.

Although the Spanish were not signatories to the Washington Treaty of 1922, they did participate in the Rome Conference of 1922. As the Royal Spanish Navy was then interested in the construction of new battleships of the 35,000-ton class, the Spaniards withdrew from the conference when it became clear that the prospective Rome Treaty would have limited the Spanish to a total of only 81,500 tons, reflecting the authorization of three ships of much less displacement than was desired.

The 1926 building program projected no capital ships, but three heavy cruisers were authorized, of which two, the *Canarias* and the *Baleares*, were eventually completed.

Several years later, during 1929–1931, plans for a "reduced *Nelson*" were seriously considered, but nothing came of the design studies.

During 1934–1935, "pocket battleship" conversions of the two old dreadnoughts *Jaime I* and *España* (originally named *Alfonso XIII*) were contemplated. The radical conversion envisioned the installation of oil-fired boilers and the lengthening of the ship. A major rearrangement of the main battery would have resulted in the four 305-mm turrets being positioned on the centerline. A new dual-purpose armament of 120-mm guns was projected.

By 1936, prior to the outbreak of the Spanish Civil War, a fleet law including a more modest modernization of the old battleships was proposed. This would have included

A 1919 view of the first *España*, with her four twin turrets trained to port. The superstructure and funnel partially obstructed the firing arcs for the two-echelon turrets near amidships for targets broad on the bow or on the quarter.

oil-fired boilers, improved internal subdivision, increased elevation for the main-battery guns, more antiaircraft guns, and the installation of a modern gunfire-control system. The outbreak of the civil war prevented any of this from being accomplished.

The fate of the two remaining Spanish dreadnoughts was a microcosm of the tragedy that was the Spanish Civil War of 1936–1939. The *España* served on the Nationalist side under General Franco. The ship served on blockade duty until she struck a floating mine on 30 April 1937; the sinking ship was bombed by Republican aircraft. The *Jaime I* served with the Loyalist (Republican) forces. The battleship was heavily damaged at Málaga on 13 August 1936 by Italian aircraft, which were supporting the Nationalist forces. On 17 June 1937, while moored at Cartagena, the dreadnought was seriously damaged by an internal explosion. The ship was scuttled to prevent further explosions. About a year later, the hulk was raised and partially stripped. The remnant of the *Jaime I* was stricken in July of 1939, and the hulk was scrapped.

The Vittorio Veneto Class. The Spanish Civil War ended officially on 1 April 1939. Within a few months, the secret naval building program signed into law by General Franco at Burgos signaled that the official Spanish interest in capital-ship construction had resumed.

The Spanish projected the construction of four battleships to the basic design developed by the Italians for their ships of the *Vittorio Veneto* class. The Spanish-built ships probably would have been very similar to this view of the *Vittorio Veneto.* (Italian Navy)

A large Spanish technical commission, headed by the new minister of industry (by profession a naval architect), visited Italy late in 1939. The Italians promised the necessary technical and material assistance to support the construction of four battleships of the Italian *Vittorio Veneto* type in Spain. With this support in mind, the Spanish concluded agreements with the Italian government formalizing these understandings.

At the insistence of the Italians, first priority was given to the necessary repair, modernization, and enlargement of Spanish shipyards. A large slipway was actually constructed at El Ferrol; the size of the new slipway was such that it would have been possible to build two battleships there at the same time.

The *Vittorio Veneto*-class battleships were designed during 1934–1935, with a total of three ships out of a projected class of four actually entering service with the Italian Navy. The basic design proved to be a very successful one. Characteristics of the *Vittorio Veneto,* as completed in 1940, are presented in table 8-2.

Armament. The main-battery guns were remarkably powerful for their caliber, with a maximum range of some 42,260 meters at only 35 degrees maximum elevation while firing an 885-kilogram shell. (In every ballistic respect, these guns were substantially superior to the 380-mm guns given the German *Bismark* and *Tirpitz.*) The mixed battery of 152-mm and 90-mm guns was typical of modern continental European capital-ship design. In service, the 90-mm antiaircraft guns proved to be notoriously unreliable, with defective loading mechanisms and poor stabilizing equipment for the gun mounts. Furthermore, the 90-mm projectiles proved to be faulty. These deficiencies radically reduced the effectiveness of the 90-mm guns, which had very good ballistic performance. As a result of the operational difficulties encountered by the Italians with their 90-mm guns, it is probable the Spaniards would have selected an alternative gun. Most likely, this would have been of 120-mm caliber, as the two heavy cruisers of the 1926 program featured such antiaircraft guns, and the postwar destroyers of the *Oquendo* type were also armed with 120-mm guns.

TABLE 8-2
The Vittorio Veneto

Displacement	
Light ship	37,613 tons (38,216 mt)
Standard	40,516 tons (41,167 mt)
Normal	42,935 tons (43,624 mt)
Full load	45,029 tons (45,752 mt)
Dimensions	
Length overall	780.05′ (237.76 m)
Length between perpendiculars	734.07′ (224.05 m)
Maximum beam	108.104′ (32.95 m)
Waterline beam	106.408′ (32.433 m)
Mean draft	34.252′ (10.44 m) @ 45,029 tons (45,752 mt)
Armament	
	nine 15″/50 (381 mm) tripled
	twelve 6″/55 (152 mm) tripled
	twelve 3.5″/50 (90 mm) single
	twenty 1.46″/54 (37 mm)
	twenty-four 0.79″/65 (20 mm)
Protection	
Belt	11.02″ over 2.76″ + 0.39″ @ 8° inclination (280 mm over 70 mm + 10 mm)
Turrets	14.96″ (380 mm) Face plates
	7.87″ (200 mm) Sides, forward
	5.11″ (130 mm) Sides, rear
	7.87″ (200 mm) Roof
Secondary	11.02″ (280 mm) Face plates
	5.11″ (130 mm) Sides, forward
	3.15″ (80 mm) Sides, rear
	5.91″ (150 mm) Roof, forward
	4.11″ (105 mm) Roof, rear
Deck	6.65″ (169 mm) Amidships, centerline TOTAL
	6.12″ (156 mm) Amidships, outboard TOTAL
	8.62″ (219 mm) Magazines, centerline TOTAL
	6.52″ (166 mm) Magazines, outboard TOTAL
Machinery	
Shaft horsepower	128,222 (130,000 mhp) normal maximum
	138,085 (140,000 mhp) overload
Maximum speed	30 knots at normal displacement
	29 knots at full-load displacement
Endurance	4,700 nautical miles @ 14 knots
	3,900 nautical miles @ 20 knots

Protection. The battleships proved to be generally well protected, although their scale of horizontal protection was only marginally satisfactory for combat service during World War II.

The novel Pugliese system of side protection was comprised of a large hollow cylinder installed along the length of the citadel on both sides of the ships. This cylinder was to be kept void at all times while completely immersed in either fuel oil or water, or a mixture of the two. The supporting structure bounding this protective system was much more heavily constructed than the hollow cylinder, which was designed to absorb energy while being hydraulically crushed after a torpedo detonation. The resulting turbulence as the cylinder was crushed would further dissipate the destructive forces of the underwater explosion. The Italians considered this to be a very effective and lightweight system as compared to more coventional multiple-bulkhead configurations.

The horizontal protection of the ships proved to be less than was desirable for wartime conditions. A Spanish version of the ships could well have featured a modest upgrading of this protection.

Similarly, the Italian structural design was predicated solely on the ships serving in the Mediterranean. A Spanish *Veneto* would very likely have been structurally strengthened. Part of this added weight could also have served ballistic purposes, enhancing the horizontal protection that was discussed in the preceding paragraph.

Propulsion plant. The relatively modest endurance of the *Vittorio Veneto* would probably have been considered adequate for service in the Spanish Navy. Any substantial increase in the fuel carried could only have been achieved by a major design modification, providing the necessary added volume by adding length or beam or by a major internal rearrangement, at the expense of other military attributes.

Probable Spanish design modifications. Overall, a Spanish variant of the *Vittorio Veneto* would have probably differed from the Italian ship the following general ways:

Displacement	Slightly increased, due to added deck armor, improved structural strength, and modified antiaircraft armament.
Dimensions	Generally the same, but added draft due to the increased displacement.
Armament	Same main battery and secondary armament, but a different antiaircraft armament, probably including 120-mm dual-purpose guns, as well as 40-mm or 37-mm and 20-mm machine guns.
Protection	Minor changes, due to added deck armor.
Machinery	No changes to power plant, but a modest reduction in maximum speed and endurance due to the added displacement.

Disposition of the Veneto project. Other than the physical improvements to the construction facilities at El Ferrol, the projected construction of the four battleships by the Spanish Navy was thwarted by the Italian involvement in the war in Europe, commencing in June of 1940, combined with limited Spanish resources.

Armored Cruiser Design Studies.

The basis for Spanish projections for new armored cruisers was a powerful heavy-cruiser design known as Project 138. This design, essentially a more powerful development of the earlier *Canarias,* was initiated at the Ferrol Shipyard in 1938. The characteristics of this "Crucero Super-Washington" are presented in Table 8-3.

This extremely powerful cruiser design provided an outstanding combination of gun power, armor protection, and high speed. The modest beam, as for all cruisers, allowed for no torpedo-defense system.

Late in 1939, an armored cruiser version of the Project 138 design was prepared. This design variant provided for slightly improved armor protection, and a main-battery armament of six 305-mm guns in three twin turrets. Table 8-4 provides the authors' *estimated* characteristics for this armored-cruiser design study.

As the Spanish had been greatly impressed with the German "pocket battleships" *Deutschland, Admiral Graf Spee,* and *Admiral Scheer,* which occasionally operated off the Spanish coast in support of Franco's forces, it is not surprising that they eventually contemplated an armored cruiser strongly influenced by the older German design.

TABLE 8-3

Crucero Super-Washington

Standard displacement	17,500 tons (17,781 mt)
Overall length	689.129′ (225.66 m)
Beam	72.007′ (23.58 m)
Draft	20.177′ (6.15 m)
Hull depth amidships	41.995′ (12.80 m)
Armament	twelve 8″/50 (203 mm) paired
	twelve 3.5″ (90 mm) paired
	sixteen 1.57″ (40 mm) paired
	twenty 0.79″ (20 mm)
Protection	
Main side belt	5.91″ (150 mm) amidships
	11.81″ (300 mm) magazines
Turret	
face plates	5.91″ (150 mm)
sides	3.94″ (100 mm)
roof plates	3.94″ (100 mm)
Armor deck	2.95″ (75 mm) amidships
	3.94″ (100 mm) magazines
Shaft horsepower	167,674 (170,000 mhp)
Maximum speed	36 knots
Endurance	7,000 nautical miles @ 18 knots

TABLE 8-4

Armored-Cruiser Design Study

Standard displacement	19,000 tons (19,305 mt)
Armament	six 12″/50 (305 mm) paired
	twelve 3.5″ (90 mm) paired
	sixteen 1.57″ (40 mm) paired
	twenty 0.79″ (20 mm)
Maximum speed	34 knots

During 1942–43, the Spanish proposed another variant of the original heavy-cruiser design. This scheme would have featured a main battery of six German 283-mm guns in triple turrets, fore and aft. Such an armament was more readily attainable within the basic scope of the original cruiser design, permitting a more balanced design with better protection than would have been possible with three twin turrets for 305-mm guns. In particular, this variation of the cruiser design would have been much more readily accommodated than the earlier armored-cruiser scheme, having fewer problems with available hull space and volume (particularly forward).

The German battlecruiser *Gneisenau* had been severely damaged by bombing in February of 1942. The forward 283-mm turret was totally wrecked, and the damage was so severe that the Germans estimated reconstruction of the ship would take two years. As a result, the Germans decided to take advantage of the available time and rebuild the *Gneisenau* with three twin turrets for 380-mm guns. It is certainly plausible to infer that the Spanish were hopeful of obtaining the two triple 283-mm turrets that such a conversion would have made available. This would have substantially accelerated the construction of the lead Spanish armored cruiser.

These three artist's conceptions by Richard Allison depict a family of cruiser designs developed by the Spanish. The Project 138 design of a *Washington*-type heavy cruiser (top) was an improved, more powerful version of the *Canarias*, with an armament of twelve 203-mm guns and a standard displacement of 17,500 tons. The middle view shows an armored cruiser version of this design, with a proposed main battery of six 305-mm guns. During 1942–1943, a different version, armed with six 283-mm guns in triple turrets (bottom), was proposed.

Summary. In the first flush of victory after a long civil war, the Spanish Navy devised an ambitious capital-ship building program, with the expectation of extensive technical and material assistance from Italy and Germany. The adverse fortunes of war soon made such assistance highly problematical at best. By the end of 1943, Fascist Italy was out of the war, Germany was reeling from a devastating series of defeats, and Spanish expectations of foreign assistance were dead.

In view of the quite modest course of Spanish naval construction over the past several decades, the extensive 1939 naval building program hardly seems realistic. It is obvious, however, that the Spanish were deadly serious in their plans to build up and modernize their navy. Had the nation's finances so permitted, and given the advantages of neutrality during World War II, the Spanish Navy could, by 1950, have ranked fourth or fifth in the world. However, events proved that the national economy simply could not support the heavy expenditures that such a program would have demanded.

CHAPTER NINE

Conclusion

The battleship has been one of the most controversial weapons systems of the twentieth century. In the years prior to World War I, naval power was measured by the number of battleships in a fleet. Therefore, a number of these ships were built or procured by the major naval powers—France, Germany, Great Britain, Russia, Japan, Italy, and the United States, as well as a number of minor naval powers such as Brazil, Chile, Argentina, and Austria-Hungary. These ships did not distinguish themselves in World War I as they were not used aggressively, with the possible exception of the German ships. Because of the amount of capital invested and their political importance, the ships were cautiously deployed. The Battles of Jutland and Dogger Bank were the most significant naval engagements of the war, and the former markedly influenced capital-ship design and strategy for some 25 years thereafter.

During the interwar years, a nation's naval power was still measured by the number of battleships in its fleet. The Washington Naval Treaty, however, prescribed the total standard displacement of ships the major naval powers could have in service at any one time.

With the development of the airplane, the aircraft carrier became increasingly important. After the Battle of Midway in June 1942, it was recognized that the carrier was the capital ship and that important naval battles would probably not be fought between ships, but between forces with both naval ships and *aircraft*. The serious torpedo damage to the *Vittorio Veneto* in November 1940 and the loss of the *Bismarck* in May 1941—and particularly, the losses of the *Prince of Wales* and *Repulse* in December 1941—showed that the battleship, like other surface ships, needed air cover to survive engagements at sea. Thus, the expensive, heavily protected, and massively armed battleship gradually became subordinate to the aircraft carrier.

The battleship did prove useful in her new role of providing fleet antiaircraft defense, but it was shore bombardment that proved to be her most useful role. After World War II these ships slowly disappeared from the naval scene, but the U.S. Navy maintained four of the superlative *Iowa* class in reserve until 1981. The *Iowa, New Jersey*, and *Missouri* were ordered reactivated to serve in a limited role and given a missile armament. The last battleship to leave the naval scene, other than those of the

Gneisenau is shown in harbor shielded by barrage balloons to discourge low-flying aircraft attack. The most important factors in the final years of the battleship era figure in this composition. The threat of air attack was a major factor in battleship/battlecruiser operations. Radar, shown here on the forward director, profoundly influenced naval ship design, naval operations, and naval tactic. High-elevation main-battery guns reflected advances in ordnance technology that led the battleship to its peak of offensive power at the time air power was rendering the battleship-type ship obsolete. (Albrecht Schnarke Collection)

United States, was the French *Jean Bart* in 1971. It would appear that the American battleships will remain in commission for the duration of the twentieth century.

Before World War II it was thought that the main function of a battleship was to engage similar ships in a fleet action like the one at Jutland. For the Axis powers, few such gunnery engagements were fought; only those involving the *Bismarck* and *Scharnhorst* were close to the classic gunnery engagements perceived before the war. Before being committed to the war in 1944, the giant *Musashi* and *Yamato* spent much of their time at anchor. Even during the Battle of Leyte Gulf, which was the largest naval engagement fought in modern history, these ships played a minor role because of the ascendancy of the aircraft carrier and the airplane. The Italian *Vittorio Veneto*-class ships were so cautiously deployed that there were only brief encounters with British ships.

All of the capital ships described in this volume were projected or built by the navies of right-wing military dictatorships. Despite the resulting absence of the domestic controversy that so often attended the construction of capital ships for the various Allied navies, controversy over the allocation of limited funds and resources for capital-ship construction nonetheless bedeviled the naval authorities of Japan, Germany, Italy, and Spain.

Japan. During the interwar period, the Imperial Japanese Navy was one of the most powerful navies in the world, generally being considered secondary only to the Royal Navy and the United States Navy. In the Pacific during the interwar years, the Imperial Japanese Navy was predominant, as the British naturally concentrated their naval strength in European waters in two major fleets, the Home Fleet and the Mediterranean Fleet. American strength was divided between the Pacific and Atlantic Fleets. Despite the paramount importance of its navy to an island nation such as Japan, the fact remains that the Japanese Army was politically by far the more powerful. As a

Both Japanese battleships of the *Yamato* class were destroyed by overwhelming numbers of American carrier-based aircraft. The *Yamato* is shown here during the Battle for Leyte Gulf in October 1944 (Fahey Collection)

The desperate plight of the Imperial Japanese Navy in 1945 is typified by this 28 July photograph of the battleship *Haruna* under air attack in port. Extreme shortages of petroleum products had virtually immobilized the Japanese Navy, and Japanese air power was virtually nonexistent by this time of the war. The American submarine campaign had strangled the Japanese war economy. (Fahey Collection)

matter of fact, General Hideki Tojo led the Japanese as premier from 1940 until the waning days of the war, when a Japanese defeat was all too obvious.

Internally, there was a very strong faction within the Imperial Japanese Navy that advocated naval aviation as the primary agent of naval power, and Japanese naval building programs, even before World War II, provided for more aircraft carriers than battleships, although the latter were much larger. After the disastrous loss of four carriers during the Battle of Midway in June 1942, the Japanese were convinced of the importance of naval aviation and gave near-absolute priority to aircraft-carrier construction, including the conversion of the third battleship of the *Yamato* class, the incomplete *Shinano*, to an aircraft carrier.

The Imperial Japanese Navy completed only two new battleships during the World War II era, the *Yamato* and *Musashi*. Both ships had relatively inactive operational careers—their great cost and prestige led the Japanese to avoid their commitment to the long, sanguinary campaigns in the southwestern Pacific. When the world's largest battleships finally were committed to battle in 1944, U.S. naval air power had become overwhelming. As a result, these two ships accomplished little other than absorbing impressive numbers of bomb and torpedo hits from hundreds of American carrier-based aircraft before sinking. Admiral Yamamoto's prophesy concerning a confrontation between a battleship and torpedo-bearing aircraft had come true.

450

Germany. The German Navy actually commissioned more new capital ships—four—than either of its allies, although both Japan and Italy possessed much larger surface navies than did the Germans in the years prior to World War II. This achievement, despite the relatively low priority given to naval construction from 1935-1939, is impressive testimony to the great industrial resources possessed by Nazi Germany.

As in Japan, the German Navy was quite inferior politically to both the Army and the Air Force. The German Air Force benefited from the great political influence of its leader, Hermann Göring, who was the single most influential figure in the German armed forces in prewar Germany and the closest military official to Hitler, who also did not understand the strategic concepts of sea power. These Nazi leaders concentrated on enhancing the power of the German Army and Air Force, the latter as a supporting arm for the Army. Inevitably, the small prewar German Navy could never hope to confront its potential adversaries, the navies of France and Great Britain, on anything approaching an equal basis.

As a result of the strength of Allied naval power, a vocal faction in the German Navy, led by Admiral Dönitz, strongly urged that the naval building program be focused soley on submarine construction. Fortunately for the Western Allies, the chief of the German Navy, Admiral Raeder, favored the construction of a balanced navy, including heavy capital ships and aircraft carriers. The limited industrial resources available to the German Navy were thereby divided between the two competing arms of the fleet. In fairness to Admiral Raeder, it must be pointed out that he had been personally assured by Hitler that his naval construction plans could be predicated on a war occurring no sooner than 1944-45. By that time, the Germans would have possessed a formidable surface navy as well as a large submarine fleet.

The German invasion of Poland in September 1939 found the German Navy ill-prepared for war. The submarine force was still relatively small, and the new battleships *Bismarck* and *Tirpitz* were far from complete, while the keels of battleships "H" and "J" had only been laid down. The German naval planners had not expected the outbreak of war so soon. Had the early outbreak of war been foreseen, it is quite possible that a greater proportion of the industrial resources available to the German Navy would have been allocated to submarine construction. Capital-ship design and construction is a very complex, tedious procedure, only partially susceptible to acceleration in time of emergency (refer to the *Jean Bart* in Volume II). Submarines, on the other hand, can be built in a relatively short time.

As events were to show, the only naval asset available to the Germans in 1939 that could possibly have influenced the outcome of a major war was their submarine fleet. Had the resources devoted to battleship and battlecruiser construction been allocated to submarine construction and planning in the late 1930s, the German submarine force would have been even larger and more efficient than it was during the critical period of 1939–1941. Such construction would have been recognized in Great Britain, and the Royal Navy would have possibly increased its destroyer strength to counter the German threat. As it was, German submarine warfare posed a grave threat to Great Britain, and such an increment of strength might have proved decisive. It was not until 1942, when U-boat strength was substantially increased, that absolute priority was given to submarine construction. In January 1943 Admiral Dönitz was appointed by Hitler to succeed Admiral Raeder, and still more emphasis was given to submarine construction and operation. Fortunately for the Allies, this was much too late in the course of the war to have any effect on its outcome. It should be noted, however, that the small

The *Admiral Graf Spee,* the last of the pocket battleships and the largest in terms of displacement. The scuttling of the ship by her crew on 17 December 1939 in Montevideo Harbor was the first great defeat suffered by the German Navy in World War II.

German surface navy made possible the invasion of Norway in April 1940, albeit at a ruinous cost. Without the surface fleet, it is unlikely that Norway could have been taken.

Italy. Italian interwar planning was based on a commerce war with France, but by 1934 it was apparent that Italy would have to contend with a combined French and British fleet in the Mediterranean. By 1939, the Italian Navy posed a major threat to Allied maritime lines of communication between Gibraltar and Suez. Although proportionately greater resources were allocated to Italian naval development, the relatively limited Italian industrial base precluded the development of the Italian Navy appreciably beyond the strength levels attained by the German Navy.

Interestingly, although the major efforts of both European Axis navies were devoted to the interdiction of Allied maritime traffic, their approaches to the problem were quite different. The Germans, inhibited by the Royal Navy's geographical advantage of lying astride the German approaches to the open sea, opted for a raider strategy. German surface ships and submarines were all designed with an emphasis on commerce warfare, and the German plans for the expansion of their fleet were oriented more towards the dissipation and dispersion of British maritime strength than its defeat in a Jutland type of major engagement. The Italians, on the other hand, developed a powerful, balanced fleet (less aircraft carriers) that was capable of engaging the French Mediterranean Fleet or the Royal Navy's Mediterranean Fleet with a reasonable chance of success. Both the German and Italian navies sorely lacked the crucial ingredient of naval power that would play a key role in their defeat at sea—naval aviation.

Naval aviation—the complex of land-based and shipborne aircraft designed for naval service and operated by specialist naval personnel—was shown by events during World War II to be sufficiently unique that air force units rarely operated successfully with naval forces on a sustained basis. The Channel Dash was a rare exception to this. The Italian Navy, with the advantage of operating near its homeland in the central

Both the Germans and the Italians were singularly unfortunate in failing to develop effective naval aviation. Neither navy succeeded in completing any aircraft carriers in World War II, although work was begun on such ships. The incomplete conversion of the Italian *Aquila* is shown here. (R. O. Dulin Collection)

Mediterranean, often found itself at a disadvantage when engaging the Royal Navy fleet units that had aircraft-carrier support because air support was usually not there at critical times. Problems in training and communications often proved insuperable obstacles to the effective application of air force assets to naval missions. Nowhere was this better shown than in the lack of critical air support during the Battle of Cape Matapan.

Geography dictated the general nature of the missions of the German and Italian navies. Effectively, the German role was a strategic one—the interdiction of maritime support of the United Kingdom by submarines and surface raiders—with minimal aviation support. The Italians tailored their ships to operations in the Mediterranean, thereby accepting a tactical mission. With the defeat of France, the Italians could at best achieve a local superiority that would threaten British operations in the Mediterranean and the Middle East. The Germany Navy, on the other hand, could contribute directly to the winning of the war by effectively isolating the United Kingdom.

Spain. Although events of later decades have shown the inadequacies of Spanish resources for the construction of capital ships, the fact remains that the victorious Franco regime was serious in its plans to construct several new capital ships, with major assistance from Italy. These ships would have been part of an ambitious program to rebuild and modernize the Spanish Navy.

After years of a particularly brutal civil war, the Spanish Navy consisted of only two old dreadnoughts, the smallest ever built, a few cruisers, and numerous smaller craft. The limited scope of actual construction during the Franco era makes the idea of Spanish battleship construction seem almost outlandish. The construction of battleship ways at El Ferrol, however, is eloquent testimony to the serious intentions of the Spanish Navy to build such ships.

License-built variants of the Italian *Vittorio Veneto* class would have been reasonably well suited to Spanish requirements, as their relatively high speed would have been helpful for operations in Spanish waters, while their modest endurance would not have been a major liability for the limited area of Spanish warship operations.

Whether or not the new Spanish government needed additional capital ships is debatable, but the fact remains that the ships were seriously contemplated, and they merit consideration as more than mere paper studies.

Evaluation and Comparison. This volume has discussed the characteristics and predominant technical features of modern Japanese, German, Italian, and Spanish capital ships. The authors will now very briefly summarize the background to the development of the various designs in this text and compare the ships with those described in the two previous volumes. This discussion will be grouped in the following order:

- Battlecruisers
  - *Scharnhorst* class
  - "0" class
  - Number 795 class
  - *Alaska* class
  - *Dunkerque* class
  - Dutch battlecruisers
  - *Kronstadt* class

- Treaty battleships
  - *King George V* class
  - *North Carolina* class
  - *South Dakota* class
  - *Lion* class
  - *Vanguard*
  - *Iowa* class
  - *Bismarck* class
  - *Vittorio Veneto* class
  - *Richelieu* class
- Large battleships
  - *Yamato* class
  - *Montana* class
  - *Sovetskii Soyuz* class
  - "H" class

Battlecruisers. Modern battlecruisers, because of speed requirements and displacement limitations, often had their armament limited to gun calibers ranging from 280 mm to 330 mm, although the projected German battlecruiser "0" was to feature 380-mm guns. The *Gneisenau* and *Scharnhorst* were scheduled to be rearmed with 380-mm guns in 1940–41, but such a program was canceled with the outbreak of war in September 1939. If this program had been implemented, these ships would have outclassed all other battlecruisers. Since the conversion was not effected, these ships, and a similar Dutch design that was to be armed with the 283-mm gun, had the weakest main battery of all battlecruisers built after 1930. The *Alaska* class, the Japanese Number 795 class, and the Russian *Kronstadt* class all projected a main battery of nine 305-mm guns in triple turrets. The French *Dunkerque* and *Strasbourg* had the most powerful main-battery gun for a battlecruiser, and in fact, their guns were more powerful than the 356-mm battleship guns projected for the *North Carolina* and *King George V* classes and the old 381-mm/42 gun used on HMS *Vanguard.*

Table 9-1 presents the key characteristics of modern heavy naval guns. The relative performance of the German 283-mm gun was excellent, reflecting the efficiency of the design, but the fact remains that the *Scharnhorst* and *Gneisenau* were outgunned by all

TABLE 9-1
Modern Heavy Naval Guns

Gun (mm)/cal	*Nation*	*AP Shell lbs(kgs)*	*Muzzle Velocity fps(mps)*	*Range @ 30° Yards (Meters)*	*Armor Muzzle*	*Penetration—Inches (mm) 20,000 yds*	*30,000 yds*
18.1"(460)/45	Japan	3,219(1,460)	2,559(780)	39,000(35,662)	34.0 (864)	20.5(521)	15.4(391)
16" (406)/50	Germany	2,272(1,031)	2,657(810)	40,245(36,800)	31.7 (805)	18.8(478)	13.6(345)
16" (406)/50	USA	2,700(1,225)	2,500(762)	36,700(33,559)	32.6 (828)	20.0(508)	15.0(381)
16" (406)/45	USA	2,700(1,225)	2,300(701)	32,300(29,444)	29.7 (754)	17.6(447)	12.8(325)
16" (406)/45	UK	2,375(1,077)	2,475(754)	35,000(32,004)	29.0 (737)	15.3(389)	11.5(292)
15" (381)/50	Italy	1,951(885)	2,789(850)	42,738(39,079)	32.1 (815)	19.8(503)	15.0(381)
15" (380)/47	Germany	1,764(800)	2,690(820)	39,589(36,200)	29.2 (742)	17.3(439)	11.9(302)
15" (380)/45	France	1,949(884)	2,575(785)	38,058(34,800)	29.4 (747)	18.5(470)	13.2(335)
15" (381)/42	UK	1,938(879)	2,400(732)	32,100(29,352)	27.1 (688)	11.7(297)	9.0(229)
14" (356)/50	USA	1,500(680)	2,700(823)	36,800(33,650)	28.0 (711)	13.8(351)	9.3(236)
14" (356)/45	UK	1,590(721)	2,475(754)	32,300(29,535)	25.4 (645)	11.2(284)	8.5(216)
13" (330)/52	France	1,235(560)	2,854(870)	42,104(38,500)	28.1 (714)	16.0(406)	11.5(292)
12" (305)/50	USA	1,140(517)	2,500(762)	33,600(30,724)	24.5 (622)	12.7(323)	9.1(231)
11.1"(283)/54	Germany	728(330)	2,920(890)	40,000(36,576)	23.8 (605)	11.5(292)	8.1(206)

other new-construction capital ships of the World War II era. Nowhere was this better shown than in the desperate last gun battle the *Scharnhorst* fought with the *Duke of York* at the Battle of the North Cape.

The most versatile battlecruiser secondary armaments were mounted on the *Alaska* and *Dunkerque* classes, although German capital ships were given larger guns. Because of space and weight considerations, the American ships were limited to twelve 127-mm guns in twin mounts, no better than U.S. cruisers. The number of 150-mm gun positions for the *Gneisenau and Scharnhorst* was limited also by space considerations. The largest and most powerful battlecruiser secondary armament belonged to the French *Dunkerque* and *Strasbourg*, which featured 130-mm guns of dual-purpose capability. The sixteen 130-mm guns were concentrated into three quadruple turrets and two twin mounts to save weight and conserve space. The magazines that served these guns featured a loading hoist with horizontal and vertical movement to move the ammunition to the handling rooms. Although the French secondary armament was

The two battlecruisers *Gneisenau* and *Scharnhorst* were very active capital ships. However, their main battery of nine 283-mm guns was a disadvantage in an era when some battleships could match the speed of these ships and bring heavier guns to bear. The *Scharnhorst* is shown in the lower view and *Gneisenau* in the upper. (Upper view: Albrecht Schnarke Collection; lower view: Foto Druppel)

tested only once—in the pursuit of the *Strasbourg* by British forces in July 1940—the quadruple mounting was not considered a success and was not repeated.

The secondary armament of Japanese battlecruisers would have been better than that of some battleships, but their 100-mm gun barrels had a very short service life of only 450 rounds; however, these guns had exceptional ballistics. They illustrate the trade-offs that have to be made in the design of warships to achieve a compromise that combines the attributes sought by a particular navy.

Since the new battlecruisers were by nature smaller than the new battleships, their deck areas were smaller, and this limited the number of antiaircraft weapons. Thus, no battlecruiser of the last generation of dreadnoughts had a particularly outstanding antiaircraft battery. The growing threat of air attack, however, would have forced a significant proliferation of machine guns (20 mm, for example). This did happen in the case of the *Scharnhorst* and *Gneisenau*, particularly before the Channel dash, when a number of permanent and temporary 20-mm gun mounts were added. The light machine-gun batteries of the *Alaska*-class battlecruisers were substantially different as completed from their contract design and did reflect the need to combat the air menace.

In World War I, swift and well-protected German battlecruisers carried smaller-caliber guns than comparable British ships. This concept was used in the French

The USS *Alaska* (CB 1) on 11 September 1944 was one of two battlecruisers completed for the U.S. Navy. Although often referred to as large cruisers, these ships had a battleship armament of nine 305-mm/50-caliber guns, but an American heavy-cruiser style secondary armament and aircraft arrangement. Neither the *Alaska* nor the *Guam* had a significant role in World War II, as they were not assigned to combat status until February 1945.

The USS *Alaska*, shown in Pearl Harbor in February 1945, had a more powerful main armament than the German *Gneisenau* or *Scharnhorst*, but their armor and underwater protection systems were inferior to those of their German counterparts.

Strasbourg, which, with her powerful main armament, adequate armor, and the broadest side-protection system ever used in a ship of her size, was the best-protected modern battlecruiser. The *Dunkerque,* her sister ship, with somewhat lesser armor protection and an equal side-protection system, was outclassed by the German *Gneisenau* and *Scharnhorst,* which had better protection. The shell damage sustained by the *Scharnhorst* from the 356-mm guns of the *Duke of York* proved that she could take as much punishment as the *Bismarck* had two years previously. Her most glaring weakness was the deck-armor system, as the events at La Pallice with the *Scharnhorst* and with the *Gneisenau* at Kiel proved. The underwater protection was more gravely weak. Because of the hull taper required for the 32-knot speed and the beam of only 30 meters, the underwater protection at the stern end of the citadel on these German battlecruisers was very weak. The vulnerability was due to the passage of the outboard shafts through the side-protection system and to a poor structural arrangement in way of the shaft passage. The torpedo bulkhead was also not continued to the main deck, and the riveted connection was so weak that large stress concentrations and poor stress flow occurred. This contributed to substantial structural internal damage whenever the stern area was struck by a torpedo.

One of the best underwater protection systems for a modern capital ship was provided the *Dunkerque* class. The torpedo bulkhead extended the full depth of the hull girder, becoming a splinter bulkhead above the armor deck, with no discontinuities along its entire length. The French naval constructors also paid strict attention to structural details of this bulkhead to ensure that no stress concentrations could occur that would mar its performance. In areas where the protection was diminished in breadth by the hull form, the bulkhead was gradually curved inboard (rather than stepped as in British battleships), and thicker plates were used to compensate for the loss in transverse depth at the ends of the armor box. In summary, this was an excellent system and withstood an unusually large explosion at Mers el Kébir, enabling the *Dunkerque* to remain afloat.

The *Alaska* and "0"-class battlecruisers had weak armor protection and in the case of the American battlecruisers, no torpedo defense system, making them the most vulnerable battlecruisers to enter service during the World War II era. The concept of providing the *Alaska*s with a torpedo defense system was explored in their preliminary design, and it was found that the weight and space of such a system could not be tolerated within the displacement sought. An alternative was to sacrifice some of the guns in the main battery. The adoption of a torpedo defense system with all other features remaining unchanged would have meant a larger ship. The German "0"-class battlecruisers were planned to have an armor system to withstand only the gunfire of heavy cruisers. It is fortunate that they were not built because they not only would have been vulnerable to long-range gunfire, but also to aerial assaults.

Speed had been traditionally considered the best defense for a battlecruiser, and the ones built after 1930 had excellent speed properties. Unfortunately, these ships were developed in the same era as the fast battleship and, therefore, did not have the speed superiority that battlecruisers enjoyed over battleships in World War I. In fact, with their weaker armament and similar speed, these ships were extremely vulnerable in a clash with a fast battleship, such as a *Bismarck, Iowa,* or *Richelieu.* The most advanced type of propulsion plant, featuring double reduction gears, high-temperature-and-pressure steam, and high-speed turbines was fitted to the *Alaska* class. Similar plants were provided for all American cruisers, destroyers, carriers, and battleships built after 1934. The *Scharnhorst* and *Gneisenau* were almost equal to the American battlecruisers in terms of propulsion, but the German ships were equipped with single reduction gear, which led to lower turbine speeds and inefficient use of the very efficient high-temperature-and-pressure boiler. As a result, the fuel consumption was high. The German boilers featured higher pressures and temperatures than the American equipment. Flaws in their superheater tubes caused some early problems, finally requiring the *Scharnhorst* to be taken out of service in April 1941 for a period of four months. This came at a crucial moment in the war, when the German Navy had taken the initiative, and forced a replanning of the *Bismarck* operation. German marine engineers were not satisfied with the reliability of the power plants of these ships.

Diesel engines of the type and power required to propel German battlecruisers at speeds of 30+ knots were not ready in 1935. Developments of diesel propulsion systems continued. By the time the design of battlecruiser "0" was being finalized, diesel engines were used in conjunction with a steam power plant to combine the economy of diesels at cruising speeds with the technical efficiency of steam power for high speed. All other battlecruisers attained their speeds with conservative steam pressures and temperatures, with the possible exception of the *Alaska*-class ships. These American battlecruisers had the same type of steam plant that was used in the battleships, and their engine and boiler spaces were very compact in terms of volume use per horsepower. The *Dunkerque* and *Strasbourg* also exhibited compact propulsion plants. The *Dunkerque* had an unusually long endurance of some 16,000 to 17,000 miles at a speed of 17 knots.

Of all the ships to enter service, by far the best battlecruiser design was that of the *Strasbourg*, which could be considered a small-displacement battleship. The German *Scharnhorst* and *Gneisenau* featured the best armor protection given to a battlecruiser type; in many respects it was comparable to that of the *Bismarck* and *Tirpitz*. The *Strasbourg*, however, had a more powerful main-battery gun, better underwater protection, good armor protection, and was equal to the German battlecruisers at least in speed.

Treaty Battleships (35,000 to 45,000 tons). The outstanding element in any comparison of "treaty" battleships is the balance of speed, protection, endurance, and armament to satisfy the treaty limits on gun power and displacement. Varying schemes were used in these ships for deployment of armor, electric power, propulsion-plant arrangement, steam conditions, and the concentration and type of armament. Since treaty battleships varied in displacement, it is difficult to compare them. For example, the 45,000-ton *Iowa*s, being 10,000 tons heavier than the *King George V*s, could afford more armament, protection, and machinery than the first-generation treaty battleship. This compounds the difficulty in comparing ships.

In French, German, and Italian treaty battleships, speed was an important design factor, coupled with the recognition that the 380-mm gun was the most suitable gun for the ship size sought. The Italian gun was a remarkably powerful weapon for its size; at normal ranges, both the French and Italian guns were ballistically superior to the American 406-mm/45-caliber (Mark VI) guns mounted on the *North Carolina* and *South Dakota* (see table 9-1). There was a price to pay—the barrel life was shorter.

This left the *King George V* class, the last generation of dreadnoughts, with the smallest gun caliber, 356 mm, of all. This was the least powerful main armament for any of the new battleships. Nonetheless, two shells from the *Prince of Wales* were sufficient to inflict enough damage to the larger and more powerful *Bismarck* to cause the German battleship to abort her mission. Later, the shellfire from the *King George V* would help sink the *Bismarck*. It is important to remember that even the most inferior of the new battleship guns was still a very powerful weapon.

HMS *Renown* is shown with the new French battleship *Richelieu* and HMS *Valiant* during exercises of the 1st Battle Squadron of the Far Eastern Fleet in early 1944. By this time, battleships and battlecruisers had become subordinate to the aircraft carrier. (National Archives)

The *Bismarck* and *Tirpitz* were the largest battleships completed by a European naval power. Under the bold leadership of Admiral Raeder, the *Bismarck*, shown here, was deployed to the Atlantic Ocean in May 1941. Her short wartime cruise came to an abrupt and tragic end through the use of aircraft and radar. No longer would the vast expanses of the oceans be safe havens for surface raiders. (Imperial War Museum)

Another important comparison is the burster-charge weight and overall projectile design. Due to the muzzle velocity and weight, the Italian 381-mm/50 and the American 406-mm/45 guns appear to have had relatively high penetration capability. The American armor-piercing shell had a burster-charge weight of 15.23 kilograms, thicker caps, and a more oval-shaped nose, while the Italian shell had a charge of 10.54 kilograms and a strong projectile body. The British 356-mm shell had a burster charge of 18.1 kilograms, while the French 380-mm shell had one of 21.7 kilograms. Because of the heavier steel mass, the American and Italian shells could pierce thicker armor, but both the French and British shells, with their heavier burster charges, could cause more serious damage if they pierced the armor. In summary, the British 356-mm shells were designed for optimum penetration of face-hardened armor at 30°–35° obliquity, while those of the United States were designed for maximum penetration of any armor, however thick, at any obliquity with the maximum velocity—even if that was achieved by decreasing the effectiveness of the penetration. American projectiles could penetrate thicker armor than any foreign gun. The German 380-mm shells were designed to penetrate thick face-hardened armor of inferior quality to the Krupp armor used in the *Bismarck* and *Tirpitz.* The most interesting outcome of the *Bismarck-Hood-Prince of Wales* encounter was the poor performance of the German 380- and 203-mm projectiles that struck the *Prince of Wales.* Poor fuze action resulted in low-order detonations and dud projectiles, sparing this British battleship from more serious damage. Nothing, however, was wrong with the shell that exploded in the aft 4-inch magazine of the *Hood.* Its explosion ignited the cordite to produce a massive explosion that destroyed this British battlecruiser much like her three predecessors during the Battle of Jutland some 35 years earlier.

In an effort to save construction time, the Royal Navy chose a 356-mm main battery for the *King George V* ships. Despite the potential for a change in the armament limitation to 406 mm if the Japanese failed to ratify the 1936 London Naval Treaty, the British confined their efforts to the 356-mm gun and turret designs, even in view of the fact that Germany, France, and Italy favored the 380-mm gun, since twelve guns looked better than eight or nine. Available resources were insufficient to permit the parallel development of 356- and 406-mm guns and turrets that was achieved in the United States. The Royal Navy's acceptance of a smaller armament in the interest of saving time was a conscious decision. Later, when confronted with overweight problems forced by the need to upgrade armor protection, one of the forward quadruple turrets was changed to a twin in order to adhere to treaty displacement limitations. This added almost one year to the construction time of the ships, thereby frustrating the earlier efforts to minimize construction time, and in a sense, undermining the entire rationale for accepting a smaller gun caliber.

All these treaty battleships, with the exception of the *Vittorio Veneto* and *Bismarck* classes, featured a dual-purpose secondary armament. The choice of a single-purpose battery in the Axis ships reflected the need to counter the large number of destroyers and cruisers in the French and British navies. Only American and British secondary gun projectiles were equipped with a proximity fuze; the effectiveness of such a dual-purpose armament was clearly demonstrated during the Battle of Santa Cruz Island, when the *South Dakota* downed 26 planes. Dual-purpose guns could be mounted for more effective and better-sustained antiaircraft fire than was possible with a mixed-caliber secondary armament—a feature that ballistic tables will not reveal. Contrast this performance with that of the *Tirpitz,* which required barrage fire from her 15-inch guns to supplement her antiaircraft defense, all of which was not effective in defense against the bombing attack that finally destroyed her. The weakest secondary armament (in terms of antiaircraft defense, although it was rather effective against surface

The British *King George V* and the American *North Carolina* classes of battleships were built to the terms of the 1936 London Naval Treaty, with designs featuring 356-mm guns. However, an anxious U.S. Navy had specified that its ships be able to mount 406-mm guns in case Japan did not ratify the 1936 treaty by 1 April 1937, and the *North Carolina* and *Washington* (shown here) were completed with 406-mm guns. (Top: U.S. Navy; bottom: W. H. Garzke Collection)

ships) was that of the *Vittorio Veneto* class. These ships had a mixed battery of 152-mm and 90-mm guns that were optimized for surface and antiaircraft fire. Unfortunately, the 90-mm mounts were mechanically unreliable, and defective projectiles detracted from the excellent ballistic performance of the gun. As a result, these Italian ships were extremely vulnerable to air attack.

Armor protection in treaty battleships was arranged somewhat differently than in earlier dreadnoughts. The new ships had to be able to withstand longer-range shellfire (larger immunity zones) and bombing attacks from aircraft—the latter deemed a more serious threat by naval staffs in 1935 when they considered the anticipated length of service (20–26 years). This forced a greater relative emphasis on horizontal armor protection than in earlier battleship designs. Another aspect influencing the armor protection sought in these ships was the limitation on displacement. Because of weight constraints, the armor would have to be concentrated around the magazines and machinery, and greater concentrations of gun power and machinery were sought to permit more efficient deployment of armor. This led to the universal acceptance of the "all or nothing" concept, in which the armor was concentrated in the armor citadel with the areas outside of it—with limited exceptions, such as the steering-gear room and the conning tower—being lightly protected or not protected at all by armor. In Germany, Italy, and Japan, treaty restrictions were not governing criteria, and the best possible protection was sought against all forms of attack from a destroyer to a battleship. In the design of the *Yamato*, however, the Japanese had to concentrate their armament and propulsion plant to "limit" these ships to a displacement of 72,000 tons.

The French battleship *Jean Bart* was the last battleship to be completed. Mounting eight 380-mm/45-caliber guns in quadruple turrets, she and her sister ship *Richelieu* were of an outstanding design. The reconstructed *Jean Bart* carried an impressive electronics array and a potent antiaircraft battery (nine 152-mm dual-purpose, twenty-four 100-mm dual-purpose, and twenty-eight 57-mm guns). (R. J. Scott Collection)

HMS *Vanguard* in June 1960, just before she was to be towed away to the breakers. This British battleship, built to rival the *Bismarck* and *Tirpitz*, never saw combat, as she was completed in 1946. The last of the British battleships, she mounted a very successful 381-mm gun.

In the *North Carolina, South Dakota, Vittorio Veneto*, and *Richelieu* classes, the main side belt was sloped and the armor thickness reduced. In the *King George V, Lion, Vanguard*, and *Bismarck* classes, a vertical armor belt arrayed at 90° was used. When the preliminary design of the *King George V* commenced in 1935–36, the director of naval construction in the Royal Navy, Sir Arthur Johns, did not want to repeat the sloping belt arrangement that had been used in the *Nelson* and *Rodney*. He surmised that diving shells would be deflected downward into the bilges, causing severe damage and flooding. This decision imposed a severe weight penalty on the last generation of British dreadnoughts, but was carried forth by his successor, Sir Stanley Goodall. For example, to resist a 356-mm shell that weighed 681 kilograms at a distance of 17,374 meters, a 381-mm vertical belt would be required instead of a 340-mm belt sloped at 10 degrees. For the same target height presented to the oncoming shell, the 10-degree sloping armor permitted a total weight saving of some 250–300 tons. Of course, the greater the slope, the thinner the armor could be for the same protection. This was a principle conscientiously followed by American, French, and even Japanese naval constructors. To make up for the disadvantage of the vertical side armor, the British and the Germans improved the metallurgy of their face-hardened armor to achieve superior resistance, despite the inefficient arrangement of the armor from a weight viewpoint. The French *Richelieu* and the German *Bismarck* had the best protection for the spaces within the citadel, due to the use of armor slopes to reinforce their main side belts. The weakest armor protection in this region of the hull was in the *Vittorio Veneto* class ships. Their main side belt extended only some 0.75 meters below the design waterline, making these ships vulnerable to diving shell sttack. The importance of such protectoin was clearly demonstrated in the *Prince of Wales-Bismarck-Hood* battle where the *Prince of Wales* and *Bismarck* were damaged by projectiles with underwater trajectories. It is also possible that the fatal shell hit to the *Hood* could have come from a shell with such a trajectory.

Heavy deck protection in the treaty battleships was greater than in previous battleships, reflecting the fact that in 1935 there were two potential dangers—bomb attack and longer-range shell trajectories that would cause projectiles to fall at angles more nearly normal to the armor surface. Most of the designs featured a light armored weather deck and a thick armor deck just above the neutral axis of a given structural section. The Italians, however, used a multi-deck protection scheme that was not as effective as that provided other ships. In general, the strongest citadel was provided to the *South Dakota* and *Iowa* classes, while the weakest was that of the *North Carolina* class.

With the development of the submarine and, in particular, torpedo-carrying aircraft, the threat of torpedo attacks on battleships had grown during the interwar period. Admiral Yamamoto of the Imperial Japanese Navy was often quoted about the development of torpedo-bearing aircraft attacking battleships: "The fiercest serpent can be overcome by a swarm of ants." Torpedo attack was, therefore, a greater threat to battleships of the pre–World War II era. Thus, more effective side-protection systems were needed to contain the effects of larger charges. The use of sloped armor in the *North Carolina* led to the decison to use an underwater blister hull form. The system in the *North Carolina*, however, was designed to resist 318-kilogram charges, despite the fact that minority opinion in the U.S. General Board regarded this as inadequate for torpedo attacks of the future. The torpedoing of the *North Carolina* in 1942 verified this minority opinion. Near-miss bombs and diving underwater hits of large projectiles were also perceived as problems.

The most effective side-protection system fitted to a treaty battleship was that in the *Richelieu* class, where the widest torpedo protection layer fitted to a capital ship was 9.45 meters in the *Jean Bart* and 8.25 meters in the *Richelieu* at a point near amidships. The *King George V* class had the weakest underwater protection, revealed during the torpedoing of the *Prince of Wales* in the South China Sea on 10 December 1941. The damage to the ship from torpedoes that had charges substantially less than the design charge revealed flaws in the original design: poor structural details, the absence of an expansion zone over the system, and the failure to provide a strong

The *Duke of York*, one of the *King George V*-class ships, profited from the loss of her sister ship the *Prince of Wales* in that a number of beneficial changes were made. Shown here in 1942 before her encounter with the *Scharnhorst*, more antiaircraft guns were to be added and the amidships aircraft facility was to be removed. Although mounting ten 356-mm/45-caliber guns, these ships had a very potent 134-mm/50-caliber dual-purpose gun that was powerful against aircraft, particularly when the shell had a proximity fuze.

structural foundation and continuity of structure above the system that would allow the elastic deformation of bulkheads to occur. These defects were largely corrected in extensive refits of her sister battleships during the course of World War II. Similar defects were corrected in the *Lion* design and the *Vanguard.*

The most unusual side-protection system was fitted to the *Vittorio Veneto* class. The Pugliese underwater-protection system featured a cylindrically shaped hollow structure designed to be kept immersed in either water or oil. The inherent strength of circular structure, combined with the absorption of the charge's energy by the hydraulic crushing of the hollow cylinder, was expected to provide excellent protection against torpedo damage. Herein exists one of the real potential weaknesses of the Pugliese system. Once the cylinder was crushed, there was no material other then the torpedo bulkheads to resist any subsequent torpedo hits. Although such a potential weakness existed in the multi-layer system when subjected to damage, this system had more mass of steel to absorb the forces of one or two torpedo hits. Although the system performed reasonably well in those cases of torpedo damage, there were some problems with flooding beyond the torpedo bulkhead when the *Roma, Italia* (ex-*Littorio*), and *Vittorio Veneto* sustained underwater damage. This can be attributed to a weak riveted

This is an aerial view of the USS *North Carolinia* in June 1942. Although originally conceived to carry nine 356-mm/50 (Mark 7) guns, Japanese refusal to sign the 1935 London Naval Treaty by 31 March 1937 brought about a contract plan changeover to nine 406-mm/45 (Mark 6) guns. The *North Carolina* sustained torpedo damage early in the war, but her sister ship the *Washington* sank the Japanese battelship *Kirishima* in one of the fierce night naval encounters off Guadalcanal.

The Italian *Vittorio Veneto* class mounted very powerful guns that had exceptional range for their caliber. The lack of aircraft support frequently hindered the operations of these ships. The *Vittorio Veneto* is shown here at high speed. (Ufficio Storico)

joint failing because of unusually high shear-stress leads and the fact that forces created by the explosions took the path of least resistance. The cylindrical section was too strong to accomplish its objective. Welded connections between the torpedo bulkhead and shell would have been more effective for this system, since the major portion of the energy from an underwater explosion would then have been absorbed in the destruction of the cylinder and the elastic deformation of surrounding structure.

Perhaps the most important difference between treaty battleships and those built during the World War I era involved their propulsion plants. Because of vastly improved marine-engineering technology, the newer ships had much higher speeds and superior endurance. The emphasis on weight savings profoundly influenced modern capital-ship design and made it possible to arrange more powerful units in smaller spaces. Although steam plants with electric propulsion were considered to be more flexible in operation and permitted better subdivision of propulsion plants, the steam turbine with mechanical reduction gearing was less expensive and lighter. A diesel engine for main propulsion rated at 40,000 horsepower was considered a possibility. The acquisition cost of such diesels would have been enormous and, from the standpoint of reliability in service, would have been a step in the dark for many of the navies, with the exception of Germany. Moreover, when considering diesel propulsion for the treaty battleship, it should be recognized that fuel is not considered part of the standard displacement, so that the greater weight of the diesel plant cannot be offset (from the standpoint of treaty displacement) by its smaller fuel requirement. Table 9-2 outlines key attributes of representative propulsion plants for World War II capital ships. The German plants compare very favorably with all but the American installations, which were by far the most efficient. This was due to the introduction of high-pressure-high-temperature steam combined with a higher turbine speed and the locked-train double-reduction gear. Diesel plants, not suprisingly, are much heavier than similarly powered steam installations. The diesel, however, is more economical in terms of fuel burned,

TABLE 9-2
Propulsion Plant Comparisons*

Ship	*North Carolina*	*Bismarck*
Nation	U.S.A.	Germany
Year completed	1941	1940
Steam pressure	575 (40.44)	853 (60.00)
Steam temperature	850 (454)	869 (465)
Reduction gear	double	single
Number of shafts	four	three
Shaft horsepower	115,000 (116,595)	136,112 (138,000)
Machinery weight	3,286 (3,340)	4,723 (4,800)
SHP/ton	35.0 (34.9)	28.8 (28.75)
Ship	**"H" Class**	***Yamato***
Nation	Germany	Japan
Year completed	—	1941
Steam pressure	diesel main propulsion	355.6 (25.00)
Steam temperature		618 (325)
Reduction gear		single
Number of shafts	three	four
Shaft horsepower	147,948 (150,000)	147,948 (150,000)
Machinery weight	5,858 (5,961)	6,559 (6,707)
SHP/ton	25.26 (25.16)	22.4 (22.37)

**Note:* Machinery weight includes electric plant.

and thus its advantage in terms of specific fuel consumption partially offsets the machinery weight disadvantage.

Another measure of propulsion-plant efficiency is that of volume. Table 9-3 lists a number of ships and combines the main propulsion turbine with the electric plant because these units coexist in the machinery box. It can be seen that American ships are extremely efficient in volume use. The *King George V* class was somewhat inefficient, having the smallest power plant in a relatively large machinery box. The *Richelieu* had a very compact propulsion plant and not surprisingly the lowest volume-to-power ratio in those ships compared. The Indret-Sural boiler was an extremely compact unit for its power; the concentrated propulsion-plant arrangement chosen by the French helped to produce the results shown in table 9-3. The *Vittorio Veneto*-class ships show a remarkable concentration of machinery, and this may be due to the Italian concept of doing maintenance in port, unlike the machinery arrangements in American or British ships, where access to equipment for maintenance and repair at sea was a very important consideration.

The *Vanguard, Iowa, Richelieu, Bismarck,* and *Vittorio Veneto* were very fast ships. It would appear that the *Iowa*s had the most advanced propulsion plant afloat and attained very economical fuel rates that permitted a large radius of action.

The *Vanguard, Bismarck,* and *Tirpitz* were very seaworthy. Since speed was a mandatory factor in the development of the *Iowa*, her hull form was extremely fine, and the very slender bow tended to bury itself in rough seas. During the joint NATO exercises in the Norwegian Sea in 1953, the *Iowa* had much trouble in shipping water due to the heavy weather, while the *Vanguard* fared better, as her breakwaters and flare and fuller bow form dispersed the same seas. This is a tribute to British designers, who recognized the shortcomings of seakeeping in the *King George V* design after evaluating the performance of the *Prince of Wales* during her battle with the *Bismarck* and *Prinz*

TABLE 9-3

Machinery Box Volume Comparisons

Ship	*North Carolina*	*Iowa*	*King George V*
Nation	U.S.A.	U.S.A.	U.K.
Volume of machinery box (ft^3)	308,000 (8,723)	569,150 (16,118)	397,760 (11,265)
Shaft horsepower (shp)	115,000 (116,595)	212,000 (214,940)	100,000 (101,387)
kw	5,000	10,000	2,800
shp + kw (in shp)	121,705 (123,393)	225,410 (228,536)	103,755 (105,194)
Volume/shp + kw	2.53 (0.0707)	2.52 (0.0705)	3.83 (0.1071)

Ship	*Vanguard*	*Richelieu*	*Tirpitz*
Nation	U.K.	France	Germany
Volume of machinery box (ft^3)	395,139 (11,191)	308,406 (8,734)	452,937 (12,826)
shp	120,000 (121,664)	147,950 (150,000)	147,950 (150,000)
kw	3,720	6,000	7,910
shp + kw (in shp)	124,991 (126,725)	158,046 (160,238)	158,555 (160,754)
Volume/shp + kw	3.16 (0.0883)	1.95 (0.0545)	2.86 (0.08089)

Ship	*Vittorio Veneto*	*Yamato*
Nation	Italy	Japan
Volume of machinery box (ft^3)	312,917 (8,861)	440,048 (12,461)
shp	128,222 (130,000)	147,950 (150,000)
kw	3,600	2,400
shp + kw (in shp)	133,050 (131,230)	153,218 (155,344)
Volume/shp + kw	2.35 (0.0675)	2.87 (0.0802)

Eugen. The *Bismarck* and *Tirpitz* demonstrated good seakeeping with their "Atlantic bow."

The *Bismarck* and *Tirpitz* were very formidable opponents. Due to the exploits of the *Bismarck* during her very short career, these ships have gained legendary status. The battle between the *Hood, Prince of Wales, Prinz Eugen*, and *Bismarck* during the morning of 24 May 1941 has brought the *Bismarck* much recognition of her strengths. A careful analysis of her short career, however, also reveals many weaknesses. One compensating fact overlooked by many authors was the quality and training of the crew. The crew of the *Bismarck* was exceptionally well trained, and they fought their ship with great valor, but the *Bismarck* was simply outgunned. Technically, the emphasis on the mixed secondary battery and the four-turret arrangement was the major reason why this battleship weighed in at 42,000 tons. Triple turrets with 380-mm guns and a dual-purpose battery would have produced greater firepower and also permitted more armor protection. Although superior to their British contemporaries, these German ships were somewhat inferior to comparable American, French, and even Italian ships in many respects. In fact, had Winston Churchill been successful in bringing the *Richelieu* over to the Allied side in 1940, she would have been a more serious threat to the *Bismarck* in a naval engagement than the newly completed *King George V* and *Prince of Wales*. These British battleships had repeated mechanical failures in their quadruple turrets, and full salvos were impossible during their battles with the *Bismarck*. These difficulties were later corrected. A battle group of the *Hood* and *Richelieu* could have dealt more effectively with the *Bismarck*. The severe damage sustained by the *Bismarck* before her sinking on 27 May 1941 indicates a good protection system, but it also attests to good damage control and crew discipline.

Large Battleships. The *Yamato* and *Musashi* were the most heavily gunned battleships ever constructed. Their nine 460-mm guns were the largest ever mounted on a dreadnought, but were only marginally superior in performance to the smaller American 406 mm/50, as table 9-1 shows. The Japanese gun would have had a slight advantage in close-range actions; however, at ranges of 20,000 yards or more, the 406-mm gun was almost as powerful. This reflected American advances in gun technology vis-à-vis the Japanese during the interwar period.

The German "H"-class ships were a logical evolution of the *Bismarck* design, with more powerful 406-mm guns, slightly improved protection, and a powerful diesel main propulsion plant that would have provided an exceptional endurance.

The only ships to equal the size of the *Yamato* and the "H" class were the American giants of the *Montana* class, projected but never laid down, and the Russian *Sovetskii Soyuz* class, laid down but never completed. Forty years after World War II, details of these Russian giants are still not known with confidence.

For surface engagements, the "H" and *Sovetskii Soyuz* classes had secondary guns that were superior to those projected for the *Montana*s, but the American battleships would have been provided with the best dual-purpose gun developed during the battleship era. Their twenty 127-mm/54-caliber dual-purpose guns, with semiautomatic loading and twin mountings, had better range and more rapid fire than earlier 127-mm/38-caliber guns used in three previous American capital-ship classes. The Japanese battleships were equipped with a mixed battery when the ships were first completed, but as the war progressed the Japanese recognized the need for better antiaircraft coverage, and two of the four triple 152-mm turrets were removed from each ship in favor of added antiaircraft firepower. A battery of 127-mm guns was installed in their place on the *Yamato*, and heavy antiaircraft machine-gun batteries were added to the *Musashi*. There was also a substantial increase in light machine guns, almost 150 being on the *Yamato* at the time of her sinking. Without central fire control for these weapons, however, the effectiveness of such a mass of gun power against aircraft was limited. One important point should be made about the antiaircraft battery on the *Yamato* and *Musashi*; the concentration of gun power was achieved by removing boats and aircraft from the central superstructure. This arrangement permitted wide arcs of fire for the antiaircraft battery unimpeded by topside clutter of either boats or aircraft.

The armor protection of the large battleships was massive, with extensive subdivision and torpedo-protection systems that could not be installed in many of the treaty battleships. The deck-armor system of the *Montana* was the strongest, while the side protection of the *Yamato* and *Musashi* was marginally superior. The distribution of armor in the "H"-class battleships was slightly different than in the *Bismarck*. Many of the features of necessity surrendered or compromised in the *Bismarck*'s design were provided in the larger "H" class. Based on the experiences of the battlecruiser *Lützow* during the Battle of Jutland, more splinter protection was used in the bow region. The armor belt was thinner, and the upper citadel belt was retained in both arrangement and armor thickness. The absence of bow splinter armor was one of the sacrifices the Germans had to accept to keep down the displacements of the *Scharnhorst* and *Bismarck* classes. The armor deck of the "H"-class battleships was thinner and one deck lower than in any of the other comparable large battleships.

The horizontal armor protection of the *Yamato* and *Musashi* was more than sufficient to prevent shell or bomb damage from reaching the vitals; the "Achilles heel" of these Japanese ships was their torpedo-defense system. A joint between the upper and lower belt was gravely weak, and its failure during torpedo attacks contributed to the sinking of both battleships and the converted aircraft carrier *Shinano*, which, like her battleship cousins, had the same joint detail. Proposals were made by Japanese naval constructors to correct the deficiency, but the penalty was severe—a 5,000-ton increase in displacement and the removal of the ships from war duty at a time when Japan desperately needed them. In addition, their modifications would have delayed other urgent ship construction. The volumes outside the armored box were also too great, permitting large flooding and consequently large trims after damage. The compartmentation within the machinery box was extensive, and flooding of outboard spaces led to large lists. The combination of heavy trims and large lists resulted in progressive flooding and the ultimate loss of the battleship versions. The carrier *Shinano* was lost through a combination of progressive flooding through non-tight penetrations in watertight structure (due to the incompleteness of the ship), poor damage-control techniques on the part of an undisciplined crew, and finally a large list.

The "H" ships and the *Montana*s were to be provided with extensive splinter protection and good subdivision. The Germans used a splinter belt forward and aft of the armored citadel to limit flooding from fragmentation damage. The torpedo-defense system of the German battleship was a repeat of that provided the *Bismarck* and *Tirpitz*. The diesel-propulsion and electric plants would require greater space than a comparable steam plant. To increase the side-protection system would have required a larger beam and more displacement; it was decided to retain the *Bismarck* torpedo-defense system, which was calculated to be proof against 250 kilograms of TNT. Furthermore, a greater beam and displacement would have caused draft problems in ports like Wilhelmshaven, and the "H"-class ships could just navigate in the Jade with their H-39 characteristics. On the other hand, the underwater-defense system of the American *Montana* class was very deep and well conceived to absorb torpedo damage. This system was also much improved over earlier torpedo-defense systems in American battleships.

The protection provided the *Sovetskii Soyuz* is not precisely defined due to Soviet secrecy. The size of the ship, 62,000 tons full load, indicates that heavy armor protection must have been used in these ships. The torpedo-defense system was probably a copy of the Pugliese type and no doubt had the same strengths and weaknesses, since the Russians were not as advanced in welding as other European maritime powers.

All of the large battleships featured steam propulsion with the exception of the German "H" class. The all-diesel electric and propulsion plant featured twelve diesel engines powering three shafts through a complex gear arrangement. The defects in the marine diesels and the vibrations experienced in the armored ships of the *Deutschland* class had been overcome. This would have assured quieter and more efficient operation and economy in the burning of fuel. The Japanese had wanted to use diesel propulsion in the *Yamato* class, but they encountered difficulties with the diesel prototype in a seaplane tender and were reluctant to install an unreliable plant in a ship where engine replacement would be very difficult. The *Yamato* class finally had a steam propulsion plant that was not as efficient. The *Montana*-class ships would have had the best propulsion plant of the larger battleships. The double reduction gear and high-pressure-and-high-temperature steam cycle was a proven success in earlier U.S. battleships and yielded large cruising radii. The *Sovetskii Soyuz* probably did not have as efficient a propulsion plant as the American ships, as it featured single reduction and large, inefficient boilers.

The *Montana*-class ships would have had better gunnery and protective systems than the *Iowa*-class ships that preceded them. Their maximum speed was reduced to 28 knots to realize these improvements, since even greater propulsive power, displacement, and ship size would have been required to maintain the same speed as the *Iowa*s. The U.S. Navy's General Board was unwilling to authorize a battleship in excess of 72,000 tons full load, and thus discussions before the board in 1940 resulted in a decision to limit the standard displacement of America's last battleship design to 60,000 tons. The propulsion plant featured the same type of equipment being used in other U.S. warships of the time, but there was better subdivision than had been employed in previous American dreadnoughts. In fact, the layout of the engineering plant was the basis for that of the large carriers of the *Midway* class. This extensive subdivision was only possible because the U.S. Navy had decided that the requirement for battleship transit of the Panama Canal would be waived so that these ships could have a greater beam with an improved side-protection system.

The *Yamato*-class battleships had only a fair propulsion plant. The power was developed by outmoded boilers and turbines, which permitted a speed of 27 knots. A combination of steam and diesel power had been sought in the original design, but serious flaws in the design of the diesel engines forced the choice of an all-steam propulsion plant. The low speed of 27 knots, therefore, conferred a tactical advantage on faster opposing battleships, but most of the ships discussed in these volumes would have had to close range to seriously damage these Japanese ships by their shell fire. At such close ranges, gunfire from the *Yamato* or *Musashi* would have become devastatingly effective. The steam power plant also led to a rather large fuel rate, and during their careers these ships had rather limited fighting roles. The endurance of these ships was half that sought in the *Iowa*- and *Montana*-class ships.

The best design of the larger battleships was that of the *Montana* class, although some of the other ships may have exceeded some characteristics of this class. There was a more effective distribution of gun power, a more efficient propulsion plant, and the much improved secondary battery featured twin-mounted 127-mm/54-caliber guns with a dual-purpose capability. These guns were used for effective antiaircraft defense in postwar American destroyers and have survived to the missile age.

Although complete details are lacking on the *Sovetskii Soyuz* class, the authors consider that these Russian ships would have been formidable warships in terms of gun power and protection. Only their propulsion plant, underwater protection, and antiaircraft capabilities probably would have been liabilities.

Epilogue. It is difficult for anyone to compare and contrast all the "modern" battleships and battlecruisers discussed in these three volumes and determine which one was the best. Each ship, whether designed or built, was an attempt to achieve a solution to a different set of requirements within limiting conditions of cost, treaty limitations, staff requirements, industrial capabilities, and geography. Desirable features had to be blended into a workable design through the skill and insight of engineers. Compromise was an important ingredient, because the "optimum" armament, protection, and propulsion could not be integrated into a single design. Compromise was usually prompted by the need to observe certain limiting conditions, which, like differing desirable features, varied from navy to navy, and the success of a design hinged on how successfully these compromises were accomplished. For example, the U.S. Navy considered substantial endurance important in its pre–World War II warship designs and achieved this through the development of economical propulsion units that could also provide the necessary power within a restricted volume and weight.

Fortunately, American industry was able to produce such machinery to meet the requirements. On the other hand, most other countries that built battleships found it difficult to adopt the American solution. A comparison of ship costs would show that American capital ships were far more expensive in terms of dollars per ton than comparable foreign designs, but the United States had greater resources, which enabled

This aerial view of the *North Carolina* in June 1942 from an elevation of 340 meters above sea level was taken during antiaircraft practice. The arcs of fire of the antiaircraft battery, including the dual-purpose 127-mm–38-caliber gun, was an important consideration in the design of these and other last-generation dreadnoughts, as the experience of the *Bismarck* was to show. The development of an adequate light machine-gun battery was equally important. This unusual early view of the *North Carolina* shows a mixture of forty 20-mm and twenty .50-caliber machine guns, and four 20-mm machine-gun quadruple mounts that later evolved into a battery of 20- and 40-mm guns.

better designs to be produced. Great Britain was unable to achieve as efficient an electric and propulsion plant in her *King George V* class as the U.S. Navy was able to install in its *North Carolina* and *South Dakota* classes.

The reasons for this American advantage are many, but the most important was that it was possible to make greater industrial investment, provide more resources, and appropriate more funds for research and development for battleship development in the United States than was possible in the United Kingdom in the late 1930s, despite the fact that both these industrial giants were recovering from a serious economic depression. This meant that high-temperature-high-pressure boilers, locked-train double-reduction gears, high-speed turbines, and alternating-current machinery could be produced in American industry without seriously impairing other industrial activities. In addition, there were significant and enthusiastic supporters of the high-temperature-and-pressure concept in Rear Admiral Harold Bowen, USN, who headed the U.S. Navy's Bureau of Engineering, and William Francis Gibbs of Gibbs and Cox, Inc., who successfully pioneered that concept in American destroyers of the *Mahan* class in the early 1930s. The technical advantages of higher steam conditions were recognized by certain parties in the Royal Navy, as seen by the design of the propulsion plant for the destroyer *Acheron*, but there was a failure to allocate the required resources to overcome the problems in the subsequent operation of this ship, none of which were fundamental. On top of this, in the 1930s Parliament was unwilling to grant to its military establishment the necessary funds to conduct the propulsion-plant evaluations that were undertaken in the United States and Germany. Also, there was the lack of a farsighted naval designer in the United Kingdom who could give the necessary support to such a project, like Gibbs or Bowen. These differences matter, because needed improvements in any one characteristic may sometimes be achieved only at the expense of another unless there is a technological breakthrough. Such a breakthrough did occur in the United States with the use of double reduction gears, high-pressure-high-temperature boilers, high-speed turbines, and alternating current in new ships built after 1935. The *King George V* had better armor protection than the *North Carolina*, but the British ship was hampered by a smaller main-battery gun and an inferior propulsion and electric plant. The only way to partially make up the difference was to make the ship larger, a solution followed by the French in the *Richelieu*, the Italians in the *Vittorio Veneto*, and the Germans in the *Bismarck*. Bigger should always be better, but it also implies greater cost and more controversy in securing the funds to build the ship. The latter was particularly true at a time when the airplane was in the ascendancy.

One has to measure a ship's capability not only against the resources available, but also against the strategic and tactical demands of a nation's defense policy. Great Britain certainly needed the five ships of the 1936/1937 *King George V* class as soon as it was possible. She did not have the resources to build a second class of battleships (*Lions*) until 1938. By 1939, Great Britain no longer had the resources to construct the *Lion*-class ships in the face of all the competing claims of an all-out war effort. The better is often the enemy of the good, and the Royal Navy would have been ill-advised to replace the three *King George V* ships authorized in 1937 with two *Lion*-class ships of "better" design or greater military potential, even at the same total cost, because they could not have been completed before 1943. To have done so might have created serious difficulties in the overall war effort.

When a design has been completed and the plans sent to a shipyard, post-design depression is a familiar condition for those naval architects and marine engineers engaged in its preparation. If they only knew at the start what they knew at the conclusion of the design, the ship design would have been much better. This learning process is a key element in the comparison of capital ships discussed in these three volumes. Each capital-ship class was an improvement over another of basically similar and earlier design. The battlecruiser *Strasbourg* owes her place as the most formidable battlecruiser to lessons learned in the development of her sister ship, the *Dunkerque*. The treaty battleship *South Dakota* represented a chance for American designers to do the *North Carolina* over again, while the *Iowa* and *Montana* were evolutionary developments of the *South Dakota*. It should not surprise one to learn that the later American dreadnoughts were truly formidable opponents because they were well designed and because many of the compromises that had to be made in the earlier designs did not have to be made in these larger ships. The Germans completed the *Bismarck* and *Tirpitz*, which were products of experiences with the *Scharnhorst, Gneisenau* and *Deutschland*-class armored ships. The "H"-class battleship designs resulted from the lessons learned in the design and operation of the previous two capital-ship classes. Although there is little official information known about Russian capital ships, there is some indication that the design of the *Sovetskii Soyuz* reflected lessons learned in the design of the battlecruiser *Kronstadt*.

Because of internal political strife prior to 1930 and the lack of experience in capital-ship design and construction, Russia had a very difficult time in assembling a design team. It was necessary to have some outside technical assistance, but once this was gained, a very fine design emerged. In Italy, a design team was trained in the redesign and reconstruction of two old battleship classes, and its services were sought by the navies of Spain, the Netherlands, and Russia. The fine design of the *Vittorio Veneto* was the final product of this team's experience. In other words, the way to learn capital-ship design is to design and build capital ships.

The Japanese never went beyond the first design, although some preliminary design was done on a 70,000-ton battleship armed with 500-mm guns. They suffered, to some degree, because in the process of design they had to train a design force to be skilled in making compromises and design decisions. The *Yamato* and *Musashi* represented a "great leap forward" approach to capital-ship design. By sheer will and talent, the Japanese proceeded from paper studies to the actual contract design of new battleships, twice as large as any previous Japanese one. These ships were the embodiment of the cliché, "more of the same is better." This daring design approach, however, resulted in ships with serious design flaws in structural arrangements, which impaired the damage stability and ultimately contributed to the loss of all three ships.

Without a viable naval air force, the Germans and Italians can be excused for underestimating the importance of a dual-purpose secondary battery. Their lack of a good antiaircraft battery was compensated for to a degree by the fact that between 1939–1943 the Royal Navy lacked aircraft with the speed and maneuverability of those used by the United States and Japan in the Pacific phase of World War II. The escape of the *Tirpitz* on 9 March 1942 was in part due to the slow-moving Albacore torpedo bomber. This was no longer the case in 1944 when the *Tirpitz* was attacked by the highly mobile and fast Barracuda dive bomber.

The most successful battleship or battlecruiser design in the last generation of dreadnoughts was the American *Iowa* class. They were not to be the battle-tested veterans of naval warfare such as were the *Bismarck* and *Prince of Wales*. Their combat careers consisted primarily of antiaircraft defense of aircraft carriers and support of ground action with naval bombardment. Nonetheless, these ships were exceptional battleships and have survived over forty years, much of that time in a deactivated state. The veteran USS *New Jersey* has been called to duty an unprecedented four times during her career. Shortly after entering service in 1943, she became the flagship of Admiral William Halsey for the duration of her World War II service. Admiral Halsey was a carrier admiral and until that time had preferred to command his fleet from a carrier. The activation of this ship and her sisters in the Korean War and her reactivation for the Vietnam Conflict were done principally to provide shoreside fire support. Their record is impressive in this regard. The decision to reactivate the *Iowa*-class ships in 1981 was brought about by an imbalance of naval strength and the fact that aircraft carriers have become so large and complicated that it requires almost ten years to build and outfit one. Placing cruise missiles aboard the reactivated *Iowa*-class ships provides them with a limited strategic-warfare capability, while the ships still retain unmatched capabilities for heavy gunfire support of ground troops in the coastal zone.

Battleships, despite their size and extensive clear deck areas, are relatively difficult ships to convert or modernize. Much of the apparent available deck area is denied any useful alternative service by the gun blast of the main armament and, to a lesser degree, by the secondary armament. Moreover, the conversion process is greatly complicated by the massive armor protection. In the 1981–84 conversion of the *Iowa*-class ships, the placement of some modern missile armaments proved to be difficult, and this led to some fire-arc restrictions on the main armament to avoid damage to sensitive electronics and weapons systems. The modern encapsulated liferaft launchers and cases were incompatible with the overpressure from a 406-mm gun. Four 127-mm/38 gun mounts were removed to avoid cutting into the armor to add new cables and pipes for the new armament and habitability requirements. The *Iowa*-class battleships were able to achieve a successful conversion from the battleship era to the missile age because of their balance of speed, armament, and protection. This balance was accomplished without an undue sacrifice in any one category of gun power, armament, or propulsion. The ship design allowed changes to be made at a later time without undue consequences. This is a tribute to the design team that developed these ships. Nonetheless, there were other "modern" battleships that excelled the *Iowa*s in certain categories, but no other ships had such a balanced design. For example, the Japanese *Yamato* had thicker armor protection and a larger main-battery gun, but it had less speed, less endurance, inferior torpedo protection, and an inefficient and voluminous propulsion plant. Although the main battery was larger in gun bore in the *Yamato*, the gun performance of the American 406-mm/50-caliber (Mark VII) gun was almost equal in performance to that of the Japanese 460-mm/45-gun at long ranges. This was achieved despite the fact that the gun was lighter in weight than earlier American 406-mm guns. There are good reasons why these ships have outlasted their contemporaries and why the authors consider them to be the classic battleship design of all time. Some will argue that the concept of the *Iowa*-class, as built, emphasized speed at the expense of an armament of 457-mm guns. The use of this larger gun would have required an increase in protection as well as an increase in the weight of main-battery ammunition (2,700 pounds for 406-mm vis-à-vis 3,850 pounds for 457-mm shells) and the necessary auxiliary machinery and equipment for the larger turrets. All of this would have resulted in exceeding the 45,000-ton limit, which the United States felt obliged to observe, even if there would be a decrease in speed to 27 knots. This criticism, however, overlooks the fact that American strategists were looking beyond 1938–1945 and were in the process of creating a fleet that would feature the battleship and aircraft carrier as the nucleus of a battle fleet. To be effective, such a fleet had to be mobile and cover large areas of the ocean.

The role of the battleship in the combat fleets of the world was challenged from the very inception of the ships. The debate raged until the Battle of Midway when the importance of the battleship in daylight battles was overtaken by modern naval aviation. At night or in very bad weather, however, when aircraft were lashed to the deck, surface ships fought the war, as the naval battles off Guadalcanal can attest. Three battleship actions were fought after Midway under non-flying conditions. When radar was made compact enough to fit within carrier-borne aircraft in 1945, the aircraft carrier became the main combatant around the clock. Battleships were then employed to defend the carrier from marauding planes and ships. These concepts led to the task force, with the carrier as the core, its mobility limited by screen and logistics craft. Battleships did, however, perform one extraordinary mission as force bombardment ships in support of ground forces ashore. This was carried forth by the *Iowa* class during the Korean and Vietnam wars as well as to the shores of Lebanon in 1983–84. This is a main reason why these ships have been retained to this day by the U.S. Navy.

This book and its two predecessors have dealt in detail with the technical aspects of the design and construction of battleships and battlecruisers. Careful design, well-conceived programs of research and development, and massive industrial resources are critical to the construction of capital ships. Competition with other phases of armament will lead to delays or cancellation of a project—such as happened to the British *Lion* class and the German "H" class.

Despite this emphasis on technology, it is only fair and proper to conclude with the observation that the most critical ingredient in the success of a ship in combat is its crew. Well-led, well-trained personnel have frequently triumphed over theoretically superior foes. Paramount attention must be given to the crews who take these ships into battle. The best-designed battleship could be seriously damaged or lost through the efforts of a poorly trained crew, especially when confronted by a better-trained adversary. On the other hand, the corollary, "a poorly designed battleship could achieve victory through the efforts of a well-trained and well-led crew," could and did happen in World War II. The *King George V*-class battleship was not as good a design as the German *Bismarck*, but Captain John Leach and his crew aboard the *Prince of Wales* achieved remarkable success in their encounter with the superior *Bismarck*. His leadership inspired a newly formed crew to accomplish remarkable gunnery accuracy, and the damage from the 356-mm guns of his ship forced the Germans to abort the *Bismarck*'s mission. The *Bismarck*'s sister ship, the *Tirpitz*, was already completed and in service at the time of the *Hood-Prince of Wales-Bismarck* battle, but despite the pleas of Captain Karl Topp to be included in the mission, the German Admiralty felt that more time was required for ship and man to become acquainted with one another. There can be no exception to the necessity for good design and well-trained crews. Second best in any naval battle spells defeat.

The colorful but controversial era of the battleship lingers on in the *Iowa*-class ships. These ships now have very different missions from those envisioned at the time of their conception. The airplane and guided missile have brought about a new era of naval strategy and warship design, where even the aircraft carrier is no longer the main combatant of the fleet. Despite their vintage and the enormous advances in technology since their inception, the reactivated *Iowa*-class battleships are valued members of the U.S. Navy of the 1980s, impressive testimony to the skill and dedication of the men who designed and built them more than four decades ago.

T.B

APPENDIX A

Stability Characteristics of the *Bismarck* Class

While analyzing the loss of the *Bismarck*, the authors came upon some of the German naval constructors' original considerations as to the damaged stability analysis of this class. This analysis formed the basis of sample calculations given to the damage-control officer. Principally, these and other German capital ships achieved damage resistance by means of comprehensive çompartmentation, proper spacing of transverse bulkheads, and relatively large beams (the *Bismarck* had a length-to-beam ratio of approximately 7, while that of the *Iowa* was 8). On the average, ships the size of the *Bismarck* and *Tirpitz* were subdivided into 20 to 26 major watertight compartments that in turn were divided into a large number of smaller watertight compartments. Also important was the provision for an adequate amount of reserve buoyancy, which represents the excess of available hull buoyancy over the total displacement, and the range of stability, which is a range of heel angles through which the ship has a positive righting arm. The upper bound of this range indicates the theoretical extreme angle to which the ship can list before it will capsize.

The Germans identified three ways in which a ship could be lost through flooding:

- Bodily sinkage. The added weight of flooding water could sink the ship on an even keel, or nearly so.
- Capsizing. The reduction in transverse stability from flooding water could cause the ship to heel over athwartships.
- Plunging. The weight of water forward or aft could reduce the longitudinal stability and cause the ship to sink by its bow or stern.

On the basis of these considerations, six cases of damage were considered for the four loading conditions tabulated in table A-1. Table A-2 summarizes the conditions that will be discussed in this appendix.

In all damage cases, it was assumed by the German naval constructors that the armor deck was not tight and that the compartments flooded were those that would produce the largest trim and list and the most adverse effect on the vertical center of gravity of the ship. The flooded portions of the ship, exposed to damage, were considered as lost buoyancy, not as added weight; only water occurring from progressive flooding was considered as added weight.

The battleship *Bismarck* is shown in the Blohm & Voss Shipyard in Hamburg during her final phases of outfitting. Work is still in progress on Turrets Anton and Bruno, and to protect the wood deck, protective wood planking has been placed around these guns. The heavy armor protection for these turrets is clearly evident—220-mm sides, 360-mm face plates, and 180-mm roofs. These were the most heavily protected parts of the ship. These battleships required a very large beam of 36 meters and had excellent stability characteristics. Note also the very large hammerhead crane used for lifting large sections of the ship or equipment. Such a crane would have been necessary to reposition Turret Caesar after the X-Craft attack in September 1943. (Blohm and Voss)

TABLE A-1

Summary of Stability—Bismarck

Loading condition	1	2	3	4
Displacement	52,359 (53,200)	46,454 (47,200)	43,009 (43,700)	39,565 (40,200)
GM (feet)	14.4 (4.4)	13.1 (4.0)	11.65 (3.55)	11.81 (3.60)
Angle of deck immersion	14°	17°	18°	19°
Range of stability	65°	59°	55°	53°

(1) Ship completely equipped—USN capacity load
(2) Ship at design displacement—USN ½ Consumed Condition
(3) Ship at standard displacement
(4) Ship, light but with all permanently provided installations—light ship

TABLE A-2

Summary of Damaged Stability Conditions

Stability Case	*Description*	*Figure*
a	Mine damage to bow	A-1
b	Stern damage	
c	Midship damage—symmetrical flooding	A-2
d	Midship damage—asymmetrical damage	A-3
e	Collision	A-4
f	Mine damage (amidships)	A-5

Damage cases (a) and (b). World War I experiences, especially that of the battleship *Bayern* when she was mined in the Baltic, showed that when a ship's bow or stern sustains contact-mine damage, the hull will be destroyed over some 30 meters of length. In such a case, a ship was assumed to be in the capacity-load condition since this produced the largest draft.

In damage case (a), the *Bismarck* was assumed to be in the capacity-load condition and to have struck a mine. This caused the flooding of the foremost watertight compartment for an extent of 36 meters. Since modern battleships were much longer than World War I types, this example showed that due to a larger moment of longitudinal inertia, the trim caused by such flooding was comparatively small. The calculated draft was 12 meters at the bow, which was not serious, especially when one considers that the bow freeboard was increased with the addition of the Atlantic bow. An alternate case (a2) was considered, whereby additional compartments in the region of turret Anton were assumed flooded, although the armor deck behind the forward armor bulkhead was still assumed to be watertight. The draft at the bow increased to 14.8 meters, with a water intrusion of some 5,200 metric tons. This demonstrated that even with large flooding in the bow region there was still sufficient freeboard to keep the ship on a forward heading and enough residual stability to prevent the ship from plunging. This was important in their analysis because the Germans were very mindful of the experiences of the battlecruiser *Seydlitz* after the Battle of Jutland. The *Seydlitz* had to be towed to port stern first due to extensive bow flooding, which also threatened her reserve buoyancy.

Parallel deliberations apply to case (b), which is the damage condition involving the stern. In this case, however, a very large amount of stern flooding could have endangered the operational effectiveness of the ship by hampering the battle readiness of the propulsion plant. The Germans considered a large draft at the stern (12.2 meters) to be of secondary importance in the case of damage.

Damage cases (c) and (d)—midship damage. Damage case (c) assumed that the *Bismarck* had flooding in three large compartments, which could be adjoining or divided by undamaged ones, as well as in tanks and voids between the torpedo bulkhead and the shell on *one* side, with the double bottoms filled with fuel oil and thus intact. It was further assumed that the location of the damaged compartments did not endanger the operation of the propulsion plant or the ship as a whole and that the ship had depleted part of her fuel, leaving the ship's bunkers some 75 to 85 percent full, with the exception of fuel compartments in the side-protection system. It was further assumed that the midship compartments between the torpedo bulkheads had sprung leaks. It is probable that empty compartments on the port and starboard side remained intact and unflooded, thereby not further aggravating the ship's damaged condition. In this case, the metacentric height (GM) increased from an initial value of 4.4 to 5.25 meters. This increase was attributed to an upward shift in the center of buoyancy.

Damage case (d) assumed the ship to be at the design waterline condition, with fuel and other consumables 50 percent expended. The flooding water was distributed over approximately three large, adjoining, and watertight compartments, as well as in the tanks and voids between the longitudinal bulkheads on one side. The opposite side was assumed to have remained intact. It was also assumed that a large part of the fuel in the wing tanks had been consumed prior to damage, but the double-bottom tanks were still considered to be full. These assumptions resulted in a list of 14 degrees, which brought the deck edge to the waterline. The water intrusion in this case amounted to 8,600 metric tons. The upper deck edge would not be submerged if the engine room on the opposite side of the damage was assumed to have been flooded by opening a valve in the centerline longitudinal bulkhead. In this case, approximately 10,800 metric tons of water would have entered the ship.

Collision—case (e). In this case, the ship was assumed to have been in a condition approximating that of the U.S. Navy minimum operating condition, that is, with two-thirds of consumables expended. It was further assumed that the collision occurred at a main transverse bulkhead and that two adjoining compartments flooded. With regard to the penetration of the colliding ship, it was assumed that damage extended to the centerline so that the maximum amount of off-center flooding would occur. Of importance was the arrangement of the longitudinal bulkheads and their influence on the magnitude of the listing moment. Side longitudinal bulkheads were installed in all large German ships to subdivide the compartments and increase the vessel's strength. Only in cases where this was ruled out by the width of an installation, such as boilers and engines, were side longitudinal bulkheads omitted. Instead, if possible, a centerline bulkhead was installed, despite its unfavorable influence on the list resulting from the flooding of wing compartments. Thus, the stability was decreased by this use of longitudinal bulkheads, although the strength of the hull was increased. Nevertheless, the Germans accepted the disadvantages of fore-and-aft bulkheads subdividing the machinery box in order to obtain smaller watertight compartments, accommodate a three-shaft propulsion plant, and fully implement the concept of using separate compartments for boilers and turbines. The adoption of a large beam reduced the dangers inherent in the potential asymmetrical flooding of the machinery box.

Assuming that the armor deck was no longer watertight, there was the possibility of an 8- or 9-degree list. The stability-reducing influence of free water surface was also considered in the analysis. Taken absolutely, this influence was not large, since the free water surface was broken down by the longitudinal bulkheads and by the armor deck with its longitudinal girders.

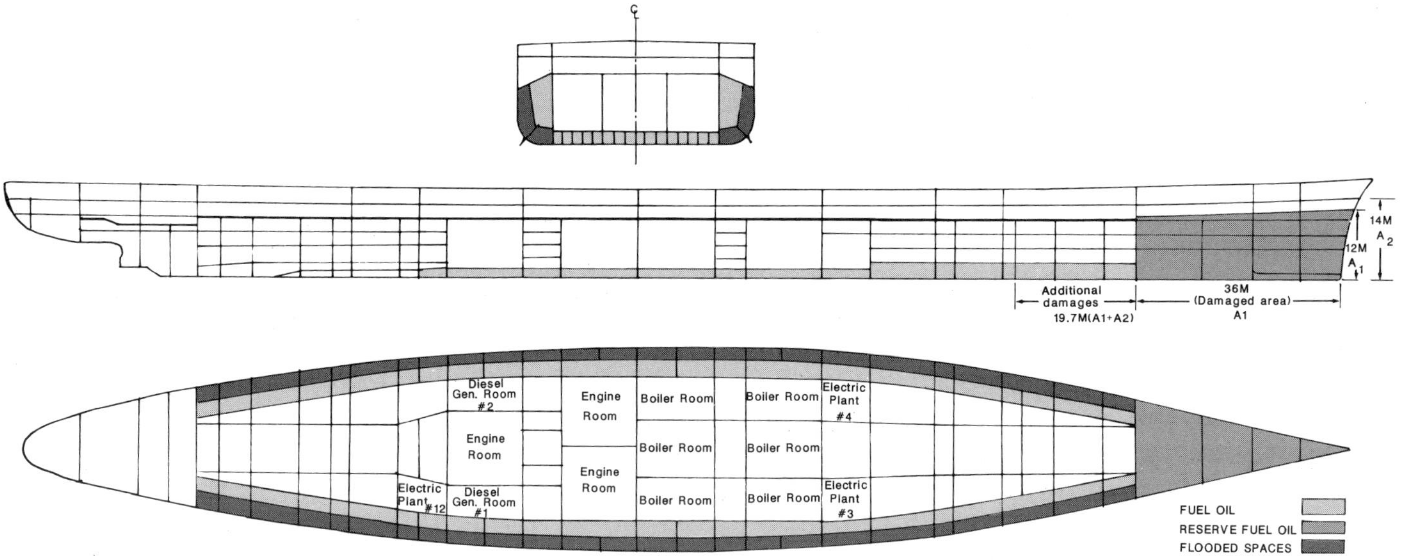

Figure A-1. Damage condition (a)*—ship with full stores, full fuel, full water (△ = 53,200 mt).*

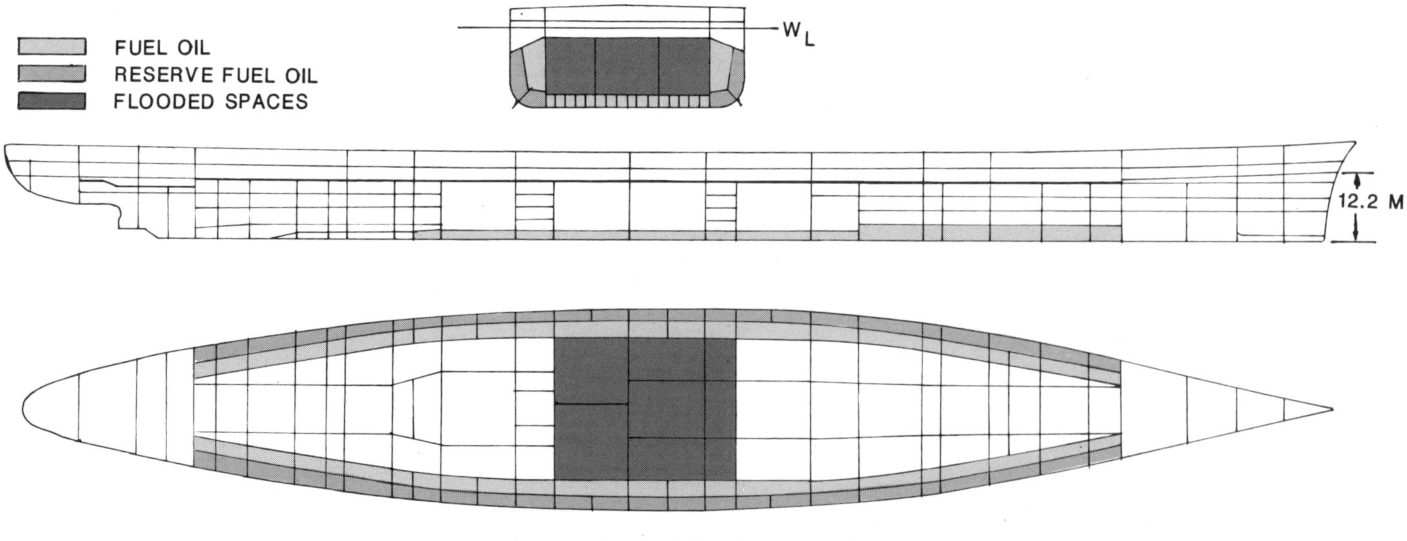

Figure A-2. Damage condition (c)*—ship with full fuel, full stores, full water (△ = 53,200 mt).*

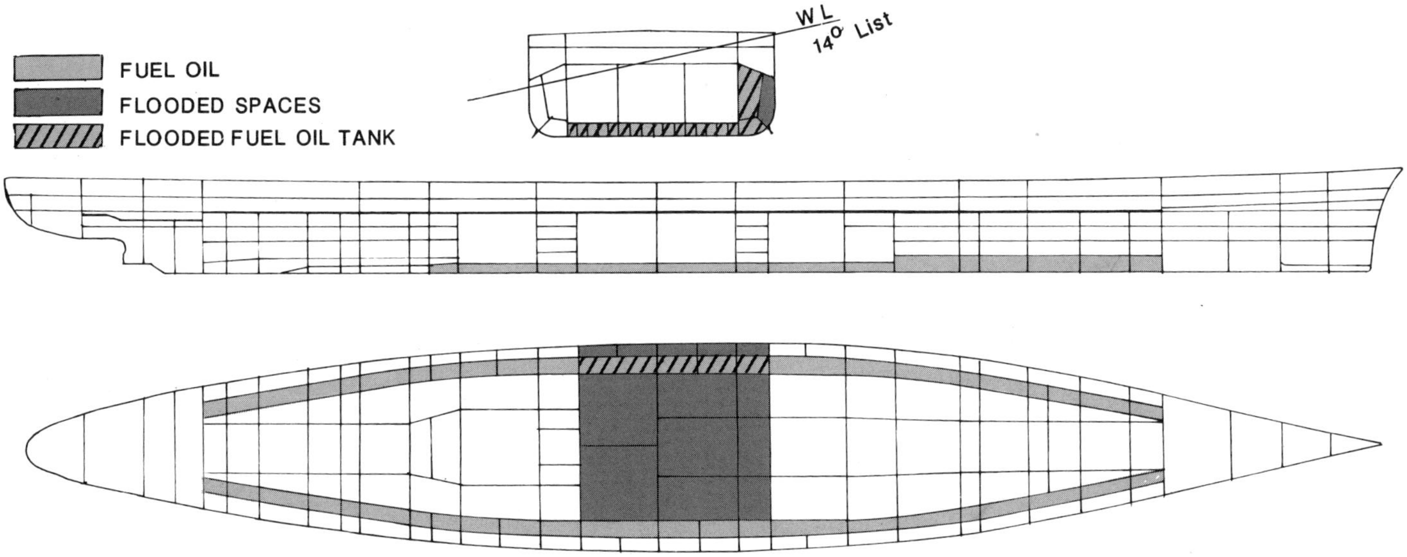

Figure A-3. Damage condition (d)*—ship with full stores, normal fuel, (no reserve fuel oil), and full water.*

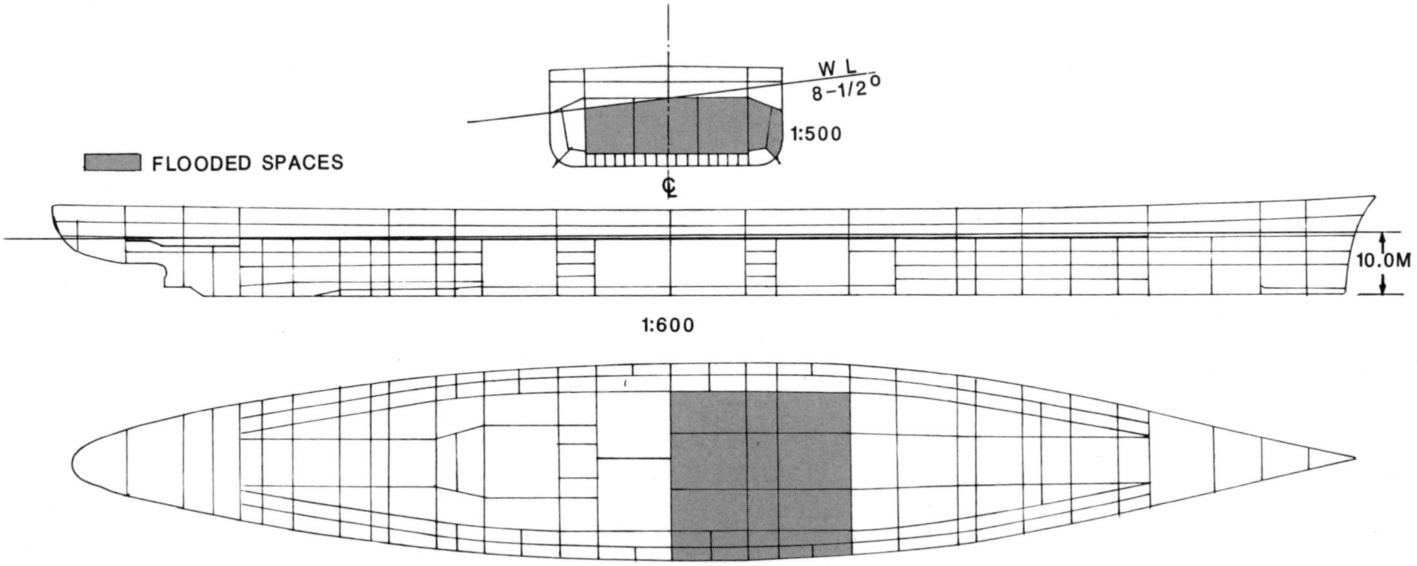

Figure A-4. Damage condition (e)*—ship in minimum operating condition.*

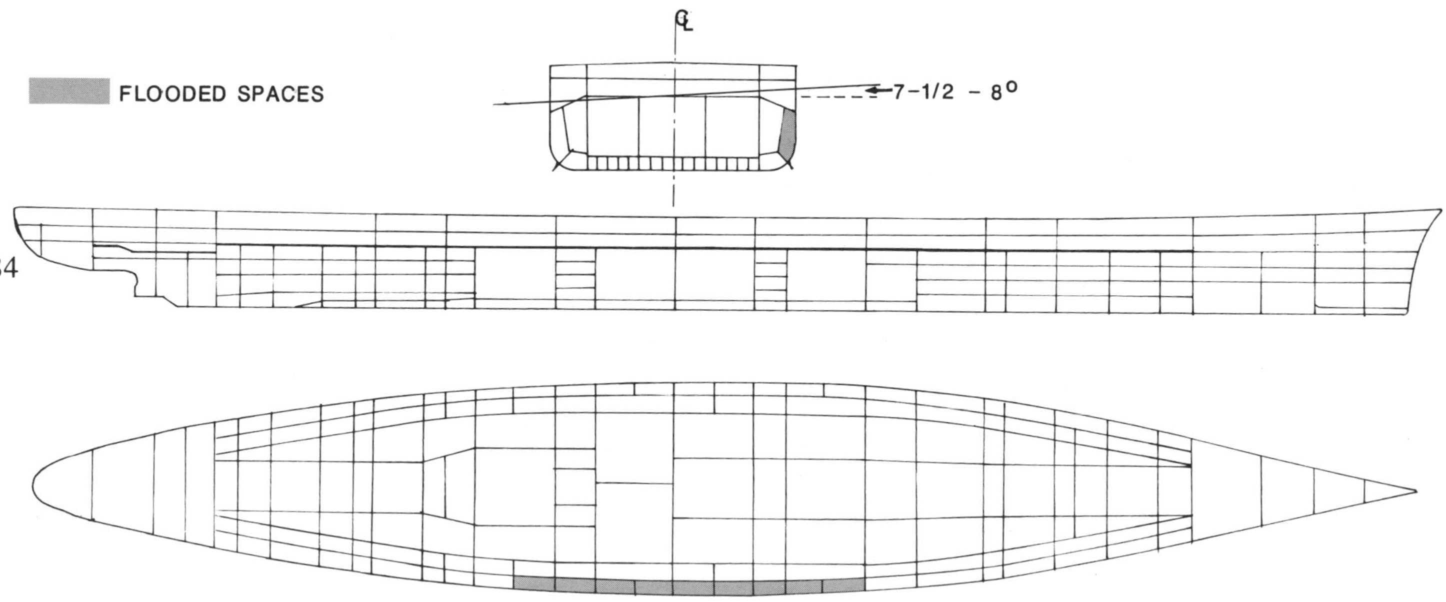

Figure A-5. Damage condition E—standard displacement condition.

Amidship damage—case (f). It was assumed here that a mine had exploded near amidships and that the side shell was torn open for a distance of 60 meters. To determine the expected list, the standard displacement loading condition was assumed, and that prior to the rupture all tanks and voids were considered to be empty in the area of damage (the worst case). In this case a list of 7.5 to 8 degrees resulted.

For the damage cases (a2) through (e), changes in the metacentric height were not large, as can be seen from the following tabulation:

Case	a2	b	c	d	e
Ship intact	14.4′ (4.4 m)	14.4′ (4.4)	13.1′ (4.0)	11.8′ (3.6)	11.64′ (3.55)
Ship damaged	16.4′ (5.0 m)	15.6′ (4.75)	11.2′ (3.4)	10.8′ (3.3)	10.2 ′ (3.1)

The flooding experienced by the *Bismarck* in the battle of 24 May 1941 as a result of 356-mm shell hits from the *Prince of Wales* can be equated to a combination of conditions (a) and (c), with several notable exceptions. First, the ship had a displacement far different from those assumed in the calculations, and second, the amount of flooding in the bow caused by one shell hit was not as extensive as had been postulated in the stability analysis. More important, the rate of flooding was much less than that from a mine or torpedo. The 356-mm shell hit outside the forwardmost boiler room on the port side produced asymmetrical flooding and further bow trim, so that during the passage of waves, there was a momentary emersion of the starboard propeller. Prompt counterflooding of aft starboard voids alleviated this problem.

As far as can be determined, the flooding caused by the two shell hits and the counterflooding water necessary to correct the ship's list and trim still existed when the two torpedoes struck the *Bismarck* well aft on 26 May. The fatal torpedo hits in the stern crippled the *Bismarck* and caused a significant amount of flooding on the port side; however, there was still sufficient freeboard and residual stability at the commencement of the artillery engagement on 27 May 1941. As far as evidence has shown, the *Bismarck* had a slight port list and no trim when her final action commenced around 0850.

Bismarck Wreck. In early June 1989, an expedition led by Dr. Robert Ballard, using a deep underwater exploration technique, located the German battleship *Bismarck* in 4,700 meters of sea water about 600 nautical miles west of Brest. The battleship was in a remarkable state of preservation with its teak decks intact except where damaged by fire and British shellfire. There was impeccable evidence, however, that the *Bismarck* had sustained much topside damage from British shellfire. Although the ship is buried in approximately 10 meters of mud, some damage to the upper and lower armored belts is visible. The starboard side of the *Bismarck* sustained less damage than the port side because the *Rodney* and *King George V* spent much of the action on the port side of the *Bismarck,* while the heavy cruisers *Norfolk* and *Dorsetshire* remained to starboard. The *Rodney* spent only a very brief period on the *Bismarck*'s aft starboard quarter at a very close range of 2,500 to 3,300 meters.

A close-up of the *Bismarck*'s starboard superstructure in March or April 1941 shows the foremast, stack, forward command tower, and the starboard boat and aircraft crane from a position just aft of the aftermost 150-mm starboard turret. The 105-mm gun in the foreground was struck with a 203-mm shell and most of the covering plates of its gun carriage were shot away leaving the guns pointing skyward. The cradle for the boat just forward of this gun mount remains with the keel rest twisted downwards. The searchlights on the platform just aft of the mainmast and those mounted on the funnel were shot away. The funnel and the battered forward command tower separated from the main hull and sank independently to the seabed below when the *Bismarck* capsized. The main battery rangefinder atop the forward command tower was toppled over the port side from an early 203-mm shell hit from the *Norfolk*. (Bibliothek für Zeitgeschichte)

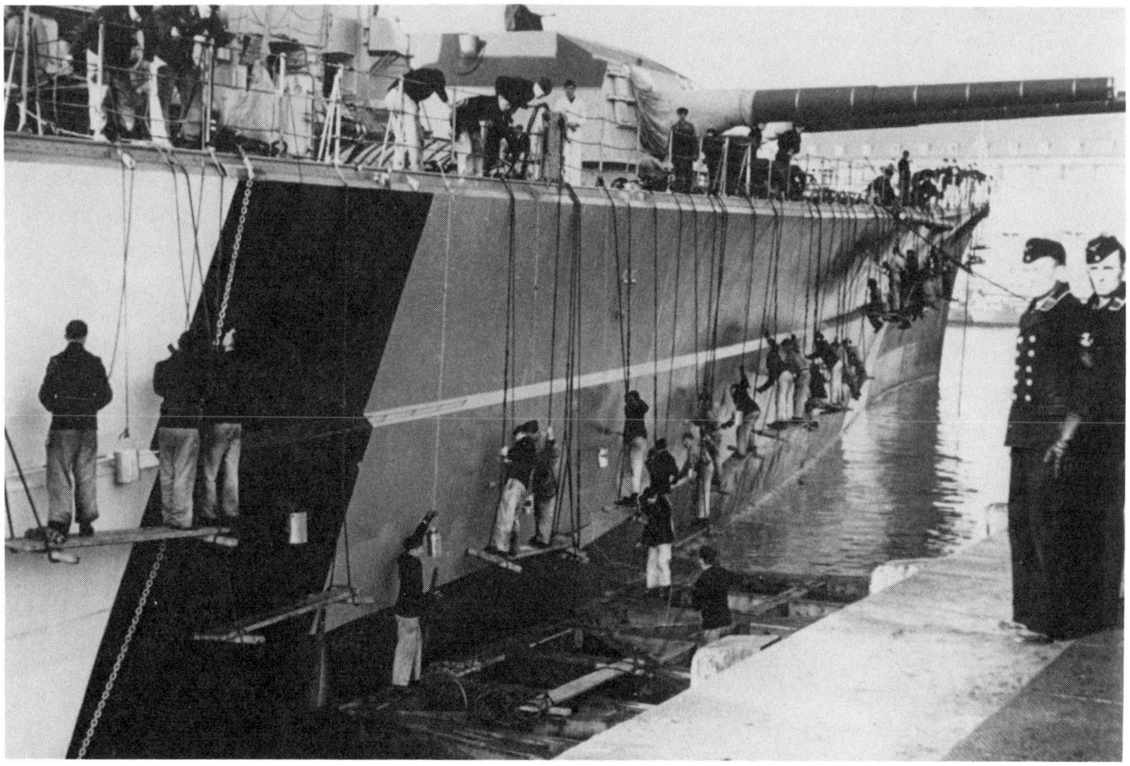

The upper photo of the *Bismarck* was taken in the Baltic during the fall of 1940. The main battery rangefinders on top of the conning tower and forward command tower are missing but the protective dome for the secondary battery directors on the port side forward has been fitted. Note the long bow wave. This was caused by a short bow entrance to a very wide beam. Although the *Bismarck*-class battleships had a significant improvement in their frictional resistance, due to a shallow draft effect, their wave-making resistance was increased because of the short bow transition to a full beam. In the lower photo this can be seen dramatically in the *Bismarck* when she was docked in Kiel. She made a brief stopover here on her way to the Baltic in March 1941. Dazzling stripes were used by the Germans to make it more difficult for Allied submarines to estimate the range and heading. The crew is busily painting the stripes as well as the sea grey on the hull sides. The false bow wave is barely visible and beyond it the extreme bow sides were painted a dark grey. Turret Anton is trained to starboard and its top was painted carmine red.

Torpedo hits—Evidence of two torpedo hits could be identified on the wreck. The torpedo hit from the *Victorious* aircraft on the starboard side abreast of the bridge was indeed a shallow hit and occurred on the 320-mm main side belt. The torpedo's explosion caused the belt armor to be displaced inboard 25-50 mm. The greater portion of its energy was vented into the air, while the remainder was expended in pushing one section of the main armor belt inboard. There would have been little or no flooding from such a hit, but the shock response from the detonation is known to have damaged equipment not properly secured and some auxiliary machinery with improper foundation design to withstand such shock effects. The shock response also caused the seals in the transverse bulkhead, damaged by a 356-mm shell hit from the *Prince of Wales* to come loose, flooding the forward fireroom on the port side, thus causing the loss of two boilers which reduced the maximum speed of the *Bismarck* by 2–3 knots.

The torpedo hit in the stern from the *Ark Royal* aircraft was the lethal blow the *Bismarck* sustained in this action with British forces. The torpedo hit on or near the port rudder at or just forward of the aft armored bulkhead of the steering gear room. The detonation tore a large hole in the *Bismarck*'s shell plating above the rudder and was vented directly upward into the ship, weakening ship structure and destroying or heavily damaging any equipment in its path. The port rudder was probably destroyed for steering purposes and combined with damage to the shell plating prevented it from being turned. The starboard rudder involvement is not clear, but probably it sustained some damage from the explosion as well as damage to its linkage. Repair crews had no success in freeing the two rudders. Since the hull is buried in 10 meters of mud the damage to the rudders will never be precisely known. We can only be guided by survivor reports and past damage to other major German warships. Based upon stern torpedo damage sustained by the armored cruiser *Lützow* in 1940 and the heavy cruiser *Prinz Eugen* in 1942, it is concluded that the *Bismarck* had a similar defect. Both the *Scharnhorst* and *Tirpitz* had corrective measures taken in their stern structure to prevent the type of structural response to torpedo hits in the stern. Also note that cracks had developed in German warships after damage, and some of these extended over quite a distance (see *Gneisenau,* page 160). A number of cracks developed in the *Tirpitz* after her mining by British midget submarines in September 1943 and from near-miss bombs in an air attack on 3 April 1944. One of the reasons for such cracks developing was that German structural detailing did not require reinforcement of cuts in way of openings in the shell plating. We believe that large cracks developed in the shell plate along the aft armored bulkhead of the steering gear room and that the torpedo damage extended a considerable distance up this plating. It probably involved the riveted seam which acted as the crack arrester. This was located at the neutral axis of the transverse section. The seas were also very rough in the area the *Bismarck* passed through after she was disabled (wave heights estimated to be 6 to 8 meters). Once the rudders were jammed, the *Bismarck* kept turning northwest into the direction of the gale, also the direction from which the British battleships *King George V* and *Rodney* were coming! All that Captain Lindemann could do was use the propellers to keep the *Bismarck* from broaching to the oncoming sea to minimize the rolling forces that would impede his gunnery. However, this maneuver also increased the pitch and heave forces, and from survivor testimony there was serious slamming of the stern as the ship rose and fell in the waves. This only would have caused the cracks in the damaged area to propagate further up the shell and across the decks. Later during the gunnery action, the *Rodney* was at very close range and hit the stern area with several shells in one of her salvos. This caused large chunks of main deck and stern side plating to be thrown into the air and overboard. Finally as the stern dipped below the water surface and the *Bismarck* capsized to port there was a clean break across the main deck reminiscent of a fatigue failure and the stern section aft of

The heavy cruiser *Prinz Eugen* (shown here), the armored ship *Lützow*, and the battleship *Bismarck* all sustained severe damage to their stern structures from torpedo hits that then led to spectacular failures. Characteristic of these failures was a clean break in the deck and side plating just aft of the heavy armor protection for the steering gear rooms. The premature termination of longitudinal bulkheads, a very poor detail structural design practice, was a major contributor. In the *Prinz Eugen*, a torpedo hit from the British submarine *Trident* caused the structure just aft of Frame 6½, which was aft of the rudder stock, to collapse on top of the rudder. As the ship was maneuvering to starboard, the rudder was at 10 degrees and it, too, absorbed some damage from the torpedo explosion. This was quite similar to the type of damage that the *Bismarck* sustained. The damage was temporarily repaired in Trondheim and jury-rigged rudders were installed with a manually or power operated capstan shown in the photograph. The clean stern break on the *Prinz Eugen* shown here is identical to the structural failure on the wreckage of the *Bismarck*. The German naval constructors recognized a serious fault in their structural design on the basis of this incident and that of the *Lützow* and corrected the problem on the other big ships—the *Admiral Hipper*, *Tirpitz*, and *Scharnhorst*. It was also to be done on the reconstructed *Gneisenau*.

the steering gear room separated from the ship. The hull and the stern section sank independently to the sea bed below.

There was survivor testimony that removal of the damaged stern section by cutting torches or explosive charges was seriously discussed between Captain Lindemann and Chief Engineer Lehman. It is our opinion that such a measure was discussed as a means to clear the damage from the rudders. Explosive charges had a risk of damaging the propellers, while it was impossible for divers to work under the prevailing sea conditions. It also would not have removed the main problem—the damaged port rudder. The rudders were housed within a stout armored box that would have been impossible to sever from the main hull with the means aboard the *Bismarck*. Captain Lindemann and Chief Engineer Lehman recognized the impracticability of such measures and saw the plight of their ship. They did evidently agree on trying to counter the rudders by the use of propellers, which failed.

Shell damage—Close examination of the wreck shows that the port side of the *Bismarck* was seriously damaged by numerous large caliber shell hits.

- Two shells hit the bridge of the *Bismarck* at or near the overhead and approximately at an angle of 60–70 degrees of each other. One shell came from abeam and the other from astern. The structure at the deck level has been pushed in about 1–2 meters and the bridge front was torn open. The port side of the conning tower was riddled with some fifty shell or fragment hits. Some large chunks of 380-mm armor were punched out of the vertical surface. The roof of the forward conning tower is virtually intact.
- A shell hit near the barbette top of turret Bruno and penetrated halfway through the 340-mm barbette armor whereupon it exploded. It curled up the armor similar to the 406-mm shell hit on the barbette of one of the *Jean Bart*'s 380-mm turrets in her action with the American battleship *Massachusetts*. This jammed the French battleship's turret in train for more than a day until the damaged face-hardened armor could be removed with special cutting torches. In the case of the *Bismarck*, however, another shell in this salvo evidently hit the turret itself and may have penetrated, causing the loss of the turret's rear armor plate. Certainly the ventilation ducts on the port side were sheared off, the turret would have been jammed in train, and fragments were sent upward and aft toward the bridgefront, killing or injuring any personnel in exposed areas. There is also the probability of damage to the sidewalls of the turret. According to Josef Statz, the rear portion of turret Bruno was uplifted so that this turret was tilted slightly forward on its barbette. We believe that this was caused by the burning of cartridges and a magazine deflagration within the barbette structure. Mr. Statz has also told us that the rear portion of turret Bruno was blown out and its debris flung against the bridgefront.
- A shell hit abreast of the forward hangar on the port side. It exploded on the main deck and caused the detonation of 105-mm ready-service ammunition or the bomb stowage for the aircraft.
- There were 406-mm shell hits in compartments IX and/or X between the Aufbaudeck and the main deck. The crew messroom (or canteen) was located here and was being used as a first aid station and assembly point for survivors near the end of the gunnery action. This also destroyed the port catapult and ripped a large hole in the deck structure.
- A 356-mm shell struck the turret face plate of turret Caesar. The shell probably exploded on contact but sent shell fragments into the Aufbaudeck and possibly several decks below. There is a very good chance that fires were ignited in combustible material along the trajectory of these fragments.
- Several 406-mm shells struck the barbette of turret Dora. They penetrated through the 340-mm face-hardened armor and exploded on contact with the inner support plate for this armor. Shell fragments entered the lower turret substructure and started a fire. This was extinguished with the prompt flooding of the magazines.
- A shell (unknown caliber) hit the bow of the *Bismarck* and carried away the bow mooring and spare anchor casting, leaving a jagged hole in the bow structure.
- The forward anchors were both missing, having their chains carried away by shell hits.
- The aft roof plate of the forward 150-mm turret on the port side was ripped away by an internal explosion. This turret was pointed aft toward *Rodney* or *King George V* when hit. There were heavy casualties in the turret crew. The superstructure adjacent to this turret was riddled with shell fragments and direct hits. There were holes in the 145- and 320-mm armor belts, just aft of this turret's position. It is also

possible that 152-mm shells from the *Rodney* could penetrate the 80-mm armor for this turret as well as the 145-mm splinter belt at the range of 3,500 meters at target angles from 60 to 90 degrees. Mr. Statz has confirmed that there was a fire in this turret's magazine which was then flooded. A fire, however, probably occurred in the turret's cupola, where charges were being loaded into the breeches of the guns. These ignited and the pressure created by burning cartridges either blew off the rear turret roof or so weakened it structurally that it came off during the sinking sequence.

- There are numerous holes in the 01 level (Aufbaudeck) from a number of shells of different calibers, supporting our contention that some 300 to 400 shells struck the *Bismarck* during the 27 May morning battle. Also it is very likely in the last thirty minutes of the battle that if one shell in a salvo hit, others in the same salvo would have likely struck as well. Such a hail of shell hits would have ripped away many exposed fitting from the decks and vertical structure, including bitts, chocks, railings, hose reels, paravane gear, tools, foundations, and boats. The wooden boats would have been good sources of the fires as would the teak decks. The latter, though intact, do show fire and shell damage.
- The port side of *Bismarck* has considerable damage to vertical and deck surfaces from the close fire from *Rodney* and *King George V,* who spent most of the action on the *Bismarck*'s port side. The starboard side, although damaged, was primarily victim to 8-inch gunfire of the *Norfolk* and *Dorsetshire.*

From the observation of the wreck it is apparent that most of the serious damage was concentrated in the forward superstructure on the port side as well as the funnel. The vertical surface of the port superstructure was ripped away by shell hits from 356-mm and 406-mm shells. Certainly there was damage done by the 152-mm guns of the *Rodney* and the 134-mm guns of the *King George V.* It is the authors' opinion that the *Rodney,* because of her close range and continuous fire, probably did more damage to the *Bismarck* than any other ship in this action. *King George V* had much trouble with interlocks in her quadruple turrets, and this deprived her of continuous fire. It is estimated that she was only 60 percent effective in the gunnery action. Fires would have been ignited in combustible materials in the *Bismarck*'s forward command tower. In addition, these shell hits also destroyed the supporting structure for the command tower as well as the funnel. When the *Bismarck* capsized to port, these structures were so weakened by shell damage that they separated from the ship upon impact with the water surface. The funnel was detached at the level of the Aufbaudeck, while the forward command tower was shaved off at the level of the forward conning tower.

There was a long oblong hole just outboard of the starboard anchor windlass. This was probably a 406-mm shell hit from the *Rodney* near the end of the gunnery engagement (0945–1000) when this British battleship was briefly on the *Bismarck*'s aft starboard quarter. The shell detonated between the main and battery decks and probably started a large fire in the crew berthing areas in compartment XXI and led observers on the *Rodney* to note that her hull was cherry-red at this location when the *Rodney* made her final approach on *Bismarck*'s port side around 1010.

There is no question that more British shells struck the superstructure than hit the hull or armor belt. The few shells that struck these belts could have penetrated the 320-mm or 145-mm armor because they were fired at such close range, certainly less than the *Bismarck* had been designed for. The armor belts visible on both sides of the hull show signs of shell impacts and penetrations. For those 152-mm and 134-mm shells striking these belts, they would have detonated, perhaps chipping or gouging the armor surface.

Damage to the 150-mm gun turrets was minimal except for the forward turret on the port side abreast of the forward conning tower. Almost all of the 105-mm mounts were damaged or demolished by direct hits, as were most of the 20-mm machine guns and the 37-mm semi-automatic guns. The aftermost starboard 105-mm mount just aft on the starboard side of the catapult has its guns pointed in a vertical direction as the result of shell hit(s) from 203-mm projectiles. Much of the wreckage of these 105-mm mounts was spilled into the sea when the ship capsized.

Her decks were littered with debris and men killed by shell explosions and flying shrapnel. The port side was covered with more wreckage than starboard because the 356-mm and 406-mm British shells tore the sides of the superstructure away and some of this toppled onto the forward 150-mm turret. The air intake louver on the port side for the port boilers was ripped away, and the 105-mm mount abreast of it was demolished. When the ship listed to 10 degrees to port, some of this debris slid into the sea.

We were able to interview one survivor, Josef Statz, who left his position in Damage Control Central in compartment XIV and at the base of the communication tube. He proceeded up to the conning tower. There he was exposed to a most gruesome scene. Several 406-mm shells had penetrated the port wall of the conning tower and had destroyed everything within. The inner surface was blackened, and a number of men in this space had been killed. Mr. Statz immediately took cover to starboard only to be wounded by a shell fragment from a 203-mm shell fired by the *Dorsetshire.* This was approximately at 1010, just before this British cruiser ceased fire. He managed to make his way around the conning tower and escaped from the port side by jumping into the sea from the middle 150-mm turret. He remarked to the authors that a good portion of the command tower had been badly damaged by shellfire and both 37-mm semi-automatic guns had been swept away. The forward conning tower door was open and had been damaged by a direct hit. The port aircraft and boat crane was demolished as were most of the boats. It was a scene of utter desolation. What had been a proud and beautiful ship to this man was now a smoking and burning wreck full of holes wherever one looked.

No other eyewitness or the photographic work of Dr. Ballard can deny the testimony of Mr. Statz. Photographic imagery taken above the forward conning tower shows a very large tear in the deck plating adjacent to the conning tower. The shell which caused this approached from the stern and then exploded against the bridge front, ripping plating apart, smashing windows on the open bridge, and ravaging the spaces below. Many men in exposed locations in this area were killed or severely wounded by this hit. Mr. Statz stated that he aided one of these men. He carried this man to the starboard side which was not as heavily damaged. From the meager evidence available, it appears that the main access in the forward superstructure was blocked from severe damage to its access ladders and bounding bulkheads. There appears to be one shell that struck the port side in the vicinity of this access and penetrated the entire transverse expanse of the superstructure before exploding against the starboard side. A large bulge was formed in the plating of the starboard superstructure, and at the apex of the bulge a large shell fragment emerged. This attests to the degree of damage that occurred in the forward superstructure and why interior access both vertically and fore to aft was very difficult.

Scuttling. The authors have made a careful examination of the stability of the *Bismarck* based upon the damage stability calculations captured from the Germans and conducted written interviews with the two surviving officers, Lieutenant Gerhard Junack and Lieutenant Commander Burkard von Müllenheim-Rechberg, as well as Chief Petty Officer Wilhelm Schmidt, who was in charge of the aft repair party. This was supplemented by a careful examination of the wreck which shows no signs of implosion as does the stern wreck of the *Titanic* where there was great effort to make compartments watertight. The evidence on the wreck clearly showed that the *Bismarck* had been destroyed as a fighting ship by British shellfire. Her sinking at 1040, however, was caused by a combination of shell hits, torpedoes, and scuttling charges. Removing any one of these meant that the *Bismarck* would not have sunk at this time. She would have sunk later from progressive flooding, which in the case of the shell hits, would have been at a slower rate than torpedo hits. Note too that turret Dora's magazines had been flooded to counter a fire from a 406-mm shell hit on her barbette. This shell had penetrated through the armor, and shell fragments had started a fire within the lower turret structure.

It should be explained that scuttling does not open huge holes. With the stout doublebottom of the *Bismarck,* the circulating water inlets and discharges were the few places where the modest-sized scuttling charges would have created the largest holes by destroying main valves and allowing large machinery spaces to flood. There were a large number of compartments to flood at the time the decision to scuttle the *Bismarck* was made shortly after 1000. Mr. Statz has indicated to the authors that the scuttling order as part of Measure V (German abbreviation for *Versenkung,* scuttling) came around 1005. Until the water entering the *Bismarck* brought the waterline to the level of the armor deck, progressive flooding could not have taken place. The pumprooms, auxiliary machinery spaces, firerooms, and engine rooms were the only spaces where qualified personnel could have correctly placed the scuttling charges. Flooding of the magazines of turrets Anton, Bruno, and Caesar cannot be confirmed. Only the center and starboard machinery complexes had scuttling charges set because there was a large amount of water in the port spaces that had come through the ventilators and boiler intakes from near misses which had ringed the port side and made it difficult for the British gunners to view the *Bismarck* from the *King George V* and *Rodney.* One of the port boiler rooms and an auxiliary machinery space had been flooded from shell damage from the *Prince of Wales.*

There were a number of valves in the *Bismarck,* but only a few piping systems outside the machinery spaces would have had direct access to the sea. Most of these would have been storerooms with combustibles and ammunition spaces. From testimony of survivors we do know that the magazines for the aft 105-mm guns on the starboard side and the forward 150-mm turret had been flooded due to fires in their shell and powder rooms. We also know from interviewing Mr. Statz that there was not sufficient damage control organization for such spaces to be flooded with charges and, in particular, there were not the men with the knowledge of the firemain system to accomplish this task under the conditions aboard the *Bismarck* at the time the Measure V order was given between 1000 and 1015. In fact, Lieutenant Commander Burkard von Müllenheim-Rechberg called Mr. Statz in Damage Control Central just around the time the order to scuttle had been given. After this it was every man for himself.

The scuttling order was given by the Executive Officer, Commander Hans Öels at approximately 1005 according to Mr. Statz. It would have taken the crews in the center and starboard propulsion spaces some 5 to 10 minutes to set the scuttling charges, and then there was a nine-minute delay once these charges had been set. This means that the charges would have exploded just around the time when the starboard torpedo hit from *Dorsetshire* occurred, around 1020. The survivors have clearly stated that they heard and felt those scuttling charges go off. The authors believe this and also think that the torpedo explosion was part of the response these men heard and felt. This torpedo hit, if it sufficiently damaged the splinter or main side belts or struck below the thick side belt, would have contributed to the bodily sinkage of *Bismarck* and allowed more water to pour into spaces on the port side since the armor deck would have been submerged. If the flooding from this torpedo hit were low in the ship, it would have provided not only beneficial low flooding water, which would provide additional righting arm, but also an advantageous counterflooding effect to counter the port list.

Around 1030 the port main deck of the *Bismarck* dipped under the water surface, and the damaged stern was completely immersed by this time. There were still ship motions from the heavy seas. Survivors were going overboard from the starboard side although there were a few who attempted escape from the port side, which would have waves rolling into it. The buoyant forces and ship motions on the main hull and the damaged stern finally caused a fatigue failure to occur in the main deck aft of the steering gear room. A 10-meter section of the stern broke away below the water surface and sank to the seabed below. The bow began to lift out of the water and very slowly the *Bismarck* began to capsize to port. When the *Dorsetshire* made her port approach to fire her torpedoes at 1036, the *Bismarck* was already in a sinking condition. The two torpedoes hit the lower or upper armor belt which would provide more resistance to damage than nonarmored structure. These torpedoes were armed with 340 kilograms of explosive, more than aerial torpedoes with 204 kilograms. Nonetheless, a great amount of energy in these torpedoes' explosions would have been absorbed in displacing the thick armor plate. The fact that the *Bismarck* sank only four minutes later after these torpedoes struck indicates that these two hits contributed little to the flooding that finally caused the *Bismarck* to sink.

The scuttling was only one of five contributing causes in the loss of buoyancy. The flooding from the torpedo hits from the *Ark Royal* aircraft, the shell hits from the *Prince of Wales,* and the counterflooding to correct adverse trim and list from these hits were still aboard the *Bismarck* the morning of 27 May. Lieutenant Commander Baron von Müllenheim-Rechberg observed a noticeable port list in the wardroom from glasses of liquid on the tables. The flooding from the *Victorious* aircraft–launched torpedo hit was very minor or possibly even nonexistent. The shell hits from the *Rodney, King George V, Dorsetshire,* and *Norfolk* were mostly to the upperworks, but there is the distinct possibility of a few shell hits below the waterline. This meant that to submerge the armor deck to downflood spaces not flooded from enemy action or scuttling charges, it would have been necessary to flood the machinery spaces as Lieutenant Gerhard Junack has maintained in his correspondence with the authors. It is our opinion, therefore, that scuttling charges were set by the Germans as a routine measure of abandoning ship. This caused the ship to sink at 1040. If they had not been used as some others have maintained, the ship would have sunk, but not at 1040! The evidence is clear that the British ships destroyed the *Bismarck* as a fighting element.

When the *Bismarck* capsized to port at the water surface, there was a great impact with the water surface on weakened structure. As the ship continued to roll over, her four main turrets that were held in place by gravity plunged to the sea bottom first. The upper portion of the forward command tower, the funnel, and the mainmast were sheared off and sank independently of the main hull. This sudden change to the weight of the ship and its stability caused a restoring force to bring the main hull upright in its passage to the seabed below. There she slid down an underwater sea mount to her final resting place in 10 meters of mud. The *Bismarck* now rests in 4,700 meters of seawater.

APPENDIX B

Inboard Profile and Deck Plans of the *Bismarck*

In order to provide a better understanding of the internal arrangement of a new-construction battleship, the inboard profile and deck plans of the German battleship *Bismarck* are reproduced here, with a numerical key to indicate the functions of the various compartments and spaces.

The key to the plans is organized as follows:

- Ordnance and fire control
- Ship control
- Outfit
- Personnel
- Damage control
- Miscellaneous
- Electric plant
- Propulsion plant
- Electronics
- Aircraft
- Minesweeping
- Tanks

Note that many of the external features of the ship are also depicted on the outboard profile and overhead view, and that some elements of the propulsion plant are shown in the machinery schematic. Note also that as the inboard profile depicts a centerline view of the ship, many offices, shops, medical facilities, and storerooms in the outboard areas are not keyed to the profile.

The *Bismarck* is shown under construction at the Blohm and Voss Shipyard in Hamburg. The barbette structures for turrets Caesar and Dora are in the foreground with the main deck (Oberdeck) in place forward of turret Caesar. In the area of turrets Caesar and Dora the shell frames between the third deck (Zwischendeck) and the second deck (Batteriedeck) on the port side have been erected. The large square opening in the main deck forward of turret Caesar is for the intakes to the center boiler rooms and the uptakes from all the boilers that will be part of the stack structure that will be mounted on the 01 level (Aufbandeck). The almost completed transverse bulkhead in the lower right corner is the main subdivision bulkhead between compartments II and III at Frame 32. (Bibliothek für Zeitgeschichte)

Ordnance and Fire Control

1 Main battery director
2 Turret (380-mm guns)
3 Aircraft bomb mag.
4 Gun chamber
5 Aft battery sec. ctl.
6 Ammunition hoist
7 Cartridge magazine (380-mm)
8 Shell magazine (380-mm)
9 Magazine cooling equipment room
10 105-mm magazine
11 Gunners' stores
12 Ordnance/mechanical spare parts storeroom
13 Main battery plotting room
14 Forward battery plotting room
15 Main battery control room
16 Forward battery amplifier room
17 Forward secondary battery control
18 150-mm magazine
19 After antiaircraft plotting room
20 Forward antiaircraft control room
21 Forward antiaircraft plotting room
22 37-mm magazine
23 After plotting room
24 Main battery amplifier room
25 Antiaircraft repair shop
26 AA electric station

Ship Control

27 Pilothouse
28 Chart room

Outfit

29 Pump room
29A F.O. pump room
30 Refrigeration machinery room
31 Gyro room
32 Aft master gyro room
33 Regulating machinery room
34 Booster pump room
35 Canvas locker
36 Carpenter storeroom
37 Chain locker
38 Boatswain's storeroom
39 Cordage storeroom
40 Paint storeroom
41 Ventilation trunk
42 Flammable liquid storeroom
43 Alcohol storeroom
44 Windlass-machinery room
45 Steering-gear room
46 Machine shop
47 Steering-motor room
48 Forge shop
49 Hand-steering room
50 Sound-detector room
51 Fan room and dry stores
52 Blacksmith shop

Personnel

53 Crew living space
54 Petty officers' living space
55 Petty officers' baggage room
56 Main battle dressing station
57 Sick bay
58 Technical petty officers' room and sickbay
59 Secondary battle dressing station
60 Warrant officers' staterooms
61 Warrant officer mess
62 Stewards' storeroom
63 Canteen storeroom
64 Clothing storeroom
65 Provisions storeroom
66 Cold store
67 Officer stateroom
68 Warrant officer storeroom
69 Admiral's storeroom
70 Captain's storeroom

Damage Control

71 Void (damage control)
72 Damage-control room
73 Central station and damage control central

Miscellaneous

74 Passages
75 Trunk
76 Pipe tunnel

Electric Plant

77 Motor generator room
78 Switchboard room
79 Diesel generator room
80 Electric plant spare parts
81 A.C. generator room
82 Battery room
83 Cable tunnel
84 Transformer room
85 Electric shop
86 Searchlight shop

Propulsion Plant

87 Machinery spare parts storeroom
88 Water analysis
89 Engine room
90 Auxiliary engine room
91 Boiler room
92 Boiler auxiliary machinery room
93 Boiler action station
94 Thrust bearing compartment
95 Shaft tunnel
96 Uptakes
97 Intakes
98 Propulsion-control station
99 Shaft
100 Propellers

Electronics

101 Hydrophone room
102 Radio room
103 Battle radio room
104 Radio transformer room
105 Radio room B

Aircraft

106 Catapult
107 Hangar
108 Bomb magazine
109 Airplane
110 Ammunition storeroom
111 Airplane smoke tanks

Minesweeping

112 Paravane tube

Tanks

113 Fuel oil
114 Salt-water ballast
115 Lube oil
116 Potable water
117 Reserve feed-water tank
118 Gasoline tank
119 Fuel-oil filling trunk
120 Fuel-oil settling tank
121 Peak tank
122 Trim and reserve tanks
123 Void
124 Speed-log tank
125 Condensate tank
126 Wash-water tank
127 Diesel-oil tank
128 Turbine-oil tank
129 Regulating tank

Special

130 Bow protection room
131 Water analysis room
132 Quartermaster's storeroom
133 High tank
134 Battle radio room
135 Radio room
136 Landing force equipment room
137 Radio equipment
138 Primer magazine
138A Small arms magazine
139 Radio transformer room
139A Saluting magazine

Inboard Profile *(Bismarck)*

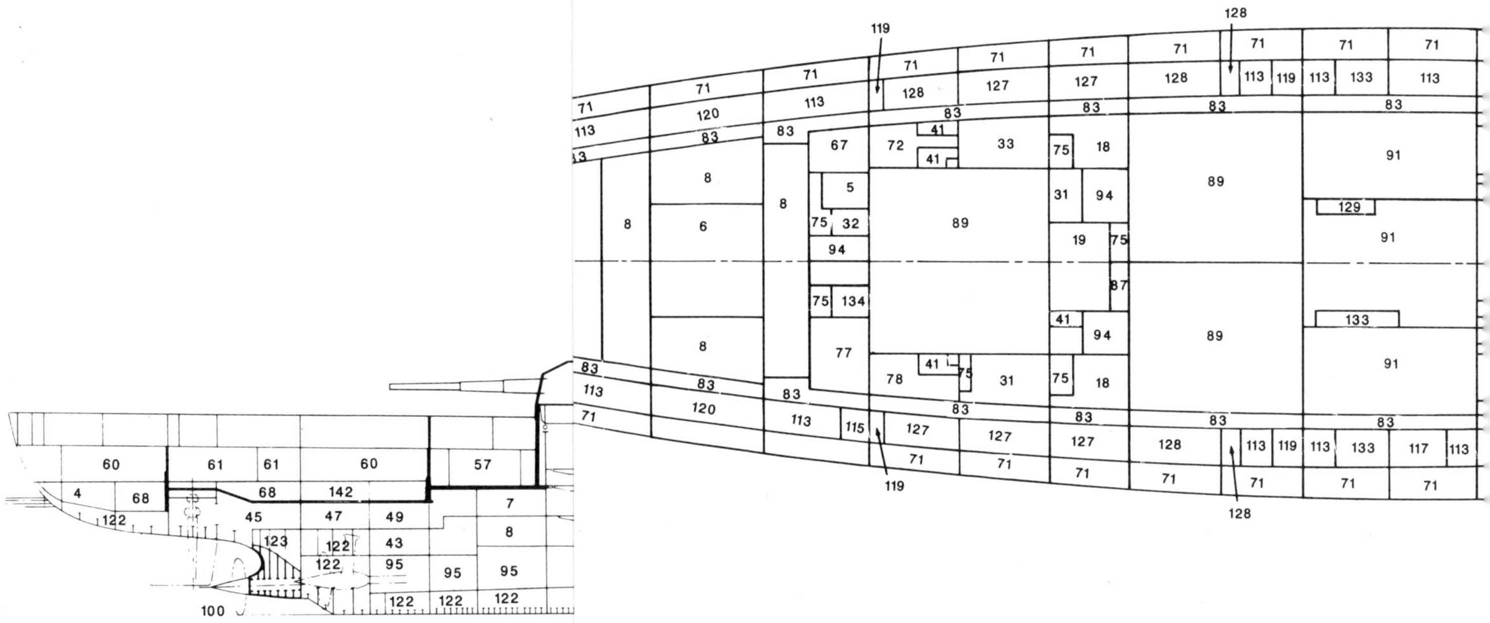

Tween Deck

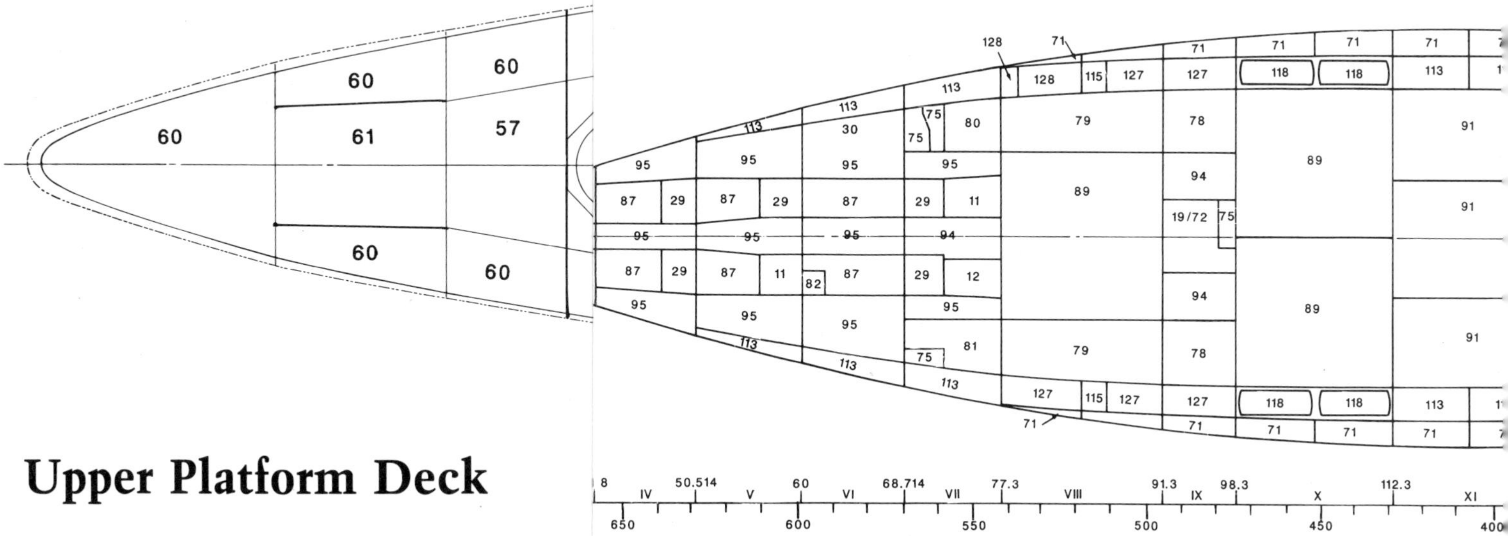

Upper Platform Deck

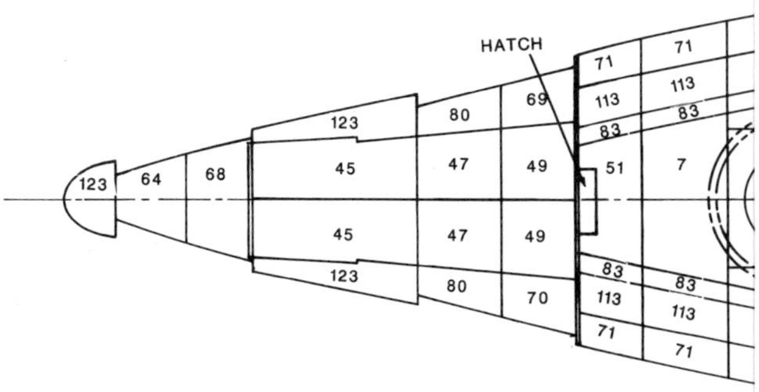

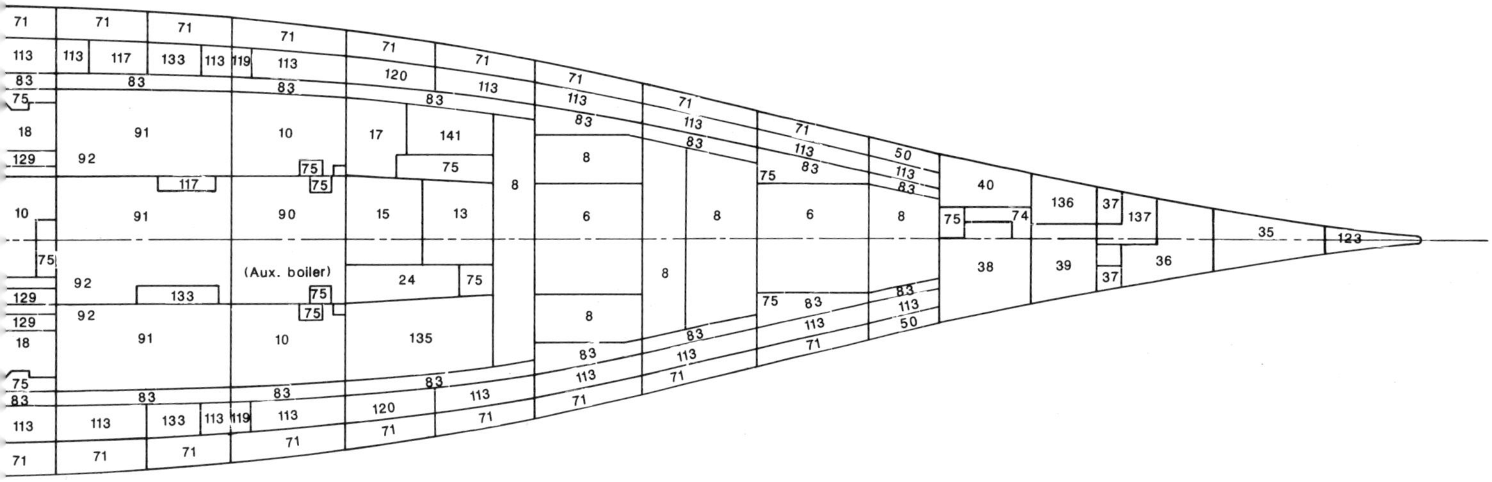
(Aux. boiler)

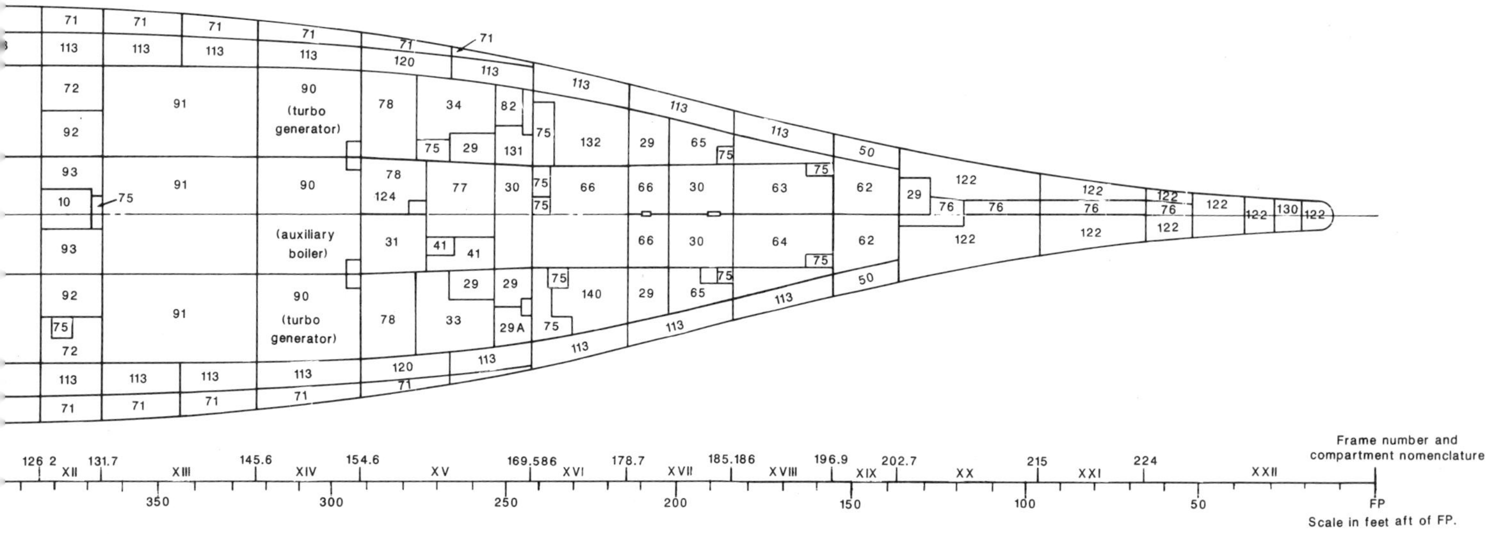
(turbo generator)
(auxiliary boiler)
(turbo generator)
Frame number and compartment nomenclature
126 2
XII
131.7
XIII
145.6
XIV
154.6
XV
169.586
XVI
178.7
XVII
185.186
XVIII
196.9
XIX
202.7
XX
215
XXI
224
XXII
350
300
250
200
150
100
50
FP
Scale in feet aft of FP.

APPENDIX C

Battleship and Battlecruiser Guns

Introduction. Battleships and battlecruisers in general have been perhaps the world's most popular warships, in terms of public interest in the ships, their characteristics, and their careers. As conventional capital ships served primarily as gun platforms, naturally their heavy guns were a major factor in their potential combat effectiveness. All other attributes were intended only to allow the ship to determine the time and place of the engagement, function effectively in combat, and withstand the destructive efforts of the enemy.

The primary weapons of battleships and battlecruisers, their guns, have been perhaps the subject of the greatest interest and speculation. Unfortunately, as is usually the case in such technical matters, one cannot merely assume that because one ship has 16-inch guns, it is necessarily more powerfully armed than another ship with 15-inch guns. Additionally, despite the potentially destructive effects of a given gun, as measured by ballistics and penetration performance, such capabilities are totally wasted if the gunfire-control equipment is unable to direct the gunfire accurately.

In order to furnish a better insight into the gunnery capabilities of the capital ships described in this volume, and to compare the main-battery guns of these ships with contemporary foreign ships, this appendix is designed first to discuss in general various aspects pertinent to the question of the effectiveness of capital-ship guns in combat, and then provide added comments by way of explaining the gun data tabulation that follows.

Main-battery guns. Capital-ship main-battery guns were primarily intended to engage and disable enemy surface ships, and particularly capital ships. For this purpose, the most important performance characteristic was the ability to penetrate the armor protecting the enemy ship. The armor-piercing shells fired by heavy naval guns relied exclusively on kinetic energy to force their way through the armor. The ability of a shell to penetrate armor is a function of a series of factors interrelated in a complex way not thoroughly understood even today. (The best formulations for estimating armor penetration are totally empirical, as opposed to being derived from theoretical considerations.) The basic factors include the following:

- Diameter of the shell.
- Weight of the shell. The greater the specific frontal density—the weight per unit frontal cross-section area—the better.

The reconstructed *Gneisenau,* as depicted in this Richard Allison painting, would have been armed with six 380-mm guns. The added weight forward would have forced the lengthening of the bow to add buoyancy. As the armor of the *Gneisenau* was quite similar to that of the later *Bismarck* and *Tirpitz,* a *Gneisenau* armed with 380-mm guns would have featured a good balance of offensive and defensive capabilities.

- Striking velocity. The greater the velocity, the greater the kinetic energy and the better the chances of penetration.
- Angle of fall. The more nearly perpendicular the angle of impact of the shell against the armor, the greater the thickness penetrated. If the angle is excessively oblique, the shell may ricochet from the armor.
- Characteristics of the armor. This is the most difficult factor to evaluate when comparing penetration tables for guns of different navies. Differing penetrations may be as much a function of varying qualities of armor as of the basic performance of the gun.
- Structural adequacy of the projectile. If the shell breaks up on striking the armor, or the fuze is rendered inert, then the projectile has been defeated by the armor, and the armor has accomplished its purpose. As a result, heavy-gun armor-piercing shells are massively constructed, with the explosive charge always comprising less than 3 percent of the total weight of the shell.
- Shape of the projectile. This affects the ballistic performance and the penetrative performance of the shell. Similarly, the adequacy of the cap, which is designed to lessen the tendency of the shell to ricochet, fundamentally influences the ability of the shell to penetrate armor.

The tendency in modern heavy naval guns was to accentuate penetrative performance by employing as heavy a shell as possible. This was a compromise judgment, partially sacrificing close-range performance in order to maximize the likelihood of penetration at longer ranges. The high-velocity, lightweight shell (tank guns are classic examples of this approach to armor penetration) is outstanding at the closer ranges, but performance deteriorates severely as range is increased. This is a consequence of the fact that velocity is a major factor in the kinetic energy of the projectile, and it deteriorates much more rapidly in the case of relatively lightweight projectiles. The heavier shells, although with a lower muzzle velocity, were far superior in retaining velocity with range, hence were superior in armor protection at longer ranges.

Capital-ship main-battery guns were also employed against lighter ships and shore targets, where some form of high-explosive shell is more suited to the target than the armor-piercing shell, which has a modest lethal radius—the distance in which any exposed personnel and unprotected equipment are subject to almost certain destruction by shell fragments. As a matter of fact, during the course of World War II, bombardment became so important to U.S. Navy capital ships that reduced charges were often used to reduce barrel wear. Barrel wear is a fundamental consideration in the performance of heavy guns. The bore erosion is so significant that the muzzle velocity and ballistic accuracy deteriorates slightly with every round that is fired. The reduced velocity was of no particular concern with high-explosive shells—assuming the available ballistic range was adequate—as the striking energy is not a factor in the destructive effects of the projectile.

Secondary-battery guns. The design concepts applied to capital-ship secondary batteries varied from navy to navy. In the navies of the United States, Great Britain, and the Netherlands, single-caliber, dual-purpose batteries were selected. As indicated by the terminology, such DP guns were expected to be heavy enough to be able to engage enemy light cruisers and destroyers successfully, while retaining sufficient flexibility and rates of fire to destroy enemy aircraft. All other nations that projected new capital ships—France, Spain, the Soviet Union, Germany, Japan, and Italy—opted for a mixed-caliber secondary battery.

The mixed-caliber secondary battery typically featured a light-cruiser gun of about

152-mm caliber, primarily designed to engage enemy surface ships (yet at times given a modest AA capability), while a lighter antiaircraft gun was used, varying from 90 mm to 127 mm in caliber. The major advantage of the dual-purpose battery was in its arrangement; more guns could be deployed, capable of engaging any one target, than would be available in the mixed battery with its single-purpose guns—each partial battery being less numerous than the total DP battery.

Antiaircraft projectiles are only as effective as their fuzes. Even under the best of circumstances, a direct hit on an aircraft was highly unlikely. As a result, the mechanical time fuze provided for a delay in detonation of the shell until it reached the predicted location of the aircraft. The inevitable time lag inherent in firing at ranges of several thousand yards made this prediction a chancy process at best. Hence, the Allied wartime development of the variable-time (proximity) fuze, which used a small radio transceiver inside the projectile to sense the proximity of the target and activate the fuze, was a major breakthrough in antiaircraft gunnery, considerably enhancing the effectiveness of heavy antiaircraft guns against air targets.

At times, the proximity or the mechanical time fuze might be used against surface targets, but normally the quick-acting point-detonating fuze was employed against "soft" targets. Projectiles with a delayed-action base-detonating fuze were used against armored or reinforced structures. The more protected location of the base-detonating fuze permitted its reliable operation—a time lag was necessary to permit the shell to penetrate the armor or heavy structure before its detonation.

Machine guns. Antiaircraft machine guns, such as the 40 mm and the 20 mm, relied on high volumes of fire to contend with high-speed targets. Normally, such machine-gun projectiles had only a primitive contact fuze at best, and the smaller-bore weapons, such as the 0.50 caliber, frequently did not have explosive projectiles.

Gunfire-control systems. The great intangibles in evaluating gunnery systems are the relative capabilities of the gunfire-control systems of the ships of the various navies. This is extremely difficult to quantify, and this study will limit itself to a few very superficial observations. In general, the Axis ships had somewhat of an advantage early in the war because of their superb optical rangefinding systems. Midway through the war, the Allies overcame this advantage with the development of radar gunfire-control systems. The Germans did have similar systems, but they did not approach the operational effectiveness of U.S. and Royal Navy systems.

Gun data table. Table C-1 is intended to give a detailed comparison of the relative characteristics of capital-ship main-battery guns. It is organized as follows:

- *Projectiles.* Types of shells and their weights.
- *Ballistic data.* Shell weight, muzzle velocity, and maximum ranges. Wherever possible, new-gun muzzle velocities are given. Frequently, range table calculations are based on an average estimated muzzle velocity over the life of the gun.
- *Armor penetration.* For a given shell weight, data for penetration of vertical and horizontal plates is given for specified ranges. To permit comparisons, wherever ballistic data permits, calculated penetrations using a detailed U.S. Navy empirical equation are given. This means such data may not be precisely correct, but it should give a very accurate measure of relative gun performance.

U.S. Navy, Royal Navy, and French Navy data is exclusively from official sources, while the information for the other guns is from a variety of official and unofficial but authoritative sources.

TABLE C-1

Capital-Ship Main-Battery Guns

Gun	*Nation*	*Gun Weight* Kg	(lbs)	*Shell Type*	*Weight* Kg	(lbs)	*Muzzle Velocity* mm/sec	(ft/sec)
18.11″ 460mm/45 (Type 94)	Japan	165,000	(363,762)	AP	1,460	(3,219)	780	(2,559)
				HE	1,360	(2,998)	805	(2,640)
16″ 406mm/52 (SKC/34)	Germany	160,000	(352,740)	AP	1,030	(2,272)	810	(2,657)
				HE	1,030	(2,272)	810	(2,657)
16″ 406mm/50 (Mark 7)	U.S.A.	121,615	(267,904)	AP	1,225	(2,700)	762	(2,600)
				HE	862	(1,900)	820	(2,690)
16″ 406mm/45 (Mark 6)	U.S.A.	100,363	(221,220)	AP	1,225	(2,700)	701	(2,300)
				HE	862	(1,900)	803	(2,635)
16″ 406mm/45 (Mark 2)	U.K.	120,656	(266,000)	AP	1,077	(2,375)	732	(2,400)
15″ 381mm/50 (Model 1934)	Italy	118,000	(260,145)	AP	885	(1,951)	850	(2,789)
				COM	824	(1,817)	870	(2,854)
15″ 381mm/42 (Mark 1)	U.K.	101,605	(224,000)	AP	879	(1,938)	732	(2,400)
				HE	879	(1,938)	732	(2,400)
14.96″ 380mm/52 (SKC/34)	Germany	111,000	(244,713)	AP	800	(1,764)	820	(2,690)
				HE	800	(1,764)	820	(2,690)
14.96″ 380mm/45 (Modele 1935)	France	94,130	(207,522)	AP	884	(1,949)	785	(2,575)
				HE	884	(1,949)	785	(2,575)
14″ 356mm/45 (Mark 7)	U.K.	93,986	(207,200)	AP	721	(1,590)	732	(2,400)
				HE	721	(1,590)	732	(2,400)
12.99″ 330mm/52 (Modele 1931)	France	70,535	(155,503)	AP	560	(1,235)	870	(2,854)
				HE	522	(1,151)	885	(2,904)
12″ 305mm/50 (Mark 8)	U.S.A.	55,262	(121,856)	AP	517	(1,140)	762	(2,500)
				HE	426	(940)	808	(2,650)
11.14″ 283mm/56 (KC/13)	Netherlands	47,100	(103,818)	AP	330	(728)	890	(2,920)
				HE	315	(694)	900	(2,953)
11.14″ 283mm/54 (SKC/34)	Germany	53,250	(117,396)	AP	330	(728)	890	(2,920)
				HE	315	(694)	900	(2,953)

Maximum Range			*Armor Penetration (Calculated)*					
m.	(yds)		mm/m	(in/yds)	mm/m	(in/yds)	mm/m	(in/yds)
41,400	(45,276)	Side	864 @ 0	(34 @ 0)	494 @ 20,000	(19.43 @ 21,872)	360 @ 30,000	(14.19 @ 32,808)
41,696	(45,600)	Deck	—		109	(4.30)	189	(7.43)
43,000	(47,025)	Side	806 @ 0	(31.72 @ 0)	479 @ 18,288	(18.84 @ 20,000)	345 @ 27,432	(13.60 @ 30,000)
43,000	(47,025)	Deck	—		80	(3.15)	127	(4.99)
38,720	(42,345)	Side	829 @ 0	(32.62)	509 @ 18,288	(20.04 @ 20,000)	380 @ 27,432	(14.97 @ 30,000)
38,059	(41,622)	Deck	—		99	(3.90)	169	(6.65)
33,741	(36,900)	Side	755 @ 0	(29.74 @ 0)	448 @ 18,288	(17.62 @ 20,000)	324 @ 27,432	(12.77 @ 30,000)
36,741	(40,180)	Deck	—		109	(4.29)	194	(7.62)
34,766	(38,200)	Side	737 @ 0	(29.03 @ 0)	389 @ 18,288	(15.3 @ 20,000)	292 @ 27,432	(11.5 @ 30,000)
		Deck	—		82	(3.24)	146	(5.73)
44,899	(48,884)	Side	814 @ 0	(32.07 @ 0)	510 @ 18,000	(20.06 @ 19,685)	380 @ 28,000	(14.93 @ 30,621)
46,533	(50,899)	Deck	—		73	(2.86)	130	(5.11)
33,558		Side	687 @ 0	(27.07 @ 0)	*297 @ 18,288	(11.7 @ 20,000)	*229 @ 27,432	(9.0 @ 30,000)
33,558		Deck	—		*79	(3.1)	145	(5.7)
36,200	(39,589)	Side	742 @ 0	(29.23 @ 0)	419 @ 18,000	(16.50 @ 19,685)	304 @ 27,000	(11.98 @ 29,527)
36,200	(39,589)	Deck	—		75	(2.96)	126	(5.02)
37,550	(41,065)	Side	748 @ 0	(29.43 @ 0)	393 @ 22,000	(15.49 @ 24,060)	331 @ 27,000	(13.12 @ 29,528)
37,550	(41,065)	Deck	—		104	(4.15)	138	(5.44)
33,924	(37,100)	Side	668 @ 0	(26.29 @ 0)	*285 @ 18,288	(11.2 @ 20,000)	— @ 25,603	(@ 28,000)
33,924	(37,100)	Deck	—		*72	(2.83)	*121	(4.75)
41,700	(45,604)	Side	713 @ 0	(28.08 @ 0)	342 @ 23,000	(13.46 @ 25,153)	292 @ 27,500	(11.48 @ 30,114)
40,600	(44,401)	Deck	—		87	(3.47)	110	(4.32)
35,271	(38,573)	Side	622 @ 0	(24.48 @ 0)	323 @ 18,288	(12.73 @ 20,000)	231 @ 27,432	(9.08 @ 30,000)
34,766	(38,021)	Deck	—		77	(3.02)	130	(5.11)
41,900	(45,822)	Side	604 @ 0	(23.79 @ 0)	291 @ 18,288	(11.47 @ 20,000)	205 @ 27,432	(8.08 @ 30,000)
42,500	(46,479)	Deck			48	(1.87)	76	(2.99)
41,900	(45,822)	Side	604 @ 0	(23.79 @ 0)	291 @ 18,288	(11.47 @ 20,000)	205 @ 27,432	(8.08 @ 30,000)
42,500	(46,479)	Deck	—		48	(1.87)	76	(2.99)

* Based on official data, no calculation possible.

Gneisenau—After the *Gneisenau* had been seriously damaged in a British air raid in Keil on 26 February 1942 while undergoing a brief overhaul, the proposed conversion of the *Scharnhorst*-class battlecruisers to mount 380-mm guns was authorized to proceed. This made her 283-mm guns available for use in coastal artillery. The three guns of turret Anton were to be sent to the Hook of Holland, outside of Rotterdam. The turrets and guns of Bruno and Caesar were sent to Norway. These guns were located in Austråt (Batterie Oerlandat) which was near Trondheim and at Fledt, near Bergen. The emplacements were nearly 10 meters below ground level and imbedded in solid rock and concrete. The turrets were able to traverse 360 degrees in 22 seconds. The gun emplacements were not equipped with radar fire control, and it was necessary to use stereoscopic rangefinders near the shore line to sight and track a target. Both turrets were installed and operational in Norway in 1943 after their delivery from Gydnia where the *Gneisenau* was undergoing conversion. Neither turret saw action. In 1945 German forces surrendered these and other emplacements to the Norwegians. It appears the Norwegian Navy made limited use of both batteries. In 1953 they were deactivated, but both turrets are still there. With the lack of maintenance, their usefulness remains questionable.

This is one of the 283-mm turret emplacements in Norway that was taken from the battlecruiser *Gneisenau*. This turret was mounted at Austråt on the coast of Trondheim in central Norway and is termed Batterie Oerlandet. The emplacement was excavated to a depth of 10 meters to allow the undercarriage of the turret to be placed as well. Although the guns were last used in late 1953, the battery was inspected by VIPs from Norway and abroad.

Bibliography

The technical details in this volume are based upon official documents of those nations concerned, primarily those documents found in the design archives, supplemented by official reports and material from books or articles by noted authorities on the subject of warship design, construction, or operation. The authors and their designated representatives were permitted access to formerly classified material, and this included:

- design histories
- ship characteristics
- ship performance data (trials, proving-ground tests, etc.)
- battle damage reports

Such material is of fundamental importance in a book of this type, as it is this largely unpublished data that permits a systematic investigation and evaluation of the various designs. A detailed correspondence with naval constructors and recognized experts, familiar with the design evolution, was of great assistance to the authors.

The technical data was supplemented by battle-damage reports, special-incidents reports, and other operational information made available by the archives or navy department of the country concerned.

The actual characteristics of the ships as built were obtained from official sources—such as the ships' operational booklets and numerous official plans—as well as from noted authorities in Germany, Italy, Japan, and Spain. The information from the latter was cross-checked against data from shipyards and official data to verify its authenticity.

The nature of the source material available on Japanese, German, Italian, and Spanish capital-ship design and construction projects of the World War II era differs radically. In general, there is a large amount of literature available on the operational careers of these ships, but relatively limited authoritative technical data. The latter was particularly evident in Japan, where American bombing raids, and the Japanese themselves after the surrender announcement, had destroyed much of the Imperial Japanese Navy's official files and archives on the design and operation of the *Yamato, Musashi,* and *Shinano.* Emphasis was placed on the procurement of technical material so that the ships could be evaluated more thoroughly from a performance aspect. For convenience, this bibliographic essay is grouped by country.

Japan. Few published sources were available on the design or operation of the *Yamato*-class battleships and the conversion of the *Shinano* to an aircraft carrier. This was complicated by the incomplete surviving Japanese archives on the design, construction, and operation of the *Yamato*-class capital ships. Most data that remained was confiscated by American intelligence after the war, but was in a poor state for long-term preservation.

In the absence of official data, the U.S. Navy had key Japanese naval constructors re-create these ships from memory and from what little information that had not been destroyed. The intelligence unit (Naval Technical Mission to Japan—NAVTECHJAP) was headed by Admiral E. C. Holtzworth, USN, and Captain Nathan Sonnenshein, USN. The list of documents generated by NAVTECHJAP that were analyzed by the authors appears under the heading U.S. Government Documents. Both Admiral Holtzworth and Captain (later Admiral) Sonnenshein were interviewed by Mr. Garzke in person or by letter. A key Japanese participant in the NAVTECHJAP group was Lieutenant Commander Shizuo Fukui, IJN, who gave direct assistance to the authors.

A number of letters were exchanged with Mr. Fukui from 1960–66, and during this time he also interviewed key personnel on the design and construction of the *Yamato*-class battleships, as well as those involved in the conversion of the *Shinano* from a battleship to an aircraft carrier. Mr. Fukui was also instrumental in arranging letter interviews with Captain Makino, Captain Matsumato, and Admiral Fukuda, all of whom were involved in the design and construction of the *Yamato*. Captain Makino was later personally interviewed by Mr. Garzke in New York and by Mr. Dulin in Washington. Preliminary manuscripts were reviewed for completeness and accuracy.

The authors are responsible for the accuracy of the text. All opinions and conclusions are theirs.

Books, Articles, and Individual Manuscripts

Blair, Clay, Jr. *Silent Victory*. New York: J.B. Lippincott Co., 1975.

Breyer, Siegfried. *Battleships and Battlecruisers, 1905–1970*. Garden City, New York: Doubleday & Co., Inc., 1973.

Matsumoto, Kitaro. *Design and Construction of the Battleships Yamato and Musashi*. Tokyo: Haga Publishing Co., Ltd., 1961.

Potter, John Deane. *Admiral of the Pacific: The Life of Yamamoto*. New York: Viking Press, 1965.

Sakai, Michio, "Medium-sized Battleships After the World War I." *Ships of the World* 9(1958).

Saunders, Mel. "Get the Yamato." *Saga* Magazine (January 1968).

Spurr, Russell, *A Glorious Way to Die, The Kamikaze Mission of the Battleship Yamato*. New York: Newmarket Press, 1981.

Toland, John, *The Rising Sun*. New York: Random House, 1970.

Watts, Anthony J. *Japanese Warships of World War II*. London: Ian Allan, Ltd., 1966.

Interviews and Letters

Fukui, Shizuo. Letters to Robert O. Dulin, Jr., 1954–1965.

Fukui, Shizuo. Letters to William H. Garzke, Jr., 1960–1965.

Holtzworth, Rear Admiral E. C. Interviews with William H. Garzke, Jr., 1963–1964.

Matsumoto, Kitaro. Letters to William H. Garzke, Jr., 1965.

Makino, Shigeru. Interview with Mr. Garzke in New York and Mr. Dulin in Annapolis, Feb. 1961.

Sonnenshein, Rear Admiral Nathan. Letters to William H. Garzke, Jr., 1965–1966.

Yano, Shizuo. Letters to Robert O. Dulin, Jr., 1960–1961.

U.S. Government Documents

U.S. Department of the Navy. Bureau of Ordnance and Division of Naval Intelligence. *A Statistical Summary of Japanese Naval Material*. Ord-ONI Technical Intelligence Bulletin 9, 1 December 1944.

U.S. Department of the Navy. U.S. Naval Technical Mission to Japan. *Reports of Damage to Japanese Warships—Article 3—Japanese Records of Major Warship Losses*. Report S-06-3, January 1946.

U.S. Department of the Navy. U.S. Naval Technical Mission to Japan. *Torpedo and Mine Tests*. Report S-01-9.

U.S. Department of the Navy. U.S. Naval Technical Mission to Japan. *Japanese 18″ Gun Mounts, Intelligence Targets Japan*. Report 0-45(N), February 1946.

U.S. Department of the Navy. U.S. Naval Technical Mission to Japan. *Japanese Naval Guns.* Report 0-54(N), February 1946.
U.S. Department of the Navy. U.S. Naval Technical Mission to Japan. *Ships and Related Targets; Reports of Damage to Japanese Warships*—Article 2—*Yamato* (BB), *Musashi* (BB), *Taiho* (CV), *Shinano* (CV). Report S-06-2, 6 January 1946.
U.S. Department of the Navy. U.S. Naval Technical Mission to Japan. *Japanese Interior Ballistics.* Report 0-21, February 1946.

Plans and Engineering Calculations

Yamato

Protective Arrangements, U.S. Naval Ship Repair Facility, Yokosuka, Japan, Drawing No. 91-48, 2 June 1948.
Rough Profiles, U.S. Naval Ship Repair Facility, Yokosuka, Japan, Drawing No. 90-48, 4 June 1948.

Shinano

Stability Chart, U.S. Naval Technical Mission to Japan Document No. ND-1003.1.
Results of Inclining Experiment, U.S. Naval Technical Mission to Japan Document No. ND-50-1003.2.
General Arrangement Plans, U.S. Naval Technical Mission to Japan Document No. ND-50-1003.5 (Four Plates).
Armor Arrangement, U.S. Naval Technical Mission to Japan Document No. ND-50-1003.6.
Lines, U.S. Naval Technical Mission to Japan Document No. ND-50-1003.7.
Midship Section, U.S. Naval Technical Mission to Japan Document No. ND-50-1003.11.

Germany. The publications available on German World War II battleship and battlecruiser operations and characteristics are many and varied. There are also many conflicting and incomplete accounts that have been published. These could only be resolved by an extensive correspondence with the key naval officers who sailed in these ships and with the designers who planned them. Instrumental in this systematic and thorough research was Rear Admiral Ernest Holtzworth, USN, who gave detailed suggestions to Mr. Garzke when he embarked on this research in 1964. This was further supplemented by an extensive search for original plans, reports, and survivor testimony.

There is little material published outside of Germany on German capital-ship design philosophy. To better understand the successes or failures of these ships in combat, a firm understanding of their design was necessary. The following men have helped give the authors that insight:

- Mr. Otto Reidel—Chief civilian naval constructor in the Construction Office of the OKM. He had intimate knowledge of the design and major characteristics of all modern German capital ships, with particular expertise in the preliminary design of battleship "H." He reviewed the final drafts of the manuscript and gave expert advice from 1961–1982.
- Admiral Werner Fuchs—Naval officer in charge of battleships "H" and "O" who served as Admiral Raeder's aide in matters of ship design and construction in conferences with Adolf Hitler on naval construction.
- Admiral Karl Witzell—Chief of Marine Waffenamtes (U.S. equivalent—Bureau of Ordnance) in the OKM, who had a general knowledge of weapon development in the German Navy prior to and during World War II.
- Dr. Erwin Strohbusch—Naval constructor in the Construction Office who worked on the designs of the *Scharnhorst*- and *Bismarck*-class ships and was later the project engineer in charge of the redesign and reconstruction of the *Gneisenau* in 1942.

- Mr. Wilhelm Hadeler—A naval constructor in the Construction Office who worked on the designs of the aircraft carrier *Graf Zeppelin* and carrier "B." He authored several books on warship design after World War II, the most notable being *Kriegschiffbau* in 2 volumes. He was also familiar with battleship and battlecruiser designs and those who were involved with them.
- Dr. Hermann Burckhardt—Former chief naval constructor in the German Navy, who prior to and during World War II was a consultant to the Construction Office on the design of German cruisers, battleships, and battlecruisers.

The following naval officers provided information on ship operation and battle damage:

Scharnhorst—Captain Helmuth Giessler (chief navigating officer)

Gneisenau—Admiral Wolfgang Kähler (the ship's last commanding officer) Captain Hans Eberhard Busch

Tirpitz—Captain Albrecht Schnarke (gunnery officer) Captain Hans Meyer (commander during X-craft attack)

Bismarck—Captain Gerhard Junack (lieutenant—machinery division) Lieutenant Burkard von Müllenheim-Rechberg (senior surviving officer—gunnery division) Warrant Officer Wilhelm Schmidt (damage control)

Prinz Eugen—Captain T. Gerhard F. Bidlingmaier (aboard during the *Bismarck* operation)

All of the above officers and naval constructors reviewed preliminary versions of the manuscript and with the assistance of Dr. Jürgen Rohwer and Captain Albrecht Schnarke, who both translated these manuscripts into German, obtained critiques from noted German naval constructors and officers. Of particular assistance to the authors was Mr. Reidel, who provided the necessary insights into the designs and the activities of the Construction Office. Captain Schnarke provided Mr. Reidel with translated copies of the final manuscripts, and numerous corrections and additions were made to incorporate Mr. Reidel's suggestions.

The book *Battleship Bismarck* by Baron von Müllenheim-Rechberg served as a final check in the research on the loss of this ship. The authors corresponded with the baron and interviewed him personally in Annapolis. Mr. Strafford Morss also conducted a transatlantic telephone interview with the baron, asking many questions suggested by the authors on the operation and loss of the *Bismarck*. Correspondence with the baron, Captain Junack, and Wilhelm Schmidt brought about a greater understanding of the ship's operation and the damage sustained in her various actions. Data on the trials of the *Bismarck* in the Baltic, provided by Otto Reidel, provided the authors with the missing link in evaluating her fuel situation on 24 May 1941. Mr. Richard von Hooff provided the authors with expert advice on the likely seakeeping responses of the *Bismarck*, based on published data and data from the authors' research.

Material on the *Scharnhorst* and *Bismarck* was prepared with the assistance of some excellent data from the writings of Vulliez and Mordal, Busch, Korotkin, and Bredemaier (*Scharnhorst* and *Gneisenau*) and Brennecke (*Bismarck* and *Tirpitz*). In researching the latter two ships, the most important sources were the individuals involved in the ships' operations. Of particular assistance was Captain Schnarke, who served on board the *Tirpitz* from 1942–1944. Mr. Reidel and Dr. Strohbusch contributed much information on the ships' characteristics (dimensions, armor thicknesses, displacements, etc.). Dr. Strohbusch also provided assistance on the wartime alterations to the *Gneisenau* and *Scharnhorst*, while the Bundesarchiv in Freiburg found plans of the converted ship. Mr. Reidel also provided the authors with a naval constructor's war damage report on the mining (by the X-craft) and bomb damage (3 April 1944) suffered by the battleship *Tirpitz*.

The texts for battleship "H" and battlecruiser "O" were first developed from U.S. Department of the Navy, U.S. Naval Mission in Europe, reports and then were thoroughly checked in correspondence with Admiral Fuchs, Admiral Witzell, and Mr. Reidel. Captain Albrecht Schnarke provided the authors much insight into the design and service performance of the seven 406-mm guns produced by Krupp for these ships. Much of the material about these ships has not been previously published. The plans of battleship H-39 were developed from authentic copies of the original plans.

Books, Articles, and Individual Summaries

Bekker, C.D. *Defeat at Sea*. New York: Henry Holt and Company, 1956.

Bekker, Cajus. *Die Versunke Flotte, Deutsche Schlachtschiffe und Kruezer, 1925–1945*. Hamburg: Gerhard Stalling, 1964.

———. *Hitler's Naval War*. Translated by Frank Ziegler. Garden City, New York: Doubleday and Company, Inc., 1974.

Bidlingmaier, Captain Gerhard. "Exploits and End of the Battleship *Bismarck*." U.S. Naval Institute *Proceedings* (July 1958): 77–89.

———. "Unternehmen *Cerebus*—der Kanaldurchbruch. *Marine Rundschau* 59 (1962):19–40.

Bradford, Ernie. *The Mighty Hood*. Hodder and Stoughton, 1959.

Bredemaier. *Schlachtschiff Scharnhorst*. Jugenheim, 1962.

Brennecke, Jochen. *Schlachtschiff Tirpitz*. Munich: Koehlers, 1953.

———. *Schlachtschiff Bismarck*. Munich: Koehlers, 1960.

Breyer, Siegfried. *Schlachtschiffe und Schlachtkreuzer, 1905–1970*. Munich: J.F. Lehmans, 1970.

Brickhill, Paul. *The Dam Busters*. New York: Ballantine Books, Inc., 1955.

Brown, Anthony C. *Bodyguard of Lies*. New York: Harper and Row, 1975.

Brown, David. *Tirpitz, the Floating Fortress*. Annapolis, Maryland: Naval Institute Press, 1977.

Burckhardt, Hermann, "Die Entwicklung des Unterwasserschutzes in der Deutschen Kriegsmarine." *Marine Rundschau* 58 (1961): 151–69; 204–18.

Busch, F.O. *Das Geheimis der Bismarck*. Hannover, 1950.

———. *Tragödie am Nordkap*. Hannover, 1951.

———. *The Drama of the Scharnhorst*. New York: Berkley Publishing Corporation, 1956.

———. *Kreuzer Prinz Eugen*. Hannover, 1953.

Cook, Gervis Frere. *The Attacks on the Tirpitz*. Annapolis, Maryland: Naval Institute Press, 1973.

Dönitz, Admiral Karl (translation by R.H. Stevens). *Admiral Dönitz Memoirs*. London: Weidenfeld and Nicolson, 1958.

Evers. *Kriegschiffbau*. Berlin, 1943.

Grenfell, Captain Russell. *The Bismarck Episode*. New York: Macmillan and Company, 1954.

Groener, Erich. *Der Deutschen Kriegschiffe, 1815–1945*. Vol. 1. Munich: J. F. Lehmans, 1966.

Hadeler, Wilhelm. *Kriegschiffbau*. Vols. 1 and 2. Darmstadt, 1968.

Herzog, Bodo. *Schlachtschiff Bismarck*. Freiburg: Podzun Verlag, 1975.

Hitler, Adolf. *Mein Kampf*. Eher Verlag, 1926.

Hough, Richard. *Dreadnought, History of the Modern Battleship*. New York: Macmillan and Company, 1964.

Hubatsch, Walter. "Schiffbauplanung, Technischer Rustungsstand und Politsche Zielsetzung beim Aufbau der Deutscher Marine 1848–1955. "*Marine Rundschau* 60 (1963): 67–79.

Janes Fighting Ships. London: Samson and Low, 1935–1949.

Kemp, Peter. *Escape of the Scharnhorst and Gneisenau*. Annapolis, Maryland: Naval Institute Press, 1975.

Kennedy, Ludovic. *Pursuit: the Chase and Sinking of the Battleship Bismarck*. New York: Viking Press, 1974.

Korotkin, I.M. *Battle Damage to Surface Warships During World War II*. Carderock, Maryland: U.S. Joint Publications Research Service for the David Taylor Model Basin, February 1964.

Macintyre, Captain Donald. "Shipborne Radar." U.S. Naval Institute *Proceedings* (September 1967): 70–83.

Martienssen. *Hitler and His Admirals*. New York: 1947.

Müllenheim-Rechberg, Baron Burkard von. *Battleship Bismarck, A Survivor's Story.* Annapolis, Maryland: Naval Institute Press, 1980.
Polmar, Norman. *Aircraft Carriers.* Garden City, New York: Doubleday and Company, 1969.
Porten, Edward P. Von der. *German Navy in World War II.* New York: Thomas Crowell Company, 1969.
Raeder, Grand Admiral Erich. *My Life.* Translated by Henry W. Drexel. Annapolis, Maryland: United States Naval Institute, 1960.
Rohwer, Dr. Jürgen and G. Hummelchen. *Chronology of the War at Sea, 1939–1945.* Vols 1 and 2. New York: Arco, 1973–1974.
Roskill, Captain S.W., DSC, RN,(Ret.). *The War at Sea.* 4 vols. London, 1954.
———. *The White Ensign, the British Navy at War, 1939–1945.* Annapolis, Maryland: United States Naval Institute, 1966.

Ruge, Vice Admiral Friedrich. *Der Seekrieg.* Annapolis, Maryland: United States Naval Institute, 1957.
Schmalenbach, Paul. *Die Geschichte der Deutscher Schiffartillerie.* Hereford, 1968.
Schofield, Vice Admiral B.B. *Loss of the Battleship Bismarck.* Annapolis, Maryland: Naval Institute Press, 1972.
Stevenson, William. *A Man Called Intrepid—The Secret War.* New York: Harcourt Brace Jovanovich, 1976.
Vulliez, Albert, and Jacques Mordal. *La Tragique Destinée du Scharnhorst.* Paris: Amiot Dumont.
Warren, C. E. T., and James Benson. *Midget Raiders.* New York: William Sloane Associates Publishers, 1954.
Watts, Anthony. *The Loss of the Scharnhorst.* Annapolis, Maryland: United States Naval Institute, 1970.
Woodward, David. *The Tirpitz.* London: William Kimber, 1953.
———. *The Tirpitz and the Battle of the North Atlantic.* New York: Berkley Publishing Company, 1956.
Zienke, Earl F. *The German Northern Theater of Operations, 1940–1945.* Washington, D.C.: U.S. Department of the Army, 1959.

Technical Papers

Ballard, Dr. Robert. "Quest for the *Bismarck.*" (London) *Sunday Express Magazine,* no. 443 (29 October 1989) and no. 444 (5 November 1989).
———. "Finding the *Bismarck.*" *National Geographic Magazine* 176 (November 1989): 622–37.
Parkes, Dr. Oscar. *German and Japanese Battleships.* Transactions of the Institute of Naval Architects, 1949.
Schade, Commodore Henry A. *German Wartime Developments, 1939–1945.* Transactions of the Society of Naval Architects and Marine Engineers, 1946.

Periodicals

Brassey's Naval Annual, 1949. "Fuehrer Conferences on Naval Affairs."
Life.
Marineblad.
Marine Rundschau.
Sea Classics.
Ships of the World.
Soldat und Technik
U.S. Naval Institute *Proceedings.*
Warships International.

U.S. Government Documents

U.S. Department of the Navy. United States Naval Mission in Europe. *Loss of the Battleship Tirpitz.* NavTech MisEu No. 222–45.
U.S. Department of the Navy. United States Naval Mission in Europe. *Bismarck and Tirpitz.* NavTech MisEu No. 224–45.
U.S. Department of the Navy. United States Naval Mission in Europe. *Projected Battleships, Types O and H.* NavTech MisEu No. 526–45.
U.S. Department of the Navy. Division of Naval Intelligence. *The Conduct of the War at Sea,* Admiral Karl Doenitz, January 1946.
U.S. Department of the Navy. Division of Naval Intelligence. *Reflections Attending the End of the Last Four German Battleships,* Vice Admiral H. Stiegel, January 1948.

Stability Characteristics of Former German Warships. Accurate Translation Service, Inc., Washington, D.C.
U.S. Department of the Navy. Preliminary Design Branch, Bureau of Ships. *Study of Bismarck.* December 1941.

German Navy Documents

Maintenance Trials of Battleship Tirpitz, March 1944.
Official War Damage Report on Damage Sustained by Battleship Tirpitz, September 1943 and April 1944.

Interviews and Letters

Bidlingmaier, Captain Gerhard. Letters to William Garzke, 1964–1966.
Burckhardt, Dr. Hermann. Letters to William Garzke, 1961–1967.
Campbell, John. Letters to Robert Dulin, 1968–1978.
Fuchs, Werner. Letters to William Garzke, 1967–1971.
Giessler, Helmuth. Letters to William Garzke, 1968–1975.
Hadeler, Wilhelm. Letters to William Garzke, 1966–1971.
Junack, Gerhard. Letters to William Garzke, 1966–1967.
Kahler, Wolfgang. Letters to William Garzke, 1969.
Müllenheim-Rechberg, Baron Burkard von. Letters to William Garzke and Robert Dulin, 1966 & 1980–1990.
———. Interview, Annapolis, Maryland (Sept. 1980).
———. Telephone interview, July 1980.
Reidel, Otto. Letters to William Garzke and Robert Dulin, 1966–1981.
———. Interview by Albrecht Schnarke, 1981.
Rohwer, Jürgen. Letters to William Garzke and Robert Dulin, 1964–1970.
Schnarke, Albrecht. Letters to William Garzke, 1967–1985.
Statz, Josef (*Bismarck* survivor). Letters to William Garzke, 1989–1990.
Strohbusch, Erwin. Letters to William Garzke, 1966–1978.
Witzell, Karl. Letters to William Garzke, 1967–1971.
Wagner Hochdruck Dampfturbinen Gesellschaft M. B. H. Letters to William Garzke, 1965–1966.
Maschinen Fabrik Augsburg-Nürnberg (M.A.N.). Letters to William Garzke, 1966, 1985.
Okun, Nathan. Letters to William Garzke, 1980–1982.
Der Bundesminister der Verteidigung, Bonn. Letters to William Garzke, 1965.

Plans and Engineering Calculations

OKM, Schlachtschiff *Gneisenau,* Geschützstände der Leichten Artillerie (SKC/30), 1 June 1938.
  Takelriss, 22 July 1938.
  Geschützstände der Schwere Artillerie.
  Art—Vordere Türme.
  Geschützstände der Mittelartillerie, 12 May 1938.
Deutsche Werke Kiel A.G., *Gneisenau* K235/N4
  Langschnitt, SK2-RA90.
  Obere Ansicht, SK2-RA88.
  Hintere Querschnitte, SK2-RA93.
  Vordere Querschnitte, SK2-RA91.
  Mittlere Querschnitte, SK2-RA92.
  Aufbaudeck und Brücken, SK2-RA89.
  Batteriedeck, SK2-RA63.
  Panzerdeck, SK2-RA87.
  Mittleres Plattformdeck, SK2-RA77.
  Unteres Plattformdeck, Stauung u. Ausserer Wallgang Unten, SK2-RA78.
  Doppelboden u. Stauung, SK2-RA79.
  Zwischendeck, SK2-RA64.
  Oberdeck, SK2-RA86.
  Ober Plattformdeck, SK2-RA65.
OKM, Sclachtschiff *Gneisenau,* Geschützstände der Leichten Artillerie (SKC/33), 11 May 1938.
Deutsche Werke Kiel A.G., Schlachtschiff *Gneisenau,* Geschütztürme A, B, und C der Schwere Artillerie.

OKM, Schlachtschiff *Gneisenau*, Geschützstände der Schwere Artillerie Hintere Turm, 10 June 1938.
Deutsche Werke Kiel A.G., *Gneisenau* K235/N4, Doppellungen auf dem Oberdeck.
Marinewerft Wilhelmshaven, Schlachtschiff *Scharnhorst*, Ruder und Schraubenanordnung, November 1938.
OKM, Schlachtschiff *Scharnhorst*, Luftungsplan, 6 January 1939.
Aufbaudeck, Brücken und Aufbauten, August 1942.
Innenboden und Stauung, August 1942.
Panzerdeck, August 1942.
Batteriedeck, August 1942.
Zwischendeck, August 1942.

Oberdeck, August 1942.
Mittleres Plattformdeck, August 1942.
Obere Ansicht, August 1942.
Oberes Plattformdeck, August 1942.
Langschnitt, August 1942.
Takelriss, August 1942.
Unteres Plattformdeck und Stauung, August 1942.
Hauptspant, February 1939.
Vordere Querschnitte, August 1942.
Mittleres Querschnitte, August 1942.
Hintere Querschnitte, August 1942.
Raumbezeichnungen für den Leckwehrdienst, June 1938.
Isolierungsplan, 3 September 1940.
OKM, Schlachtschiff *Bismarck*, Langschnitt, 27 June 1941.
OKM, Schlachtschiff "F" (Ersatz Hannover), Schema für die Einrichtungen für Wasser, Zwischendeck, 21 January 1937.
Schema für die Einrichtungen für Wasser, Unteres Plattformdeck, 21 January 1937.
Schema für die Einrichtungen für Wasser, Mittles Plattformdeck, 21 January 1937.
Schema für die Einrichtungen für Wasser, Querschnitte, 21 January 1937.
Schema für die Einrichtungen für Wasser, Oberes Plattformdeck, 21 January 1937.
Blohm & Voss, Schlachtschiff *Bismarck*, Stauung und Doppelboden, 28 April 1941.
Oberes Plattformdeck, 28 April 1941.
Panzerdeck, 28 April 1941.
Linienriss, 17 April 1940.
OKM, Schlachtschiff *Tirpitz*
Langschnitt.
Takelriss.
Obere Ansicht.
Aufbaudeck, Aufbauten und Brücken.
Oberdeck.
Batteriedeck.
Zwischendeck.
Panzerdeck.
Oberes Plattformdeck.
Mittleres Plattformdeck.
Unteres Plattformdeck und Stauung.
Doppelboden und Stauung.
Vordere Querschnitte.
Mittlere Querschnitte.
Hintere Querschnitte.
Linienriss.
Querschnitte.
Hinterschiff.
Seitenpanzer.
Panzerquerschott Spt 202.7.
Übersicht der Raderlange.
Druckhöhen plan F.d. Rüfung.
Wellenabschlussbuck der Mittelwelle.
Querschnitte zur Beslimmung ober Widerstandsmomente für die Nachrechnung der Längsfestigkeit.

Marinewerft Wilhelmshaven, Schlachtschiff "G", Vorentwurf dur Ballastwasser, Entölungsanlage der Schlachtschiffe "F" und "G", 1 April 1938.
OKM, Schlachtschiff "H", Linienriss, K.Z. No 132.
Panzerabwicklungszeichnung (Panzerquerschnitte), 1940.
Panzerabwicklung (Barbetten der M. A.), 1940.
Panzerabwicklung (Seiten- , heck- , und Bugpanzer), 1940.
Vordere Kommandostand, 1940.
Langschnitt.
II u. III Aufbaudeck, Brücken und Aufbaudeck.
I. Aufbaudeck.
Oberdeck.
Batteriedeck.

Oberes Plattformdeck.
Mittleres Plattformdeck.
Unteres Plattformdeck.
Stauung.
Stauung und Innenboden.
Vordere Querschnitte.
Mittlere Querschnitte.
Hintere Querschnitte.
Deutsche Werke Kiel A.G., Schlachtschiff Typ "O", K265-A81, Innenboden, 3 March 1941.
——————, ——————, K265-A80, Langschnitt Obere Ansicht, 3 March 1941.
OKM, Schlachtschiff "O," Druckhöhenplan, 1940.
Langschnitt, 1 October 1941.
Aufbaudeck und Brücken, 1 October 1941.
Obere Ansicht, 1 October 1941.
Oberdeck, 1 October 1941.
Batteriedeck, 1 October 1941.
Zwischendeck, 1 October 1941.
Oberes Plattformdeck, 1 October 1941.
Mittleres Plattformdeck, 1 October 1941.
Stauung, 1 October 1941.
Innenboden, 1 October 1941.
Vordere Querschnitte, 1 October 1941.
Mittlere Querschnitte, 1 October 1941.
Hintere Querschnitte, 1 October 1941.
OKM, Unterlagen und Richtlinien zur Beftimmung der haptkampfentfernung und der Geschokmahl, Sextband, Berlin 1940.
———. Durchschlagsangaben fremder Schlachtschiffe, Berlin 1940.
———. Durchschlagsangaben fremder Kreuzer, Berlin 1940.
OKM, Unterlagen und Richtlinien zur Beftimmung der Hauptkampfentfernung und der Geschokwhal, Eigene Durchschlagsangaben für Leichte Kreuzer und Beifpiele, Berlin, 1940.
———. Eigene Durchschlagsangaben für Panzerschiffe und Beifpiele, Berlin 1940.
———. Eigene Durchschlagsangaben für Schlachtschiffe *Scharnhorst, Gneisenau,* und Beifpiele, Berlin, 1940.
———. Eigene Durchschlagsangaben für Schlachtschiffe *Bismarck, Tirpitz,* und Beifpile, Berlin, 1940.
OKM, Durchschlagskurven 15–28 cm Pzgr gegen KC u. hom. Panzerplatten.

Italy. The technical details in this chapter are based principally upon official Italian Navy documents, primarily from materials provided by the Historical Office of the Italian Navy (Ufficio Storico Marina Militare). The office of the Italian Naval Attaché in Washington was also helpful in forwarding the authors' requests to Rome for official replies from the Italian Navy. Materials provided included:

- Outline design histories
- Battle damage data—plans and narratives
- Detailed engineering analyses and calculations (stability, weight, propulsion, etc.)
- Gun data and specifications
- Ship plans/photographs

The exceptional cooperation of the Historical Office of the Italian Navy was of fundamental importance. This largely unpublished data, when combined with published source materials, permitted a thorough analysis of the design and operational careers of the battleships of the *Vittorio Veneto* class and aided in the description and location of damage to the *Italia* (ex-*Littorio*), *Vittorio Veneto*, and *Roma*.

The characteristics of the ships were obtained from official data provided by the design plans and engineering data for the ships.

The admirable *Italian Battleships—1861–1969 (Le Navi di Linea Italiane)* is an excellent starting point for any study of the ships of the *Vittorio Veneto* class.

There are a number of helpful published sources regarding the Italian naval campaigns in the Mediterranean that provide useful insights into the operational careers of these battleships. Of these, perhaps the most relevant are *The Italian Navy in World War II, Night Action off Cape Matapan*, and *The Attack on Taranto*.

The spectacular destruction of the *Roma* by German radio-guided bombs is perhaps best described in Charles H. Bogart's Naval Institute *Proceedings* article, "German Remotely Piloted Bombs." The Italian Navy also provided the authors with some additional information on the loss of this ship.

The authors have been most fortunate in having several authorities on these ships review drafts of the present text, and provide a number of helpful comments and corrections/amplifications: the Historical Office of the Italian Navy, Aldo Fraccaroli, and Guiseppe Sitzia. The two shipyards that built the four ships, Cantieri Ansaldo, Genoa, and Cantieri Riuniti dell'Adriatico, Trieste, also provided some assistance to the authors during their research.

The authors and illustrators remain fully responsible for the accuracy of this text. All opinions and conclusions, unless specifically identified otherwise, are theirs, and are not to be construed as official or necessarily reflecting the views of the Italian Navy or any other organization or individual.

Books, Articles, and Individual Summaries

Bargoni, Franco. *Orizzonte Mare, Navi Italiane Nella 2ª Guerra Mondiale, Corazzate Classe* Vittorio Veneto. Rome: Edizioni Bizzarri, 1974.

Baumbach, Werner. *The Life and Death of the Luftwaffe*. New York: Ballantine Books, 1960.

Bragadin, Marc' Antonio. *The Italian Navy in World War II*. Annapolis, Maryland: U.S. Naval Institute, 1957.

Bogart, C. H. "German Remotely Piloted Bombs." U.S. Naval Institute *Proceedings* (November 1976): 63–68.

Breyer, Siegfried. *Battleships and Battlecruisers, 1905–1970*. Garden City, New York: Doubleday and Company, Inc., 1973.

Fraccaroli, Aldo. *Italian Warships of World War II*. London: Ian Allen, 1968.

Hough, Richard. *Dreadnought—A History of the Modern Battleship*. New York: MacMillan and Company, 1964.

Jane's Fighting Ships. London: Sampson and Low, 1939, 1942, 1943–1944, 1944–1945, 1950–1951.

Pack, S.W.C. *Night Action off Cape Matapan*. Annapolis, Maryland: Naval Institute Press, 1972.

Polmar, Norman. *Aircraft Carriers*. Garden City, New York: Doubleday and Company, Inc., 1969.

Potter, E.B., and Chester W. Nimitz. *Sea Power—A Naval History*. Englewood Cliffs, New Jersey: Prentice Hall, 1960.
Ross, Frank, Jr. *Guided Missiles: Rockets and Torpedoes*. New York: Lothrup, Lee and Shepard Company, Inc., 1951.
Schofield, Vice Admiral B.B. *The Attack on Taranto*, Annapolis, Maryland: Naval Institute Press, 1973.

Unpublished Material

Italian Navy. Historical Office (Ufficio Storico M.M.). *Littorio*, Battle Damage Sketch and Descriptive Text, Damage Sustained November 11, 1940.
———. *Littorio*, Battle Damage Sketch and Descriptive Text, Damage Sustained June 15, 1942.
———. *Vittorio Veneto*, Battle Damage Sketch and Descriptive Text, Damage Sustained March 28, 1941.
———. *Vittorio Veneto*, Battle Damage Sketch and Descriptive Text, Damage Sustained December 14, 1941.

Interviews and Letters

Fraccaroli, Aldo. Letters to Robert Dulin, 1965–1978.
Gates, Thomas F. Letters to Robert Dulin, 1966–1969.
Italian Navy, Historical Office (Ufficio Storico M.M.). Letters to William Garzke and Robert Dulin, 1961–1981.
Nani, Augusto. Letters to William Garzke and Robert Dulin, 1964–1966.
Sitzia, Guiseppe. Letters to Robert Dulin, 1969–1970.

Plans and Engineering Calculations

Ansaldo, S.A. *Ship Armed with Six 343-mm Guns* (Nave Armata Di VI 343 mm), Number 770 (Plates 1–30), dated September–December 1932.
R. N. *Impero*, Forward Superstructure Tower—Middle Part, Number L.301/2293 (Plates 1–2), dated July 19, 1940)
R. N. *Impero*, Forward Superstructure Tower—Upper Part, Number L.301/2648 dated July 22, 1940.
R. N. *Impero*, Forward Superstructure Tower—General, Number L.301/1717, dated July 17, 1940.
R. N. *Impero*, Forward Superstructure Tower—Lower Part, Number L.301/2186 (Plates 1–2), dated July 11, 1940.
R. N. *Italia*, Scheme of Protection, Number 57761 (Plates 1–3), dated May 1945.
R. N. *Littorio*, Midship Section, Number L.301/4436, dated October 18, 1940.
R. N. *Littorio*, Armor Plans, Number L.301/4520 (Plates 26–29).
R. N. *Littorio*, Arrangement Plans, Number L.301/4573 (Plates 1–12), dated February 19, 1941.
R. N. *Littorio*, Subdivision Booklet, Number L.301/4577 (Plates 1–13), dated August 18, 1940.
R. N. *Littorio*, Fuel, Water, and Ballast Tanks Booklet, Number L.301/4578 (Plates 1–10), dated August 18, 1940.

Spain. Mr. Ricardo de Sabrino, a former Spanish naval officer now residing in the United States, was the first individual to make the authors aware of Spanish efforts to develop designs for the construction of modern battleships for the Spanish Navy. Mr. de Sabrino, in addition to providing much helpful information, was able to put Mr. Dulin in contact with a number of individuals who were familiar with the design proposals or knew where the information existed in the archives. Among those individuals he suggested were:

- Admiral Pedro Nieto Artinez—A commanding officer of the Spanish cruiser *Galicia* and a personal naval aide to General Franco from 1954 to 1957. He provided information on the negotiations with the Italians on the battleships of the *Vittorio Veneto* type, on the designs of the armored cruisers of the Super-*Washington* type, and on the 1939 fleet law signed by General Franco.

- Andres Gutierez—Reviewed early versions of the Spanish chapter and made some constructive critiques.
- M. Ramirez Gabarrús—Provided background information regarding the design of the *España* class, details on the *España*-class dreadnoughts as finally completed, and information on the Rome Conference of 1922.
- Juan Genova Sotil—Helped authors to evaluate Spanish efforts in battleship design, including details of those his father made in a Spanish technical journal in 1942.
- Vicente Elias—Gave the authors contacts on new Spanish battleship products, including that of the license-built *Vittorio Veneto* type.
- J.I. Taibo—Sent authors information on the Super-*Washington* designs.

Other Spanish contacts who provided information on Spanish battleship design efforts were Alfredo Aguilera and Commander Usatorre of the Spanish Navy.

Assistance also came from Sr. L. Mazarredo of the Associación de Investigación de la Construcción Naval, who provided the authors with information on the Spanish *Vittorio Veneto* version as well as the necessary changes that would have to be made at the El Ferrol Shipyard to build these ships.

The Office of the Naval Attaché to the Spanish Embassy in Washington, D.C., was helpful in obtaining technical information on Spanish battleship design efforts.

Index

Key to Abbreviations
Australia (HMAS)—His or Her Majesty's Australian Ship
France (FNS)—French Naval Ship
Germany (RM)—Reichs Marine
(KM)—Kaiser Marine
Italy (IINS)—Imperial Italian Naval Ship
(IIN)—Imperial Italian Navy
Japan (IJNS)—Imperial Japanese Naval Ship
(IJN)—Imperial Japanese Navy
Russia (USSR)—Union of Soviet Socialist Republics
Spain (SNS)—Spanish Naval Ship
(SS)—Steam Ship
United Kingdom (HMS)—His or Her Majesty's Ship
(RN)—Royal Navy
United States (USS)—United States Ship
(USCG)—United States Coast Guard
(USN)—United States Navy

The Naval Institute Press is the book-publishing arm of the U.S. Naval Institute, a private, nonprofit, membership society for sea service professionals and others who share an interest in naval and maritime affairs. Established in 1873 at the U.S. Naval Academy in Annapolis, Maryland, where its offices remain today, the Naval Institute has members worldwide.

Members of the Naval Institute support the education programs of the society and receive the influential monthly magazine *Proceedings* and discounts on fine nautical prints and on ship and aircraft photos. They also have access to the transcripts of the Institute's Oral History Program and get discounted admission to any of the Institute-sponsored seminars offered around the country.

The Naval Institute also publishes *Naval History* magazine. This colorful bimonthly is filled with entertaining and thought-provoking articles, first-person reminiscences, and dramatic art and photography. Members receive a discount on *Naval History* subscriptions.

The Naval Institute's book-publishing program, begun in 1898 with basic guides to naval practices, has broadened its scope in recent years to include books of more general interest. Now the Naval Institute Press publishes about one hundred titles each year, ranging from how-to books on boating and navigation to battle histories, biographies, ship and aircraft guides, and novels. Institute members receive discounts of 20 to 50 percent on the Press's more than eight hundred books in print.

Full-time students are eligible for special half-price membership rates. Life memberships are also available.

For a free catalog describing Naval Institute Press books currently available, and for further information about subscribing to *Naval History* magazine or about joining the U.S. Naval Institute, please write to: